COMPOST ENGINEERING

Principles and Practice

COMPOST ENGINEERING

Principles and Practice

by
Roger Tim Haug

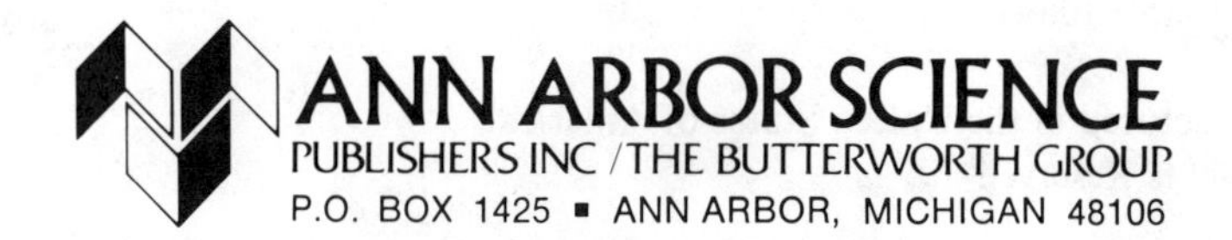

230 Collingwood, P.O. Box 1425, Ann Arbor, Michigan 48106

Library of Congress Catalog Card No. 79-56119
ISBN 0-250-40347-1

Manufactured in the United States of America

Paris throws five millions a year into the sea. And this without metaphor. How, and in what manner? day and night. With what object? without any object. With what thought? without thinking of it. For what return? for nothing. By means of what organ? by means of its intestine? its sewer. . . .

Science, after long experiment, now knows that the most fertilising and the most effective of manures is that of man. The Chinese, we must say to our shame, knew it before us. No Chinese peasant . . . goes to the city without carrying back, at the two ends of his bamboo, two buckets full of what we call filth. Thanks to human fertilisation, the earth in China is still as young as in the days of Abraham. Chinese wheat yields a hundred and twenty fold. There is no guano comparable in fertility to the detritus of capital. A great city is the most powerful of stercoraries. To employ the city to enrich the plain would be a sure success. If our gold is filth, on the other hand, our filth is gold.

What is done with this filth, gold? It is swept into the abyss.

We fit out convoys of ships, at great expense, to gather up at the south pole the droppings of petrels and penguins, and the incalculable element of wealth which we have under our own hand, we send to the sea. All the human and animal manure which the world loses, restored to the land instead of being thrown into the water, would suffice to nourish the world.

These heaps of garbage at the corners of the stone blocks, these tumbrils of mire jolting through the streets at night, these horrid scavengers' carts, these fetid streams of subterranean slime which the pavement hides from you, do you know what all this is? It is the flowering meadow, it is the green grass, it is marjoram and thyme and sage, it is game, it is cattle, it is the satisfied low of huge oxen at evening, it is perfumed hay, it is golden corn, it is bread on your table, it is warm blood in your veins, it is health, it is joy, it is life. . . .

Put that into the great crucible; your abundance shall spring from it. The nutrition of the plains makes the nourishment of men.

from *Les Miserables*
by **Victor Hugo** (1862)

PREFACE

My introduction to composting came in 1972 when I was an engineering faculty member at Loyola University. My father then worked for the Los Angeles County Sanitation Districts and was instrumental in implementing a windrow composting facility on anaerobically digested sludge. Until that time, dewatered sludge was placed on the ground in thin layers (about 0.5 m depth) to air dry before reuse. Septic conditions often developed beneath the surface layer and odors were a constant problem. Various techniques had been tried over the years to control odors with only moderate success.

The Districts had conducted considerable research on refuse composting, but it was commonly thought that dewatered sludge cake could not be composted by itself because of the high moisture content. A crisis of odor complaints in late 1971 convinced District management to try the composting approach, even though some concern was expressed that periodic compost turning would only aggravate the odor problem. A key factor which made the operation successful was my father's idea of recycling dry product and blending with dewatered cake to adjust conditions of the starting mixture. The experiment began in February 1972. Within the first few days of composting, previously odorous material was converted to an aerobic condition. Odors were markedly decreased and complaints from the surrounding neighborhood dropped sharply. By July 1972 all dewatered sludge was being windrow-composted. Results were so impressive that the Districts were commended by the county for solving a major community problem.

A number of lessons were learned during this time. First, dewatered sludge cake could be composted using recycled product as the sole amendment. Second, the importance of odor and nuisance control in sludge management was emphasized. No other factor, including economics, seemed as important as odor control, particularly if odors were out of control. Composting was also demonstrated to reduce odor problems compared to certain other management alternatives. Because of this and other experiences with odors, an entire chapter of this book has been devoted to nuisance control.

In 1975 I removed my full-time academic gowns to become senior engineer for the LA/OMA project. The latter was a facilities planning effort to develop a long-range plan for management of sewage sludges generated in the Los Angeles and Orange County metropolitan areas. The project gave me the opportunity to observe sludge management systems throughout the United States, to study alternative systems and to direct technical studies and field experiments designed to gather information in areas where knowledge was lacking. All feasible management alternatives, including sludge composting, were analyzed in depth by the project, and several field-scale demonstration projects were conducted.

It began to appear that our basic knowledge of composting was becoming complete. Just as complacency was about to set in, a sequence of events occurred which profoundly altered my thinking. The first was a letter from a colleague which had been prompted by a presentation on composting I had made recently. The letter asked a rather simple question regarding the effect of product recycle on the energy budget achieved during composting. My inability to answer the question adequately triggered thought processes which eventually led to detailed thermodynamic analysis of the problem. It's interesting to speculate on the small things, such as a question asked at the right time, which trigger sudden insights and result in extensive human endeavors.

A second event was the startup of expanded dewatering facilities by the Los Angeles County Sanitation Districts in 1977. It was intended that all dewatered sludge continue to be windrow composted. Since 1972 the windrow system had successfully processed about 90 dry metric tons per day of digested sludge, dewatered to 30-35% solids. On completon of the new dewatering facilities, sludge tonnage increased to about 270 dry ton/day, but cake solids decreased to only about 23%. When combined with effects of wet weather and other operational difficulties, odor emissions and complaints increased dramatically. Water load on the system increased by a factor of about five as a result of the decrease in cake solids and the increase in dry tonnage. Although it was not known at the time, thermodynamic constraints had been exceeded. This in part led to process failure with subsequent high odor emissions and reduced composting temperatures

This experience emphasized the need for more fundamental knowledge of the compost process. Indeed, millions of dollars had been spent and much research conducted in designing and constructing the sludge dewatering facility. Unfortunately, similar attention was not given the downstream compost process. I must point out that this judgment is made with the clear vision provided by hindsight. Nevertheless, the experience highlighted the fact that solids content produced from dewatering is an important variable in determining the successful composting of sludge. As it turns out, moisture and volatile solids control and the energy budget for the system are largely influenced by this parameter. Sludge composting has since been described as a problem of moisture control. The reader will note that several chapters of this book are devoted to the subjects of moisture control and system thermodynamics.

Another event which sparked my interest was development of the aerated static pile compost system at Beltsville, MD. This was an entirely different approach to composting compared to the windrow system. Wood chips were used as a "bulking agent" and periodic turning was not used. This prompted several questions. What was the function of the bulking agent? What physical factors influenced the required ratio of wood chips to sludge? What changes in physical properties resulted from use of bulking agents? Were these changes the same as those resulting from use of recycled compost in the windrow system? These intriguing questions spurred further investigations into feed conditioning requirements to produce a starting material with the proper combination of moisture content and free airspace.

Finally, in my position with the LA/OMA project, I had the opportunity to investigate the many enclosed reactor systems available for sludge composting. Some of these systems were designed from the start for sludge composting, while others were originally used on refuse or other relatively dry material. Questions immediately arose in trying to provide a technical assessment of these systems. As one example, required detention times quoted by various manufacturers ranged from as low as 1 day to as high as 14 days. The literature on refuse composting clearly indicated that a one-day detention time was not adequate for stabilization of most organic components. Beyond this, however, the available literature was not particularly helpful. Further analysis of process kinetics was clearly necessary to determine detention time requirements and to identify tradeoffs between detention time, organic stabilization and reactor stability.

It is against the backdrop of these and other experiences that I undertook the study and analysis which culminated in this book. My goal has been to produce a more fundamental engineering approach to the analysis of composting. I hope I have been at least partially successful in achieving this goal. Perhaps the crowning achievement of this effort has been the integration of thermodynamics and process kinetics into a unified model of composting. Although much work remains to improve and further verify the model, its use can guide the analysis and design of present day systems. Preliminary answers can be provided to questions for which experimental data are not available.

In compiling this book, I have drawn heavily on the work of early pioneers and present workers in the field of composting. The reader of this text will become quite familiar with their names. I have tried to represent their data as accurately as possible. Any errors in the analysis are solely my own, as are the comments and opinions expressed in the book.

I owe a great debt of gratitude to those who trained me in the disciplines of engineering. The efforts of Dr. James Foxworthy of Loyola University and Dr. Perry McCarty of Stanford University are most appreciated. I owe a debt to my father, Mr. Lester Haug, which can never fully be repaid. He introduced me to this subject, provided fresh ideas and insights when I needed them, and was the sounding board for numerous theories. In the past eight years, I hardly recall a family meeting which did not end up in a discussion of fundamental aspects of composting, much to the dismay of wives and children. My wife deserves special thanks for the support she provided during the long hours necessary to complete this text and for her proofing of the manuscript. Mrs. Alma Rios and Mrs. Jan Tanori receive my warmest thanks for typing the bulk of the manuscript and Mr. Greg Jowyk for supplying many of the graphics.

Roger Tim Haug

To Peg, David, Jim

To Les Haug, the best practical engineer I have known

mbma

Roger Tim Haug is Senior Engineer for the Regional Wastewater Solids Management Program for the Los Angeles Orange County Metropolitan Area (LA/OMA Project) where he is responsible for all technical planning and engineering aspects of a five-year program to develop a management plan for wastewater solids produced in the region.

Dr. Haug earned his MS and PhD in Environmental Engineering from Stanford University and a BS in Civil Engineering, *magna cum laude*, from Loyola University. He is a registered Professional Civil Engineer in California.

He was formerly Assistant Professor of Civil Engineering and Environmental Sciences at Loyola Marymount University, where he still maintains a position as Adjunct Professor. He has worked as a consulting engineer. His responsibilities now include technical program development, supporting documentation and advice on all technical aspects of the $3.4 million LA/OMA program.

A member of the American Water Works Association, Water Pollution Control Federation, and the California Water Pollution Control Association, Dr. Haug has published extensively in professional journals.

CONTENTS

CHAPTER 1

INTRODUCTION

COMPOSTING–DEFINITION AND OBJECTIVES

There is no universally accepted definition of composting. In this text, composting is defined as the biological decomposition and stabilization of organic substrates under conditions which allow development of thermophilic temperatures as a result of biologically produced heat, with a final product sufficiently stable for storage and application to land without adverse environmental effects. Thus, composting is a form of waste stabilization, but one that requires special conditions of moisture and aeration to produce thermophilic temperatures. Most biological stabilization and conversion processes deal with dilute aqueous solutions, and only limited temperature elevations are possible. Thermophilic temperatures in aqueous solutions can be achieved if substrate concentrations are high and if special provisions for aeration are employed. Aside from such special cases, composting is usually applied to solid or semisolid materials, making composting somewhat unique among the biological stabilization processes used in sanitary and biochemical engineering.

Aerobic composting is the decomposition of organic substrates in the presence of oxygen (air). The main products of biological metabolism are carbon dioxide, water and heat. Anaerobic composting is the biological decomposition of organic substrates in the absence of oxygen. Metabolic end products of anaerobic decomposition are methane, carbon dioxide and numerous intermediates such as low-molecular-weight organic acids. Anaerobic composting releases significantly less energy per weight of organics decomposed compared with aerobic composting. Anaerobic composting has a higher odor potential because of the nature of many intermediate metabolites. For these reasons almost all engineered compost systems are aerobic. Mass transfer limitations, however, may cause anaerobic zones in an otherwise aerobic system. Such subtleties aside, this book will deal primarily with aerobic systems because of their commercial importance to man.

The objectives of composting have traditionally been to biologically convert putrescible organics to a stabilized form and to destroy organisms pathogenic to humans. Composting is also capable of destroying plant diseases, weed seeds, insects and insect eggs. Odor potential from use of compost is greatly reduced because organics that remain after proper composting are relatively stable with low rates of decomposition. Composting can also effect considerable drying, which is of particular value with wet substrates such as municipal and industrial sludges. Decomposition of substrate organics together with drying during composting can reduce the cost of subsequent handling and increase the attractiveness of compost for reuse or disposal.

Compost can be disposed of in a sanitary and usually convenient manner. If the product is reused, it can accomplish a number of additional purposes including:

1. to serve as a source of organic matter for maintaining or building supplies of soil humus, necessary for proper soil structure and moisture-holding capacity;
2. to improve growth and vigor of crops in commerical agriculture or home-related uses; and
3. to reclaim and reuse certain valuable nutrients including nitrogen, phosphorus and a wide variety of essential trace elements.

The nutrient content of compost is related to the quality of the original organic substrate. However, most composts are too low in nutrients to be classified as fertilizers. Their main use is as a soil conditioner, mulch, top dressing or as an organic base with fertilizer amendments. On the other hand, nutrients such as nitrogen are organically bound and slowly released throughout the growing season, making them less susceptible to loss by leaching compared to soluble fertilizers.

PRESENT LIMITATIONS

The most common engineered application of microbes is to treat or convert substrates in aqueous solution. Suspended growth reactors, such as the activated sludge process, or fixed film reactors, such as the trickling filter and rotating biological contactor (RBC), are widely used for treatment of municipal and industrial liquid wastes. Biological engineering is well developed and it is possible to design and operate such systems using a reasoned, engineered approach.

There are a number of biological processes used on solid or semisolid materials including fermentation and ripening of cheese, production of silage and, of course, composting. At least in the case of composting, a reasoned, fundamental approach to analysis has not been fully-developed.

Almost every book on the subject of composting begins with the statement that composting is an ancient art, probably practiced by man since before the dawn of recorded history. Although the evidence suggests that man has had a long affair with composting, fundamental scientific studies of the process have generally occurred in the past three decades. Our ability to engineer the process and to understand the numerous competing forces within a composting material is even less well developed. In other words, the theory of composting may be understood and most of the forces involved may be known, yet engineering of systems is still often conducted using a "handbook" approach with little knowledge of how to control these forces to achieve the final end product. It is a goal of this book to develop a more fundamental approach to analysis and design, one that would rely as much as possible on first principles of physics, chemistry, biology, thermodynamics and kinetics.

COMPOSTING SUBSTRATES

The quantity of substrates potentially suitable for composting is indeed large. One estimate of solid and semisolid organic wastes generated and collected in the United States is shown in Table 1-1. Urban refuse, manure and agricultural wastes represent major components of the collected fraction, which totals over 100 million ton/yr. Problems encountered in managing waste materials depend not only on quantity, but also on their characteristics. Thus, municipal and industrial sludges, because of their high moisture

Table 1-1. Estimates of Organic Wastes Generated and Collected in the U.S. in 1971 (10^6 ton/yr, dry wt) [1][a]

Waste Type	Generated	Collected
Urban Refuse	115	65
Manure	180	24
Agricultural Crops and Food Wastes	355	21
Industrial Wastes	40	5
Logging and Wood-Manufacturing Residues	50	5
Miscellaneous Organic Wastes	45	5
Municipal Sewage Solids	11	2
Total	796	127

[a]Values rounded from original reference.

content, may present management difficulties far in excess of their relative tonnage.

There appear to be essentially two basic approaches to resource recovery from organic wastes: (1) use of the organics (and associated nutrients) either directly in the soil or after production of compost material, and (2) conversion of the organics to alternative energy forms. Both paths to resource recovery have noble objectives. Compost is a proven organic supplement which, by supplying humus and nutrients to deficient soils, can greatly improve crop yields. Alternatively, the energy potential of organic wastes is considerable. Development of the resource as an alternative fuel has received economic impetus from worldwide increases in energy prices. Which of these reuse possibilities will win out? To answer this, the characteristics of waste organics must be examined.

Relatively dry wastes, such as municipal refuse, are probably more valuable as energy resources. Energy can be extracted efficiently by thermal processes (e.g., incineration, pyrolysis and gasification) because of the dry nature of the material. Costs of extraction may be high and certain components of a heterogeneous waste, such as refuse, will not be amenable to thermal processing. As moisture content increases, thermal processing becomes much less efficient. For combustion to be self-supporting, it is usually necessary for moisture levels to be less than 60-70%, although the exact value depends on the nature of the organic being burned. If the waste is in the form of a liquid slurry or suspension, anaerobic digestion is the only practical energy recovery method. But what about residues remaining after anaerobic digestion, or other organic wastes too wet for efficient thermal conversion to energy? In the past, fossil fuels were often added to such wastes either to support combustion of the organics or to remove moisture (heat drying). But such processes are energy-intensive, and the use of fossil fuels in this manner is falling into increasing disfavor, as well as becoming very expensive.

High-moisture (greater than about 60%) organic wastes represent a rather unique management problem. Direct application to land is possible, but such practice is usually limited to rural areas where sufficient land is available. Composting can be particularly effective in converting wet materials to a more usable or easily disposable form. At the same time, composting can stabilize putrescible organics, destroy pathogenic organisms and provide significant drying of the wet substrate. All of these advantages are obtained with minimal outside energy input; the major energy resource being the substrate organics themselves. Furthermore, composting is a flexible process: it can be viewed as a conversion process to produce a material suitable for reuse or simply as a stabilization and drying process to provide for easier disposal. Composting is also compatible with a wide variety of feedstocks.

Sludges from municipal and industrial wastewater treatment, certain other industrial processes and animal manures represent a major portion of the high-moisture organic wastes. Estimates of past and future municipal sludge production in the United States are presented in Figure 1-1. Amounts are expected to increase as treatment plants are expanded and upgraded to higher levels of treatment. The present annual production of municipal sludge is about 6-7 million metric tons dry weight. Suler [2] presented the following partial list of organic industrial sludges:

- The food industry generates about 650,000 dry ton/yr of organic sludges which are mostly readily degradable.
- The textile industry produces about 300,000 dry ton/yr, mostly organic and composed of cotton, wool, synthetic fibers, dyes and sizing.

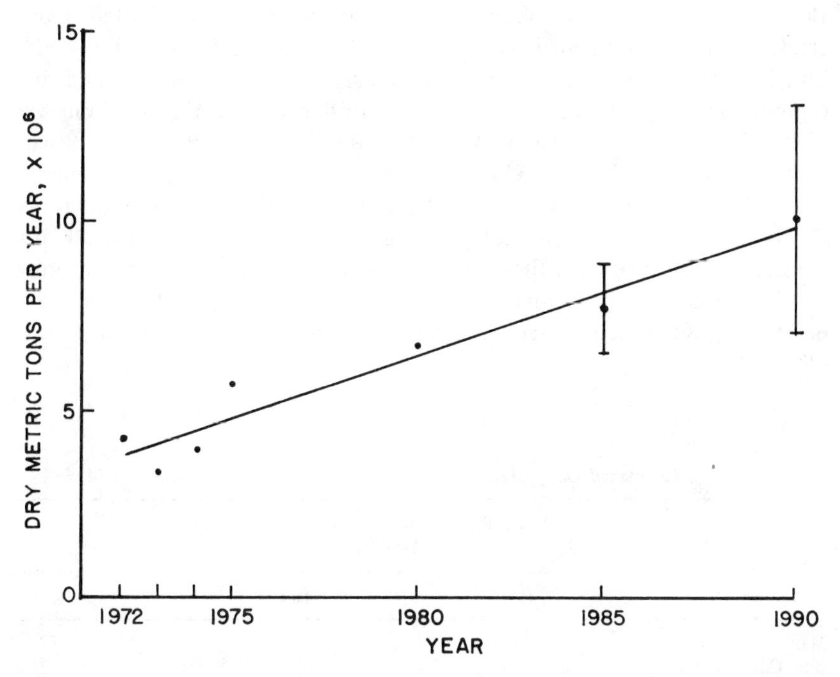

Figure 1-1. Estimates of present total U.S. sludge production and predictions for future sludge production. Values for metric tons dry weight produced in the years 1972-1975 are based on estimates of population served by wastewater treatment and per capita solids production. Projections for 1980, 1985 and 1990 reflect the increase of sludge expected to arise from institution of secondary wastewater treatment at all facilities where it is not now in effect and from construction of new facilities. Ranges are given for 1985 and 1990 estimates. Data presented here indicate that between 1972 and 1990 the amount of sludge produced per capita in the U.S. may more than double. Data compiled by National Research Council of the National Academy of Sciences [3]. See cited reference for original sources of information.

- The pulp and paper industry generates about 2 million dry ton/yr, mostly carbonaceous containing cellulose fibers and other biomass.
- The pharmaceutical industry produces about 200,000 dry ton/yr.
- The petroleum industry generates about 850,000 dry ton/yr of petroleum-related sludges.

Composting has even been applied on laboratory- and pilot-scale to hazardous industrial wastes such as TNT, petroleum sludges and certain pesticide residues. Manures represent the other major class of wet organics. Production per head and representative moisture contents are presented in Table 1-2.

The question posed previously regarding resource recovery by composting (or direct land application) or thermal processing can now be answered. It is likely that both avenues of resource recovery will be practiced in the future. Energy recovery will find its greatest potential with dry waste organics. High-moisture materials present unique problems and are often more conducive to management by land disposal, land application or composting. This is not to imply that a distinct boundary exists between wet and dry organics. Indeed it does not. For example, drying systems are now available to remove water from wet organic slurries with less thermal input than conventional heat drying. With such equipment the feasibility of energy recovery from wet organics is enhanced. However, the technology is sophisticated and certainly not applicable in all cases. Also, not all dry organics will be processed thermally because of costs, air pollution concerns, competing disposal alternatives, etc. Although each case should be evaluated on its own site-specific factors, it would seem that composting will be a

Table 1-2. Estimated Daily Manure Production by Various Farm Animals [4][a]

Animal	Daily Raw Manure Production, Wet Weight (lb)	Daily Raw Manure Production, Wet Weight (kg)	Moisture (%)
Man	0.33	0.15	77
Beef Cattle	53.8	24.5	82
Dairy Cattle	102.5	46.6	87
Chicken	0.32	0.15	73
Broiler	0.28	0.13	75
Turkey	0.89	0.40	68
Duck	0.74	0.34	
Sheep	3.8	1.7	73
Horse	44.2	20.1	60
Swine	7.1	3.2	84

[a]Values are averages of those reported by several different authors.

favored process, particularly with wet organic substrates and in some cases even for dry organics. Furthermore, there appears to be a considerable quantity of such wastes, which makes engineering analysis of the compost process worthwhile.

The focus of attention in this text is on composting wet organic substrates, particularly municipal and industrial sludges. Nevertheless, the principles developed will apply to composting of all materials.

PROBLEMS OF HIGH-MOISTURE WASTES

Generally, composting has been applied to nonfluid, dry materials. Municipal refuse, agricultural residues, dried animal manures and forest wastes are examples of substrates commonly used as composting materials. Composting of dewatered sludges differs uniquely from composting of such relatively dry materials in that dewatered sludge may still be 70–80% water. Water is often added while composting dry substrates to avoid moisture limitations, whereas wet sludges usually require the addition of a drying agent. The presence of much water in sludge can result in reduced composting temperatures and inefficient operation if moisture is not controlled. An understanding of the thermodynamics of composting is essential for designing workable composting systems.

As a rule, the higher the moisture content of the organic material, the greater is the need to maintain a large void volume to ensure adequate aeration. Dewatered sludge cake is not a bulky material, and it lacks the porosity of materials such as straw or refuse. Because of its plastic nature, sludge also tends to compact under its own weight which further reduces the void volume.

The high moisture content, lack of porosity, tendency to compact and the need to dry dewatered sludge during composting make sludge composting somewhat unique and often difficult. In recent years several composting systems have been developed or modified to overcome problems associated with sludge composting.

Before beginning the text, a description of some additional differences between composting wet organic sludges and dry organic solids is in order. Sludge, both liquid and dewatered, has been added to dry composting material as a source of moisture and nitrogen. For example, the addition of sludge to composting refuse adjusts moisture and nutrients. The focus of this book is on the opposite case, where the purpose is to compost wet sludge either alone or with small amendment additions. In the latter case, amendment addition is for the purpose of adjusting conditions for proper sludge composting and not primarily for the purpose of composting the amendment.

APPROACH TO ANALYSIS

A fundamental approach has been used throughout this text. Whenever possible the first principles of physics, chemistry, biology, thermodynamics and kinetics have been applied to problem analysis and solution. In particular, a heavy emphasis has been placed on process thermodynamics. No other science can so unify the approach to analysis. Application of thermodynamic principles is perhaps the most fundamental way of analyzing composting systems, just as it has been a fundamental tool in analyzing and coordinating physics, chemistry and biology.

The book contains a heavy emphasis on theory, but guidance is provided to the engineer in applying the theory to design of practical compost systems. The goal has been to develop a working model of the process based on theoretically sound arguments, supported when necessary by experimental data. The model allows one to study the relationship of process variables; to determine limits within which composting must be conducted; to provide answers to questions for which field data are lacking; and to provide guidance for improved design of future systems.

Design of compost facilities is a challenge, but one that is made considerably easier by a firm grasp of the engineering principles of composting. To develop these principles, background chapters on compost systems, chemical thermodynamics, microbiology and heat inactivation kinetics have been provided (Chapters 2 through 5). From this point, the analysis centers on physical aspects of composting such as moisture and volatile solids control, the relationships between density, porosity and free airspace, and aeration requirements (Chapters 6 through 8). Thermodynamic principles are then applied to a generalized compost system in Chapter 9, followed by a discussion of process kinetics in Chapter 10.

At this point, if the reader is not exhausted, he should have sufficient background to tackle the development of process simulation models presented in Chapter 11. These models are based on combining thermodynamics and kinetics to produce useful mathematical models of the compost process. Application of the simulation models to various compost systems, the interplay of process variables and implications for design and operation are presented in Chapters 12 through 14.

Composting can be subject to nuisance problems such as odors and dust. Therefore, measures to control such conditions should be a part of any process design. Solutions to these problems are discussed in Chapter 15. The book then concludes with a discussion of the future of composting, design aspects of future composting systems and the need for reasoned, fundamental analysis of the process as well as other alternatives on a case-by-case basis (Chapter 16).

BASIC UNITS

To the extent possible, SI (metric) units have been used in this text. However, much of the composting literature is in English units and, at times, it was more appropriate to maintain the original units. In cases where the the nature of the unit is not obvious, a designation will appear in parentheses next to the unit. If no specification is given, the reader should assume the units to be metric. The most common example of this is use of the term "ton." This should always be interpreted as a metric ton, which is 1000 kg, unless specifically written as "ton (short)," which is 2000 lb.

GREAT EXPECTATIONS

It is hoped that the reader will come to appreciate the ordered complexity of what appears on the surface to be a simple process. Indeed, composting can be a very contradictory process. For example, rapid and extensive organic stabilization is desired. But this leads to rapid and extensive heat evolution, which in turn can elevate temperatures to the point where biological reaction rates become temperature-limited, slowing down the organic stabilization. With "energy-rich" sludges, the system thermodynamics may impede kinetics of the process once started. In such a case, it may be very difficult to achieve optimum values for more than one or two variables at a time. For example, a high level of organic decomposition, high temperature elevation and high level of drying may not be simultaneously achievable. As another example, low ambient temperatures can lead to kinetic limitations which impede development of the available energy resource. It is much like placing a large log on a fire, slow to start but once started, hard to control. Exploring such interrelationships has been a fascinating experience and I hope to convey a sense of this fascination to you the reader.

CHAPTER 2

COMPOSTING SYSTEMS

INTRODUCTION

This chapter provides a brief description of the available types of composting systems. Emphasis will be placed on the fundamental concepts which underlie all composting systems and their operation. New composting systems will likely be proposed at an increasing rate in response to the growing interest in sludge composting, and any text that simply describes available systems would be obsolete within a few years of publication. The basic principles, however, remain unchanged regardless of how new or novel the composting system or how eloquent and loud the demands of the entrepreneur to the contrary.

Nevertheless, it is necessary to have some understanding of available composting systems before proceeding too far into fundamental concepts. Therefore, this chapter will introduce the general nature of composting systems, describe their basic similarities and differences, introduce terminology used in later chapters, and effect transition into the more fundamental aspects of composting presented in later chapters.

GENERALIZED COMPOSTING SCHEMATIC

A generalized schematic diagram for composting is presented in Figure 2-1. Wet feed substrates are difficult to compost alone because of the high moisture content. Dewatered sludge cake has never successfully been composted alone except in small pilot-scale facilities where constant mixing can be applied. This is caused by the high moisture content of the sludge cake which saturates all of the void space. Oxygen transfer into the composting mass is effectively prevented unless high levels of agitation are used continually to expose new surfaces for oxygen transfer. Three approaches have

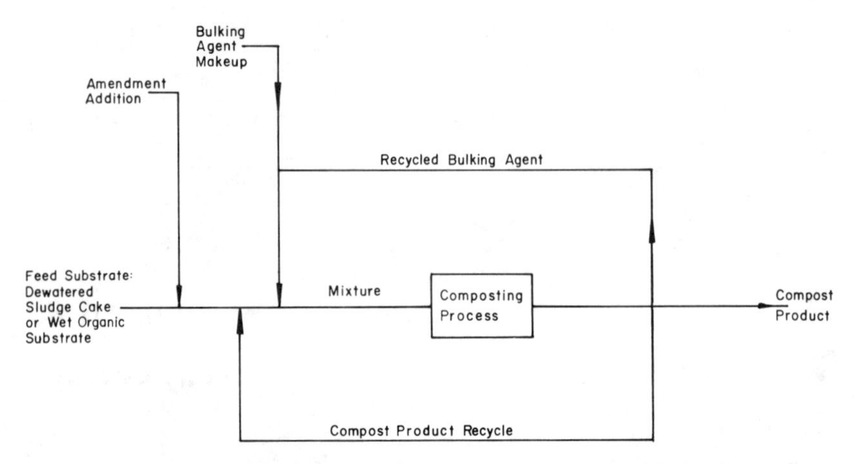

Figure 2-1. Generalized diagram for composting showing inputs of feed substrate, compost product recycle, amendment and bulking agent.

been used to overcome the problem of oxygen transfer: (1) recycle of compost and blending with the dewatered cake before composting (note that this approach is often referred to as sludge-only composting because the recycle is entirely of sludge origin with no outside amendments); (2) addition of an organic amendment to the sludge; and (3) addition of a bulking agent such as wood chips and subsequent screening of the bulking particles from the compost product.

At this point it would be advisable to define "amendment" and "bulking agent" as used in this text. An amendment refers to an organic material added to the feed substrate primarily to reduce bulk weight and increase air voids allowing for proper aeration. Amendments can also be used to increase the quantity of degradable organics in the mixture. Amendments that have been used with dewatered sludge include sawdust, straw, peat, rice hulls, manure, solid wastes such as refuse and garbage, tree and lawn trimmings, and a variety of other waste organics. The ideal amendment would be dry, have a low bulk weight and be readily degradable. Compost recycling can be used to accomplish the reduction in bulk weight and in this sense should be considered as an amendment. It has been distinguished from other amendments listed above because it allows the feed substrate to be composted without any external organic additives. Compost recycling along with amendment addition has been used in some cases.

A bulking agent is a material, organic or inorganic, of sufficient size to provide structural support and maintain airspace when added to wet sludge. "Bulking particle" may be more descriptive and is the term preferred by this author although use of the term "bulking agent" is irreversibly ingrained in the literature. Bulking particles form a three-dimensional matrix of solid particles capable of self-support by particle-to-particle contacts. Sludge cake can be viewed as occupying part of the void volume in the interstices between particles. If the bulking particle is organic, an increase in the quantity of degradable organics may be a secondary benefit. Wood chips are most commonly used as bulking agents although the use of pelleted refuse, shredded tires, peanut shells, tree trimmings, rocks and other materials has been reported.

A variety of different composting systems has been used in the "compost process" block of Figure 2-1. Although details of these systems may vary, basic principles governing the composting process remain unchanged, and all systems use either recycled product, amendments or bulking agents to overcome problems of excessive feed substrate moisture.

CLASSIFICATION OF COMPOSTING SYSTEMS

Attention will now be turned to the various processes that have been used in the "composting process" block of Figure 2-1. A variety of techniques can be used to classify composting systems. A chemical engineering approach to classification will be used here which emphasizes reactor type, solids-flow mechanisms, bed conditions in the reactor and manner of air supply. Perhaps the most basic distinction is between those systems in which the composting material is contained in a reactor and those in which it is not. Systems which use reactors are popularly termed "mechanical," "enclosed" or "within-vessel" compost systems, whereas those which do not are often termed "open" systems. The term "mechanical" is a misnomer, however, because all modern compost systems are mechanical to some extent. Some employ mobile equipment while others use stationary reactors and conveyors. One system might be considered more mechanical than another but the basic distinction is illusive.

The terms "enclosed" and "open" are also poorly defined. For example, composting material might be housed under a roofed structure and thereby be enclosed but not contained in a reactor. Would such a system be "open" if it is under a roofed structure? To avoid such confusion, we will make the basic distinction between reactor and nonreactor systems. Reactor systems are those in which composting material is placed in a reactor, and nonreactor systems are those in which it is not. Both types of systems can employ mechanical equipment and both may or may not be enclosed under protective housings. Simplified sketches of the basic reactor and nonreactor systems are presented in Figures 2-2 through 2-4. Brief descriptions of representative compost systems are given in Tables 2-1 and 2-2.

Table 2-1. Composting Processes Developed Primarily for Mixed Solid Wastes as Main Feed Component (Adapted and Expanded from References 5 and 6)

Classification	Process Name	General Description
Nonreactor		
Agitated Solids Bed	Bangalore (Indore)	Developed by Sir Albert Howard (1925) in India. Trench in ground, 2-3 ft deep. Material placed in alternate layers of refuse, night soil, earth, straw, etc. No grinding. Turned by hand as often as possible. Detention time of 120-180 days. Used extensively in India.
	Conventional and Forced Aeration Windrow	Open windrows with a haystack cross section. Refuse ground. Aeration by turning windrows. Detention time depends upon number of turnings and other factors. Once used in Mobile, AL (270 ton/day); Boulder, CO (90 ton/day); and Johnson City, TN (47 ton/day); currently used in Israel and Mexico City.
	Van Maanen	Raw refuse without any pretreatment except moisture adjustment placed in open windrows for 120-180 days. Turned once by grab crane. Must be conducted in remote areas because of odors, fly and rodent problems. First used in Netherlands in 1931.
	Others	Numerous process names have been applied to the windrow system, often taking the name of the front-end equipment manufacturer.
Static Solids Bed	Brikollari (Caspari) (Briquetting)	Ground material is compressed into blocks and stacked for 30-40 days. Aeration by natural diffusion and airflow through stacks. Curing follows initial composting. Blocks are later ground. Sludge can be added to mixture up to moisture content of about 53%. Plants in Schweinfurt, Germany and Biel, Switzerland. No present U.S. installations.
Vertical Flow Reactor		
Moving Agitated Bed	Earp-Thomas	Possibly the oldest reactor system. Silo type with 8 decks stacked vertically. Center shaft drives a plow which agitates the compost and moves it downward from deck to deck. Air passes upward through the silo. Digestion of 2-3 days followed by windrowing. Installations reported in Seoul, Korea; Verona, Italy; and Basel, Switzerland. No present U.S. installations.

Table 2-1, continued

Classification	Process Name	General Description
Vertical Flow Reactor		
Moving Agitated Bed	Frazer-Eweson	Ground refuse placed in vertical bin with 4 or 5 perforated decks and special arms to force composting material through perforations. Air is forced through bin. Detention time of 4-5 days. Problems with bridging of solids across perforated deck. An 18-ton/day facility operated on garbage in Springfield, MA from 1954 to 1962. No known facilities presently in operation.
	Jersey (John Thompson)	Structure with 6 floors, each equipped to dump ground refuse onto the next lower floor. Aeration effected by dropping from floor to floor. Detention time of 6 days; 6-8 weeks of additional curing in unturned piles. A 300-ton/day plant constructed in Bangkok, Thailand.
	Naturizer (International)	Five 9-ft wide steel conveyor belts arranged to pass material from belt to belt. Each belt is an insulated cell. Air passes upward through digester. Detention time of 6-8 days. Plants once located in Norman, OK (1959-64), San Fernando, CA (1963-64), St. Petersburg, FL (1966-?). The last was a 90-ton/day facility, but was closed because of uncontrolled odor.
	Riker	Four-story bins with clam-shell floors. Compost is dropped from floor to floor to provide agitation. Forced aeration. Total detention time of 20-28 days. Treated mixture of ground garbage, corn cobs and sludge. Problems reported in maintaining aerobic mixture. A 4-ton/day facility operated in Williamston, MI, from 1955 to 1962. No known facilities presently in operation.
	T. A. Crane	Two cells consisting of three horizontal decks. Horizontal ribbon screws extending the length of each deck recirculate ground refuse from deck to deck. Air introduced in bottom of cells. Three days composting followed by curing for 7 days in a bin. 18-ton/day pilot refuse and sludge system was installed in Kobe, Japan.
	Varro	Ground refuse placed in 8-deck digester and moved downward from deck to deck by plows. Each deck pair had own recirculating air supply to control CO_2 level. Output dried, reground and used as base material for fertilizer, soil conditioner, wallboard, etc. Digestion time 40 hr. ~55-ton/day system constructed in Brooklyn, NY in 1971.

Table 2-1, continued

Classification	Process Name	General Description
Vertical Flow Reactor		
Moving Packed Bed	Triga	See Table 2-2 for description. As of 1978, two plants were in operation in France using mainly municipal solid waste (MSW) with some sludge.
Horizontal and Inclined Flow Reactor		
Tumbling Solids Bed	Dano	Dispersed flow rotating drum, slightly inclined from the horizontal, 9-12 ft in diameter, up to 150 ft long. Drum kept about half full of refuse. Drum rotation of 0.1-1.0 rpm. One to five days digestion followed by windrowing. No grinding. Forced aeration into drum. Probably the most popular reactor process for MSW with 160 plants worldwide as of 1972. Plant in Rome, Italy handles over 450 ton/day. Plant in Leicester, England composts refuse and sludge. U.S. installations once located in Sacramento, CA (1956-1963) and Phoenix, AZ (1963-1965).
	Fermascreen	Hexagonal drum, three sides of which are screens. Refuse is ground and batch-loaded. Screens are sealed for initial composting. Aeration occurs when drum is rotated with screens open. Detention time about 4 days.
	Geochemical-Eweson	Cells-in-series type. Unground refuse placed in rotating drum, 11 ft diam, 110 ft long, slightly inclined from horizontal. Three compartments in drum. Refuse transferred to next compartment every 1-2 days for total digestion time of 3-6 days. Screened output cured in piles. 35-ton/day facility constructed in Des Moines, IA and Big Sandy, TX in 1972. Latter composted a mixture of 27 ton/day refuse and 9 wet ton/day sludge.
Agitated Solids Bed	Fairfield Hardy	Circular tank. Vertical screws, mounted on two rotating radial arms, keep ground material agitated. Forced aeration through tank bottom and holes in screws. Continuous flow type. Detention time about 5 days. Original plant constructed at Altoona, PA in 1951 with capacity of 25 ton/day is still in operation. 135-ton/day facility constructed in San Juan, PR in 1969. In 1978 a 45-ton/day facility for composting garbage and sludge was constructed in Toronto, Ontario and is currently operating.

Table 2-1, continued

Classification	Process Name	General Description
Horizontal and Inclined Flow Reactor		
Agitated Solids Bed	Snell	Rectangular tank about 8 ft deep with porous floor equipped with air ducts for forced aeration. Tank inclined on a 6° slope. Traveling bridge with vertical paddles provides agitation and movement of material along incline of tank. Detention time 5-8 days. 275-ton/day facility constructed in Houston, TX in 1967. Reported to operate successfully but local problems forced closure.
	Metro-Waste	Rectangular tanks about 20 ft wide, 10 ft deep, 200-400 ft long. Refuse is ground. Residence time of about 7 days. "Agiloader" moves on rails mounted on bin walls and provides periodic agitation by turning. One of the more successful reactor types. 275-ton/day MSW facility operated in Houston, TX from 1966-1970. Other refuse installations once operated in Largo (45 ton/day) and Gainesville (135 ton/day), FL. A similar facility in Ohio (1972-present) composts 360 ton/day of cattle manure.
	Tollemache	Similar in design to Metro-Waste system. Installations reported in Spain and Rhodesia in 1971.

Nonreactor Systems

The nonreactor systems sketched in Figure 2-2 are divided between those which maintain an agitated solids bed and those which employ a static bed. An agitated solids bed means that the composting mixture is disturbed or broken up in some manner during the compost cycle. This may be by periodic turning, tumbling or other methods of agitation. Agitation is not synonymous with mixing. If a system is well mixed it means that the infeed will be dispersed throughout the bed volume as a result of agitation. In a completely mixed system no concentration gradients exist throughout the bed volume. Thus, compost withdrawn will have the same characteristics as material within the bed. For a system to be well mixed it must be agitated in some manner. However, an agitated system may be either well mixed or not. In fact, in some agitated systems very little, if any, mixing occurs as a result of the energy input.

The windrow system is the outstanding example of a nonreactor, agitated solids bed system. Mixed compost material is placed in rows and turned

A. Agitated Solids Bed (Windrow System)

1. Conventional: Solids are agitated by periodic turning without forced aeration. Batch feed of solids. Plug flow of solids with some dispersion from agitation.
2. Forced Aeration: Same as conventional windrow but with provision for forced aeration.

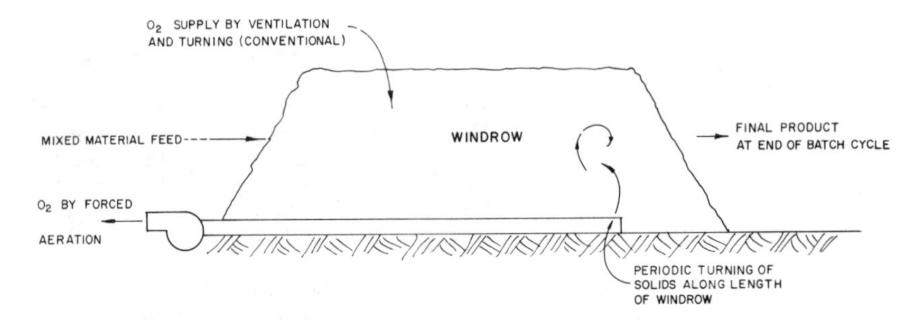

B. Static Solids Bed–No agitation or turning of static bed. Batch feed of solids. No dispersion or mixing of solids in bed.

1. Forced Aeration: Example–aerated static pile process.

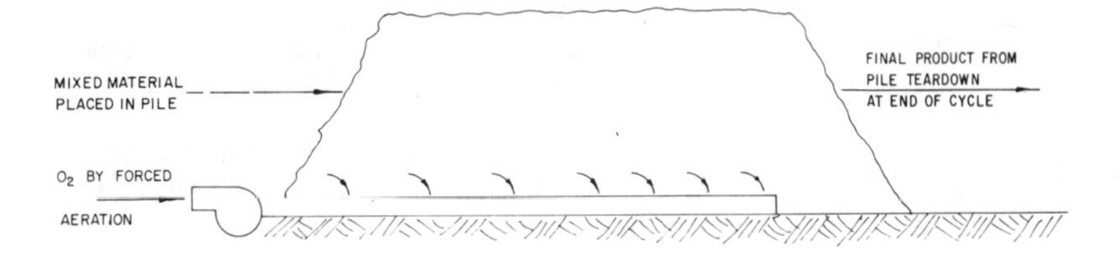

2. Natural Ventilation and Diffusion of O_2: Example–Brikollari Process.

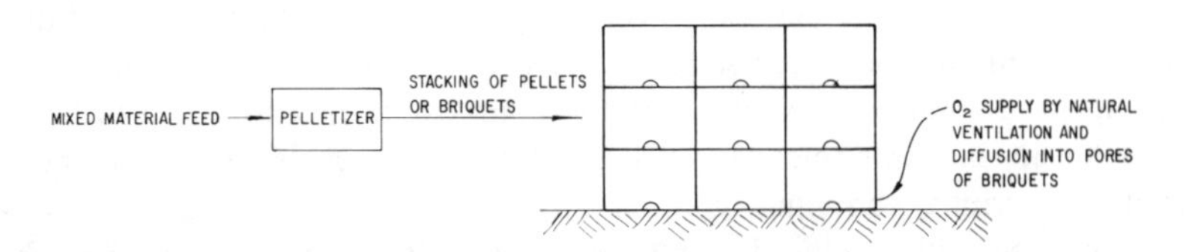

Figure 2-2. Nonreactor composting systems.

periodically, usually by mechanical equipment. Height, width and shape of the windrows will vary depending on the nature of the feed material and the type of equipment used for turning. Oxygen is supplied by gas exchange during turning and by natural ventilation resulting from buoyancy of the hot gases in the windrow. In the forced aeration windrow system, oxygen transfer

Table 2-2. Composting Processes Developed for or Adapted to Wet Substrates as Main Feed Component

Classification	Process Name	General Description
Nonreactor		
Agitated Solids Bed	Conventional and Forced Aeration Windrow	Open windrows of generally triangular cross section. Requires use of recycled compost product and/or amendment for moisture control. Conventional process used by Los Angeles County Sanitation Districts (1972-present).
Static Solids Bed	Aerated Static Pile (Beltsville or ARS process)	Open piles constructed of a mixture of dewatered sludge cake and bulking agent. Oxygen supplied by forced aeration. Used at a number of municipal U.S. facilities mainly on the East Coast.
	Pelleting Process	Dewatered sludge and recycled product used for moisture control are extruded into pellets with a diameter of about 1 cm. Pellets are then stacked into piles for composting. Oxygen is supplied by natural ventilation through the large pore spaces which result from the pelleting. System reportedly developed in Germany [7].
Vertical Flow Reactor		
Moving Agitated Bed	Thermax	Similar to Earp Thomas system. Pilot operation conducted at Orange County Sanitation Districts, CA in 1976. Used digested, dewatered sludge with recycled product and other amendments.
	Schnorr "Biocell" System	Reactor consists of a vertical tower with 10 floors one above the other. Each floor contains a hydraulically operated valve which allows material to be discharged to the next floor. Oxygen is introduced by forced aeration. Feed consists of dewatered sludge, recycled compost and ground bark in proportions of 2:2:1 by volume. Composting mixture is about 1 m deep on each floor. Residence time is about 3 days on each floor, giving a total reactor time of 30 days. Three European facilities as of 1978.
Moving Packed Bed	Kneer (Later versions known as BAV)	Reactor consists of a cylindrical tower with no interior floors or other mechanisms. Feed usually consists of dewatered sludge, recycled compost and some amendment such as sawdust or bark. Feed is introduced at the top and flows downward as product is removed by mechanical scraper from the bottom. Oxygen is supplied by forced aeration from the bottom. Residence times of 7-12 days. Over 25 installations in Germany with reactor volumes as large as 375 m^3.

Table 2-2, continued

Classification	Process Name	General Description
Vertical Flow Reactor		
Moving Packed Bed	Triga	Reactor is a concrete tower called a "Hygiensator" which is divided into four separate vertical compartments. Residence time of 4-10 days depending on feed. Air is pulled out of top of reactor. Screw extractor removes and agitates material from bottom of reactor. Extracted material is recycled 3-5 times during compost period to avoid compaction at bottom. Feed usually consists of sludge and amendments such as MSW, bark or sawdust. Curing period of 2-4 months used after reactor. French facilities at Dreux, Montargis and St. Palais. Latter facility uses dewatered sludge mixed with sawdust or bark which accounts for 1/3 by weight of the mixture.
	Euramca (Roediger, Fermentechnik)	Reactor is of the tower type with a special extraction and agitation mechanism at the bottom. The reactor is batch-operated with dewatered sludge and recycled product as the feed. Once the reactor is loaded, material is frequently recycled from the bottom to the top to provide agitation and assure uniform exposure to temperature. Residence time is 6 days. After composting, the material is pelleted and air-dried in windrows or an enclosed forced-air reactor. Pellets are broken before recycling to the composting reactor. Has been demonstrated on both raw and digested sludges. In December 1978 a facility was commissioned at Wutöshingen, West Germany, with a reactor volume of 50-m^3. Pilot unit tested by LA/OMA project in California.
	ABV	Similar to Kneer and BAV systems. At least two facilities in Sweden as of 1979.
Horizontal and Inclined Flow Reactor		
Tumbling Solids Bed	HKS	Reactor is of the complete-mix, rotating drum type. Feed consists of dewatered sludge and recycled compost product. As of 1978 the only facility was a European pilot plant at Aachen.
	Dano	See Table 2-1 for description. A number of facilities process mainly MSW along with sludge. Reactor should be capable of processing mainly sludge but no operation in this manner has been reported.

Table 2-2, continued

Classification	Process Name	General Description
Horizontal and Inclined Flow Reactor		
Agitated Solids Bed	Fairfield-Hardy, Aerotherm Processes	See Table 2-1 for descriptions. In 1969 the Fairfield-Hardy composter at Altoona, PA, was used to compost raw dewatered sludge. Feed consisted of sludge, heat-dried recycled compost and shredded paper in a dry weight ratio of 1:1.2:0.07. Material discharged from the reactor was about 50% solids and was heat-dried to 87% before recycling.
	Metro-Waste	See Table 2-1 for description. System has processed manure. Composting by forced aeration and turning of sludge/recycled compost mixtures in 10-ft-deep bins was demonstrated by the LA/OMA Project. Since this is similar in concept to the Metro-Waste system, the latter should be applicable to sewage sludges.

into the windrow is aided by forced or induced aeration. In most composting literature the term forced aeration is used regardless of whether the aeration is forced or induced. In either case, periodic agitation by turning is used to restructure the windrow. As a result, considerable mixing can be expected along the height and width of the row, but little mixing will occur along the length.

Two static solids bed systems can be defined by the method of aeration. The first uses forced aeration and is typified by the aerated static pile process which will be described later. The second relies on natural ventilation and diffusion of oxygen (e.g., the Brikollari process). In either case, no agitation or turning of the static bed occurs during the compost cycle and the piles or briquets are formed on a batch basis. Because there is no bed agitation, no mixing occurs once the pile is formed. Some recent designs for the aerated static pile process have included provisions for tearing the pile down, delumping solids and reforming for additional composting. If this provision is used the system could be classified as an agitated solids bed system even though the agitation is considerably less than that normally associated with the windrow system.

Reactor Systems

Reactor systems have been divided into vertical flow reactors (Figure 2-3), and horizontal and inclined flow reactors (Figure 2-4). Vertical flow reactors

A. Moving Agitated Bed Reactor–Solids are agitated during movement down the reactor. Forced aeration. Continuous or intermittent feeding. Some mixing in reactor.

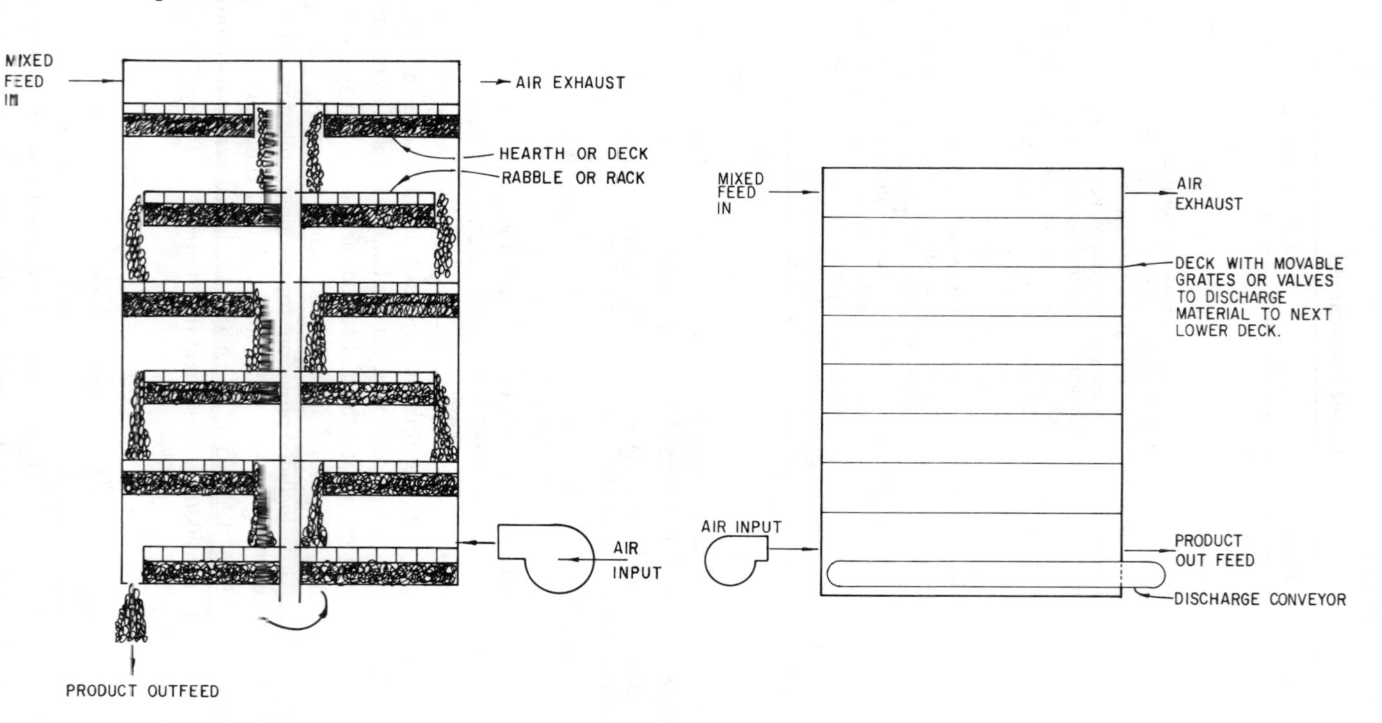

Example–Earp Thomas Reactor

Example–Schnorr Reactor

B. Moving Packed Bed Reactor–Solids are not agitated during movement down the reactor. Forced aeration. Plug flow of solids in reactor.

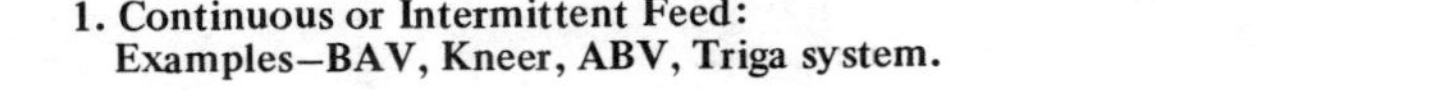

1. Continuous or Intermittent Feed:
Examples–BAV, Kneer, ABV, Triga system.

2. Batch Feed: Example–Euramca Process.

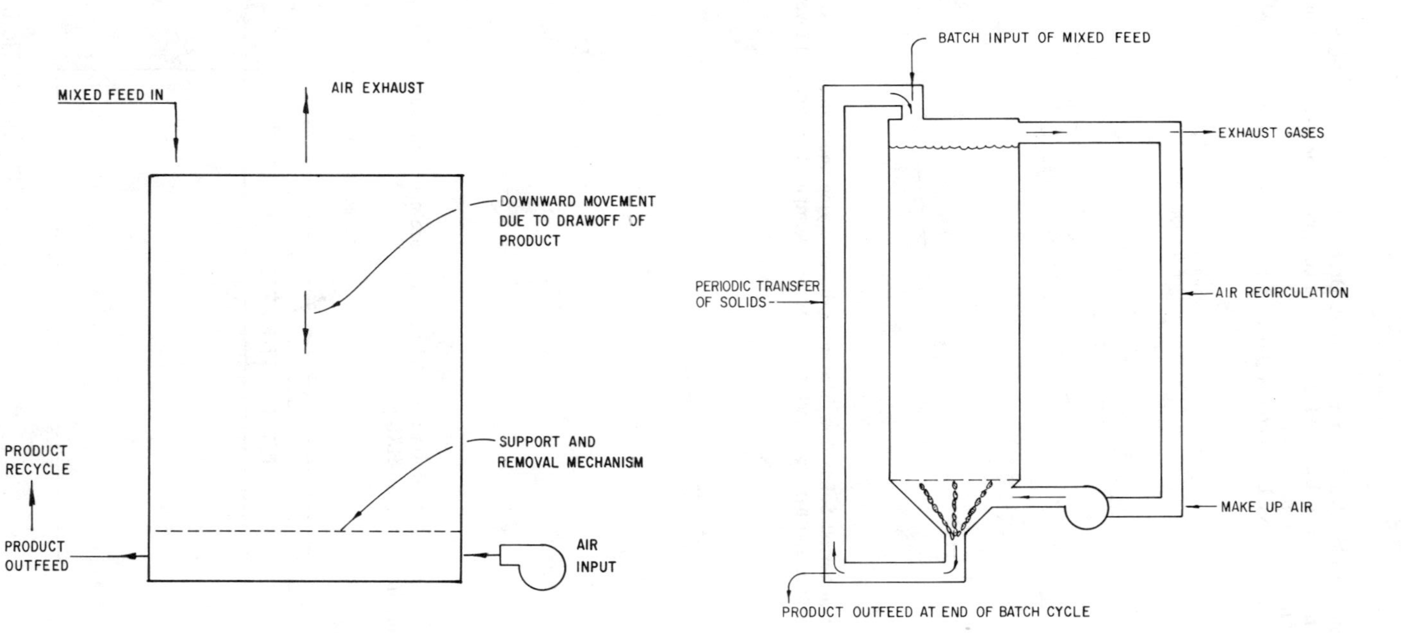

Figure 2-3. Vertical flow reactor systems (tower or silo reactors).

A. Tumbling Solids Bed Reactor (Rotating Drum)–Solids are agitated by nearly constant rotation of a drum and are fed on a continuous or intermittent basis. Forced aeration is usually provided.
   1. Dispersed Flow: Dispersion is provided by constant tumbling action. Example–Dano process.

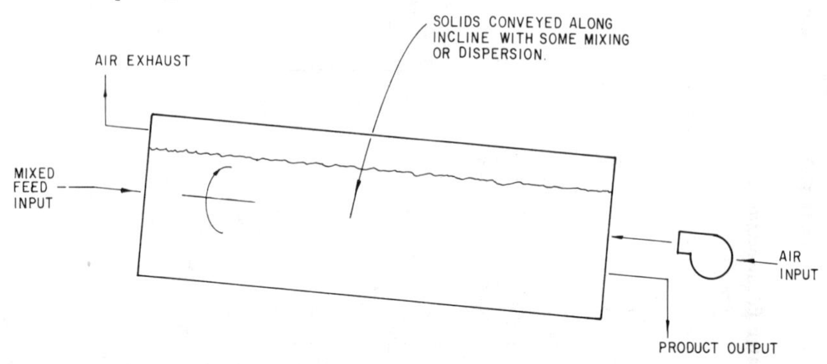

   2. Cells in Series: Solids flow is by periodic emptying and transfer of material from one cell to another. Each cell is well mixed. Example–Geochemical-Eweson.

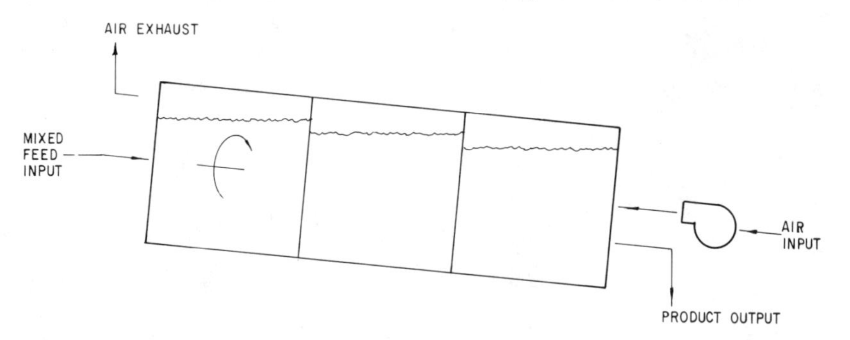

   3. Complete Mix: Uniform feed and discharge are maintained along with a high level of mixing. Example–HKS process.

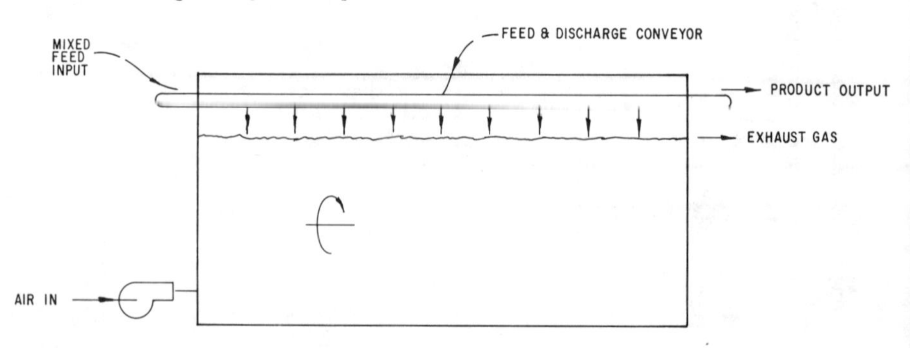

Figure 2-4. Horizontal and

B. Agitated Solids Bed Reactors (Bin Reactors)–Solids are agitated by mechanical turning devices and are fed on a continuous, intermittent or batch basis. Forced aeration is provided.

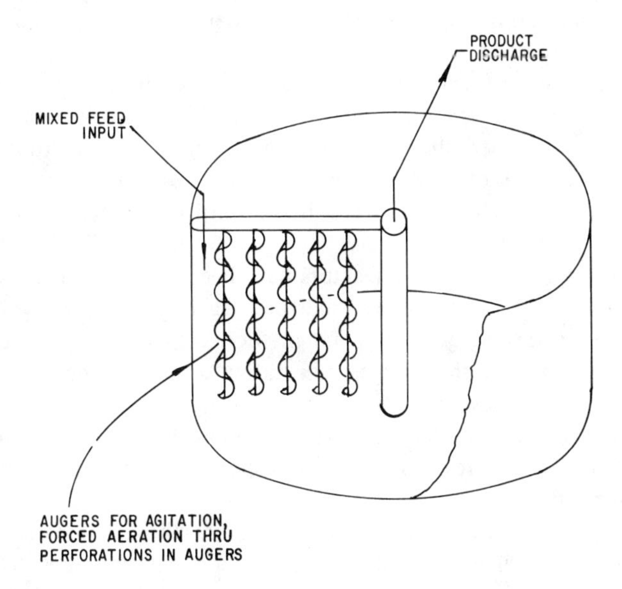

Examples: Fairfield-Hardy, Aerotherm, Snell (rectangular tankage).

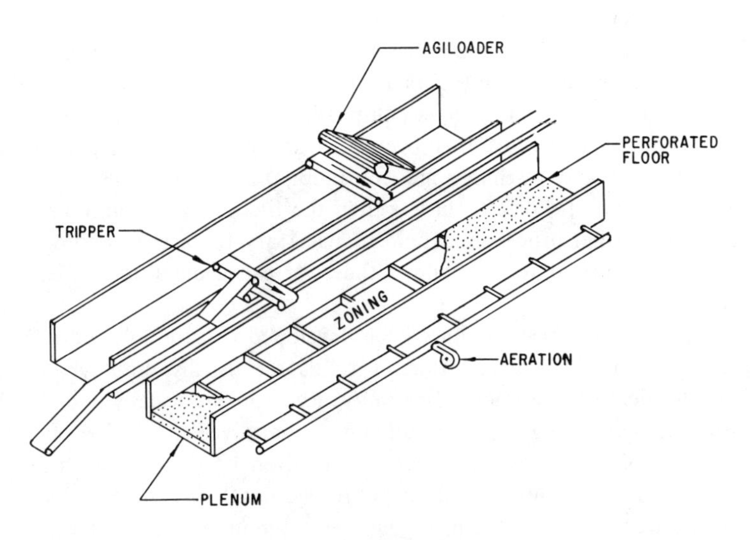

Example: Metro-Waste System.

inclined flow reactors.

can be defined further according to bed conditions in the reactor. Some systems allow for agitation of solids during their transit down the reactor, and are termed moving agitated bed reactors. These are usually fed on either a continuous or intermittent basis. In other systems, the composting mixture occupies the entire bed volume and is not agitated during any single pass in the bed. These are termed moving packed bed reactors and can be fed on either a continuous, intermittent or batch basis. The batch system shown in Figure 2-3 allows for periodic transfer of solids from the bottom to the top of the reactor. Agitation of solids occurs as a result of this transfer, but on any single pass the bed solids remain unagitated until they are again withdrawn from the bottom for transfer.

Horizontal and inclined flow reactors have been divided into those which employ a rotating drum (tumbling solids bed reactor) and those which use a bin structure of varying geometry and method of agitation (agitated solids bed reactors). At least three types of rotating drums can be distinguished based on the solids flow pattern within the reactor. In the dispersed flow case, material inlet and outlet are located on opposite ends of the drum. Plug flow conditions would result except for dispersion of material induced by the tumbling action.

To ensure that material does not short-circuit through the reactor, the drum can be compartmented into a number of cells in series. The most common operation is to discharge product from the last cell. Each cell is then consecutively discharged into the next. Feed material is added to the first cell once it has been emptied. Thus, feed is added on an intermittent basis and the reactor can be viewed as a number of well-mixed cells in series.

In the last case an attempt is made to mix completely the contents of the drum to produce a homogeneous composting mass. This requires that both feed and discharge of material occur uniformly along the length of the reactor. If feed and discharge are continuous, a considerable quantity of material will exit the reactor in less than the theoretical detention time. In such a case, further processing would be necessary to ensure a pathogen-free product. Intermittent feeding and withdrawal can avoid this problem to a certain extent.

A number of bin reactors have been developed (Figure 2-4). Both forced aeration and mechanical agitation of solids are usually employed, which allows considerable operational flexibility. Reactors are usually uncovered at the top and may be housed in a building for improved all-weather operation and control of nuisances. Bin reactors can often be constructed using less expensive building materials such as poured concrete or concrete block. The Fairfield-Hardy and Aerotherm systems are commonly operated on a continuous- or intermittent-feed basis. Augers used for agitation will also induce some mixing of the material as it is transferred from inlet to outlet. The Metro-Waste system may be operated on a batch or intermittent basis and less mixing is likely because of the high length-to-width ratio of the reactor.

Available Systems–Yesterday and Today

A brief overview of some of the nonreactor and reactor systems developed for solid wastes (primarily refuse and garbage) is presented in Table 2-1. Many of these systems are of historical significance only and are no longer pursued because of a number of factors including high costs, mechanical difficulties or the general decline in refuse composting in the late 1960s. This decline was particularly evident in the U.S., which witnessed a nearly total elimination of refuse composting, even though some of the systems were mechanically successful. Sanitary landfilling was often less expensive and refuse composting was often plagued by lack of markets for the final product. Compounding this, compost produced from solid wastes usually contained glass fragments, rubber and plastic material, which limited its attractiveness in the specialty and home markets.

Renewed interest in composting has been spurred by the pressing need to dispose of sludge and other wet organic substrates. Competing disposal alternatives for sludges are expensive and economic limitations against composting are not as serious as was the case with solid wastes. Furthermore, the product of sludge composting is often more aesthetically attractive and major markets for home and specialty products exist. For these reasons, manure and agricultural wastes have for a long time been viewed as suitable composting materials. In some cases, solid wastes may again enter the picture as the utility of processed refuse as an amendment or bulking agent for sludge composting is realized.

As a result of the interest in sludge composting, new processes have been developed and certain older processes adapted to composting of wet feedstocks. A number of these systems are briefly described in Table 2-2. A select number of the systems in Table 2-2 will be described in more detail in the next section.

At this point it would be advisable to reiterate that sludge, both liquid and dewatered, has commonly been added to composting material as a source of moisture or nitrogen. Addition of sludge cake to composting refuse is an example and many of the systems outlined in Table 2-1 used this approach. Primary emphasis was on refuse composting with sludge serving as an amendment to adjust moisture and nutrients for composting. Use of sludge in this manner is not the main theme of this text. Instead, attention is focused on cases where sludge is composted either alone or with small amendment additions. In this case, the amendment addition is for the purpose of adjusting conditions for proper sludge composting and not primarily for the purpose of composting the amendment.

SLUDGE COMPOSTING SYSTEMS

The Windrow System

In the windrow composting system, wastes are stacked in piles that can be arranged in long parallel rows or windrows. In large systems, such windrows are turned at regular intervals by mobile equipment. In cross section, the windrows may range from rectangular to trapezoidal to triangular, depending largely on characteristics of the composting material and equipment used for turning. The windrow system has been used successfully for composting a wide variety of organic residue. In general, windrow composting is relatively low in cost because of the productivity of the mechanical equipment involved, but can also be relatively land-intensive.

A major adaptation that has allowed the windrow system to be applied to wet organic substrates is the concept of recycling dry compost to blend with wet feed. The quantity of recycled material is adjusted to obtain a mixture moisture content of $\leqslant$60% [9]. By so doing, the structural integrity of the mixture is increased so that a properly shaped windrow can be maintained. Friability or porosity of mixed material is also greatly improved, which, in turn, increases the effectiveness of windrow turning for aeration. Amendments such as wood chips, sawdust, straw or rice hulls can also be added to the dewatered cake, either with or without dry compost recycle, and will accomplish the same purposes.

The windrow composting technique is thought to be the oldest "sludge only" composting system. Since 1972 it has been used by the Los Angeles County Sanitation Districts to compost approximately 90 dry ton/day of digested, dewatered sludge cake. Typical composting time for the windrow system has been reported to be 3-4 weeks during most of the year. Final compost solids contents of 60-70% are consistently achieved during favorable weather conditions [9]. The Upper Occoquan Sewer Authority, VA, operates a windrow composting facility that processes about 8 dry ton/day of digested primary, secondary and advanced wastewater treatment sludges, combined. The system is completely covered by a roofed structure (Figure 2-5). A concrete slab flooring with provision for forced aeration of the windrows is also provided to improve operational characteristics during wet weather [10].

A schematic diagram of a windrow sludge composting system is shown in Figure 2-6. Temporary storage is typically provided for dewatered sludge

Figure 2-5. Enclosed windrow compost facility operated by the Upper Occoquan Sewage Authority, VA. Note the copious steam cloud as a result of turning.

cake. This allows sludge to be held during periods of wet weather and provides a transition between the operating schedule of the dewatering facilities and that of the composting operation. All materials to be added to the feedstock must then be mixed together to achieve as homogeneous a feed as possible. Mixed materials are then formed into windrows.

A roughly trapezoidal cross section has been reported [9], by use of digested sludge cake and recycled compost (Figure 2-7), with a loading of about 2650 m^3 mixed material/ha (1400 yd^3/ac). The actual cross section is largely a function of the feedstock and the turning equipment used. More fibrous feedstocks can generally be formed into deeper windrows of rectangular cross section. This is illustrated in Figure 2-9 for a composting operation producing bedding material for mushroom growing. A major component of the mixture is straw which allows a rectangular windrow shape with depths of 2.4-3.0 m (8-10 ft) without excessive compaction of the bottom material. At present there is considerable competition among equipment manufacturers to develop machines capable of forming and turning larger windrows. Therefore, considerable improvement in the loading per hectare can be expected in the future.

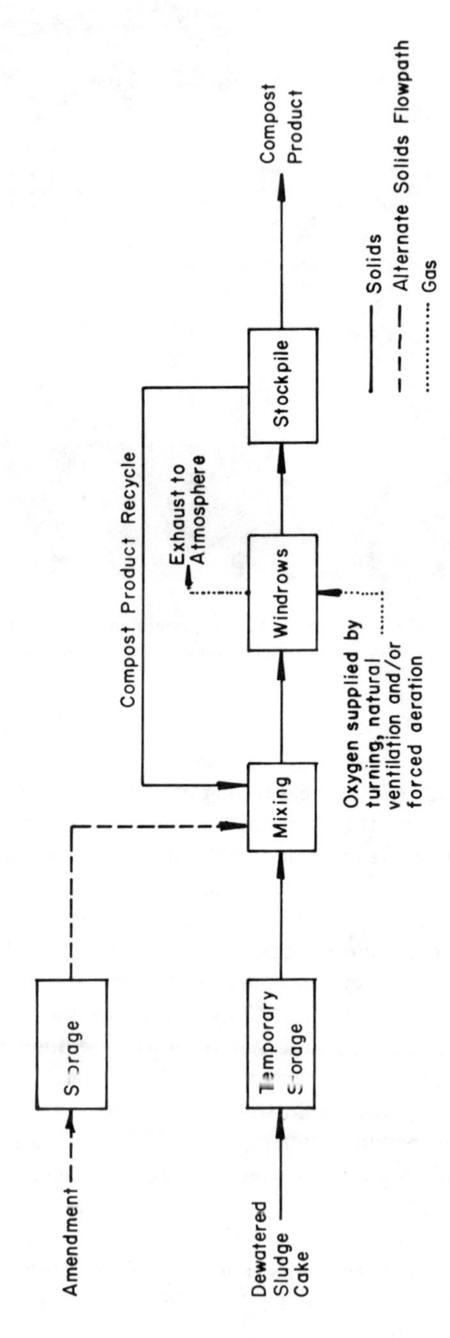

Figure 2-6. Process flow diagram of a windrow sludge composting operation.

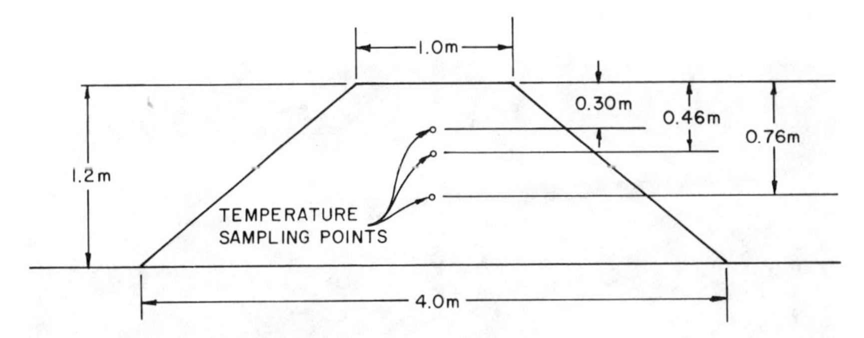

Figure 2-7. Typical windrow formed from a mixture of sludge and recycled compost [9]. Actual dimensions are largely a function of the feedstock and turning machine.

Figure 2-8. Windrow composed of dewatered, digested sludge and recycled compost product being turned by mobile equipment. Note steam cloud rising from windrow after turning.

A materials balance reported by Horvath [9] for windrow composting of digested sludge cake is shown in Figure 2-10. It is interesting to note that even though the feed substrate was anaerobically digested, a considerable quantity of remaining organics were biodegradable under the composting conditions. For each 1.9 tons of volatile solids (VS) in the feed sludge about

Figure 2-9. Windrow composting of a mixture of straw, manure and corn cobs. Product is used as a mushroom bedding material. Note the rectangular shape of the windrows.

0.9 ton is lost during composting, nearly a 50% reduction of feed VS. VS content of the final product is typically less than 40%. The significant loss of water during the windrow process should also be noted.

Figure 2-11 contains plots of temperature, total solids (TS) and VS against time for an experimental windrow of the cross section shown in Figure 2-7. Feedstock consisted of digested sludge with a solids content of about 35% blended with recycled compost to achieve about a 40% solids mixture. Results are typical of the process during periods of favorable weather and are intended to provide the reader an introduction to the type of information available. Further discussion of windrow composting data will be deferred until later chapters.

The Aerated Static Pile System

An aerated static pile process for composting dewatered sludge has been developed at the U.S. Department of Agriculture (USDA) Agricultural Research Service (ARS) experimental station at Beltsville, MD, and is often referred to as the Beltsville process or the ARS process. The term "aerated static pile" will be used here because it is more in keeping with the classification scheme presented earlier. Both digested and raw dewatered sludges have been composted by this technique. Several treatment plants are

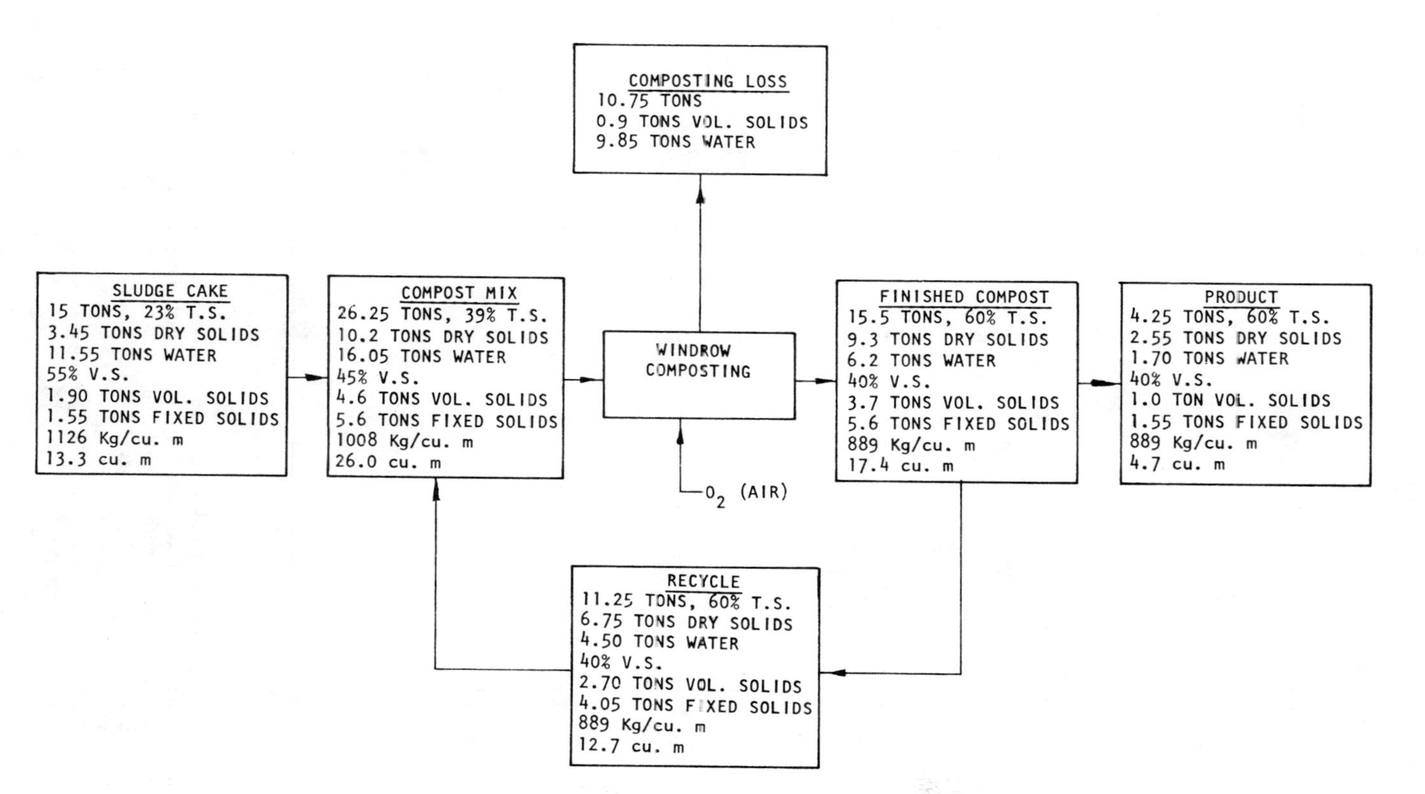

Figure 2-10. Materials balance observed during windrow composting of digested sludge cake blended with recycled product [9].

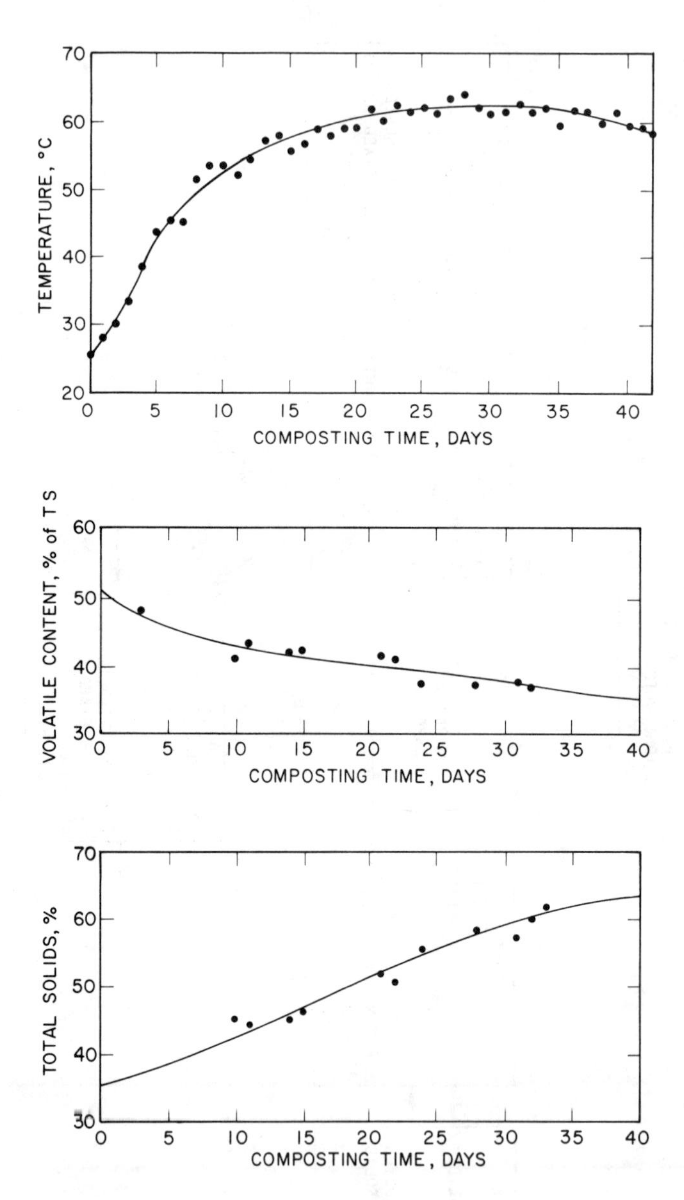

Figure 2-11. Temperature, total solids and volatile solids as a function of time for a windrow compost system. Feed consisted of dewatered, digested, primary sludge at 30–35% cake solids blended with recycled compost at about 65–70% solids. Windrow cross section corresponds to that of Figure 2-7. Temperatures are average of recorded readings at sampling points shown in Figure 2-7. Compost facility is located in southern California and data were recorded during April and May with a once per day windrow turning frequency [9].

evaluating or have implemented the process including Beltsville, MD (12 dry ton/day); Washington, DC (55 dry ton/day); Bangor, ME (1.3 dry ton/day); Camden, NJ (24 dry ton/day); Windsor, Ontario (22 dry ton/day); and Durham, NH (3.2 dry ton/day). As of 1979 several large-scale facilities were under design including a 110-dry ton/day facility for the Washington Suburban Sanitary District [11-13]. Obviously, the system is enjoying considerable popularity on the East Coast.

The aerated static pile process differs from the windrow process in that composting material is not turned. Aerobic conditions are maintained by mechanically drawing air through the pile. Another difference is that previously composted material is usually not recycled to produce a friable mixture or to adjust the starting moisture content. Instead, dewatered sludge is mixed with a bulking agent such as wood chips, which serves as a moisture absorbent and provides porosity to the material. The required ratio of sludge to wood chips has been reported to be in the range of 1:2 to 1:3 on a volumetric basis [11-13]. Most of the experience at Beltsville has been with wood chips, although other bulking particles should also be suitable. As mentioned earlier, use of pelleted refuse, shredded tires, peanut shells and other materials has been reported. Obviously, both size and quantity of the bulking agent must be controlled to maintain porosity throughout the pile and assure adequate airflow without excessive blower head loss.

A process flow diagram for the aerated static pile system is shown in Figure 2-12. Sequential steps in the formation of a pile are as follows:

1. mix sludge with the bulk agent;
2. lay a prepared base of wood chips or other bulking agent along with perforated aeration piping;
3. place the sludge/wood chip mixture in a deep pile on the prepared bed;
4. cover the outer surface of the pile with a layer of screened or unscreened compost; and
5. attach a blower to the aeration piping.

At this point the pile is ready to begin operation. The blower is operated to either pull or push air into the pile. Blower operation is controlled to maintain aerobic conditions throughout the pile. An on-off sequence is usually employed to avoid excessive cooling of the pile. Exhaust air from the blower can be deodorized before discharge. There is a variety of techniques which will accomplish deodorization to varying degrees (see Chapter 15). A common practice is to vent the gas through a pile of screened compost.

Detention time in the aerated pile is usually about 21 days [11,13], after which the pile is dismantled. The mixed pile, compost cover, prepared base and deodorization pile are usually placed in a temporary stockpile. Subsequent drying may be required before screening the mixture to separate the bulking agent. It has been found, for example, that a moisture content $<50\%$ greatly facilitates separation of wood chips from sludge with vibrating screens

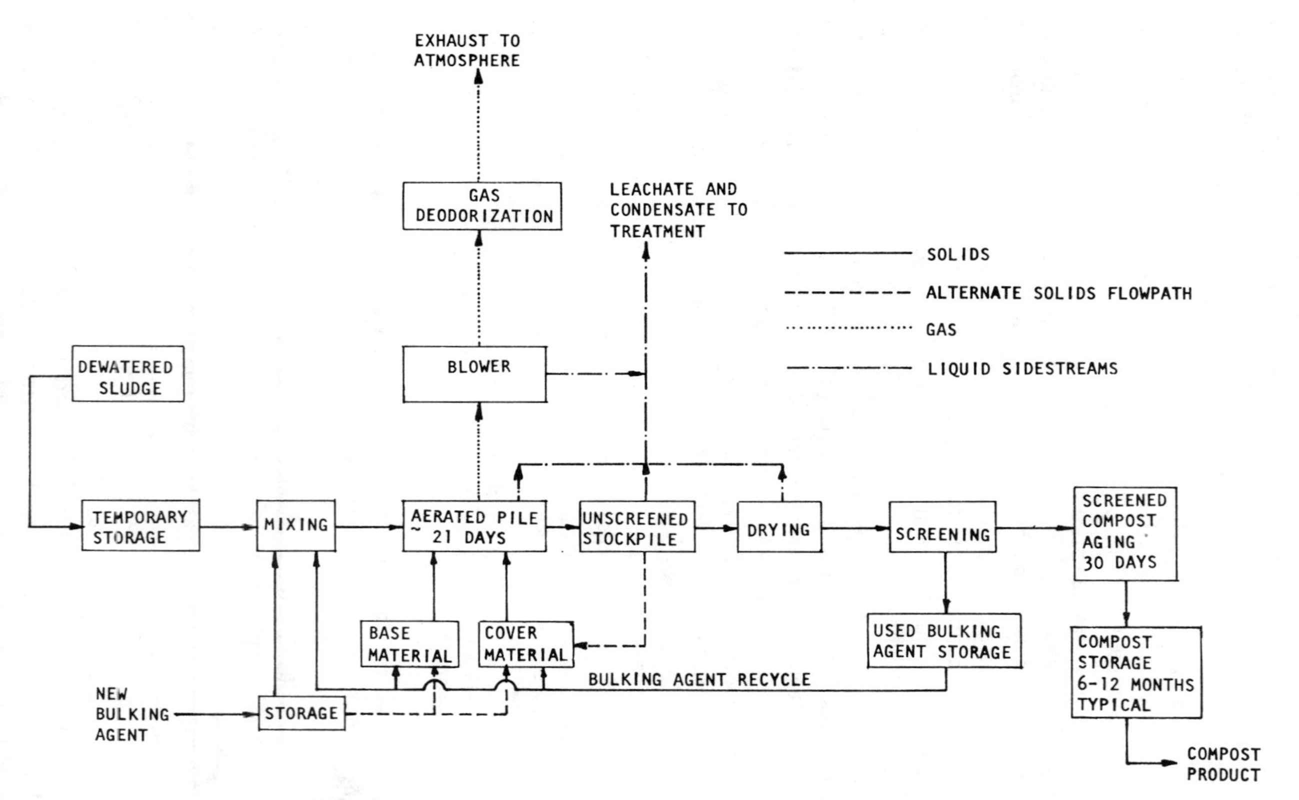

Figure 2-12. General process flow diagram of the aerated static pile compost system.

and trommels [14]. Separation and reuse of bulking agent is usually required because of the large volumes normally used and the high cost of materials such as wood chips. Drying can be accomplished by maintaining high aeration rates during composting, by forced aeration in the stockpile and/or windrowing to accomplish open-air drying.

If wood chips or other degradable material is used as the bulking agent, some degradation and physical breakdown can be expected during composting. Eventually the size reduction will be such as to allow bulking material to pass through the screen with the composted sludge. Thus, continual makeup of bulking agent is necessary to balance that lost in the final product.

It is common to collect some leachate from the bottom of the piles. Also, a water trap must be placed on the air piping to collect condensate formed when hot, moist exhaust gas is cooled. Both leachate and condensate must be collected and treated.

A cross section of a typical aerated static pile is shown in Figure 2-13. Several process modifications have been suggested to reduce the land requirement when single piles are constructed. In one modification, the "extended pile method," new piles are built on the structure of the preceding piles as

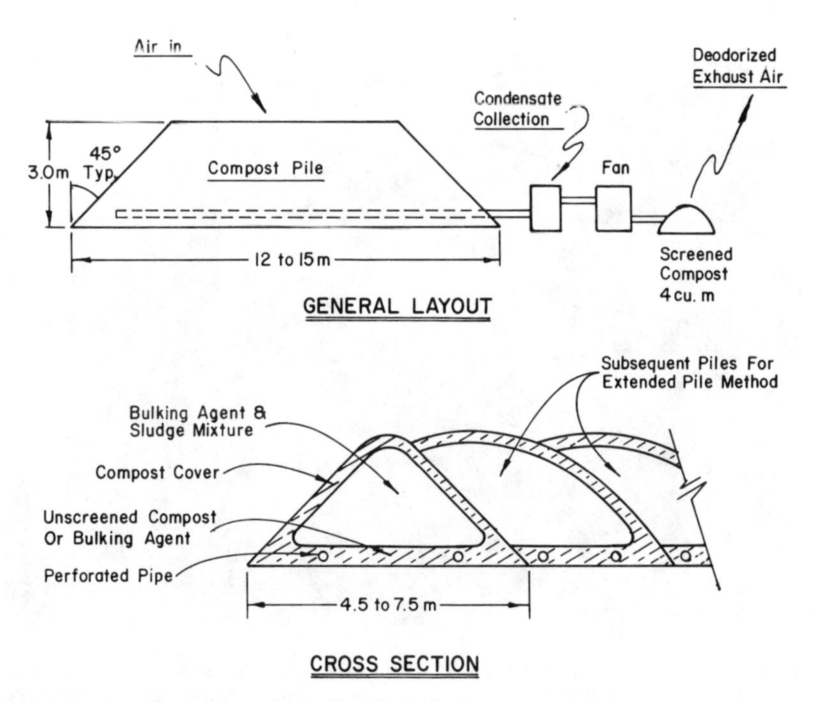

Figure 2-13. Typical forced aeration compost pile for 40 m^3 of dewatered sludge [15].

shown in Figure 2-14. In another, piles as high as 18 ft were constructed using a crane and the modification was termed the "extended high pile method" [Figure 2-15]. Neither of these modifications represent fundamental changes and would be used simply to reduce land requirements or operational problems.

Typical temperatures recorded during the composting of raw sludge using wood chips as a bulking agent are shown in Figure 2-16. In general, good temperature elevations have been observed with aerated static pile techniques on raw sludge in cold and wet climates. As seen from Figure 2-16, temperature increases rapidly during the first three to five days, holds relatively constant, and then begins to decrease after about three weeks. Temperature profiles observed during composting of digested sludge blended with various amendments is shown in Figure 2-17. Good temperature elevations were achieved but air channeling was more of a problem compared to use of bulking particles such as wood chips.

A process flow diagram of a proposed 544-wet ton/day aerated static pile design is shown in Figure 2-18. Most of the process elements of Figure 2-12 are incorporated with the addition of heat exchange between inlet ambient

Figure 2-14. Composting operations at the USDA Agricultural Research Station at Beltsville, MD, in 1976. Note single aerated static piles at top left and extended pile modification at bottom. Windrow operation is at right of photo.

Figure 2-15. Aerated static pile compost system in operation at Blue Plains Treatment Plant, Washington, DC.

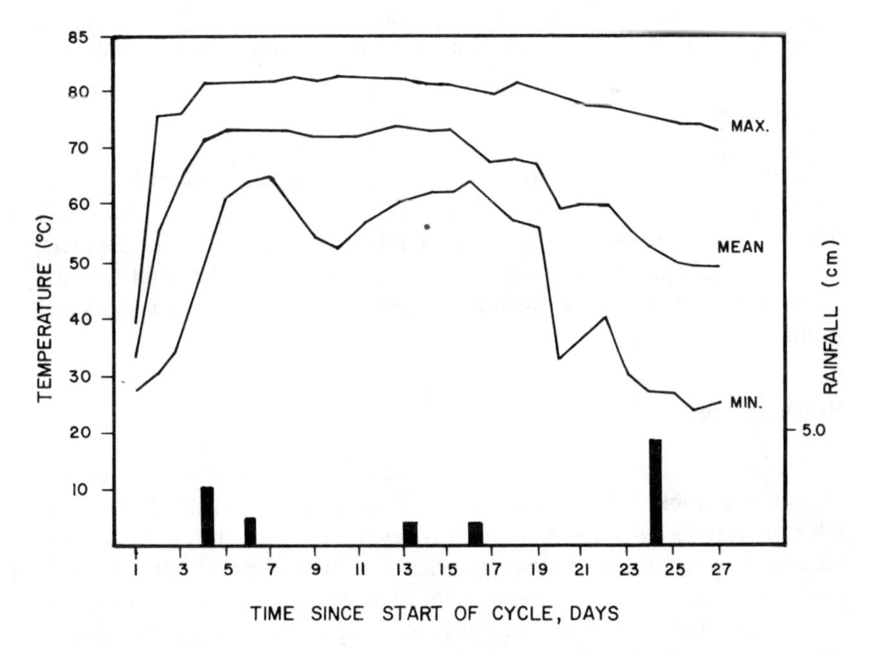

Figure 2-16. Typical temperatures recorded during aerated static pile composting of a raw sludge-wood chip mixture. Bars at bottom indicate rainfall events [16].

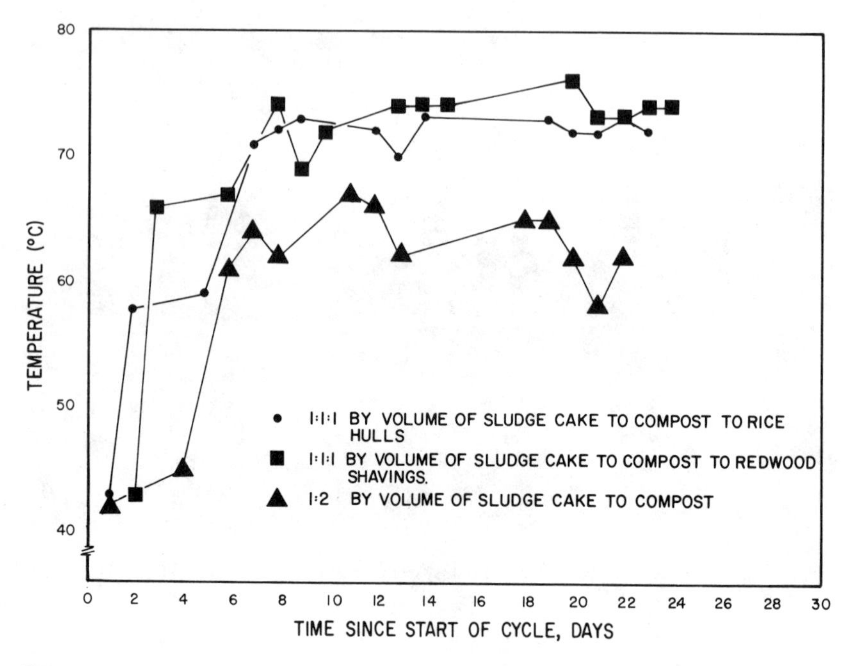

Figure 2-17. Temperature profiles observed during aerated static pile composting of digested sludge blended with various amendments. Each data point is the average of 10 readings made over the pile cross section [17].

air and exhaust gases from the compost pile. A detailed materials balance for the proposed design is presented in Table 2-3. This is one of the better balances presently available and will be referred to in succeeding chapters. Process design criteria for the proposed facility is presented in Table 2-4 and summarizes a considerable quantity of practical information necessary for facility design.

Reactor Systems

Operational data are becoming available on a number of reactor systems described in Table 2-2. However, few of these data are available in the published literature. Even if they were, it is beyond the present scope to provide material balances for the numerous reactor systems. Instead, attention will be focused on a few documented tests of reactor composters operating on wet organic substrate. The reader will learn what can generally be accomplished with such systems and some of the operating variables important to achieving proper composting.

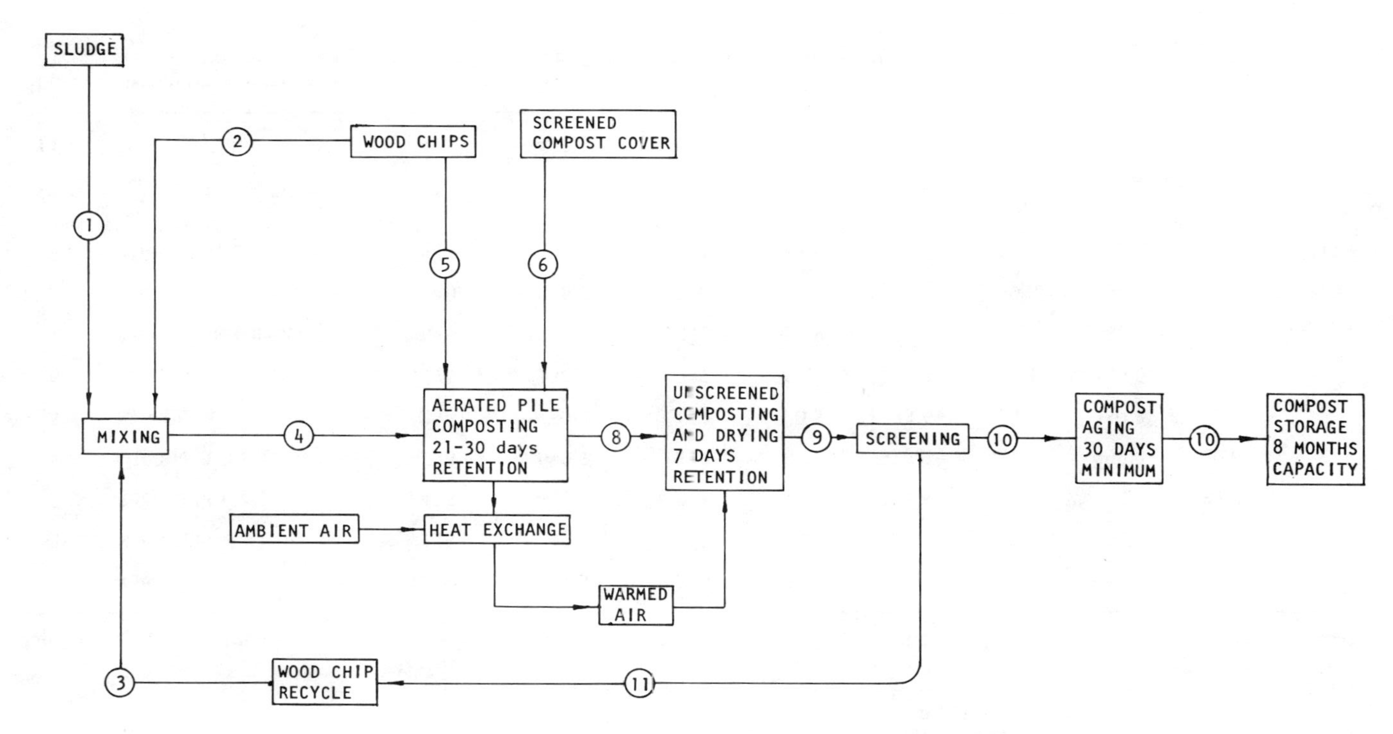

Figure 2-18. Process flow diagram for a proposed 544-wet ton/day aerated static pile compost system designed to process raw sludge. Circled numbers correspond to line numbers in Table 2-3. Seven day/week operation is assumed. Developed by Toups and Loiederman Engineers [18].

Table 2-3. Material Balance for the Process Flow Diagram Shown in Figure 2-18, Developed by Toups and Loiederman Engineers [18]

Line[a]	Station	Density (kg/m^3)	Total Flow (ton/day)	Total Flow (m^3/day)	Material Flow (avg ton/day) Fixed Solids	Material Flow (avg ton/day) Volatile Solids	Material Flow (avg ton/day) Water	Material Flow (avg ton/day) Total	Percent Water	Cumul. Water Loss
1	Raw Sludge[b]	1067	544.3	510	32.7	76.2	435.4	544.3	80.0	
2	New Chips	296	52.6	178	0.5	34.9	17.2	52.6	32.9	
3	Recycled Chips	415	402.8	971	47.1	205.8	149.9	402.8	37.2	
4	Mix/Blend Output	726	999.7	1377	80.3	316.9	602.5	999.7	60.0	
5	New Chips (Pad Bed)	296	50.8	172	0.5	33.6	16.7	50.8	32.9	
6	Compost (Top Cover)	474	136.0	287	57.1	30.8	48.1	136.0	35.0	
7	Aerated Pile (Input)	646	1186.5	1836	137.9	381.3	667.3	1186.5	56.2	
8	After 21-Day Composting	440	808.3	1836	137.9	331.1	339.3	808.3	42.0	328.0
9	After 7-Day Secondary Composting	545	705.2	1294	137.9	311.6	255.7	705.2	36.3	411.6
10	Screened Compost	512	302.4	591	90.7	105.8	105.9	302.4	35.0	
11	Screened Chips	415	402.8	971	47.1	205.8	149.9	402.8	37.2	

[a] Line numbers refer to numbers in circles in Figure 2-18.
[b] Average sludge input per day: 544.3 wet tons, 20% solids, 80% water, 30% inert, 70% volatile.

Table 2-4. Process Design Criteria for Aerated Static Pile Composting, Developed for an Average Sludge Load of 544 Wet Ton/Day [18]

Unit Operation	Design Criteria
Mixing (Enclosed Building)	Blended Mix: 2.1 m^3 wood chips/wet ton of sludge. One day of sludge input = 544 wet tons
	Operating Cycles: sludge unloading and mixing 7 day/week during daylight work shifts
Aerated Pile Composting Area	Mixed Volume: 2.5 m^3/wet ton of sludge
	Effective Mix Depth: 2.44 m
	Wood Chip Bed Depth (Pad): 0.3 m
	Compost Cover: 0.46 m
	Overall Pile Width: 61 m
	Detention Time: 21 days
	Composting Area: 0.98 m^2/wet ton of sludge in place
	Aeration: 0.78 m^3/dry ton-min; aerate minimum 25% of cycle time by intermittent induced-draft aeration
	Operating Cycles: mix emplacement 7 days/week during daylight shifts: pile removal 7 days/week and 7.5 hr/day
Odor Absorption Pile	Volume Dry Compost: 0.84 m^3/dry ton
	Volume Wood Chips: 0.08 m^3/dry ton
Second Stage Composting and Drying	Pile Depth: 3.2 m
	Pad Area: 0.7 m^2/wet ton of daily sludge input
	Storage Time: 7 days
	Operating Cycle: removal 7 day/week and 7.5 hr/day
	Forced Air: 7 days, 24 hr/day
Screening Plant (Enclosed Building)	Screen System: two-stage (disc screen followed by vibratory screen)
	Screen Retentions: 60% by volume on first stage; 40% by volume on second stage
	Operating Cycle: screen 7 day/week (normal), with provision for 5 day/week screening.
Screened Wood Chip Recycle	Operating Cycle: recycle 7 day/week and continuous during screening
	Storage: none
Compost Curing	Aging Period: 30 days minimum
Compost Storage	Storage Period: 8 months
New Wood Chips	Storage Period: one-year supply
	Storage Area: 56,700 m^3/ha
	Average Depth: 6.1 m

Fairfield-Hardy Tests (1969)

From May 27 through July 7, 1969, the Fairfield-Hardy plant at Altoona, PA, processed a total of 416 wet tons of raw filter cake sludge at 75% moisture [19]. The Fairfield composter used in the demonstration was 11.6 m in diameter with a capacity at 1.8 m depth of 23 ton/day of material weighing approximately 450-510 kg/m^3 (28-32 lb/ft^3) at approximately 55-60% moisture. Both agitation and forced aeration were provided during the composting cycle. Agitation screws continuously moved material from the outside of the reactor to the center where it was discharged after 5.5-7 days. Outfeed from the reactor was dried in a rotary drum dryer to approximately 13% moisture. Dried compost was then stored for use as a conditioning material for the infeed sludge. Shredded paper was also added to the mixture to improve porosity and reduce bulk weight.

A material balance covering the entire test period is shown in Figure 2-19. Mixed substrate averaged about 63% moisture and outfeed from the reactor about 49% moisture. It is interesting to note that of the 306 tons of water removed, about 78% was removed in the compost reactor and 22% in the heat dryer. Thus, energy requirements for drying are significantly reduced compared to direct heat drying of dewatered cake. This is caused by the heat released as a result of biological decomposition in the compost reactor. VS destruction over the course of the study averaged about 64%, accounting for the large heat evolution.

During the last three weeks of the test period, reactor temperatures averaged 67.5°C with a range of recorded values from a low of 63.9°C to a high of 72°C. The consistency of temperature readings was attributed to the close control over aeration rates obtained with the reactor. Quite often an excess of air was required to cool the digesting mass to the desired temperature. Temperatures as high as 86°C were obtained on other occasions using lower aeration rates.

During the first 3.5 weeks of the study, shredded paper was the only amendment used. It was blended with dewatered cake in the approximate ratio of 4.0 dry tons of cake to 1.6 dry tons of paper. Bulk density of the mixture averaged about 396 kg/m^3 (24.8 lb/ft^3). After this period compost which had been heat-dried to 13% moisture was used as the sole amendment. Recycling was adjusted to achieve a mixture moisture content of about 56%. However, mixture bulk weight increased to about 673 kg/m^3 (42 lb/ft^3). The drive mechanism for the agitation screws was originally designed for a 560-kg/m^3 (35 lb/ft^3) material and the mechanism soon stalled. The problem was entirely mechanical and recycled product could be used as the sole amendment if the reactor were designed for a material of greater bulk weight [20].

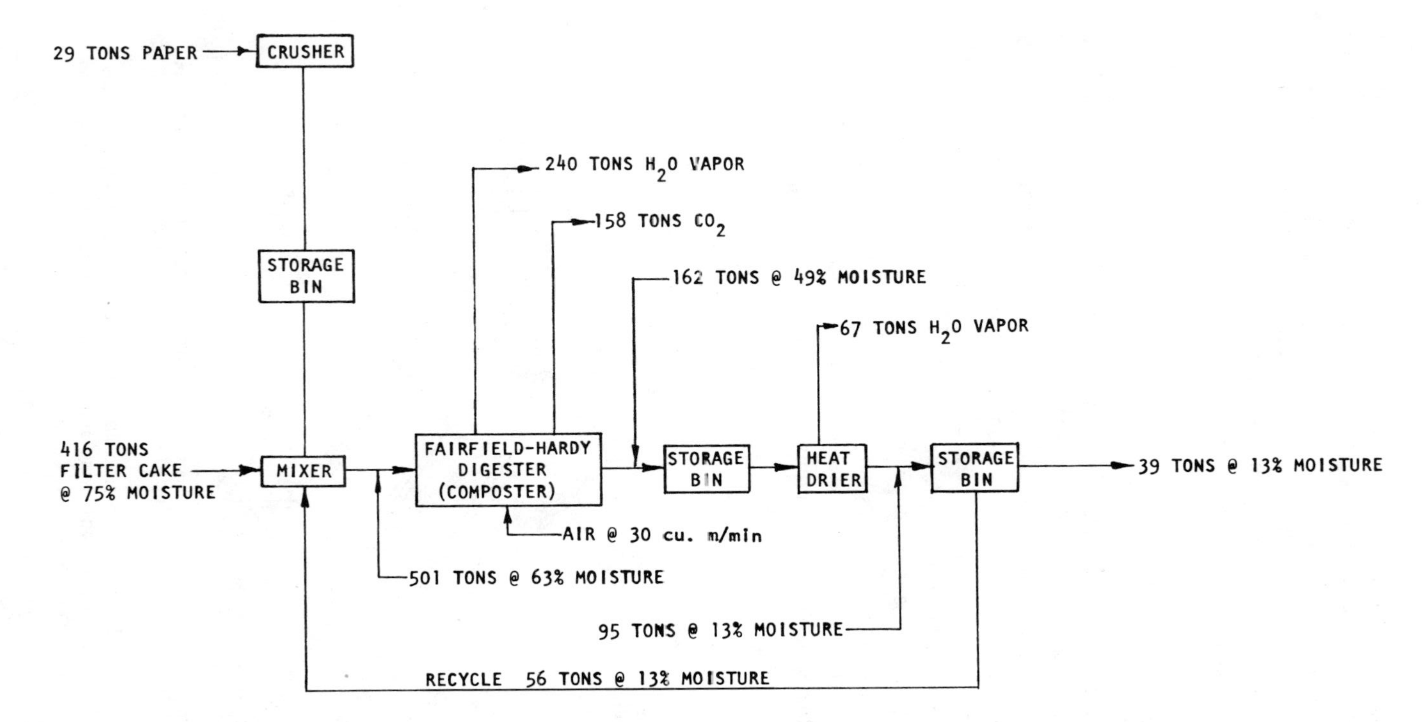

Figure 2-19. Process flow diagram and material balance observed during composting of raw, dewatered filter cake in the Fairfield Hardy digester (composter) in Altoona, PA. Material balance covers the entire test period from May 27 through July 7, 1969. The system includes a pelleter ahead of the heat-dryer which was not used in the sludge composting study [19].

Stalling was corrected by addition of a small amount of shredded paper to the mixture. Throughout the remainder of the test period, reactor infeed consisted of a 1:1.2:0.07 mixture of cake, dried compost and shredded paper on a dry weight basis. Mixture moisture content was about 56% with a bulk weight of 505 kg/m^3 (31.5 lb/ft^3). The reactor obtained a satisfactory steady-state operation under these conditions.

Bin Composting Experiments

At least three pilot studies have been conducted using bin-type composters (horizontal flow, agitated solids bed reactor) with wet organic feed materials. Senn [21] reported the use of such a reactor for composting dairy manure. A diagram of the test facility is shown in Figure 2-20. Manure was brought to the test facility from adjacent corrals and mixed with compost and/or air-dried manure to achieve the desired initial moisture content. Because of the nature of the loading equipment, bins were staggered to allow transfer of material from one bin to another. This provided agitation of the material during the compost cycle. A forced aeration system was provided in the floor of the bins. Normal maximum aeration rates were about 6.7 m^3/h/m^3 of

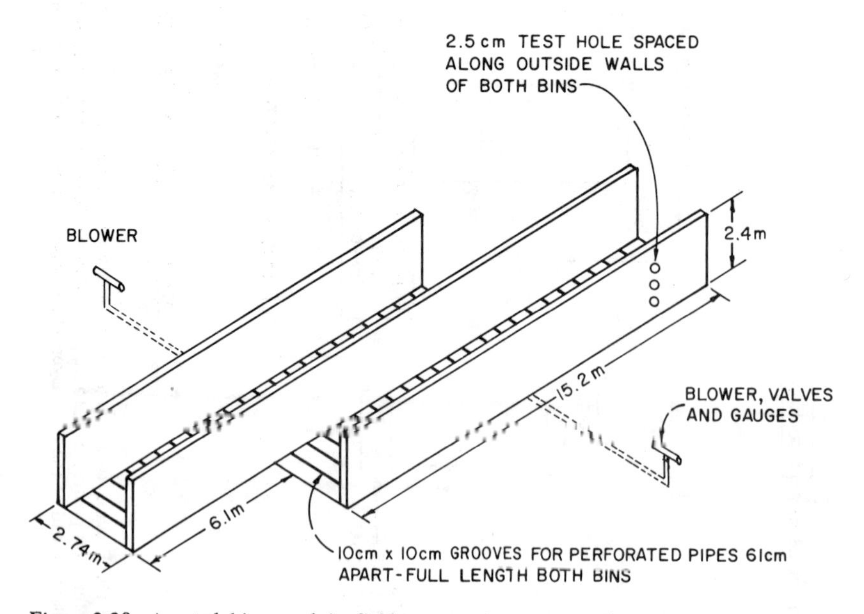

Figure 2-20. Aerated bins used in field composting tests on dairy manure. Bins were staggered and sized for transferring and unloading by conveyor buckets and belt of "eagle loader" [21].

manure. Maximum static pressures were about 15 cm water gauge. A composting time of 2-7 days was provided in each bin followed by a minimum 30-day period of aging in stockpiles.

A typical temperature profile developed under near optimum conditions of air flow and moisture is shown in Figure 2-21. Rapid temperature elevations within the first 24 hr were characteristically observed. The most rapid production of attractive, stable compost occurred at temperatures between

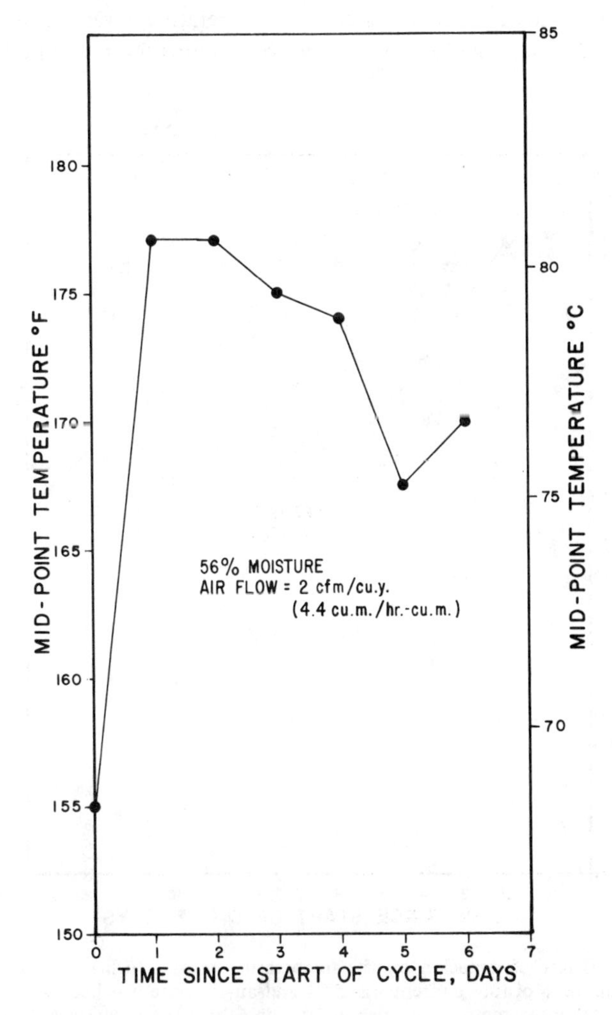

Figure 2-21. Typical temperature developed during bin composting of dairy manure under conditions of constant airflow and optimum moisture [21].

71 and 79°C. Temperatures as high as 87°C were noted but above about 79°C the compost had a nearly black color, a "cooked" odor, and was not considered a desirable product. As pointed out by Senn [21], the range of optimum temperatures found in these studies was higher than that often reported in the literature.

The material was successfully composted if loaded at moisture contents of 45-60%. Optimum moisture content was found to be in the range of 50-55% moisture. "Packing" of material was commonly observed in the bins, particularly at higher moisture contents. Transfer of material from Bin 1 to 2 loosened the material, allowing increased temperature development as shown in Figure 2-22. Agitation was also effective in loosening the material so that

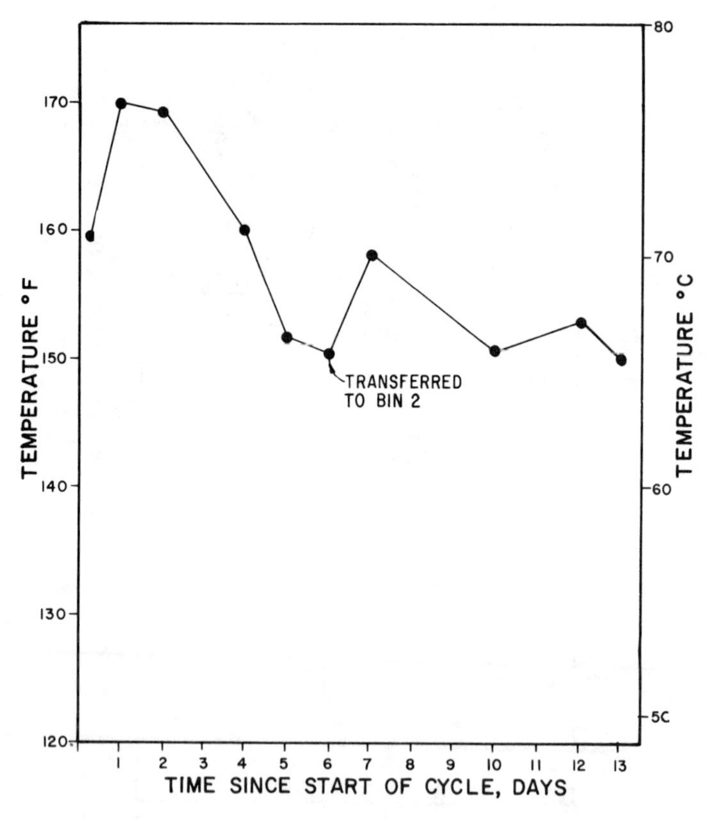

Figure 2-22. Effect of agitation on the temperature profile during bin composting of dairy manure. Moisture content was 50% and air flowrate was held constant. Points to note: (1) rapid temperature rise to 70°C in 6 hr, (2) temperature drop is graded and can be stopped by lowering air flowrate; (3) temperature rise caused by mixing and loosening from transfer [21].

higher air flowrates could be maintained. It was found that low air flowrates could lead to excessively high temperatures. The product was well "cooked" but not well stabilized in terms of organic decomposition. This demonstrates that temperatures can become so hot that rates of biological activity, and hence the process kinetics, actually decrease. Higher air flowrates resulted in a higher rate of heat removal from the bin, which lowered the temperatures to a more optimum range.

Senn [21] reported considerably more data than are summarized here, some of which will be used in later chapters. It is interesting to note that a 360-ton/day manure compost plant of the bin type is currently operating at an Ohio feedlot. The design is of the Metro-Waste type and incorporates two bin reactors, each 6 m wide and 122 m long. The reactors and the storage, screening and bagging facilities are all enclosed in one large building. The Metro-Waste system has many good points in its favor including a relatively simple design which incorporates both agitation and forced aeration. Unfortunately, no operational data were available on the full-scale manure system in Ohio, but the facility is thought to be successful. Figure 2-23 shows profiles of temperature, and oxygen and carbon dioxide content observed in a Metro-Waste system operating on shredded municipal refuse.

The LA/OMA Sludge Project sponsored tests of bin composting of dewatered digested sludge. The tests were conducted by staff of the Los Angeles County Sanitation Districts using the test bins shown in Figure 2-24. These bins were designed to simulate an elemental volume of an agitated solids bed reactor as described in Figure 2-4. Provisions for both agitation and forced aeration were provided in the design.

Temperature profiles observed for two different mixtures are shown in Figure 2-25. Proper compost temperatures were achieved using recycled compost as the sole conditioning material. It should be noted that each data point in Figure 2-25 represents an average of 17 temperature readings over the cross section of the reactor. Because of the large surface-to-volume ratio, cooling near the edges was more significant and average temperature somewhat lower than would be expected in a larger system. Temperatures as high as 80°C were observed at points within the bin.

Air was supplied to the compost bin by a blower controlled by an on-off timer. When the blower was operating, air was supplied at 11 $m^3/h/m^2$ of floor area. The air pressure drop was generally in the range of 7-18 cm of water with the bin loaded, of which about 4 cm was caused by aeration ducting and piping. Thus, about 3-14 cm of water loss was caused by the 2.5- to 3.0-m depth of mixed material in the bin.

As was the case with manure composting, it was observed that the initial mixture solid content was important to the achievement of rapid temperature elevation. This is illustrated by the data of Figure 2-26. A 50% solids starting mixture was considered near optimum.

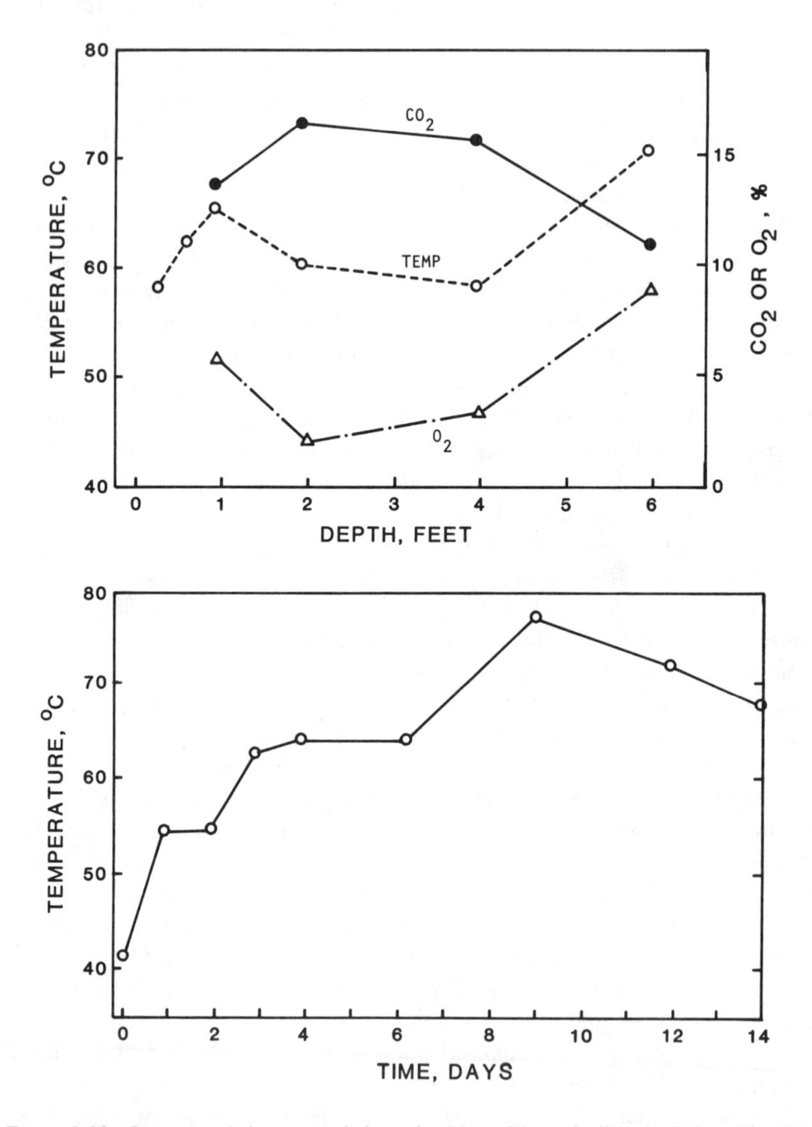

Figure 2-23. Operational data recorded at the Metro-Waste facility in Gainesville, FL. Composting material consisted of shredded municipal refuse with moisture adjusted by addition of water or sewage sludge. (Top) Vertical profiles of temperature, oxygen and carbon dioxide content observed at day 5 of the compost cycle. (Bottom) Temperature profile observed throughout the compost cycle. All measurements were taken at a depth of one foot or greater and each value is an average of three or more measurements [49].

Figure 2-24. Top. Test bins used by the LA/OMA project for composting dewatered municipal sludges. Two 3- × 3- × 3-m steel bins were mounted on a concrete slab. The bins were located side by side with doors on one side to allow access to the composting material. Agitation was provided by using a front-end loader to transfer material from one bin to another or to unload the material and reload in the same bin. A forced aeration system was located in notches in the concrete slab. Bottom. Compost material being removed from bin composter. The compost mixture consisted of digested sludge cake mixed with recycled compost product. Note the steam clouds as water vapor is released from the hot material. Temperatures as high as 80°C were observed in the composting material.

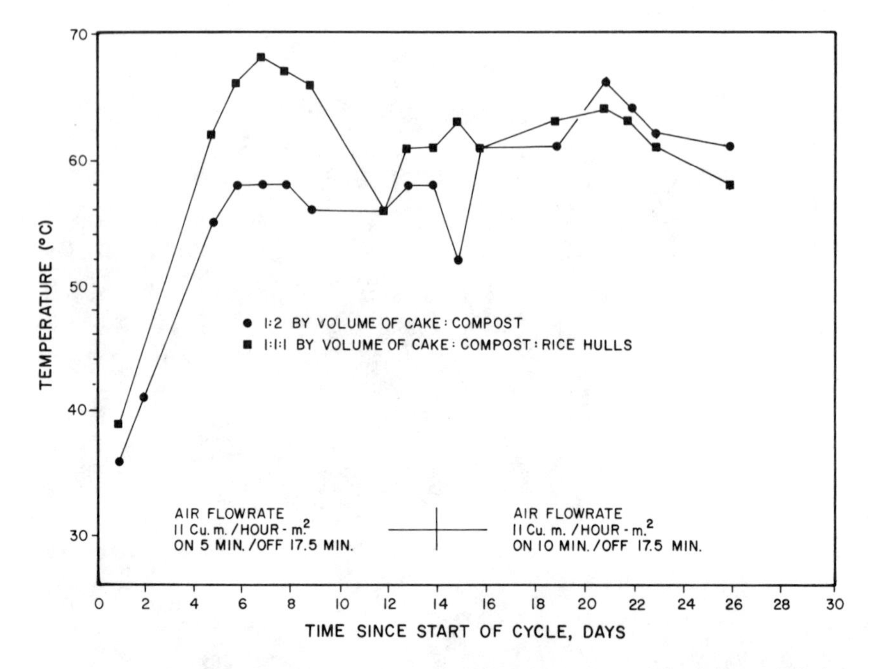

Figure 2-25. Average temperatures observed during bin composting of digested sludge cake mixtures. Sludge cake was approximately 25% solids and recycled compost about 65% solids. Each data point represents an average of 17 temperature readings over the cross section of the reactor [17].

The compost experiment sponsored by LA/OMA showed that dewatered municipal sludge can be composted in a bin-type system with bed depths of 2.5-3 m (Figures 2-25 and 2-26). Recycled compost can be used as either the sole or primary amendment and both agitation and forced aeration should be provided. Head losses during aeration are generally small even though bulking particles were not added to the mixtures in these studies.

From 1977 to 1980, a 1-m^3/day (about 0.5-wet ton/day) bin composter was operated at the Minamitama Sewage Treatment Plant, Japan. Feed material consisted of raw sludge, conditioned with lime and ferric chloride, and dewatered by filter press to about 35% cake solids. Dewatered sludge was blended with recycled compost product to achieve a 50% solids mixture which was then composted for about 10 days. Compost product was used for feed conditioning because (1) it did not increase the quantity of final product for disposal; and (2) other additives such as wood chips were not consistently available.

The Minamitama composter was about 9 m long, 2 m wide, with the compost bed depth maintained at about 1.5 m. Material was fed daily and turned

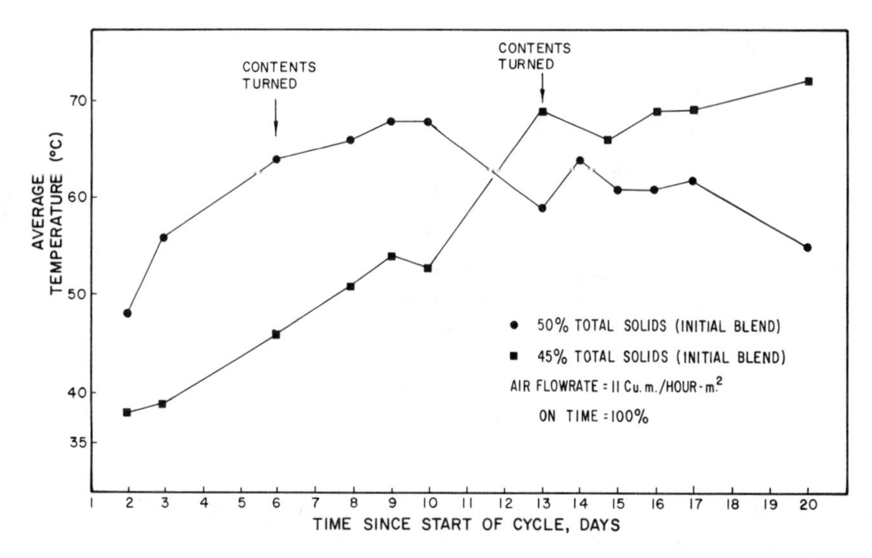

Figure 2-26. Effect of initial mixture solids content on the temperature profile during bin composting of digested sludge cake blended with recycled compost. Each data point represents an average of 17 temperature readings over the cross section of the reactor [22].

about 3 times per week using a special turning device similar to the "agi-loader" used in the Metro-Waste system. Bed temperatures above 60°C were maintained with air flowrates of 7-18 m^3/min/dry ton/day of sludge cake feed. A materials balance for the process is shown in Figure 2-27 and indicates considerable organic decomposition and moisture removal. An expanded facility (10 m^3/day) of similar design was scheduled to be completed in early 1980.

There are no full-scale facilities presently using bin reactors for composting dewatered sludge cake as the primary feed material. However, available data indicate this type of reactor to be well adapted for such purposes. Several small-scale facilities (10-20 dry ton/day) are being pursued actively in several regions of the U.S.

Additional Comments

Composting is primarily a problem of materials handling. The type of material being handled often dictates the design of the system. When municipal solid waste (MSW) was being pursued actively as a composting feedstock, reactor systems fell into disfavor in the eyes of many investigators. Although

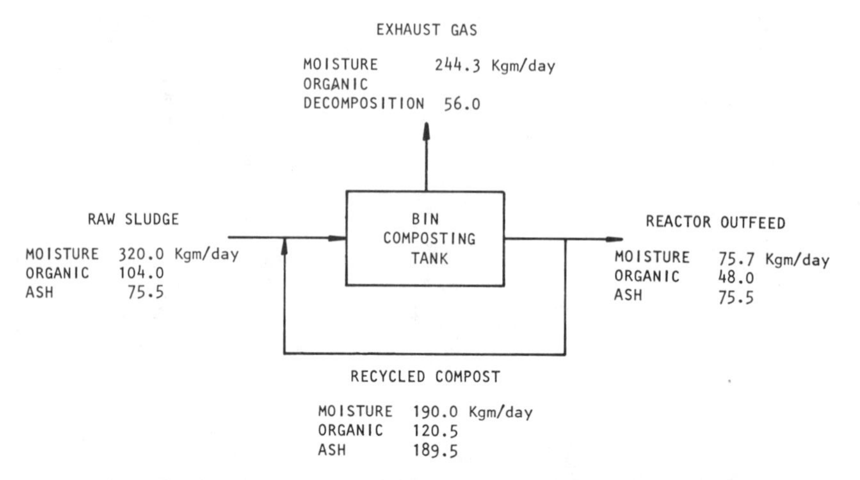

Figure 2-27. Materials balance observed during bin composting of raw municipal sludge at the Minamitama, Japan, pilot bin compost facility. Input sludge was conditioned with recycled compost to achieve a feed mixture of about 50% moisture content. Organic content of feed sludge was about 58% of total solids, reducing to about 39% in the final product. About 76% of the moisture content of raw sludge was removed in the exhaust gases from the composter [209].

Figure 2-28. The Metro-Waste system composting shredded solid wastes. Note the long bins and rail-mounted "agiloader" used for loading, unloading and agitation of material. Forced aeration is provided through floor of bin.

Figure 2-29. A Fairfield-Hardy reactor showing the circular tankage and traveling arm with vertical augers for agitation. Forced aeration is provided through the reactor bottom and through holes in the augers. Feed is introduced on a continuous or intermittent basis at the outer periphery. Augers mix and transport material to the discharge located at the center of the reactor. Reactor shown is at the Toronto Resource Recovery facility.

some systems were successful, most were not. Furthermore, windrow composting or curing was often used after the reactor because many organics in MSW are resistant and slow to degrade. Why go to the trouble of having an enclosed reactor if open windrows or piles are going to be used anyway?

As a feedstock, sludge differs in many ways from materials like MSW. Because sludge cake is mostly water the thermodynamic balance in the system can easily be stressed. Both raw and digested sludges are prone to produce odors if handled improperly. Because of the fine particulate nature of most sludges, the final compost has a tendency to dust if moisture content decreases below about 30-35%. In addition to these feedstock differences, sewage treatment plants always seem to be located close to residences and always seem to have less than sufficient land area. A clear statement of Murphy's law! Treatment plant neighbors are rightly concerned about odors and dusts from any treatment plant operation.

Some of the reactor systems offer potential for composting at reasonable cost with improved control over environmental problems such as odors, dusts and general unsightliness. Because of the rather unique feedstock

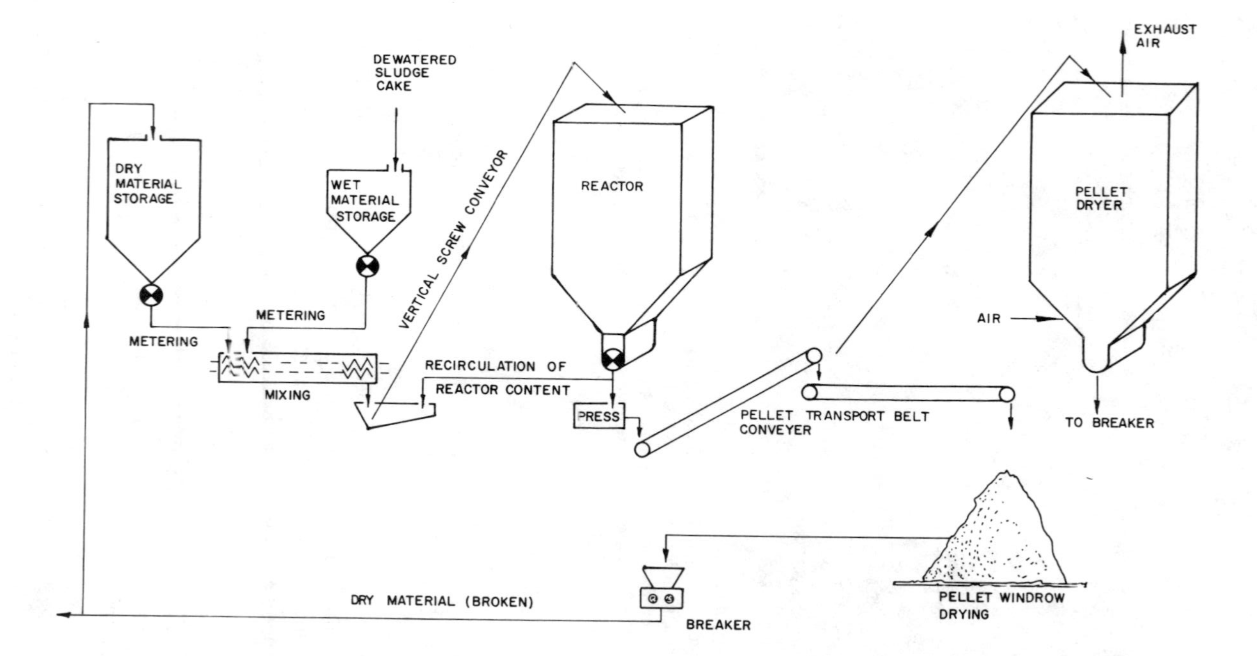

Figure 2-30. Process flow diagram of the Euramca Fermentechnik system of sludge composting. Reactor is of the vertical, moving packed bed type and is batch-loaded. Reactor contents are recirculated frequently during the compost cycle to provide agitation of the material. A unique feature of the system is the pelleting of compost before drying. Pellets are then dried either in windrows or an enclosed forced-air dryer. Pelleting can reduce much of the dust associated with dry sludge compost. A detention time of 6–7 days is provided in both the compost reactor and pellet dryer.

A pilot-scale system was tested by the LA/OMA Project [23]. Temperatures above 55–60°C were achieved with both raw and digested sludges using dry product recycle for initial mixture conditioning.

characteristics and the environmental constraints placed on modern treatment facilities, renewed interest in reactor composting seems warranted. This does not mean that there is anything wrong with open composting systems. It just means that a decision on the type of composting system must recognize the feedstock characteristics and other site-specific constraints. All available systems should be considered in such a decision. Design selection between various processes must consider the availability of amendments and bulking agents, local weather conditions, land availability, proximity of adjacent residences, whether digested or raw sludge is to be composted, the cake solids produced during dewatering, desired operational flexibility, and local environmental standards.

Many of the above factors will be discussed in detail in later chapters. After studying this text the planner or designer should be in a better position to make such a decision. However, it is obvious from the nature of many of these factors that any decision will be site-specific.

One additional observation will be offered. Many systems which appear simple on a small scale become remarkably complex when expanded to larger-scale facilities. For example, the aerated static pile system developed at Beltsville was highly advertised as a simple, low-technology composting system. Indeed, on a small scale it is. But the same concept becomes increasingly complex when expanded to meet the needs of a larger-scale facility. Operations once conducted by hand or simple mobile equipment must give way to more mechanized, automated approaches. Such an evolution is evident in the design of recent large-scale facilities for the aerated static pile system. For a facility processing 100 dry ton/day there may be little difference between open and reactor systems in terms of mechanization and reliance on automated methods of materials handling.

If reactor systems are to be considered, which systems offer the most promise? First, a composting system should provide the operator the maximum degrees of freedom possible in operating the plant. In other words, the more variables the operator can control, the more tools he has to produce acceptable compost. Systems which provide both agitation and forced aeration during the compost cycle meet this criterion. Aeration rates can be controlled by the operator in response to feed characteristics, weather conditions, desired temperature elevations and other factors. Rates of agitation can be controlled to average-out errors in the initial mixing of feed components, assure that all material experiences the proper temperature conditions, prevent consolidation of materials, break up clumps or balls of material which may form, and reduce the chances of air channeling during forced aeration. Such systems are favored because of their increased operating flexibility.

The aerated static pile system does not usually provide agitation of material once the pile has been formed. However, use of bulking particles is rather unique to this system and overcomes many operating difficulties without the need for agitation. In a sense, bulking particles are being used to reduce the need for agitation. Good initial mixing of the feed and bulking particles is required. Some recent designs have incorporated a single agitation of the pile to break up clumps of solids and average-out errors in the initial mixing. However, many static pile systems have operated successfully without any intermediate agitation.

A second factor that should be considered is the cost per unit volume of the reactor. Obviously, this should be as low as possible to minimize capital expenses. Simple designs using common building materials are probably more cost-effective on a per-volume basis than complex systems requiring extensive fabrication.

The above observations are offered as general guidelines for selection of composting systems. The designer is urged to consider the large number of site-specific factors which must weigh on such a decision. A general comparison between various composting processes is provided in Table 2-5. Many of the items listed will be covered in greater detail in later chapters. The reader should recognize that the disadvantages and operational problems listed in Table 2-5 can often times be mitigated by proper design.

SUMMARY

Three approaches have been used to overcome composting difficulties posed by high moisture content feed materials: (1) recycle of compost product; (2) addition of an organic amendment; and (3) addition of a bulking agent. All sludge composting systems are similar to the extent that one or more of these techniques are employed. Amendments and recycled compost product decrease the bulk weight of the mixture and increase the air voids allowing proper aeration. Sawdust, straw, peat and rice hulls are examples of amendments. Bulking agents are particles of sufficient size that the wet sludge can occupy void spaces while the particles provide structural support. Wood chips have been the most commonly used bulking agent.

Composting systems can be classified according to the reactor type (or lack of), solids flow mechanisms, bed conditions in the reactor and manner of air supply. The following classification scheme was developed:

1. Nonreactor (Open) Systems
   A. Agitated Solids Bed (Windrow)
      1. Conventional
      2. Forced Aeration
   B. Static Solids Bed
      1. Forced Aeration (Aerated Static Pile)
      2. Natural Ventilation

Table 2-5. Comparison Between Various Composting Processes When Used for Sludge Composting

	Nonreactor Systems		Reactor Systems	
Item	Windrow	Aerated Pile	Forced Aeration + Agitation	Forced Aeration No Agitation
Capital Costs	Generally low	Generally low in small systems; can become high in large systems	Generally high	Generally high
Operating Costs	Generally low	High, depending largely on bulking agent used	Generally low	Generally low
Land Requirements	High	High	Low, but can increase if windrow drying is required	Low, but can increase if windrow drying is required
Control of Air Supply	Limited unless forced aeration is used	Complete	Complete	Complete
Operational Controls	Turning frequency, amendment or compost recycle addition, forced aeration rate	Air flowrate, bulking agent addition	Air flowrate, level of agitation, amendment or compost recycle addition	Air flowrate, amendment or compost recycle addition
Sensitivity to Dewatered Cake Solids	More sensitive	Less sensitive	Less sensitive	Less sensitive
Need for Subsequent Drying	Drying usually occurs in windrow but depends on climate	Drying can be achieved in pile with high air supply; windrow drying may be required	Drying can be achieved in reactor; final drying in windrow or heat dryer may be required	Less drying potential from lower air flowrates; final drying in windrow or heat dryer usually required

Table 2-5, continued

Item	Nonreactor Systems		Reactor Systems	
	Windrow	Aerated Pile	Forced Aeration + Agitation	Forced Aeration No Agitation
Sensitivity to Cold or Wet Weather	Sensitive unless in housing; demonstrated mainly in warm, dry climates	Demonstrated in cold and wet climates	Demonstrated in cold and wet climates	Demonstrated in cold and wet climates
Demonstrated on Digested Sludge	Yes	Yes	Yes	Yes
Demonstrated on Raw Sludge	Yes, but odor problems observed	Yes	Yes	Yes
Need for Amendment or Bulking Agent	Well demonstrated using recycled compost only	All designs to date use bulking agents; pilot operations used recycled compost with digested sludge	Demonstrated using recycled compost only	Amendment addition along with recycled compost usually used
Control of Odors	Depends largely on feedstock; potential large-area source	Handling of raw sludge is potentially odorous; may be large-area source	Potentially good	Potentially good
Potential Operating Problems	Low cake solids will increase odor potential and cause reduced temperature in windrows; susceptible to adverse weather	Mixing and screening of bulking agent can be difficult; anaerobic balls sometimes observed; wood chips can harbor opportunistic fungal pathogens	High operational flexibility should reduce composting problems; system may be mechanically complex	Potential problems of sludge compaction and channeling or short-circuiting of air supply; system may be mechanically complex

2. Vertical Flow Reactor Systems (Tower Reactors)
   A. Moving Agitated Bed
   B. Moving Packed Bed
      1. Continuous or Intermittent Feed
      2. Batch Feed
3. Horizontal and Inclined Flow Reactor Systems
   A. Tumbling Solids Bed (Rotating Drum)
      1. Dispersed Flow
      2. Cells in Series
      3. Completely Mixed
   B. Agitated Solids Bed (Bin Reactors)

Composting systems developed for other feedstocks as well as those developed for or adapted to wet organic materials can be classified according to this scheme.

The windrow, aerated static pile and certain reactor systems have had considerable application to composting of wet organic feedstocks. All of the available systems have advantages and disadvantages relative to one another. Reactor systems may provide greater control over environmental problems such as odors, dusts and unsightliness, which may justify higher capital costs in some cases. In general, systems that are capable of providing both agitation of solids and forced aeration are considered to have greater operational flexibility. A possible exception to this is the aerated static pile process in which bulking particles are used to assure proper airflow and reduce the need for agitation.

CHAPTER 3

SOME PRINCIPLES OF CHEMICAL THERMODYNAMICS

INTRODUCTION

Thermodynamics is the branch of science which deals with energy and its transformations. Thermodynamics is normally associated with heat, but the subject deals not only with heat but all forms of energy. The principles of thermodynamics are well established and have been applied to physical, chemical and biological systems. Lehninger [24] has stated that the proper study of biology, for example, should start from thermodynamic principles as the central theme which can best systematize biological facts and theories. This same statement can be made about the study of composting systems. Application of thermodynamic principles is a fundamental way of analyzing composting systems just as it has been a fundamental method for analysis of other physical, chemical and biological processes. Because the laws of thermodynamics appear to be inviolable, the application of those laws will reveal much about the limitations and expectations of composting systems.

The subject of thermodynamics should be distinguished clearly from the related subject of kinetics. Thermodynamics deals with the energy changes that accompany a process. Kinetics deals with rates or velocities of reactions and cannot be inferred from the thermodynamics of the system. For example, organic molecules in a piece of paper contain a rather substantial amount of energy. If a match is struck to the paper the energy is released at a rapid kinetic rate. If the paper is decomposed by microbial action, as in a compost pile, the energy is released at a much slower kinetic rate. The same amount of energy is released in either case but the kinetics are quite different. The subject here is energy changes as determined by thermodynamics. Kinetic principles of composting will be deferred until Chapter 10.

For most organisms life is a constant struggle or search for energy supplies needed to power the cellular machinery. Certain higher organisms, such as man, have freed themselves from the constant search for energy. Nevertheless, the human body is constantly "burning" various stored substances for energy. If these stored reserves are not replaced at periodic intervals, death is inevitable. Microorganisms, however, spend nearly their entire life cycles in search of energy sources. Despite this difference, many of the fuels used by microorganisms are also used by man.

From a thermodynamic standpoint, all life forms can be viewed as chemical machines which must obey the laws of energy and heat as must all other nonliving processes. Thus, thermodynamics places limits on living systems just as it does on physical systems constructed by man to extract energy from his surroundings. This discussion of thermodynamics will necessarily be brief but of sufficient depth to serve the purposes of this book. The interested reader is referred to the numerous excellent texts available on chemical thermodynamics and bioenergetics for more in-depth discussion.

Although the need for an energy supply is common to all life forms, the actual source can be markedly different. The study of different energy sources available to microbes is probably the best approach to understanding the differences between life forms. It will also lead to a better understanding of reactions which organisms mediate and the useful tasks to which they can be directed in properly engineered systems. In this sense the designer of a composting plant differs little from the biochemical engineer designing an industrial fermentation, enzyme extraction or antibiotic production plant, or the sanitary engineer designing a biological waste treatment facility. All must understand the microbes involved, reactions they mediate to obtain energy, environmental conditions required for growth and metabolism and, in certain cases, conditions required to kill organisms such as pathogens.

HEAT AND WORK, THE CONSERVATION OF ENERGY

To begin the discussion, the terms "heat" and "work" need to be defined. Heat is energy that flows because of a temperature difference between two bodies. The basic unit of heat is the calorie, which is the energy required to raise 1 g of water 1°C. The energy required actually varies somewhat with temperature and it is common to specify the 15°C calorie as the heat flowing into 1 g of water when its temperature increases at atmospheric pressure from 14.5 to 15.5°C. The equivalent English unit is the 60°F Btu (British thermal unit), which is the quantity of heat flowing into 1 lbm (pound mass) of water when its temperature increases at atmospheric pressure from 59.5 to 60.5°F. These exact definitions are seldom necessary for environmental engineering purposes and it is sufficient to define the specific heat of water as 1.00 cal/g-°C. The specific heat of other materials varies significantly and is usually much lower than that of water. The specific heats of ethanol, acetone, aluminum and copper are 0.65, 0.50, 0.22 and 0.093 cal/g-°C, respectively.

Note that the previous definitions specified atmospheric pressure conditions. Specific heat can be measured under both constant volume and constant pressure conditions. c_v is the specific heat at constant volume, and c_p is the specific heat at constant pressure. There is little difference between the two for liquids and solids. With gases, however, c_p is greater than c_v because of the added heat energy required to expand a gas against a constant pressure. Most biological processes operate under conditions of constant pressure in aqueous solutions, and the distinction is not as important to liquid-phase systems. Dry gases and water vapor are important components in a composting system, however, and the distinction will be of some importance in later studies.

The concept of work is essential to the study of thermodynamics. Work may be mechanical, electrical, magnetic or of other origin. Consider the mechanical system shown in Figure 3-1. Work is the application of force through a distance. If the force on the piston is constant, the incremental work dw in moving the piston through the distance dl is given as:

$$dw = Fdl \qquad (3\text{-}1)$$

The force F is equal to the product of pressure p times area A. Thus:

$$dw = PAdl = PdV \qquad (3\text{-}2)$$

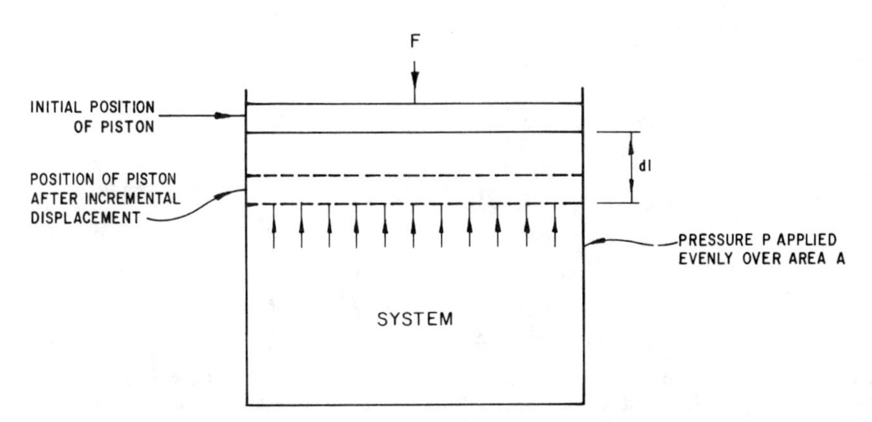

Figure 3-1. Schematic illustration of mechanical work being performed on a system.

Since dV is positive when the volume increases, work in an expansion process is positive, in agreement with the sign convention of thermodynamics. Work is measured in foot-pounds, joules or ergs.

Because heat and work are both forms of energy one should be able to equate them. James Prescott Joule verified this in a number of experimental systems between about 1840 and 1850. In each case the amount of work yielded the same amount of heat, about 4.18×10^7 ergs of work per calorie of heat. In honor of Joule, 10^7 ergs were set equal to 1 joule. Thus, today we say that 4.18 joules equals 1 calorie. In English units, 778 ft-lb is equivalent to 1 Btu.

THE FIRST LAW OF THERMODYNAMICS

Consider the system shown in Figure 3-2. If energy can be neither created or destroyed, heat energy that flows into the system, +q, must either be stored within the system, flow out of the system or appear as work done by the system. In other words, energy which flows into a system must be fully accounted for in other forms of energy. This is a statement of the law of conservation of energy which is also referred to as the first law of thermodynamics. Stated yet another way, energy can be neither created or destroyed. The German physicist Helmholtz is usually credited with first formally stating this principle in 1847, although it had been intuitively accepted as early as the eighteenth century. Notice that this concept is referred to as the first law of thermodynamics. There is no proof of this

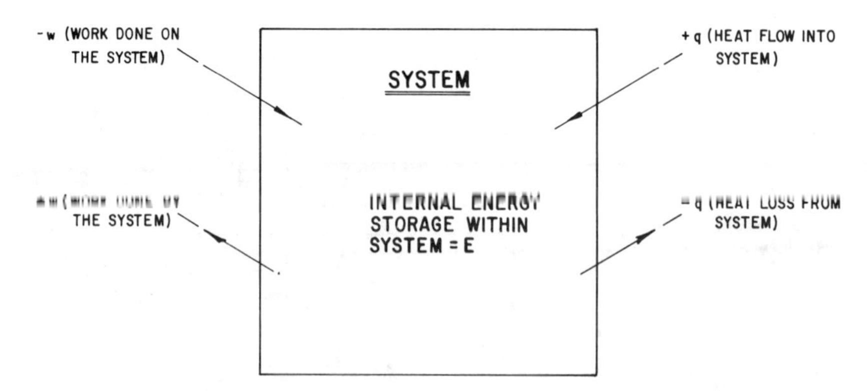

Figure 3-2. Schematic illustration of the principle of conservation of energy also known as the first law of thermodynamics. Note the sign convention on the terms for heat and work.

concept. It is accepted as a first principle, a fundamental concept which describes observed phenomena and which has never been shown to be violated. We will place considerable reliance on the concept of an "energy balance" in later discussions of composting systems.

Returning to Figure 3-2, the first law for the system shown can be stated as:

$$q = \Delta E + w \tag{3-3}$$

where ΔE = the change in internal energy of the system
$+q$ = the heat flow into the system
$+w$ = the work done by the system

If the system is maintained at constant volume (isovolumetric), Equation 3-3 becomes:

$$q_V = \Delta E \quad (v = \text{constant}) \tag{3-4}$$

No work of expansion or contraction can be performed in a constant volume system and heat absorbed by the system is exactly balanced by increased internal energy within the system.

The heat per unit mass flowing into a substance can be defined as:

$$dq = mcdT \tag{3-5}$$

where c is the specific heat capacity and m is the mass. Integrating this expression:

$$q = m \int_{T_1}^{T_2} cdT \tag{3-6}$$

The specific heat must be known as a function of T to complete the integration. Fortunately, values of specific heat for most substances remain relatively constant over small temperature ranges. Assuming a constant-volume process with constant specific heat, Equation 3-6 integrates to:

$$q_V = mc_V\Delta T \quad (\Delta T \text{ not large}) \tag{3-7}$$

Many chemical and most biological processes operate under conditions of constant pressure p. Applying Equation 3-3 to the system in Figure 3-1, assuming conditions of constant pressure and movement of the piston over the incrementally small distance dl, the following results:

$$dq = dE + pdV \tag{3-8}$$

Integrating and recalling that p is constant:

$$q_p = E_2 - E_1 + pV_2 - pV_1 \tag{3-9}$$

Rearranging:

$$q_p = (E_2 + pV_2) - (E_1 + pV_1) \tag{3-10}$$

Because constant pressure systems are so important in chemical and biochemical systems, the terms in parenthesis have been defined as the enthalpy H of the system. Thus:

$$q_p = H_2 - H_1 = \Delta H \tag{3-11}$$

When chemical changes result in the absorption of heat (i.e., heat flows into the system), the reaction is termed endothermic and both q and ΔH are taken as positive values. Reactions which result in the evolution of heat (i.e., heat flows from the system) are termed exothermic, and q and ΔH have negative values. The chemical reactions which result during composting usually occur under conditions of constant pressure. Thus, energy released can be determined from the enthalpy changes accompanying the reactions.

It is common to measure elevations against a reference point. Thus, the elevation of a mountain is referenced to sea level or some other datum, the height of a ceiling to the floor beneath it, and so on. Similarly, measurement of enthalpy is simplified by establishment of a datum against which all other values can be measured. The datum is termed a standard state and is conveniently taken as the stable state of the compound at 25°C and 1 atm pressure. For example, the standard state of oxygen is a gas. By convention, enthalpy of an element in its standard state is set equal to zero. The enthalpy of a compound is determined from the heat of reaction necessary to form the compound from its elements. For reactions in solution, the standard state corresponds to a concentration at unit activity (approximately a 1 *M* solution). An enthalpy change calculated under standard state conditions is usually termed ΔH°. Standard enthalpy values for a few compounds of interest are presented in Table 3-1.

Heat flow into a substance under constant pressure can also be determined from the specific heat and temperature difference. Integrating

Table 3-1. Standard State Enthalpy and Free Energy of Formation at 25°C[a]

Substance	State[b]	ΔH° (kcal/mol)	ΔG° (kcal/mol)
CH_4	g	−17.89	−12.14
CH_3COOH	aq	−116.74	−95.51
$C_6H_{12}O_6$	aq		−217.02
CO_2	g	−94.05	−94.26
CO_2	aq	−98.69	−92.31
H_2O	lq	−68.32	−56.69
H_2O	g	−57.80	−54.64
H_2S	aq	−9.4	−6.54
NH_4^+	aq	−31.74	−19.00
SO_4^{2-}	aq	−216.90	−177.34
H^+ (pH 7)	aq		−9.67
NH_4Cl	c	−75.38	−48.73
Cl^-	aq	−40.02	−31.35

[a]Compiled from data in References 25-28.
[b]aq = aqueous, g = gas, lq = liquid, c = crystal.

Equation 3-6 under conditions of constant pressure gives:

$$q_p = mc_p\,(\Delta T) \quad (\Delta T \text{ not large}) \tag{3-12}$$

Equation 3-12 will be extremely useful in later energy balances on composting processes.

Both internal energy and enthalpy are properties of a system. This means that changes in enthalpy and internal energy between two equilibrium states of a system are the same for all paths followed between the two states. In other words, internal energy change and enthalpy change depend only on the beginning and end points of a process. For example, heat absorbed in an isobaric (constant pressure) process is equal to the enthalpy difference between beginning and end states regardless of path. Because enthalpy is a property of a system, its values can be tabulated. The reader is probably familiar with steam tables which contain tabulated values of enthalpy for water and water vapor as functions of temperature and pressure. Not all terms mentioned in this discussion are system properties. Work, for example, depends on the path followed between equilibrium states and is not a property of a system.

The distinction made between a property of a system and other functions which depend on path may seem rather subtle. However, the distinction is quite important and has direct application to thermodynamic analysis of composting systems. Often the inlet and outlet conditions about a compost process can be defined without knowledge of the path followed between end points. Fortunately, it is not necessary to know the path to determine the enthalpy change. The following example should help clarify the distinction and highlight its importance.

Example 3-1

Consider a compost process operated under constant pressure conditions in which water and vapor enter at temperature T_1 = 15.5°C (60°F). The reactor is maintained at a temperature T_2 = 71.1°C (160°F), and different proportions of water and vapor exit the reactor at T_2. Determine the heat required to accomplish the mass and temperature changes assuming two different thermodynamic paths as follows: (1) assume that water is vaporized at temperature T_1 and then water and vapor are heated to T_2; (2) assume inlet water and vapor are first heated to temperature T_2 and then a portion is vaporized at temperature T_2.

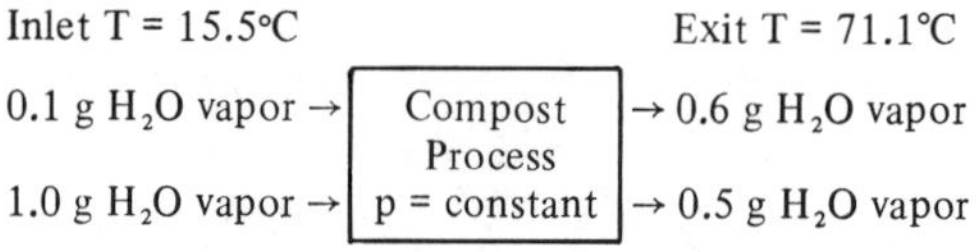

Inlet T = 15.5°C		Exit T = 71.1°C
0.1 g H_2O vapor →	Compost Process	→ 0.6 g H_2O vapor
1.0 g H_2O vapor →	p = constant	→ 0.5 g H_2O vapor

Solution

1. From steam tables the following table of enthalpy values can be constructed:

	Enthalpy (cal/g)	
Temp. (°C)	Water	Steam (Sat.)
15.5	15.59	604.28
71.1	71.09	627.89

2. For path 1 the enthalpy change can be determined as:

$$\Delta H = \underbrace{0.5(71.09 - 15.59)}_{\text{liquid to } T_2} + \underbrace{0.5(604.28 - 15.59)}_{\text{vaporize at } T_1} + \underbrace{(0.5 + 0.1)(627.89 - 604.28)}_{\text{all vapor to } T_2}$$

$$\Delta H = 336.26 \text{ cal}$$

3. For path 2 the enthalpy change is determined as:

$$\Delta H = \underset{\text{liquid to } T_2}{1.0(71.09 - 15.59)} + \underset{\text{vaporize at } T_2}{0.5(627.89 - 71.09)} + \underset{\text{input vapor to } T_2}{0.1(627.89 - 604.28)}$$

$$\Delta H = 336.26 \text{ cal}$$

4. Obviously the enthalpy change is the same regardless of path and one need only select a convenient path and be consistent in any subsequent calculations. This result will have a direct bearing on energy balances conducted in Chapters 9 and 11.

The heat release accompanying any chemical reaction is determined as the sum of the enthalpy of products minus reactants. The equation for oxidation of glucose is:

$$C_6H_{12}O_6 + 6\ O_2 \rightarrow 6\ CO_2 + 6\ H_2O$$

$$\Delta H_R^0 = -673{,}000 \text{ cal/mol}$$

The heat of reaction ΔH_R^0 is negative, which means that 673 kcal of heat energy would be released to the surroundings per mole (180 g) of glucose oxidized. If the reaction is conducted by microbes, some of the chemical bond energy will be captured for use by the microbes, reducing somewhat the amount which appears as heat.

For microbial reactions all reactants and products should be assumed to be in the aqueous state, since all energy-yielding reactions are conducted within the cellular cytoplasm. For combustion reactions the oxygen and produced CO_2 are usually assumed to be gaseous. If the produced water remains as a vapor the enthalpy change is often termed the lower heat of combustion. If produced water is condensed to a liquid, additional heat is released and the enthalpy change termed the higher heat of combustion.

Every organic molecule has a characteristic heat of combustion of significant magnitude and, in each case, of negative sign. Organic molecules contain energy in the chemical bonds which form the molecule. This energy is released as the molecule is degraded into simpler compounds, such as CO_2 and H_2O, which have a lower energy content.

Heats of reaction for the aerobic oxidation of various foodstuffs are presented in Table 3-2. It is interesting to note that the heat of reaction, expressed per g of foodstuff, varies significantly between the three major foodstuffs: proteins, carbohydrates and lipids (fats). Lipids contain over twice the energy per gram as proteins and carbohydrates, including both starch and cellulose. Because most organic wastes are composed of a mixture

Table 3-2. Heats of Combustion of Some Organic Fuels[a]

Feed	Formula	Molecular Weight	ΔH_R^0 (cal/mol)	g COD[b]/mol	g COD/g organic	kcal/g	kcal/g COD
Carbohydrate							
Glucose	$C_6H_{12}O_6$	180	−673,000	192	1.067	3.74	3.51
Lactic Acid	$CH_3CH(OH)COOH$	90	−326,000	96	1.067	3.62	3.40
Polysaccharides, e.g., Starch and Cellulose	$(C_6H_{10}O_5)_x$ Cellulose					essentially same as for glucose	
Lipid							
Palmitic Acid	$CH_3(CH_2)_{14}COOH$	256	−2,398,000	736	2.87	9.37	3.26
Tripalmitin	$(CH_3(CH_2)_{14}COO)_3(C_3H_5)$	809	−7,510,000	2,320	2.87	9.28	3.24
Protein							
Glycine (Amino Acid)	$CH_2(NH_2)COOH$	75	−163,800	48	0.64	2.18	3.41
Hydrocarbons							
n-Decane	$CH_3(CH_2)_8CH_3$	142	−1,610,000	496	3.49	11.34	3.25
Methane	CH_4	16	−210,800	64	4.00	13.18	3.29

[a]Heats of combustion were obtained from Reference 26 and corrected to products of combustion as $(CO_2)_g$, $(H_2O)_{aq}$ and $(NH_4^+)_{aq}$.
[b]Chemical oxygen demand.

of these natural foodstuffs, the heat of combustion can be expected to vary from a low of about 2100 to about 9300 cal/g (3800 to 16,700 Btu/lb). Comparative heating values of various fuels and municipal waste products are presented in Table 3-3. The organic components in domestic sludges commonly contain about 5550 cal/g of organic matter (10,000 Btu/lb).

THE SECOND LAW OF THERMODYNAMICS

In the latter part of the nineteenth century there was considerable interest among the early founders of thermodynamics as to the direction in which spontaneous chemical reactions would proceed. The French chemist Pierre Berthelot proposed that reactions would proceed spontaneously in the direction of negative ΔH. In other words, only exothermic reactions were spontaneous. This seemed to make a great deal of common sense. Organic molecules would always burn to form CO_2 and H_2O, but the latter would not spontaneously combine to form organic molecules.

Unfortunately for Berthelot, there were a few chemical reactions which did not follow this rule. For example, consider the dissolution of ammonium chloride crystals in water.

$$NH_4Cl_C \rightarrow NH^+_{4aq} + Cl^-_{aq}$$

$$\Delta H^0_R = (-31.74 - 40.02) - (-75.38) = +3.68 \text{ kcal/mol}$$

When added to water the crystals will dissolve completely provided the solubility limit is not exceeded. In other words, the reaction proceeds

Table 3-3. Comparative Heating Values of Various Fuels [29,30]

Fuel	Heating Value (cal/g)
No. 2 Oil	10,900
No. 6 Oil	9,720
Natural Gas	12,700
Bituminous Coal	7,560
Wood (Air-Dried)	3,060
Grease and Scum	9,280
Sludge (Dry Volatiles)	5,560
Digester Sludge	2,950
Digester Gas	8,560
Municipal Refuse (20% Moisture)	2,720

spontaneously to the right even though the reaction is endothermic. Something was missing from Berthelot's suggestion regarding the direction of spontaneous reactions.

The missing factor was the tendency of all systems to seek a state of maximum randomness or disorder. In the above example, the solution of ammonium and chloride ions in water is more random than the highly organized structure of the original crystal. Thus, the crystal tends to dissolve even though the reaction is endothermic, driven by the tendency toward maximum randomness.

Development of mathematical formulations to describe the tendency toward randomness dominated much of theoretical thermodynamics in the nineteenth century. The concept of entropy was developed from a search for a function that would serve as a criterion for spontaneity of physical and chemical changes. The search led to the formulation of the second law of thermodynamics, which states that all spontaneous changes in an isolated system occur with an increase in entropy or randomness. In 1876 and 1878 an American chemist, Josiah Willard Gibbs, published papers in which the two laws of thermodynamics were applied to chemical reactions. Gibbs showed that the change in enthalpy was composed of two parts: (1) a change in useful or available energy, termed free energy ΔG; and (2) a change in entropy ΔS. The relationship is:

$$\Delta H = \Delta G + T\Delta S \quad (T \text{ and } p = \text{constant}) \tag{3-13}$$

or

$$\Delta G = \Delta H - T\Delta S \tag{3-14}$$

In any spontaneous process the change in entropy ΔS must always be positive. Thus, in the case of ammonium chloride the product of $T\Delta S$ is greater than the enthalpy increase ΔH, giving a net decrease in free energy:

$$NH_4Cl_c \rightarrow NH_4{}^+_{aq} + Cl^-_{aq}$$

$$\Delta G^0_R = (-19.00 - 31.35) - (-48.73) = -1.62 \text{ kcal/mol}$$

In any spontaneous process, whether endothermic or exothermic, the entropy must increase and the free energy decrease. Final realization of this concept was one of the crowning achievements of nineteenth century thermodynamics.

It can be shown that ΔG gives the useful work that can be derived from a chemical reaction occurring under constant pressure and temperature conditions. Since these are exactly the conditions under which most biochemical reactions occur, the free energy function of Gibb's can be used to determine the energy available to microbes from various chemical reactions. The original premise of this chapter can now be restated in a more sophisticated manner: "All life forms require a source of free energy to maintain life."

Standard states must again be established to measure changes in free energy. For convenience the standard states are defined in similar manner as for enthalpy changes. For reactions which occur in solution the standard state corresponds to a concentration at unit activity (approximately 1 *M* solution). In addition, the hydrogen ion at unit activity (pH $\sim$ 0) is assigned a standard free energy of zero. Standard free energy values for a number of compounds of interest are presented in Table 3-1.

FREE ENERGY AND EQUILIBRIUM

Consider a reaction in which "a" can proceed to "b," and vice versa:

$$a \rightleftharpoons b$$

Such a reaction is termed an equilibrium reaction. Most of the metabolic reactions mediated by microbes are composed of a series of such equilibrium reactions. Thus, they are important to living systems. Let us assume that initially no "b" is present in solution. Then the reaction will proceed to the right, producing "b" and consuming "a." As soon as "b" becomes present in solution, some of it will begin to react back to the left producing "a." Eventually, a dynamic condition will be established when the rate of production of "b" is balanced by the production of "a." At that point the system is in equilibrium.

Now let us consider the free energy changes which occur in the above equilibrium. The initial movement of "a" to form "b" occurs because of a decrease in free energy for the reaction. It becomes apparent that the magnitude of the free energy change must be related to the concentration of the reactant and product. At equilibrium the rate of production of "b" is exactly balanced by production of "a." There is no tendency for the reaction to make a net movement in either direction. At this point the free energy change for the reaction must be zero. If it were negative, the reaction would move to produce more "b." If it were positive the reaction would move in the reverse direction to produce more "a."

Consider the reaction:

$$aA + bB \rightleftharpoons cC + dD$$

It can be shown that the effect of concentration of the different species can be described as:

$$\Delta G_R = \Delta G_R^o + RT \, ln \, \frac{[C]^c[D]^d}{[A]^a[B]^b} \tag{3-15}$$

where ΔG_R = reaction free energy change, cal/mol
ΔG_R^o = reaction free energy change under standard state conditions, cal/mol
R = universal gas constant = 1.99 cal/deg-mol
T = absolute temperature, °K
[A] = molar concentration of species A

Equation 3-15 can be used to adjust standard free energy values for the effect of concentration.

The reaction free energy ΔG_R must be zero at equilibrium, thus:

$$\Delta G_R^o = -RT \, ln \left(\frac{[C]^c[D]^d}{[A]^a[B]^b} \right)_{eq} \tag{3-16}$$

The term in brackets is actually the equilibrium constant for the reaction

$$K_{eq} = \text{equilibrium constant} = \frac{[C]^c[D]^d}{[A]^a[B]^b} \tag{3-17}$$

Thus

$$\Delta G_R^o = -RT \, ln \, K_{eq} \tag{3-18}$$

This equation is one of the more important results of chemical thermodynamics because it allows the prediction of equilibrium constants from standard free energy changes. Although there will be little occasion to use Equation 3-18 in this book, its significance should be appreciated.

REACTION RATES AND TEMPERATURE

Rates of chemical reactions are usually a function of temperature. A convenient method of expressing the effect of temperature is to determine the rate of activity at one temperature to the rate at a temperature 10°C lower. This ratio is the temperature coefficient Q_{10}. As a general rule of thumb, most chemical reaction rates approximately double for each 10°C rise in temperature (i.e., $Q_{10} = 2$). For biologically mediated reactions such a relationship is also observed over the limited temperature ranges suitable for living organisms. Temperatures outside this range cause inactivation of enzymes responsible for mediating the desired reactions. Diffusion-controlled reactions usually exhibit a Q_{10} value less than 2 because diffusion coefficients vary less with temperature. On the other hand, temperature coefficients for enzyme coagulation and heat inactivation of microbes are characteristically greater than 2. Therefore the effect of temperature on chemical and biochemical reaction rates is of considerable importance in the decomposition of waste materials and in thermal inactivation kinetics.

In 1889 the Swedish chemist Svante August Arrhenius developed a mathematical relationship between temperature and the rate of reaction. Based on experimental and theoretical considerations, Arrhenius proposed the following relationship:

$$\frac{d(ln\ k)}{dT} = \frac{E_a}{RT^2} \tag{3-19}$$

where k = reaction rate constant
E_a = activation energy for the reaction, cal/mol
T = absolute temperature, °K

Activation energy is interpreted as the amount of energy a molecule must have to undergo a successful chemical reaction. The integrated form of Equation 3-19 becomes

$$ln\ \frac{k_2}{k_1} = \frac{E_a(T_2 - T_1)}{RT_1T_2} \tag{3-20}$$

where k_2 and k_1 are the reaction rate constants at temperatures T_2 and T_1, respectively. Between 20 and 30°C, a Q_{10} value of 2 corresponds to an activation energy of about 12.25 kcal/mol.

There are several other commonly used forms of the Arrhenius expression developed from Equation 3-20 as follows. Taking the antilog of both sides of Equation 3-20 and rearranging:

$$k_2 = (k_1e^{E_a/RT_1})e^{-E_a/RT_2} \tag{3-21}$$

Because k_1 corresponds to temperature T_1 the term in brackets can be considered a constant for a given reaction:

$$k_2 = C\, e^{-E_a/RT_2} \tag{3-22}$$

Most biological processes operate over a limited temperature range. This is true even in composting systems. A temperature range from 0 to 80°C, which seems fairly extreme, corresponds to a range of only 273 to 353°K. Thus, the product of T_1 and T_2 changes only slightly over the biological temperature range. It is frequently assumed that E_a/RT_2T_1 is constant over this range so that Equation 3-20 becomes:

$$ln \frac{k_2}{k_1} = \theta(T_2 - T_1) \tag{3-23}$$

or

$$k_2 = k_1 e^{\theta(T_2 - T_1)} \tag{3-24}$$

Although θ should be reasonably constant, it has sometimes been found to vary considerably even over small temperature ranges. It is considered good practice to state the applicable temperature range whenever a value of θ is given. By way of comparison, a Q_{10} of 2.0 corresponds to a θ of about 0.069.

LIFE AND ENERGY

Strictly speaking, the thermodynamic principles outlined above apply only to "closed" systems, defined as those which do not exchange matter with their surroundings or across the system boundary. Living systems are constantly exchanging matter with their surroundings and hence are termed "open" systems. Furthermore, they seldom attain true thermodynamic equilibrium. Instead there is a continual flux of metabolic materials which may result in a dynamic "steady-state." Under such steady-state conditions the rate of formation of a component is balanced by the rate of subsequent breakdown or conversion to another component. Nevertheless, principles of equilibrium thermodynamics have been applied to living systems with much success, and there need not be much concern here with slight deviations from theoretical accuracy. The field of irreversible thermodynamics has been developed to deal with steady-state systems such as living microbes. One useful result of irreversible thermodynamics is the realization that steady-state is characterized by achievement of the minimum possible rate of entropy production for a given substrate use rate [31].

Another attribute of living cells is that they are highly ordered systems composed of many sophisticated molecular structures. As such, there has been a local decrease in entropy within the cell. It was observed before that any spontaneous process must occur with a net increase in entropy. Therefore, the local decrease in entropy within the cell is maintained at the expense of a larger increase in entropy in the surroundings. Consider, for example, an organism using the organic molecule glucose as a food source. The end products of aerobic metabolism are CO_2 and H_2O, which are more random than the original glucose molecule. Since the glucose is taken from the media surrounding the cell and CO_2 and H_2O are discharged back into it, an increase in the entropy of the surroundings has occurred. Lehninger [24] stated that living organisms create and maintain their essential orderliness at the expense of their environment which they cause to become more disordered and random.

Local decreases in entropy are not unique to living systems. Energy flow into a nonliving system can often cause a local decrease in entropy even though the net entropy must increase. What is unique to living systems is the use of enzyme-catalyzed chemical reactions to effect and maintain the decreased entropy within the cell. In recognition of these factors Asimov [32] distinguished life and living organisms as "characterized by the ability to effect a temporary and local decrease in entropy by means of enzyme-catalyzed chemical reactions."

One final attribute of microbial systems which should be considered is their remarkable ability to exploit available sources of chemical free energy. It is rare that an organic or inorganic reaction which yields free energy is *not* used by microbes. If it is assumed that all such reactions are capable of use by microbes, one will rarely if ever be in error. Poindexter [33] has indicated that bacteria and fungi are particularly omnivorous and as a group can use for growth every known naturally occurring organic compound. No wonder that these microbes are particularly important in composting systems.

ESTIMATING HEATS OF REACTION

It is often difficult to estimate heats of reaction for organic wastes from standard enthalpy values. Usually such wastes comprise a mixture of organics of unknown composition. In such a case, standard enthalpy values are of little use except in defining the range of probable heats of reaction. However, several experimental and empirical approaches are available for determining heats of reaction.

The heating value of an organic waste can be determined by calorimetric measurements. The quantity of heat released, however, is a function of the path followed during oxidation of the sample material. One method of determining the heat of reaction for a given chemical reaction is by means of an "open calorimeter," in which pressure is maintained constant at one atmosphere. Under constant pressure conditions, heat released is equal to the enthalpy change for the reaction. Another type of calorimeter is the "bomb calorimeter," in which reactions are conducted under conditions of constant volume. Thus, heat released in a bomb calorimeter would differ somewhat from that in a constant-pressure calorimeter. Fortunately, these differences are usually small for organic materials which release considerable energy on oxidation. Methods are also available to correct bomb calorimeter results to conditions of constant pressure [34].

The calorimetric approach is undoubtedly the most accurate way to determine heats of reaction for mixtures of organics. However, calorimetric tests are not routine for most water quality laboratories, and analysis by laboratories equipped for this purpose is usually required. As a result, a number of approximate formulas based on both theoretical and experimental approaches have been developed.

Fair et al. [35] determined the fuel values of different types of vacuum-filtered sludges. A bomb calorimeter was used in these experiments, but actual heat release under constant pressure conditions would probably not differ significantly. An empirical formula describing their results is:

$$Q = a\left[\frac{P_V(100)}{100 - P_C} - b\right]\left[\frac{100 - P_C}{100}\right] \tag{3-25}$$

where Q = fuel value, Btu/lb dry solids
a = coefficient equal to 131 for raw and digested primary sludge, 107 for raw waste activated sludge
b = coefficient equal to 10 for raw and digested primary sludge, 5 for raw waste activated sludge
P_V = percent volatile solids in sludge
P_C = percent of inorganic conditioning chemical in sludge

A formula presented by Spoehr and Milner [36] relates the heat of combustion to the degree of reduction of the organic matter. This is a rational approach because the heat of combustion has already been shown to be significantly lower for carbohydrates than for the more reduced lipids and hydrocarbons. Products of combustion are assumed to be gaseous carbon dioxide, liquid water and nitrogen gas. The degree of reduction for any type of organic matter is:

$$R = \frac{100[2.66(C) + 7.94(H) - (O)]}{398.9} \tag{3-26}$$

where C, H and O are the percentages of carbon, hydrogen and oxygen, respectively, on an ash-free basis. The heat of combustion is:

$$Q(\text{cal/g}) = 127\ R + 400 \tag{3-27}$$

A representative ultimate analysis for domestic sludge and refuse is presented in Table 3-4.

Another formula, similar to that presented by Spoehr and Milner, is called the Dulong formula and is sometimes useful in estimating gross heating values from the feed composition [37].

$$Q(\text{Btu/lb}) = 145.4(C) + 620\left(H - \frac{O}{8}\right) + 41(S) \tag{3-28}$$

Equations 3-27 and 3-28 both require an ultimate analysis of the waste to determine the percentages of carbon, hydrogen and oxygen. This may not be a routine test in many water quality laboratories. Equation 3-25 requires analysis only of the volatile solids (VS) content, which is easily handled. However, Equation 3-25 was developed only for municipal sludge and should not be extended to other organic wastes.

A rule of thumb which is reasonably accurate for most organics is that about 3.4 ± 0.2 kcal are released/g COD of the waste. Because the COD test is routinely practiced by many laboratories, it allows a relatively easy approximation of the heat of combustion. Heats of reaction were calculated

Table 3-4. Representative Chemical Analysis and Heat Content of Dry Refuse and Sewage Sludge Samples [38]

Constituent	Refuse (wt %)	Raw Sludge (wt %)	Digested Sludge (wt %)
Carbon	33.11	37.51	24.04
Hydrogen	4.47	5.54	3.98
Oxygen	25.36	22.56	12.03
Nitrogen	0.60	1.97	2.65
Chlorine	0.41	0.33	0.17
Sulfur	0.14	0.37	0.75
Metal	11.64		
Glass, Ceramics, Stone	16.23		
Volatiles @ 110°C		3.66	3.01
Ash	8.04	28.06	53.37
	100.00	100.00	100.00
Higher Heating Value, cal/g	3280	3910	2570

for the foodstuffs presented in Table 3-2. When expressed in terms of kcal/g COD, the caloric values are reasonably constant even though they vary considerably when expressed as kcal/g of organic.

The reason that the heat released per unit of COD is relatively constant lies in the fact that COD is a measure of electrons transferred. Four moles of electrons must be transferred during aerobic oxidation for each mole of substrate oxygen demand:

$$\frac{1}{4}O_2 + 1e^- \rightarrow \frac{1}{2}O^{2-}$$

Thus, substrate COD is proportional to the number of electrons transferred during aerobic oxidation. Furthermore, heat of combustion per electron transferred to a methane-type bond is taken to be about 26.05 kcal per electron equivalent [31] or, since O_2 has four such electrons, 104.2 kcal/mol O_2. This in turn is equal to 3.26 kcal/g COD. For aerobic oxidation, therefore, the COD unit turns out to be a measure of energy release and it is not surprising that energy release per electron transferred is relatively constant for a wide variety of substrates.

Example 3-2

Given the raw sludge composition in Table 3-4, compute the heat of combustion by the techniques described above.

Solution

1. Calculate Q by Equation 3-25. The VS content is 100−28.06 = 71.94%. P_c will be assumed to be zero. Using coefficients for raw primary sludge:

$$Q = 131\left[\frac{71.94(100)}{100-0} - 10\right]\left[\frac{100-0}{100}\right] = 8114 \text{ Btu/lb total solids}$$

$$= 4503 \text{ cal/g total solids}$$

2. Calculate Q by Equation 3-27. The percentages of C, H and O on an ash-free basis are:

	% from Table 3-4	Adjusted to 100%
C	37.51	57.2
H	5.54	8.4
O	22.56	34.4
	65.61	100.0

From Equation 3-26:

$$R = \frac{100[2.66(57.2) + 7.94(8.4) - 34.4]}{398.9} = 46.2$$

Therefore, from Equation 3-27:

$$Q = 127\ (46.2) + 400 = 6267 \text{ cal/g organic}$$

Because the sludge is 71.94% VS

$$Q = 6267\ (0.7194) = 4509 \text{ cal/g total solids}$$

3. Calculate Q by Equation 3-28:

$$Q = 145.4\ (57.2) + 620\ (8.4 - 34.4/8) = 10{,}859 \text{ Btu/lb}$$

$$Q = 10{,}859\ (0.7194) = 7812 \text{ Btu/lb total solids}$$

$$Q = 4336 \text{ cal/g total solids}$$

Note that the sulfur content was neglected in this calculation. Although sulfur can be oxidized by certain autotrophic organisms during composting, the contribution to the overall energy balance would be negligible for the sulfur contents found in most sludges and other organic wastes.

4. Calculate Q by rule of thumb. Consider the organic fraction to contain 5550 cal/g of organic oxidized.

$$Q = 5550\ (0.7194) = 3993 \text{ cal/g total solids}$$

5. Calculate Q based on sample COD. Average organic composition can be determined from the weight-percentages and molecular weights of the components. Consider only the C, H and O fractions:

C	57.2/12 = 4.8
H	8.4/1 = 8.4
O	34.4/16 = 2.2

which gives an average composition of $C_{4.8}H_{8.4}O_{2.2}$. COD can be determined by balancing the chemical equation for oxidation to carbon dioxide and water.

$$\begin{array}{ccc} 1\text{ g} & x & \\ C_{4.8}H_{8.4}O_{2.2} + & 5.8\ O_2 & \rightarrow 4.8\ CO_2 + 4.2\ H_2O \\ 101.2 & 5.8(32) & \end{array}$$

$$x = \frac{5.8(32)(1\text{ g})}{101.2} = 1.83\text{ g COD/g organic}$$

Assuming an average of 3400 cal/g COD:

$$Q = 3400(1.83) = 6222\text{ cal/g organic}$$

$$Q = 6222(0.7194) = 4476\text{ cal/g total solids}$$

6. Note that the range of values calculated by these techniques is 3993-4509 cal/g total solids or about ±6% of the average value. This should be sufficiently accurate for most analyses of sludge composting systems.

One final point should be made concerning heats of reaction for mixtures of organics which is characteristic of most natural waste products. The actual heat released during composting will be determined by those organics which actually degrade during composting. Thus, in the example above, heat release will be as calculated provided all organic components are equally degradable. If all components of the mixture are not degradable the energy release could vary significantly. For example, if the lipid fraction were more degradable than either the protein or carbohydrate fraction, the heat release per gram of organic decomposed would be greater because of the higher caloric value of the lipids.

SUMMARY

Thermodynamics is the study of energy and its transformations. It provides an underlying current for the understanding of physical, chemical and biological systems. It places distinct limits on the energy transformations within systems as small as a single microbe or as large as the universe. One of the central themes of this book is the application of thermodynamics to composting systems which by their nature are composed of physical, chemical and biological processes. No single science will so unify the diverse aspects of composting as will thermodynamics.

The first and second laws of thermodynamics form the foundation upon which the science is based. Both are accepted as first principles which have been repeatedly upheld by human experience with the world. The first law states that energy can be neither created nor destroyed. In engineering terms it is commonly referred to as the law of conservation of energy. The concepts of heat, work, internal energy and enthalpy are related by the first law. The second law resulted from a search to explain the direction in which spontaneous reactions would occur. This led to the realization that all spontaneous changes in an isolated system occur with an increase in entropy or randomness.

From the first and second laws the concept of free energy was developed. The free energy gives the useful work which can be derived from a chemical reaction which occurs under constant pressure and temperature conditions. This is extremely useful because most microbial processes occur under such conditions. Therefore, a measure exists of the useful energy available from the feed substrate being used by a microbial population.

All chemical reactions have a standard free energy change, measured with all reactants and products at unit activity (approximately a 1 *M* concentration). Spontaneous chemical reactions proceed in the direction of decreasing free energy. If the free energy change is zero the reaction is at equilibrium. The standard free energy change can be related to the equilibrium constant for the reaction, and can be adjusted for the effect of product and reactant concentrations which differ from the standard concentration.

The effect of temperature on the rate constant for chemical reactions can be estimated from the Arrhenius equation. Various simplified forms of the Arrhenius relationship are currently used in engineering practice.

Heats of combustion vary from ~2100 to ~9300 cal/g for the three major foodstuffs: proteins, carbohydrates and lipids (fats). Lipids generally contain about twice the energy per gram as proteins or carbohydrates. Expressed on a COD basis, however, most organics have a heat of combustion of about 3.4 ± 0.2 kcal/g COD of the organic. It is often difficult to estimate heats of reaction for organic wastes from standard thermodynamic tables because the wastes are likely to be composed of a mixture of compounds of unknown composition. Open and bomb calorimetric techniques can be used to determine experimentally the heats of combustion for such unknown materials. A number of empirical equations are also available which yield reasonably consistent results and which require only routine laboratory analysis.

CHAPTER 4

BIOLOGICAL FUNDAMENTALS

INTRODUCTION

This chapter provides a very brief introduction to the biology of organisms responsible for composting, i.e., bacteria and fungi. The discussion should acquaint the reader with sufficient terminology to allow an understanding of the fundamental biology involved in composting. It will not be a detailed discussion from the viewpoint of a microbiologist, but will stress the energy requirements and overall chemical reactions mediated by organisms.

The distinction of living things into plant and animal kingdoms is common to our experience. With the discovery of and experimentation with microorganisms, however, distinctions became less well defined. Microbes seemed to have characteristics of both the plant and animal kingdoms. In 1866 Ernst H. Haeckel, a German zoologist, proposed the kingdom Protista, which included bacteria, algae, protozoa and fungi. All unicellular (single-celled) organisms and those containing multiple cells of the same type are included in this classification. Viruses, which are noncellular, were not discovered until 1882 and so escaped Haeckel's original classification. For convenience, viruses are often included in the Protist kingdom. A breakdown of the Protist kingdom into groupings convenient for our discussion is presented in Figure 4-1.

All bacteria and blue-green algae are procaryotic cell types, while other protists (and all other living organisms) are of the eucaryotic cell type. The distinction is based on differences in cellular anatomy. In procaryotes, the nuclear substance is not enclosed within a distinct membrane and nuclear division is less complex than in eucaryotes. Procaryotes are thought to be more primitive organisms on the evolutionary scale.

Microbes of importance in composting include bacteria and fungi. All

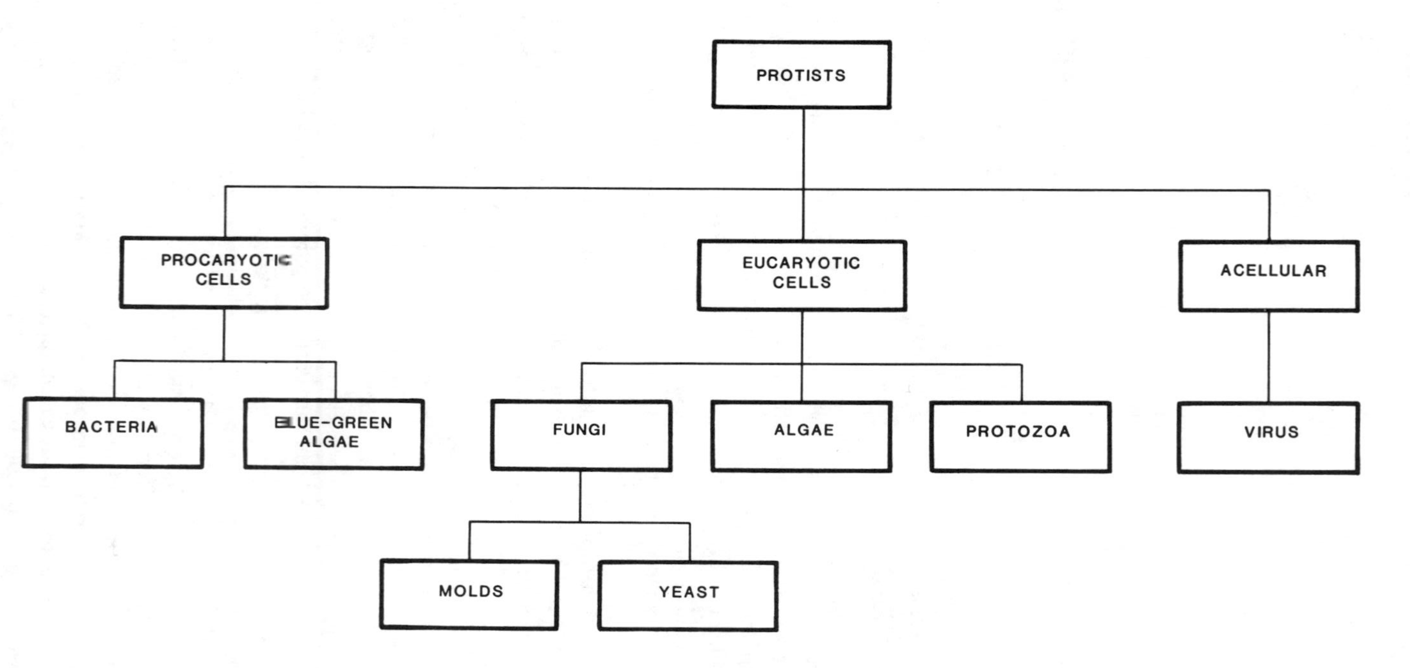

Figure 4-1. Classification of microbes in the Protist kingdom.

other groups are of minor significance. It should be noted that some workers in composting include actinomycetes as a separate group distinct from bacteria and fungi. There appears to be considerable confusion on this point, because actinomycetes have characteristics between bacteria and fungi and have been at times classified as bacteria, fungi and even as a separate phylogenetic line. As a class, they are active in degradation of insoluble high-molecular-weight organics, such as cellulose, chitin, proteins, waxes, paraffins and rubber. As such, they should be important in composting systems. For convenience, they will be considered along with bacteria, although the difficulty of classification should be recognized.

METABOLIC CLASSIFICATIONS

Metabolic distinctions between organisms provide a useful tool to understand both the effect of an organism on its surroundings and the environment necessary for proper growth of the organism. Perhaps the most basic distinction is between aerobic, anaerobic and anoxic metabolism. Aerobic refers to respiration with oxygen. Consider the aerobic oxidation of glucose as follows:

$$\overbrace{C_6H_{12}O_{6_{aq}}}^{-24e^-} + 6O_{2_g} \rightarrow 6CO_{2_{aq}} + 6H_2O_l \quad \Delta G_R^0 = -677 \text{ kcal/mol}$$

(+24e⁻ accepted by oxygen)

The reaction is of the oxidation-reduction type, because electrons are transferred from glucose and accepted by oxygen. Thus, oxygen is reduced, while carbon is oxidized. In this case, oxygen is referred to as the electron acceptor. All organisms which use oxygen as an electron acceptor are termed aerobic. All macroscopic organisms, and many microscopic ones as well, are obligate aerobes. In other words, oxygen and only oxygen will serve as an electron acceptor.

Among microbes, other compounds can be used as electron acceptors. The most notable of these are oxidized inorganic compounds of nitrogen and sulfur such as nitrate (NO_3^-), nitrite (NO_2^-), and sulfate (SO_4^{2-}). Carbon dioxide can also be used as an electron acceptor, and is usually reduced to methane. Metabolism with these electron acceptors is termed anoxic. Many organisms can function with a variety of such electron acceptors and are termed facultative. The oxidation of glucose using sulfate as an electron acceptor is given as:

$$\overbrace{C_6H_{12}O_6}_{aq} + 3SO_{4\,aq}^{=} + 6H^+_{aq} \rightarrow 6CO_{2\,aq} + 6H_2O_l + 3H_2S_{aq} \qquad \Delta G_R^0 = -107 \text{ kcal/mol (pH = 7)}$$

(−24e⁻ from $C_6H_{12}O_6$ to CO_2; +24e⁻ from $SO_4^{=}$ to H_2S)

Note that considerably less energy is available compared to aerobic metabolism. Being efficient chemical factories, microbes will first use those electron acceptors which provide the greatest energy yields. The order of use is generally O_2, NO_3^- and SO_4^{2-}. This is fortunate, because production of H_2S can be prevented by maintenance of aerobic conditions.

But what if all available electron acceptors have been used? Will electrons accumulate until the biochemical machinery grinds to a halt? The answer is no. To explain this consider the following reaction with glucose:

$$C_6H_{12}O_{6\,aq} \rightarrow 3CH_{4\,g} + 3CO_{2\,aq} \qquad \Delta G_R^0 = -96 \text{ kcal/mol}$$

(+12e⁻ from $C_6H_{12}O_6$ to CH_4; −12e⁻ from $C_6H_{12}O_6$ to CO_2)

Electrons removed from CO_2 are ultimately accepted by methane. Electrons are transferred and the reaction is of the oxidation-reduction type, but the acceptor and donor of electrons originate from the same molecule. Such reactions are termed fermentations, following Pasteur who called fermentation "life without air." Some confusion in nomenclature still persists between different disciplines. Most industrial microbial conversions are called fermentation even though many of the processes are aerobic.

Anaerobic metabolism is considerably more complex than the above fermentation reaction with glucose would indicate. Conversions are frequently mediated by a variety of organisms operating in series. In other words, the product of one organism is used as substrate by the second organism, and so on. A variety of intermediate products can be formed along the way, including low-molecular-weight organic acids (e.g., acetic and propionic), alcohols and aldehydes. In sanitary engineering practice these organisms are collectively referred to as first-stage or acid formers, although a variety of other intermediates can be formed. McCarty [39] has shown that with a complex starting substrate such as municipal sludges as much as 72% of the organics (measured as COD) pass through acetic acid. First-stage organisms are often important to industrial fermentation because of the end products formed. Ethanol is an obvious example of such an end product resulting from the fermentation of sugars by selected yeast cultures.

If fermentation is allowed to go to completion, the final end products will always be methane and CO_2. At this point as much energy has been extracted from the substrate as is possible under anaerobic conditions. Microbes which convert intermediate products to methane and CO_2 are termed methane-formers and are usually strict or obligate anaerobes.

The complexity of end products that result from anaerobic metabolism is often confusing to the student. Aerobic metabolism seems somewhat simpler because end products are usually carbon dioxide and water. In fact, the biochemical pathways are remarkably similar between the two groups. To help understand the reason for such variety of first-stage end products, a number of possible reactions were assembled in Table 4-1. In all cases, the starting material is glucose and the flow of electrons is balanced. There is a rather large range of free energy values between the reactions listed. In a mixed microbial population, reactions yielding more free energy would be expected to predominate. Production of a single intermediate such as ethanol requires sterile substrate preparation and use of pure or nearly pure cultures of selected microbes which yield the desired end product.

Metabolic classification of organisms can be continued by considering the carbon and energy sources used by the organism. Carbon accounts for about 50% of the dry mass of most organisms and is needed to synthesize the variety of organic molecules used in the structure and machinery of the cell. Two sources of carbon are available, carbon in the form of organic molecules and the carbon present in carbon dioxide. Autotrophs use carbon dioxide, whereas heterotrophs use the carbon of organic molecules.

Three distinct energy sources are available to organisms: organic oxidation-reduction reactions, inorganic oxidation-reduction reactions and the energy available in light. Organisms that use organic reactions for energy are termed organotrophs; those that use inorganic reactions are lithotrophs. Only bacteria are capable of using the energy of inorganic reactions. Organisms that use light for energy are phototrophs. By comparing the carbon and energy sources, four nutritional categories can be distinguished as shown in Table 4-2. One might suspect that with two carbon sources and three energy sources a total of six nutritional categories might be described. As far as is known today, however, all lithotrophs are also autotrophs. In other words, all lithotrophs obtain carbon from carbon dioxide. Likewise, all organotrophs will use organic carbon for synthesis. Therefore, the practical number of nutritional categories is reduced to four.

Although photoautotrophs are not significant to composting, this discussion of nutritional patterns would not be complete without a brief description of the pattern of electron flow for these organisms. Light is the ultimate source of energy for phototrophs. As such they do not depend on organic oxidation-reduction reactions in the same manner that organoheterotrophs do. However, electrons are still needed to reduce the cell carbon source, carbon dioxide, to the level needed for construction of organic

Table 4-1. Possible Anaerobic Fermentations of Glucose

End Products	Reactions	ΔG_R^0 @ pH = 7 (kcal/mol glucose)
Acetate	$C_6H_{12}O_6 \rightarrow 3CH_3COO^- + 3H^+$	−78.55
Propionate, Acetate, H_2	$C_6H_{12}O_6 \rightarrow CH_3CH_2COO^- + CH_3COO^- + 2H^+ + CO_2 + H_2$	−70.84
Butyric, H_2	$C_6H_{12}O_6 \rightarrow CH_3CH_2CH_2COOH + 2CO_2 + 2H_2$	−61.7
Ethanol	$C_6H_{12}O_6 \rightarrow 2CH_3CH_2OH + 2CO_2$	−51.14
Lactate	$C_6H_{12}O_6 \rightarrow 2CH_3CH(OH)COO^- + 2H^+$	−49.5
Methanol	$C_6H_{12}O_6 + 2H_2O \rightarrow 4CH_3OH + 2CO_2$	−21.42

Table 4-2. Metabolic Categories Based on Carbon and Energy Sources

Type of Nutrition	Principal Source of Energy	Principal Source of Carbon	Occurrence
Photoautotroph	Light	CO_2	Some bacteria, most algae, higher plants
Photoheterotroph	Light	Organics	Some algae, some bacteria
Lithoautotroph (Chemoautotroph)	Inorganic Oxidation-Reduction Reactions	CO_2	Some bacteria
Organoheterotroph[a] (Chemoheterotroph)	Organic Oxidation-Reduction Reactions	Organics[b]	Higher animals, protozoa, fungi and most bacteria

[a]In saprophytic nutrition, nonliving (inanimate) organic material is used. In parasitic or predatory nutrition, some or all of the materials used are animate.
[b]A few bacteria classified in this group are able to use CO_2 as a carbon source although this is not an energetically favorable pathway.

molecules. These electrons are usually taken from the oxygen of water as indicated by the following simplified equation for photosynthesis:

$$CO_2 + 2H_2O \xrightarrow{\text{light energy}} \underset{\text{cellular organics}}{CH_2O} + O_2 + H_2O$$

Four electrons taken from oxygen in the water molecule are used to reduce the carbon dioxide. Such a reaction is typical of algae and all higher plants.

Certain photosynthetic bacteria, notably the green and purple sulfur bacteria, are capable of extracting electrons from other sources such as sulfide, sulfur, H_2 and thiosulfate ($S_2O_3^{2-}$). A simplified equation representing the use of H_2S as an electron source is:

$$CO_2 + 2H_2S \xrightarrow{\text{light energy}} CH_2O + 2S^0 + H_2O$$

The inquisitive reader might wonder why some bacteria are classified as photoautotrophs along with the algae and higher plants. If bacteria are photosynthetic should they not then be called algae? The distinction centers on the type of chlorophyll contained in the cells. All algae, including the blue-green algae and all higher plants, contain the photosynthetic pigment chlorophyll-a, a particular type of chlorophyll, along with other light-sensitive pigments. Photosynthetic bacteria are procaryotic and do not contain chlorophyll-a. In addition, bacterial photosynthesis does not result in evolution of molecular oxygen because electrons are obtained from sources other than water. Hence, a distinction is made between photosynthetic bacteria and all other photosynthetic organisms.

Now that these metabolic distinctions have been made, we are in a position to explore the types of reactions observed in the microbial world. Respiration using various reductants and oxidants is presented in Table 4-3. The reactions are indicative of those mediated by various groups of microbes. With study of Table 4-3 the reader should be sufficiently familiar with nutritional pathways to understand subsequent material.

BACTERIA

Bacteria are the smallest living organisms known. Bacteria are typically unicellular, but multicellular associations of individual cells are also common. They may exist in a number of morphological forms including spheres (cocci), rods (bacillus), spirals (spirillum) and a variety of intermediate forms

Table 4-3. General Types of Lithoautotrophic and Organoheterotrophic Metabolism

Reductant	Oxidant	Products	Carbon Source	Respiration	Metabolic Category	Type, Representative Organism
NH_4^+, NH_3	O_2	NO_2^-, H_2O	CO_2	Aerobic	Lithoautotroph	Nitrifying bacteria, *Nitrosomonas*
NO_2^-	O_2	NO_3^-	CO_2	Aerobic	Lithoautotroph	Nitrifying bacteria, *Nitrobacter*
S^{2-}, H_2S	O_2	SO_4^{2-}	CO_2	Aerobic	Lithoautotroph	Sulfur-oxidizing bacteria, *Thiobacillus*, *Thiothrix*
Fe^{2+}	O_2	Fe^{3+}, H_2O	CO_2	Aerobic	Lithoautotroph	Iron-oxidizing bacteria, *Ferrobacillus*
Fe^{2+}	NO_3^-	Fe^{3+}, N_2, H_2O	CO_2	Anoxic	Lithoautotroph	
H_2	O_2	H_2O	CO_2	Aerobic	Lithoautotroph	Hydrogen bacteria
H_2	CO_2	CH_4, H_2O	CO_2	Anaerobic	Lithoautotroph	CO_2-reducing bacteria, *Methanobacterium*
Organics	O_2	CO_2, H_2O	Organics	Aerobic	Organoheterotroph	Many bacteria
Organics	NO_3^-	CO_2, H_2O, N_2	Organics	Anoxic	Organoheterotroph	Denitrifying bacteria
Organics	SO_4^{2-}	CO_2, H_2O. H_2S	Organics	Anoxic	Organoheterotroph	Sulfur-reducing bacteria, *Desulfovibrio*
Organics	Organics	Many intermediates, see Table 4-1 $CH_4 + CO_2$ final products	Organics	Anaerobic	Organoheterotroph	Acid-forming bacteria, methane-forming bacteria

such as comma-shaped (vibrio) and spindle-shaped (fusiform). Most bacteria reproduce by binary fission, division into two identical daughter cells. Sexual reproduction can also occur in certain cases.

Spherical bacterial cells are usually on the order of 0.5-3.0 μm in diameter. Rod-shaped cells are typically 0.5-1.5 μm in width and 1.5-10 μm in length. Because of their small size the bacteria have a very high surface-to-volume ratio. This allows for rapid transfer of soluble substrates into the cell and high rates of metabolic activity. As a result, bacteria will usually predominate over other larger microbes in aqueous substrate solutions.

Accumulations of certain unicellular bacteria often occur after cell division, particularly with spherical cells. Clumps, sheets and chains of cells can thus be formed. These accumulations are not regarded as true multicellular states because (1) they are not a regular feature of the life history of the cell, and (2) metabolism and development of an individual cell is not affected by its presence in such an association [33].

Poindexter [33] described three types of multicellular states characteristic of certain bacteria: trichomes, microcolonies containing budding filaments and mycelia. A trichome is a long, multicellular thread or chain of undifferentiated cells which remains intact indefinitely. In some cases a tubular sheath surrounds the chain of cells. Cells within the trichome continue to reproduce by cell division. Reproduction of the trichome itself occurs by cutting the thread into two subunits or by release of single cells to begin a new multicellular structure. Flagellated, nonflagellated and gliding bacteria are capable of such associations. The sulfur oxidizers *Beggiatoa* and *Thiothrix* are samples of trichome-forming, lithoautotrophic bacteria.

Certain microbes reproduce by budding (typically yeast and some bacteria). Mature cells give rise to one or more daughter cells which are initially much smaller than the mother cell, but grow and eventually divide from the mother cell. In some bacteria a filamentous outgrowth occurs from which a bud develops at the end. More filaments can develop from the newly formed bud, eventually developing into a microcolony of cells linked by filamentous appendages (Figure 4-2).

Perhaps the most elaborate multicellular arrangement occurs among the actinomycetes. A single reproductive cell gives rise to a chain of cells which branches extensively. Eventually a three-dimensional matrix of linked and branched filaments develops, termed mycelia. The general structure resembles that of molds but there are several distinctive differences. Filaments of the actinomycetes are usually 1-5 μm in diameter compared to the larger molds, which range around 10-20 μm. Furthermore, actinomycetes have a procaryotic structure compared to the eucaryotic fungi, as well as other biochemical and sexual differences. They are, therefore, usually classified as true branching bacteria and sometimes referred to as moldlike bacteria.

The subject of classification brings up the question of taxonomic

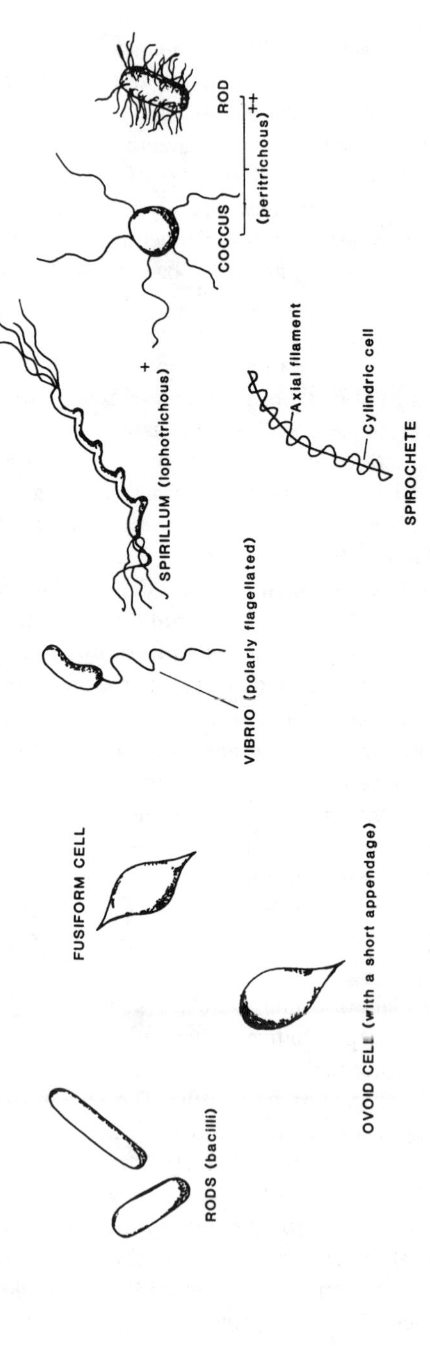

+ Tuft of flagella from end of cell.

++ Flagella protrude from entire surface of cell.

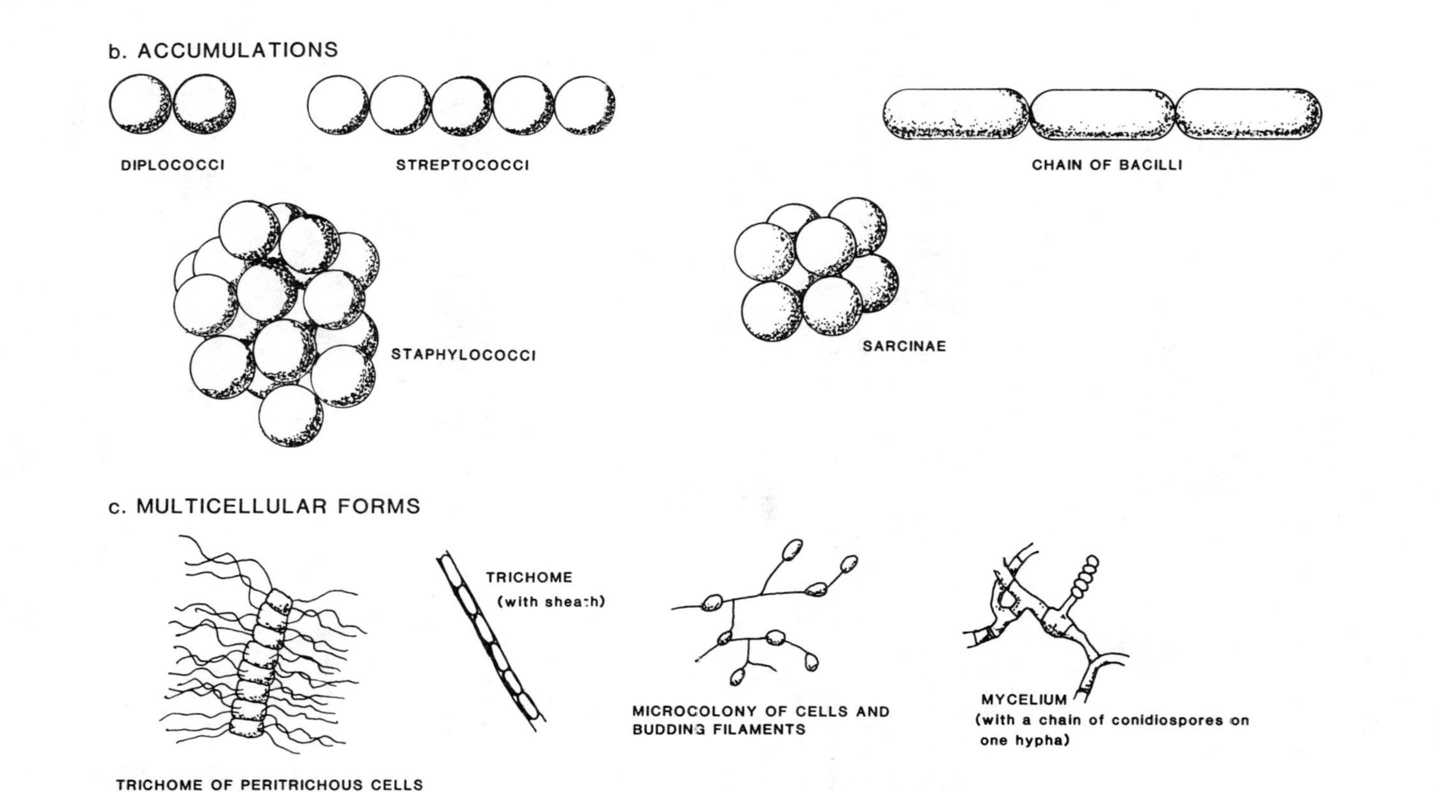

Figure 4-2. Bacterial morphology: (a) unicellular forms, (b) accumulations, (c) multicellular forms (a and c from Poindexter [33], b from Buffaloe and Ferguson [40]).

nomenclature. A sequence of taxonomic categories is employed, the idea being to group related organisms at various levels of similarity. The nomenclature commonly used in biological sciences is as follows:

Species	Organisms of one and the same kind
Genus	A group of related species
Tribe	A group of related genera
Family	A group of related tribes or genera
Order	A group of related familes
Class	A group of related orders
Phylum	A group of related classes
Kingdom	A group of related phyla

In some cases additional taxa are employed, particularly where there are a large number of species.

Classification of bacteria as to genus and species is a very difficult task partly because of the extremely small size of the microbe. This means that the form (morphology) of the organism is not sufficient for complete classification. Newer classifications take into account the type of nutrition and respiration, response to chemical tests such as the gram stain, composition of the DNA molecule, sensitivity to select bacterial viruses (bacteriophage) and other factors. The most widely accepted system of classification is that in *Bergey's Manual of Determinative Bacteriology* [41].

Table 4-4. Characteristics of

Order	Description	Motility
Class II–Schizomycetes		
Pseudomonadales	Many metabolic types; photosynthetic lithoautotrophs; organoheterotrophs; unicellular	Generally motile by flagella (polar)
Chlamydobacteriales	"Sheathed bacteria," filaments enclosed within a sheath	Cells released from sheath called swarm cells and are motile by flagella
Hypomicrobiales	"Budding bacteria," characterized by budding type of reproduction.	Some motile by flagella (polar)
Eubacteriales	"True bacteria," all are organoheterotrophs, most cocci or bacillus forms; unicellular	Some motile by flagella (peritrichous)

In this system the procaryotes are divided into three classes. Class 1, Schizophyceae, contains all of the blue-green algae. Class 2, Schizomycetes, includes the entire group of "true" bacteria. Class 3, Microtatobiotes, includes obligate intercellular parasites. Some distinctive characteristics of the orders of bacteria in Classes 2 and 3 are shown in Table 4-4.

A summary of the typical composition of bacteria, yeasts and molds is presented in Table 4-5. The cells and their associated slime layers are about 80% water and 20% dry matter of which about 90% is organic. Several empirical formulations for the organic fraction have been determined for mixed populations. For both aerobic and anaerobic bacteria these generally range from $C_5H_7O_2N$ to $C_5H_9O_3N$. The formulations do not vary significantly and will be useful in later analysis.

Some bacteria are capable of producing dormant forms of the cell which are more resistant to heat, radiation and chemical disinfection. An endospore is a thick-walled, relatively dehydrated unit formed within a bacterium and released upon cell lysis. Cysts and microcysts are formed from an entire cell that develops a thickened cell wall. Mycelial bacteria can produce spores (exospores) which can survive prolonged periods. Of the three types of dormant forms, endospores are generally more stable under adverse conditions than cysts or exospores. These forms of bacteria become significant when requirements for heat inactivation are considered (Chapter 5).

the Orders of Bacteria [41]

Mode of Reproduction	Representatives	Remarks
Binary fission	*Pseudomonas* *Nitrosomonas* *Nitrobacter* *Thiobacillus* *Acetobacter*	Gram-negative, some form trichomes; widely distributed in soil, fresh- and saltwater. Do not produce endospores. Some characterized by the variety of metabolized substrates. Includes green and purple sulfur, nitrifying, sulfur-oxidizing, sulfur-reducing bacteria.
Usually binary fission	*Sphaerotilus*	Inhabit fresh and marine waters; *Sphaerotilus natans* associated with water heavily polluted with organics.
Budding and longitudinal fission	*Hypomicrobium*	Found in all types of aquatic environments and sewage.
Binary fission	*Escherichia,* *Syaphylococcus,* *Lactobacillus,* *Azotobacter,* *Rhizobium*	Ubiquitous in nature, some produce heat-resistant endospores. Order includes many genera common to sanitary engineering such as *Escherichia*, *Enterobacter*, *Proteus*, *Salmonella*, *Shigella*, *Klebsiella*.

Table 4-4,

Order	Description	Mobility
Class II–Schizomycetes		
Caryophanales	"Filamentous bacteria," disk-like cells in trichomes	Motile by flagella (peritrichous)
Actinomycetales	"Moldlike bacteria," branching bacteria, some produce mycelia	Generally nonmotile
Beggiatoales	"Gliding bacteria"	Gliding motion without flagella caused by waves of contraction
Myxobacterales	"Slime bacteria," often termed protozoalike due to similarities with amoeba protozoa. Cells may swarm and form fruiting bodies and microcysts.	Gliding motion without flagella
Spirochaetales	"Spiral bacteria," elongate spiral cells	Rotary and flexing motion, no flagella
Mycoplasmatales	Pleuropneumonialike organisms (PPLO), extremely variant morphologic form	Nonmotile
Class III–Microtatobiotes[a]		
Rickettsiales	Obligate intracellular parasites	Nonmotile

[a]Virus particles are sometimes included in the Microtatobiotes as the order Virales.

continued

Mode of Reproduction	Representatives	Remarks
Fission and spores	*Caryophanon*	Cells usually large, measuring 20-30 μm in length. Cells occur in trichomes. Present in water and decomposing organic material.
Mostly by sporulation	*Actinomyces*, *Mycobacterium*, *Streptomyces*, *Micromonospora*	Prevalent in decomposing organic matter. *A. isarali* causes human actinomycosis. Streptomyces are major producers of antibiotics. *M. tuberculosis* causes tuberculosis.
Binary fission	*Beggiatoa*, *Thiothrix*	Order contains both chemoautotrophs and chemoheterotrophs. Filamentous with rod or coccoid cells in trichomes. Often referred to as algalike because of similar motility to blue-green algae. *Beggiotoa* and *Thiothrix* are sulfide oxidizers.
Binary fission and microcyst formation	*Myxococcus*	Found in soil, compost, manure, rotting wood. Capable of degrading complex substrates such as cellulose and bacterial cell walls. Many are predatory on other bacteria. Do not form trichomes, mycelia or endospores. All are organoheterotrophs.
Binary fission	*Treponema pallidum* (syphillis)	Large cells from 6-500 μm long. Normally occur in sewage, stagnant, fresh- or saltwater. Many are harmless saprophytes while others cause diseases of man and animals such as syphilis and leptospirosis.
Fragmentation and formation of elementary bodies	*Mycoplasma*	Extremely small cells, smallest ranging from 0.1 to 0.2 μm. Organisms lack cell wall. A few harmless saprophytic species are known. *M. pneumonia* causes primary atypical pneumonia.
Binary fission	*Rickettsia*	Very small rods or cocci. Typically parasites of arthropods such as fleas, lice, ticks. Often pathogenic to man. *R. prowazekii* causes typhus. Others cause rocky mountain spotted fever and Q fever. Order includes the smallest living cells.

Table 4-5. Typical Composition of Organic and Inorganic Fractions of Microbes

	Bacteria		Yeasts		Molds	
Component	**Avg**	**Range**	**Avg**	**Range**	**Avg**	**Range**
Organic Constituents (% dry wt)						
Carbon	48	46-52	48	46-52	48	45-55
Nitrogen	12.5	10-14	7.5	6-8.5	6	4-7
Oxygen		22-28				
Hydrogen		5-7				
Protein	55	50-60	40	35-45	32	25-40
Carbohydrate	9	6-15	38	30-45	49	40-55
Lipid	7	5-10	8	5-10	8	5-10
Nucleic acid	23[c]	15-25	8	5-10	5	2-8
Ash	6	4-10	6	4-10	6	4-10
		Bacteria		**Fungi**		**Yeast**
Inorganic Constituents (g/100 g dry wt)						
Phosphorus		2.0 -3.0		0.4 -4.5		0.8 -2.6
Sulfur		0.2 -1.0		0.1 -0.5		0.01 -0.24
Potassium		1.0 -4.5		0.2 -2.5		1.0 -4.0
Magnesium		0.1 -0.5		0.1 -0.3		0.1 -0.5
Sodium		0.5 -1.0		0.02-0.5		0.01 -0.1
Calcium		0.01 -1.1		0.1 -1.4		0.1 -0.3
Iron		0.02 -0.2		0.1 -0.2		0.01 -0.5
Copper		0.01 -0.02				0.002 -0.01
Manganese		0.001-0.01				0.0005-0.007
Molybdenum						0.0001-0.0002
Total ash		7-12		2-8		5-10

[a]Overall chemical composition: 80% water, bound and free; 20% dry matter, organic (90% of the dry weight), inorganic (10% of the dry weight).
[b]In part from Peppler [42].
[c]Values this high are observed only with rapidly growing cells.
[d]From Aiba et al. [43].

A bewildering variety of bacterial organisms can be isolated from composting material. Identification of the bacterial species present would be of academic interest only and few such studies have been conducted. It is known that mesophilic bacterial types dominate in the early stages of composting but yield to thermophilic types as temperatures increase above about 40-50°C. Species of the actinomycete genera, including *Micromonospora*, *Streptomyces* and *Actinomyces*, can regularly be isolated from composting material. Golueke [44] indicates that actinomycetes become detectable visually in undisturbed piles near the end of the composting process. They appear as a blue-gray to light green powdery or filamentous layer in the outer 10-15 cm of the pile. If the compost is mixed mechanically such large colonies will not occur. No visual evidence of filamentous colonies is usually observed in windrow sludge composting where mechanical action is frequent. Where mixing action is less intense, such as in the aerated static pile system, filamentous growths can become established to the point where visible colonies are formed.

Poincelot [45] analyzed the density of various microbes as a function of temperature during composting. His results are summarized in Table 4-6. Bacteria (including the actinomycetes) are usually present in larger numbers than the fungi in all temperature ranges tested. In most cases, bacterial numbers were about 100 or more times those of fungal species. This is consistent with a intuitive estimate by Golueke [44], who ascribed at least 80-90% of the microbial activity during composting to bacteria.

FUNGI

Fungi are of the eucaryotic cell type and are organoheterotrophic. The vast majority are saprophytic, decomposing organic matter in soil and aquatic environments. They are ubiquitous in nature. Fungi can be broadly divided between molds and yeasts. Molds are aerobic, whereas both aerobic and anaerobic metabolism is observed in yeasts. Molds tend to form filamentous structures while yeasts tend to be unicellular.

Fungi are very similar to organoheterotrophic bacteria, in that they use

Table 4-6. Microbial Populations During Aerobic Composting[a] [45]

	No./wet gram compost			
Microbe	Mesophilic Initial Temp <40°C	Thermophilic 40-70°C	Mesophilic 70°C to Cooler	Numbers of Species Identified
Bacteria				
Mesophilic	10^8	10^6	10^{11}	6
Thermophilic	10^4	10^9	10^7	1
Actinomycetes				
Thermophilic	10^4	10^8	10^5	14
Fungi[b]				
Mesophilic	10^6	10^3	10^5	18
Thermophilic	10^3	10^7	10^6	16

[a]Composting substrate not stated but thought to be garden-type materials composted with little mechanical agitation.
[b]Actual number present is equal to or less than the stated value.

most of the same organic substrates. They are distinguished from bacteria by their eucaryotic cell type, generally larger size and more sophisticated methods of reproduction. Because both types of organisms are served by similar substrates, competition between them is common. Both can use solid food materials by secreting extracellular hydrolytic enzymes. However, fungi are less affected by low moisture environments and can often grow on dry substrates nourished by moisture absorbed from damp atmospheres. Fungi can also extract moisture from materials that have high osmotic pressure such as syrups, jams and pickling brines. They can withstand a broad range of pH conditions, and often have a lower nitrogen requirement than bacteria. As such, fungi are common in soils and decaying vegetation where their lower moisture requirements give them a competitive advantage over bacteria. For example, it is common to observe development of mold on bread where the low moisture level inhibits rapid bacterial development. As a result of these distinctive properties, molds are common inhabitants of moist, dark locations where organic matter and oxygen are available. Obviously, similar conditions are found in most composting systems, and we can expect the fungi to play a prominent role in such.

McKinney [46] reported an approximate empirical formulation for the organic fraction of the fungus *Aspergillus niger* as $C_{10}H_{17}O_6N$. Assuming this formulation to be representative of other classes of fungi and comparing it with bacterial formulations given previously, the lower nitrogen content of the fungal protoplasm is immediately obvious. Thus, fungi should have a competitive edge over bacteria in nitrogen-deficient environments because the basic nitrogen requirement is lower. This can be of practical significance because many composting substrates are cellulosic and therefore tend to be low in nitrogen.

Fungi are important industrial microbes. Yeasts are the only significant microbes used in the production of alcoholic beverages such as beer and wine. Yeasts also produce industrial alcohol and glycerol, and are used as a leavening agent in baking. Molds are used industrially to synthesize a wide variety of valuable substances that cannot easily be made by artificial processes. These include antibiotics, organic acids and biological enzymes.

Over 80,000 species of fungi have been identified. Such a bewildering number of organisms presents major difficulties in classification. Fortunately, knowledge of detailed classification schemes is not necessary to design of successful composting systems. Nevertheless, a knowledge of general groupings is useful and provides considerable insight into the variety of living habits of these interesting organisms. A partial classification of fungi to the class level is presented in Table 4-7.

Fungi are classified largely on the basis of morphology and method of reproduction. In this sense the classification scheme differs from that used for

Table 4-7. Characteristics of the Classes of Fungi[a]

Class	Description	Representatives	Remarks
Kingdom: Protista Division: Mycota (Fungi) Subdivision I: Eumycotina (True Fungi)			
Class I: Phycomycetes	Primitive fungi: asexual spores contained in a sporangium, mycelia nonseptated except at reproductive site.	*Rhizopus* (black bread mold) *Mucor*	Most are saprophytic. Class includes water molds, blights, mildews and bread molds. Some species are aquatic. Terrestrial forms include *R. nigricans*, the common black bread mold. Occasionally, certain species can cause infection in man especially if large numbers of spores are inhaled.
Class II: Ascomycetes	Sac fungi: sexual spores are enclosed in sacs or asci. Mycelia septated. Includes filamentous and nonfilamentous forms (yeasts). Budding is the most distinctive method of asexual reproduction in yeasts.	*Aspergillus* *Penicillium* (blue-green mold) *Neurospora* (red mold of bread) *Candida* (a yeast)	About 30,000 species of filamentous Ascomycetes. Also includes the yeasts. Important industrial uses in fermentation, baking, brewing and biomass production. Participate in production of humus and digestion of cellulose. *A. fumigatus* can cause pulmonary infection. *P. notatum* and *P. chrysgenum* prominant as sources of penicillin.
Class III: Basidiomycetes	Club fungi: sexual spores are borne on basidia, a swollen clublike reproductive cell. Mycelia septated.		Includes macroscopic fungi such as mushrooms, puffballs and toadstools. Parasitic species such as rusts and smuts can infect plant crops. Other species aid in deterioration of wooden structures and formation of humus.

Table 4-7, continued

Class	Description	Representatives	Remarks
Class IV: Deuteromycetes	Fungi imperfecti: Heterogeneous group of about 20,000 species for which a sexual stage of reproduction has not been observed.	*Torula* *Tricoderma* *Cladosporium* *Alternaria* *Fusarium*	Vast majority are saprophytic, occurring in soils, decaying vegetable matter and foods. About 50 species are opportunists, causing disease when introduced into body under proper conditions. Infection of lungs and superficial skin infections most common. Some are carnivorous, trapping and digesting microscopic animals including nematodes in soil.
Subdivision II: Myxomycotina (Slime Fungi)			
Class I: Myxomycetes	True slime molds: In vegetative phase the microbe is a viscous mass of multinucleated protoplasm called a plasmodium. Resemble amoebas in some respects. Reproduce by spores on stalks like fungi.	*Physarum* *Badhamia*	Obtain nourishment by ingesting bacteria, small particulate matter and organic materials in soils. Some are parasitic on higher plants.
Class II: Acrasiomycetes	Cellular slime molds: Frequently occur as a pseudoplasmodia composed of many amoeba-like individual cells.	*Dictyostelium*	Diet consists mainly of bacteria, spores of fungi, and small piece of organic matter

[a] Adapted from a classification scheme by Alexopoulos [47].

bacteria, where biochemical responses are of prime importance. Fungi are usually capable of both sexual and asexual reproduction. An exception is the class Deuteromycetes, for which sexual reproductive stages have not been observed. In either case a reproductive body termed a spore is formed. Fungal spores should not be confused with bacterial spores which serve mainly as a survival mechanism. The major types of asexual and sexual spores are illustrated in Figures 4-3 and 4-4 along with much of the nomenclature common to fungal classification. Study of Figures 4-3 and 4-4 along with Table 4-7 should provide the reader with a knowledge of fungi sufficient for purposes of this text.

During the vegetative phase, virtually all filamentous fungi consist of tubular, branching filaments, called hyphae. Diameter of the hyphae typically range from 10 to 50 μm. Note that this is considerably larger than the average bacterial cell, a reflection of the eucaryotic nature of the fungi. A mass of hyphae is called a mycelium. In more primitive fungi, such as the Phycomycetes, the hyphae are not septated. In other words, individual cells in the filament are not separated by a cross wall and the hyphae can be considered as a single, multinuclear cell. Hyphae of other classes of fungi are generally septated, although the septum often contains pores which allow passage of material between cells.

Large numbers of fungal species can be isolated during both mesophilic and thermophilic stages of composting. Kane and Mullins [49] isolated 304 unifungal cultures from a Metro-Waste reactor system operating on municipal refuse. Of the total 304 isolates, 120 belonged to the genus *Mucor*, 97 to the genus *Aspergillus*, 78 to *Humicola*, 6 to *Dactylamyces*, 2 to *Torula* and 1 to the genus *Chaetomium*. As Kane and Mullins pointed out, however, the number of isolated species in each genus may not indicate the actual number of individuals of each species in the compost. Thus, the importance of each genus to composting is not indicated by the count of species. Thermophilic fungi were observe during all stages of the compost cycle, including the beginning of digestion under mesophilic conditions. No apparent succession of species was found.

One would expect the fungal population to be influenced by feed substrate, temperature, aeration, pH, moisture content and, perhaps, the mechanical agitation applied to the compost. Fungal molds are strict aerobes and would be present only where persistent aerobic conditions are maintained. Temperature conditions during composting can often reach the point where thermophilic fungi are inactivated. Some thermophilic fungi do not appear to grow well at temperatures above about 60°C, as indicated by Figure 4-5. Acclimation of the fungal species may have allowed continued metabolic activity at temperatures above these shown in Figure 4-5. Nevertheless, we can expect fungal activity to be highly variable depending on

A. SPORANGIOSPHORES: Spores inside swollen fertile structure called sporangium (limited to Phycomycetes).

B. CONIDIOSPORES (conidia): Spores supported by a specialized fertile structure, the conidiophore.

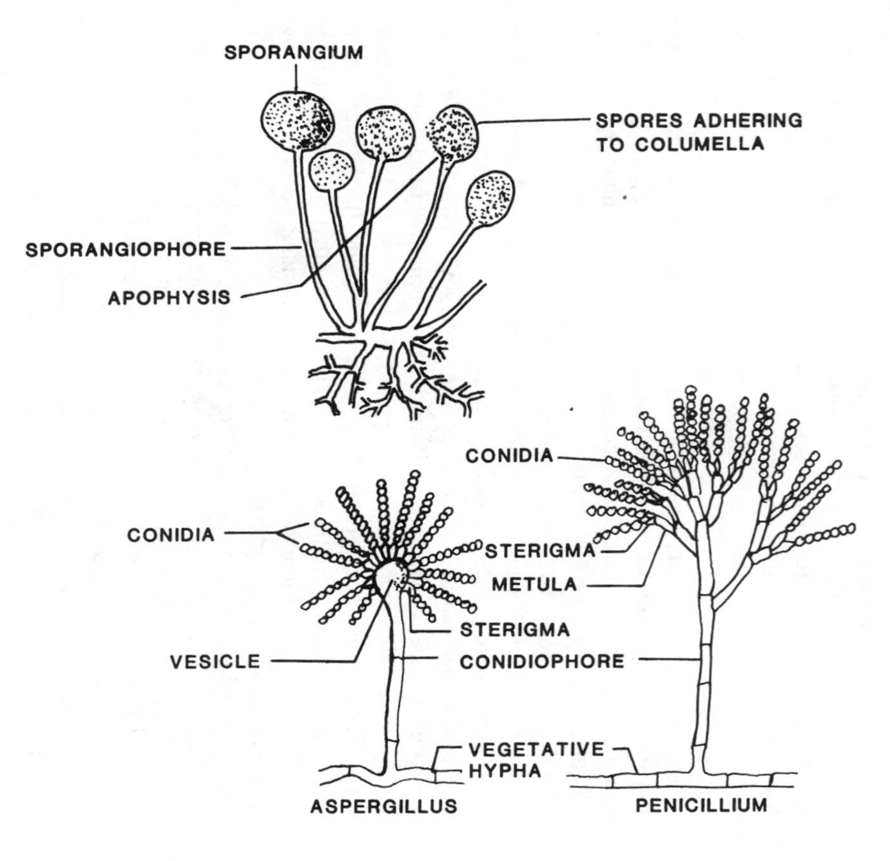

C. THALLOSPORES: Spores resulting from changes in the vegetative hyphae or thallus.

1. ARTHROSPORES (oidia): Hyphae fragment into small spores with thickened cell walls.

2. CHLAMYDOSPORES: Hyphae divide into spore-like cells with large food reserve and resistance to unfavorable environment.

3. BLASTOSPORES: Produced by budding.

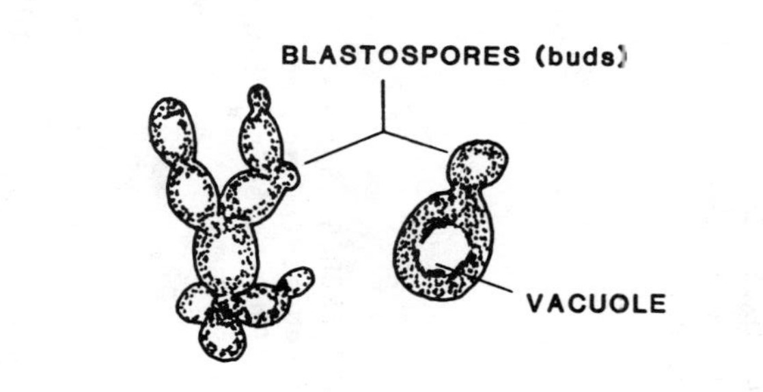

Figure 4-3. Major types of asexual spores formed by fungal microbes [48].

A. ASCOSPORES: Spores produced in a sac or ascus.

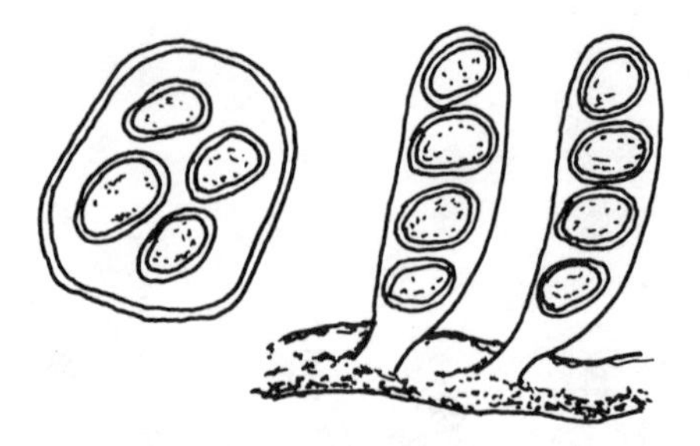

B. BASIDIOSPORES: Spores produced at the surface of a club-shaped structure, the basidium.

C. ZYGOSPORES: Spores produced by the fusion of similar-appearing gametes formed at the tips of hyphae (limited to the Phycomycetes).

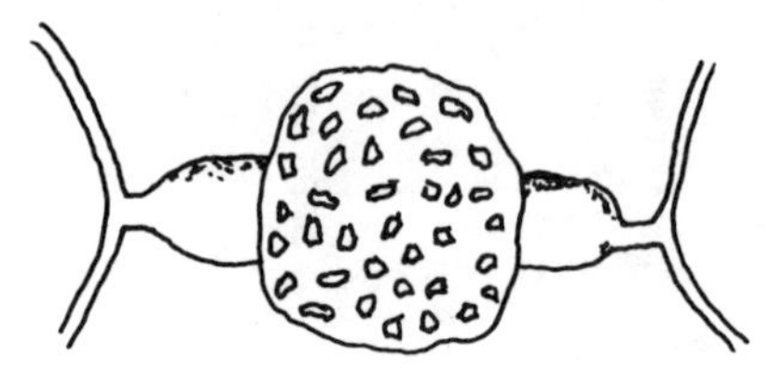

D. OOSPORES: Spores resulting from the mating of two unlike gametes (limited to the Phycomycetes).

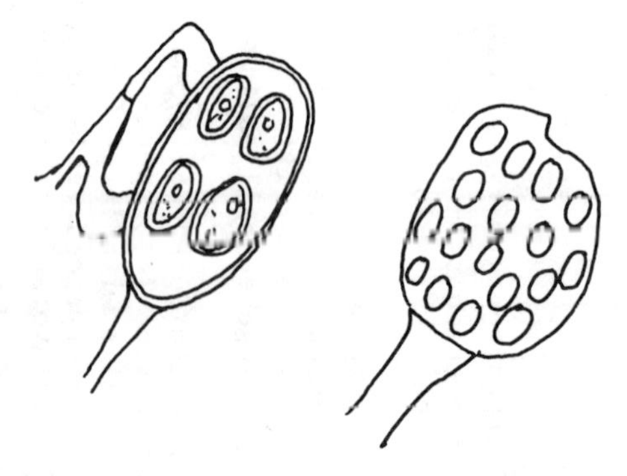

Figure 4-4. Major types of fungal sexual spores resulting from the fusion of nuclei or mating of gametes (reproductive cells) [48].

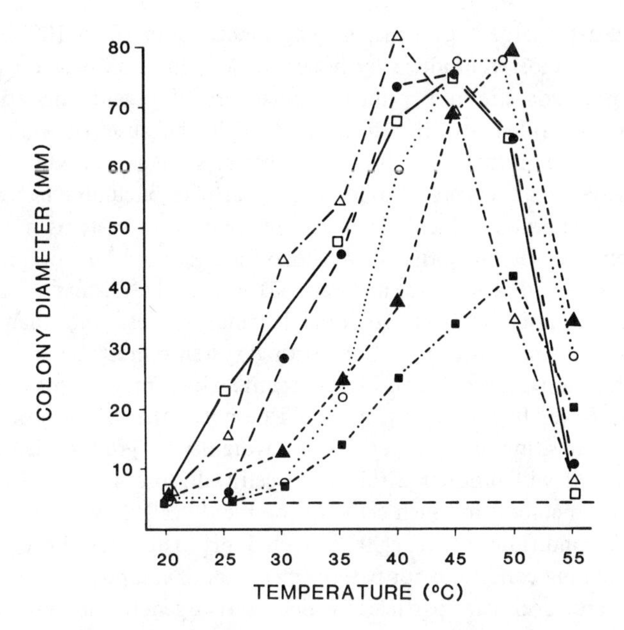

Figure 4-5. Relation of temperature to growth of the following fungi: (○) *Mucor* sp. 24; (●) *M. michei* 60; (△) *M. pusillus* 72; (▲) *Chaetomium thermophile* 48; (□) *Torula thermophilia* 48; (■) *Humicola lanuginosa* 48 (numbers are hours of growth). The microbes were isolated from a Metro-Waste reactor system composting municipal refuse [49].

conditions maintained in the compost system. Kane and Mullins [49] concluded that high temperatures, acidity and anaerobic conditions may limit fungal growth in the interior of a compost pile and restrict the role of thermophilic fungi.

OTHER PROTISTS

This discussion would not be complete without a small note of acknowledgment to the other groups of protists, i.e., algae, protozoa and viruses. Algae include both procaryotic and eucaryotic types, are photosynthetic, contain chlorophyll-a as a photosynthetic pigment and evolve O_2 as a by-product of photosynthesis. Because of their photosynthetic nature, algae are not of significance in the decomposition of organic residues.

Protozoa are eucaryotic, organoheterotrophic microbes that exhibit a tendency toward unicellular growth with elaborate intercellular organization.

They are usually large organisms, ranging in size from 10 to 100 μm. Most are motile and most reproduce by binary fission. In liquid waste treatment processes protozoa are not primary consumers of organic material. This role is usually filled by the bacteria, which, because of their higher surface-to-volume ratio, can process substrates more rapidly. Instead, protozoa serve as scavengers of solid organic particles, including bacteria, and act to polish the liquid effluent. Their role in composting systems is minor.

Although the vast majority of protozoa are saprophytic or predatory, protozoan-caused diseases are not rare. The causative agents of amoebic dysentery, giardiasis, malaria, African sleeping sickness and many other human diseases are protozoan. Composting systems must be operated to assure destruction of protozoan disease agents which may be present in the starting substrate. In this regard, it should be noted that some protozoa are capable of encysting during periods unfavorable to growth. During this process the cell will produce a thickened cell wall, lose water and maintain limited or no metabolism. Such cells are commonly called cysts and are more resistant to conditions of drought, heat and pH. This must be considered when determining conditions for pathogen control in composting.

Viruses are acellular particles which carry genetic information for reproduction but no biochemical machinery to transcribe the information or to metabolize substrates for energy. As such they are obligate parasites, using a host cell to provide the biochemical machinery which they lack. Virus particles are extremely small, ranging from about 0.01 to 0.25 μm. Individual virus types are extremely host-specific, generally invading only one type of host cell. Virus particles are known to "reproduce" in cells of almost all living organisms, including other protists. As with the protozoa, the major concern with virus is their potential for disease transmission and the conditions necessary for their destruction.

PATHOGENIC ORGANISMS

Enteric Microbes

The vast majority of protists are harmless or even helpful to man. However, a limited number are capable of causing human disease. These pathogenic microbes have thus attained a prominence far beyond their number. One of the major objectives of composting is destruction of pathogens which may be present in the original substrate. This section will present the types of pathogens common to municipal wastes and to waste composting systems. The following chapter will consider measures available to control pathogens.

Table 4-8 presents a summary of many of the enteric pathogenic organisms which potentially can be transmitted by water, sewage or sludge. The table is not presented as an exhaustive list of all potentially waterborne pathogens, but does contain those of greater importance or with a higher frequency of occurrence. Even so, the list is quite extensive. The point of Table 4-8 is not the sewage sludge (or animal manure and refuse for that matter) should be feared, but only respected. Measures must be taken to assure that pathogens are controlled to avoid public health problems. Composting, of course, has long been used to destroy such pathogens.

A wide variety of pathogenic bacterial, viral, protozoan and metazoan forms occur in sewage and, hence, in sludges of sewage origin. However, the types of pathogens present as well as the concentration may vary considerably from community to community. Bacterial pathogens include the causative agents for cholera, typhoid and paratyphoid fever, and various dysentery-related diseases. Numerous viruses are excreted by man including poliovirus (3 types), cosackievirus (25 types), echovirus (25 types), reovirus (3 types) and adenovirus (33 types) [50]. Even though numerous types are excreted, the only viral disease definitely known to be waterborne is hepatitis A (infectious hepatitis).

Various types of intestinal parasites have been reported in wastewater including *Entamoeba histolytica*, various nematodes such as *Ascaris lumbricoides*, hookworms and schistosomes. The life cycle of many intestinal parasites is quite complex, but, in general, the first stage involves excretion in the feces. It should be noted that not all metazoan disease agents are of public health concern. Although they may be present in sludge, lack of a proper intermediate host often prevents transmission of the disease. Schistosomiasis is one example. The reader should review Table 4-8 before proceeding further in the text.

Of the parasite eggs found in sewage, those of *Ascaris* species are the most common and of major concern. Geographic factors usually influence the types of parasites expected, but the ubiquity of *Ascaris* places quantities of their eggs in all sewages or sludges tested.

Ascaris was among the earliest known of human parasites. Its life cycle has been explored and well defined. The large nematode in the adult form is found in the small intestine. Females can reach a length of 20-35 cm and males 15-30 cm. Egg production of the female ascarid ranges from 200,000 to 250,000 eggs/day and is excreted in the feces of the host. The size of unsegmented eggs is 60-70 $\times$ 40-50 μm, and the eggs are covered by a thick albuminous coat. An environment with a temperature lower than the host body, a trace of moisture and a supply of oxygen are required for eggs to develop. If favorable conditions are realized, active embryos develop in about a week and reach an infective larval stage in 10-14 days. Should eggs then be

Table 4-8. Pathogenic Organisms Which Can Potentially Be Transmitted by Water, Sewage or Sludge

Causative Agent	Disease	Remarks
Enteric Viruses		
Poliovirus	Poliomyelitis	Exact mode of transmission not yet known. Found in effluents from biological sewage treatment plants. Introduction of effective vaccine has made polio a rare disease in developed areas of the world in less than a decade.
Virus	Infectious Hepatitis Type A	"Viral hepatitis" is a generic term that covers at least two distinct forms of hepatitis. Hepatitis type A, known as "infectious hepatitis," is the generally accepted term for epidemic, community-acquired disease. Type A virus is present in the feces and blood of infected persons who can contaminate milk, food and water. Hepatitis type B, also known as "serum hepatitis," is ordinarily transmitted by the parenteral route and is commonly associated with drug abuse and transfusion of blood and blood products.
Coxsackievirus	Usually Mild Infections	Responsible for aseptic meningitis, pleurodynia and infantile myocarditis. Commonly cause diarrhea in infants and young children.
Echovirus	Usually Mild Infections	These viruses (enteric cytopathogenic human orphan viruses) have been associated with illnesses of aseptic meningitis, rash, diarrhea and common respiratory diseases.
Adenovirus	Respiratory Infections	Many of the adenoviruses have been associated with a variety of respiratory diseases such as colds, influenzalike illnesses, bronchitis, croup and atypical pneumonia.
Reovirus		Diarrhea and respiratory diseases similar to those noted for adenovirus.
Virus	Gastroenteritis and Diarrhea	Causative agents not known but thought to be viral in some instances. In terms of magnitude, gastroenteritis and diarrheal disease are probably most important viral diseases transmitted by water.
Bacteria		
Coliform species	Diarrhea and Internal Infections, Gastroenteritis	Implicated in several cases of infant diarrhea in hospitals. In rare cases known to cause cardiovascular infections.

Table 4-8, continued

Causative Agent	Disease	Remarks
Vibrio cholerae	Cholera	Acute diarrheal disease transmitted by sewage and polluted waters. Disease often terminates in death. Endemic in India and SE Asia. Has occurred in Europe and N. and S. America. No reported cases in U.S. since 1913. During an epidemic in London in 1854, the causative organism was shown to be transmitted in water, making chloera the first disease for which this important fact was known. Infection results from ingestion of contaminated food or drink. Infective dose is near 10^9 vibrios.
Salmonella (many types)	Salmonellosis	Salmonellosis may range in severity from intestinal discomfort to potentially fatal diseases such as typhoid fever (see below). Food infections from salmonella are quite common. Three clinically distinct forms of the infection in man: enteric fevers, septicemias (rare) and acute gastroenteritis.
Salmonella typhi	Typhoid Fever	Most severe of enteric fever forms of salmonellosis. Occurs in all parts of the world but infrequently where good sanitation and purification of water is practiced. Can survive in water for a week or more, also transmitted by milk. Common in sewage and effluents in times of epidemics. Enteric form of infection can be produced by other *Salmonella* species, i.e., *S. paratyphi*, which causes paratyphoid fever.
Salmonella typhimurium (and others)	Salmonella Gastroenteritis	Most common form of salmonellosis. Infection occurs from ingestion of contaminated food or drink. The disease persists for 3-5 days and is usually not severe.
Shigella dysenteriae *S. sonnei* *S. flexneri*	Shigellosis (Bacillary Dysentery)	Dysentery is a clinical condition with intestinal inflamation, diarrhea, and water stools containing blood, mucus and pus. Polluted water is the main source of infection.
Bacillus anthracis	Anthrax	Anthrax is a disease of sheep, cattle, horses, goats and swine. Human infection is rare but spores can be found in sewage and are resistant to treatment. In man generally appears as a disease of the skin. Can be fatal if left untreated.

Table 4-8, continued

Causative Agent	Disease	Remarks
Brucella *abortus* (cattle) *B. suis* (hogs) *B. melitensis* (goats)	Brucellosis	Normally transmitted from animals to man by infected milk or by contact with infected meat or placentae of infected animals. Sewage also suspect. Very rare in U.S. except in Midwest. Hazard to slaughterhouse workers, farmers, veterinarians.
Mycobacterium tuberculosis	Tuberculosis	Isolated from sewage and polluted streams. Possible mode of transmission. Care with sewage and sludge from sanatoria. Deaths have been sharply reduced by early detection and treatment. Estimated 80,000 new cases per year in U.S. WHO reports death rate in Central and S. America three times greater than in N. America.
Leptospira interohaemorrhagiae (rats) *L. canicola* (dogs) *L. pomona* (cattle & swine)	Leptospirosis	Jaundice-like disease in man. Worldwide distribution and fairly common in man. Often transmitted to man by ingestion of food and drink contaminated by urine of the reservoir animal or bathing in contaminated water. Can be carried by sewer rats and documented as occurring in sewer workers in England.
Protozoa		
Entamoeba histolytica	Amoebiasis (Amoebic Dysentery)	Spread by contaminated waters and sludge used as fertilizer. Also transmitted by uncooked vegetables fertilized by sewage or sludge. Common in warmer countries. Organism can form a cyst which is resistant to disinfection.
Giardia lamblia	Giardiasis (Lambliasis)	Clinical manifestations range from asymptomatic cyst passage to severe malabsorption. Mean duration of the illness is often 2-3 months. In 1974 an outbreak occurred in Rome, NY, where over 5000 persons were affected. Giardia cysts are not destroyed by chlorination at dosages and contact times normally employed in water treatment, but it is felt that they can be removed by coagulation, settling and filtration.

Table 4-8, continued

Causative Agent	Disease	Remarks
Balantidium coli	Balantidiasis	Found throughout the world, particularly in the tropics. Illness similar to amoebic dysentery. Some persons suffer acute dysentery, but the majority are probably carriers without symptoms. Infection results from ingestion of cysts harbored in stools of man or swine. Fatalities have occurred in severe infections despite treatment.
Isospora belli *I. hominis*	Coccidiosis (Isosporosis)	Infection is usually sporadic and not severe. Most common in tropics and subtropics, but also reported in U.S. Infection results from ingestion of viable cysts.
Metazoan		
Helminths (Intestinal worms and flukes)		
Nematodes (Roundworms)		
Ascaris lumbricoides	Ascariasis	A large intestinal roundworm sometimes reaching 20-40 cm in length in the intestine. The most common of the intestinal helminths of man. Prevalent throughout the world and described as "one of man's most faithful and constant companions from time immemorial." Danger to man from sewage effluents and dried sludge used as fertilizer. Infection occurs by ingestion of mature eggs usually in fecally contaminated food or drink.
Ancylostoma duodenale *Necator americanus*	Hookworm	*Necator* is the prevailing genus in the Western Hemisphere. Formerly very prevalent in the southeast U.S. Infections developed in sewage farm workers in England. Adult worms live in intestines, fastening to walls by strong mouth parts. Eggs excreted in feces. Subsequent larval stage may enter host through skin.
Enterobius vermicularis (pinworm)	Enterobiasis	The most common cause of helminthic infection of man in the U.S. Although annoying, cure is readily effected with one of several drugs.
Tricuris trichiura (whipworm)	Trichuriasis	Common parasite of man throughout the world. New infections are acquired by direct ingestion of the infective eggs passed in the feces.

Table 4-8, continued

Causitive Agent	Disease	Remarks
Strongyloides stercoralis (threadworm)	Strongyloidiasis	Prevalent in the southeastern U.S. and tropical and subtropical areas of the world. Eggs secreted by the adult worm in the intestine develop into larvae which are passed in the feces. The free-living larvae can penetrate the skin of the next victim, enter the blood stream and eventually the small intestine where maturation to the adult stage takes place.
Toxocara cati (cat roundworm) *Toxocara canis* (dog roundworm)	Visceral Larva Migrans	Intestinal parasites found in dogs and cats. Street runoff suspected as a source of eggs. Recognized as a disease agent for children with pets.
Ancylostoma braziliense (cat hookworm) *A. canium* (dog hookworm)	Cutaneous Larva Migrans (Creeping Eruption)	Common infection of man in the southeastern U.S., particularly from contact with moist sandy soil contaminated with dog or cat feces. Larvae invade the skin surface, usually hands or feet and remain active for several weeks or months.
Cestodes (Tapeworms)		
Taenia saginata (beef tapeworm)	Tapeworm Infection	Eggs very resistant. Present in sewage, sludge and sewage effluents. Danger to cattle on sewage irrigated land or land manured with sludge. Cattle in southwest U.S. grazing on pastures treated with sludge tainted with eggs of *T. saginata* have contracted "beef measles." Require intermediate host.
Taenia solium (pork tapeworm)	Tapeworm Infection (Taeniasis)	Similar to above. No longer prevalent in the U.S.
Hymenolepis nana *H. diminuta* (dwarf tapeworm)	Tapeworm Infection	Require no intermediate host. Most common tapeworm infestation in U.S., especially in the South. Prevalent in the tropics and subtropics.
Diphyllobothrium latum (fish tapeworm)	Tapeworm Infection	Often found in Europe, Japan and Great Lakes region of U.S.

Table 4-8, continued

Causative Agent	Disease	Remarks
Echinococcus granulosus *E. multilocularis*	Echinococcus (Hydatid Disease)	Tapeworm is found in several hosts throughout the world, including sheep, dogs and other canines. S. America, Australia, Greece and other Mediterranean countries are areas of heaviest human infestation. Human infection results from ingestion of eggs passed in the host feces.
Dipylidium caninum (dog tapeworm)	Tapeworm Infection	Occasionally reported in children in Europe and the Americas who live in close association with dogs or cats. Infection results from swallowing fleas or lice which serve as the intermediate hosts.
Trematodes (Flukes)		
Schistosoma mansoni *S. haematobium* *S. japonicum*	Schistosomiasis (Bilharziasis) (Liver and Intestinal Flukes)	Eggs excreted in urine or feces of infected person. Hatch on contact with water and enter snail host. Emerging larvae (cercariae) leave the snail and can penetrate directly into human skin. Infection may continue for years as an insidious drain on body vigor. Widespread in Africa, Near East and Orient where more than 90% of population may carry the worms. Egyptian government considers disease to be major obstacle to country's economic progress. Schistosomes not transmitted in U.S. because the host snails are not present. In N. America a mild disease called "swimmer's itch" can occur. Microbes probably killed by efficient sewage treatment.
Fasciolopsis buski (giant intestinal fluke)	Fasciolopsiasis	Common parasite of man and pigs in China, Taiwan, SE Asia and India. Rare in continental U.S. Aquatic snail is the intermediate host.
Fasciola hepatica (sheep liver fluke)	Fascioliasis	Infection results from ingestion of cercariae on aquatic vegetables. Prevalent in sheep-raising countries, particularly where raw salads are eaten.
Clonorchis sinensis (liver fluke)	Clonorchiasis	Endemic in parts of Japan, China, Formosa and Indochina. Snails and freshwater fish are intermediate hosts. Infection results from eating such fish, raw or undercooked
Paragonimus westermani (lung fluke)	Paragonimiasis	Commonly infects man throughout the Far East. Snails and crabs or crayfish are intermediate hosts. Infection results from eating such crustaceans, raw or pickled.

ingested, larvae will hatch in the small intestine where they penetrate the mucous membrane and travel throughout the body to the liver, heart and/or lungs. Eventually the worm settles in the small intestine and begins the cycle again.

The shell of an *Ascaris* ovum is resistant to chemicals and desiccation. Eggs will survive for weeks in a 10% formalin solution and are very resistant to chlorination. This resistance allows the ovum to survive and remain infective for years under proper conditions.

The Center for Disease Control analyzed results of over 400,000 stool specimens examined by public health laboratories throughout the United States for the presence of intestinal parasites [51]. *Giardia lamblia* was the most commonly identified pathogenic intestinal parasite, appearing in 3.8% of all stool specimens examined. *Entamoeba histolytica* was identified in 0.6% of the specimens. Nematode ova, arranged in the order of occurrence, were: *Trichuris trichiura*, 2.7% of all samples; *Ascaris lumbricoides*, 2.3%; *Enterobius vermicularis*, 1.7%; and hookworm, 0.8%. Cestodes and trematodes appeared in about 0.3 and 0.05% of the samples examined, respectively. A total of 8.4% of all samples were positive for some form of pathogenic intestinal parasite. Geographical variations were noted in most cases. For *A. lumbricoides*, for example, the positive sample frequency ranged from a high of 9.3% in Guam to a low of 0.0% in Wyoming, Arizona and Nevada.

Because the above results are from public health laboratories, the values reflect "high risk" groups such as institutionalized patients and residents of communities with low standards of living and poor sanitation facilities. Results should not be interpreted as the prevalence of parasitic infection in the general U.S. population. Despite this limitation, the data do provide an idea of the relative prevalence of parasitic infections in a developed country. Higher incidence rates would certainly be expected in less-developed regions of the world.

Worms and eggs have a specific gravity of approximately 1.1 or greater, with settling velocities of about 3 ft/hr. Tapeworm eggs are reported to have a somewhat lower settling velocity of 0.3-0.6 ft/hr. [192]. Therefore, most worms and eggs tend to accumulate in primary sludges and also, perhaps, in waste activated sludges. Virus particles tend to be associated with particulate solids and are also concentrated in the sludge component.

Since the majority of eggs, cysts and certain other pathogens tend to occur in sludges, their fate during subsequent sludge processing is important. Mesophilic anaerobic digestion is moderately effective against certain pathogens such as virus and protozoan cysts. Thermophilic digestion (45-50°C) should be more effective, but a considerable number of all pathogenic types, particularly helminth eggs, can be expected to survive. This is further assured by the fact that most modern digesters are well mixed.

Thus, a large portion of the feed passes through the digester in less than the theoretical detention time. This alone will limit pathogen destruction achievable in a digester even if operated under thermophilic conditions.

Air drying of sludge, such as by sand drying beds, is not particularly effective, especially against eggs and cysts which can survive desiccation. It would appear, therefore, that both raw and digested sludges, whether dewatered by mechanical or air-drying techniques, are likely to contain significant numbers of pathogens. Stated another way, unless specific measures are taken to destroy pathogens, dewatered sludges produced by conventional processing can be expected to contain various concentrations of the pathogens presented in Table 4-8.

Fungal Opportunists

Although the vast majority of fungal organisms are saphrophytic, about 50 species are termed opportunistic. They are capable of causing disease in man when introduced into the body under certain conditions. Frequently, such infections occur in patients already debilitated by other diseases or in weakened conditions. Fungal infection, termed a mycosis, can be divided into systemic or dermal types. Systemic infections are those which affect or pertain to thc body as a whole. These affect deep tissues, usually the lungs, and, therefore, are regarded as serious. Dermal infections affect superficial tissues such as skin, hair or nails. They are annoying, but not generally serious. A description of the more important systemic mycotic infections and their causative agents is presented in Table 4-9.

It should be recalled that all fungal organisms, including those shown in Table 4-9, are organoheterotrophic, deriving their energy from the degrada- of organic matter. Even those that can cause human disease usually live naturally as saprophytes in soil or decaying vegetation. Thus, there are innumerable opportunities for individuals to come in contact with spores of these fungi. Infection usually occurs by inhalation of large quantities of spores into the lungs or by contamination of cuts or abrasions. Under the right conditions, the spores can develop, leading to fungal infection of body tissues. Such infections are rarely transmitted from person to person. Emmons [52] concluded that:

> the fungi that cause systemic mycoses are normal and more or less permanent members of the microflora of soil, compost, or organic debris . . . . These fungi are vigorous and self-sufficient saprophytes so long as environmental conditions are favorable. They are parasites by accident.

Because fungal organisms are active in composting piles, concern has arisen over possible growth of opportunistic species during composting. To

Table 4-9. Systemic Mycotic (Fungal) Infections

Causative Agent	Disease	Remarks
Coccidioides immitis	Coccidioidomycosis	Infection results from inhalation of spores or mycelial fragments. Fungus grows in soil in certain arid regions of southwest U.S. and Mexico. Infections commonly pass unnoticed but can be severe in certain individuals. Lung is major focus of infection. Considered an occupational hazard for archeologists in southwest U.S.
Histoplasma capsulatum	Histoplasmosis	Similar to above.
Cryptococcus neoformans	Cryptococcosis	Yeastlike fungus found in soil and pigeon nests. Human infection is worldwide. Lungs and meninges are major focus of infection.
Blastomyces dermatitides	Blastomycosis	Infections are usually severe. Lungs are major focus of infection. Geographically limited to central and eastern U.S. and Canada. Similar infections can occur in South or Central America and Mexico caused by another microbe.
Candida albicans	Candidiasis	Can be cultured from mouth, vagina or feces of about 65% of the population. Frequently a secondary invader in other types of infections. Lungs, skin, intestinal tract and internal organs can be sites of infection.
Sporothrix schenkii	Sporotrichosis	Worldwide in distribution. Most patients are people whose occupation brings them into contact with soil, plants or decaying wood. Infection takes place when an organism is introduced into a break in the skin.
Aspergillus fumigatus (and other *A.* species)	Aspergillosis	Infection usually occurs by inhalation of large numbers of spores. Lung is focus of infection and disease is not rare in man. Caused by various species of *Aspergillus*, but *A. fumigatus* causes more serious infections.

date, most concern has centered on the genus *Aspergillus*, a number of species of which can cause aspergillosis. Other fungal opportunists have not been reported to be commonly associated with composting systems. *A. fumigatus* is usually discussed in reference to aspergillosis because its infections appear to be the most common and most serious. Various members of this genus, including *A. fumigatus*, have been isolated from composting systems, particularly those with cellulytic substrates. Millner et al. [53] indicated that *A. fumigatus* has been found frequently in composting vegetative material such as grass and tree clippings, woodchip piles, refuse compost, refuse-sludge compost, sludge and moldy hay, and is often isolated from soil. The organism is also thermotolerant, growing over a range of temperatures of $<$20-50°C. The real question, then, is to what extent are individuals already exposed to opportunistic fungi and whether composting operations increase this exposure.

Aspergillus is a ubiquitous fungus to which every individual has daily contact throughout his entire life [54]. Solomon [55] sampled fungal spores in the air of 150 homes in winter. *Aspergillus* was one of the most common fungi found. *Aspergillus*-type spores in outside air probably rarely exceed 500 colony-forming units (CFU)/m^3 with an observed range of about 10-10^4 CFU/m^3 [56,57]. On the other hand, concentrations up to 21 million CFU/m^3 have been observed in farm buildings after shaking of moldy hay [58]. Hudson [208] reported concentrations of *Aspergillus fumigatus* in outdoor air ranging from 0 to 14 spores/m^3 with a mean of 3.2 spores/m^3.

Millner et al. [53] sampled for *A. fumigatus* throughout various stages of aerated static pile composting at Beltsville, MD. The composting mixture consisted of raw sludge blended with wood chips as a bulking agent. Maximum and minimum concentrations of the fungus detected in each stage of the composting process are shown in Figure 4-6. Sludge samples contained between 10^2 and 10^3 CFU/g (dry weight). Wood chips appeared to be the major source of the fungus. Samples of fresh and old wood chips contained 10^3 to 2.3×10^5 CFU/g (dry weight) and 2.6×10^6 to 6.1×10^7 CFU/g (dry weight) of *A. fumigatus*, respectively. The abundance of *A. fumigatus* was sufficient that chips were often observed to have a gray-green coloration, attributed to dense masses of dry conidia. The authors also concluded that other bulking agents of high cellulose content, such as bagasse, carob husks, corncobs and husks, grass seed straw, cereal straws and husks, refuse, peanut hulls, and licorice root, would probably be suitable for growth of the fungus. Licorice root alone was observed to contain approximately 4.7×10^6 CFU/g (dry weight) of *A. fumigatus*. Semiquantitative studies of airspora at the composting site indicated that *A. fumigatus* constituted 75% of the total viable mycoflora captured. This decreased to only about 2% at locations 320 m to 8 km from the site.

Millner et al. [53] also sampled for *A. fumigatus* in commercial potting soils, manures and mulches. Of 21 products analyzed, 5 contained no detectable *A. fumigatus*. The remainder ranged from about 10^2 to 5.7×10^5 CFU/g (dry weight). Five of the products had concentrations exceeding that found in four-month-old Beltsville stored compost (Figure 4-6).

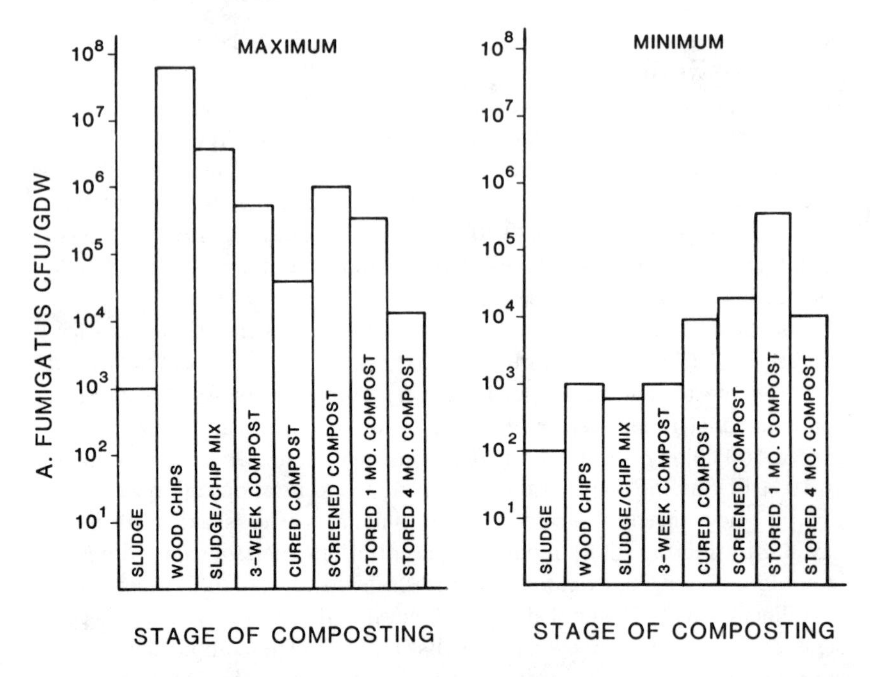

Figure 4-6. Maximum and minimum *A. fumigatus* CFU/g dry weight of substrate from each stage of the aerated static pile composting process. Numbers for stored compost represent counts from the 10-cm depth only. The composting mixture consisted of raw municipal sludge blended with wood chips [53].

For the case of aerated static pile composting using cellulosic bulking agents, it has generally been recommended that both mixing and screening operations be totally enclosed. The latter operations are major mechanisms for airborne release of spores. It has also been recommended that surrounding regions be periodically monitored for *A. fumigatus* spores and that the proximity of planned composting operations to health-care facilities be carefully considered [179]. Pahren et al. [54] recommended that personnel working in the mixing and screening building regularly be skin-tested and/or evaluated for precipitating antibodies to *Aspergillus*. Workers with bronchial asthma and positive skin tests should not be permitted in these buildings. The concensus among persons knowledgeable in the field appears to be that these measures will provide adequate protection for the general population surrounding the site and workers employed at the site.

In a separate study, the LA/OMA project sponsored field measurements of *Aspergillus* concentrations at a windrow facility and an enclosed reactor pilot system, both operated by the Los Angeles County Sanitation Districts.

Digested sludges were blended with recycled compost for conditioning in both cases. No cellulosic bulking agents or amendments were used. The starting compost material had a typical *Aspergillus* content of 10^3-10^4 CFU/g (dry weight), similar to levels reported in the Beltsville study. If internal temperatures above 60°C were attained, *Aspergillus* levels dropped off rapidly as shown in Figure 4-7. Final concentrations less than 10 CFU/g (dry weight) were readily achieved. Even if temperatures greater than 60°C were not achieved in the windrow system, *Aspergillus* concentrations in the final compost ranged from about 10-1000 CFU/g (dry weight), which is less than that of the original feed.

Aspergillus was not observed to grow during the compost cycle, unlike the experience at Beltsville where wood chips fostered growth of the fungus. Similar results were obtained with both the windrow and enclosed reactor system. Therefore, growth of *Aspergillus* is probably related more to use of cellulosic bulking agents than to the type of compost system.

An aerosol sampling program was conducted to evaluate first those portions of the windrow composting process thought to release fungal

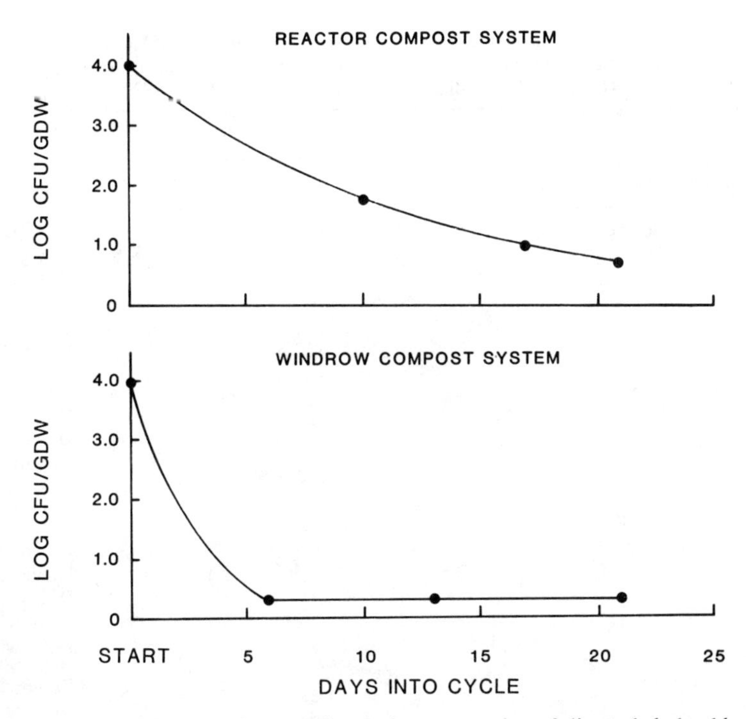

Figure 4-7. Concentration of *Aspergillus* during composting of digested sludge blended with recycled product. Temperatures >60°C were attained during each cycle [59].

spores into the air. Second, the compost field was monitored immediately upwind and downwind to evaluate the total emission of *Aspergillus.* Airborne concentrations measured during potentially dust-producing phases of the windrow operation are shown in Table 4-10. Samples were collected using an Anderson impingement sampler as close as possible to the source of emission to minimize atmospheric dilution. The sampler divided collected particles into respirable and nonrespirable size fractions. Very low levels of *Aspergillus* spores were detected during these relatively dusty operations. In some cases, the Anderson sampler was covered with dust after the sampling period. The single highest observation of nonrespirable aerosols was 60 CFU/m^3, and for the smaller respirable particles 42 CFU/m^3. For all other samples, concentrations were less than 10 CFU/m^3.

Aspergillus concentrations from the windrow compost field itself are presented in Table 4-11. Samples were taken upwind of the field, immediately downwind of (1) one windrow, (2) half the compost field, and (3) the entire field. As expected, *Aspergillus* levels were quite low. The

Table 4-10. Concentrations of Airborne *Aspergillus* Spores Measured Adjacent to Various Operations at a Windrow Facility Composting Digested Sludge Blended with Recycled Product [59]

Compost Operation	*Aspergillus* Spores (CFU/m^3)		Sampling Location
	Respirable	Nonrespirable	
Stockpiling Finished Windrows	<3	3	Downwind
	2	0	Downwind
	<1	<1	Downwind
	<1	<1	Downwind
	<1	<1	Upwind
	<1	<1	Upwind
Loading Finished Compost into Trucks	7	60	Downwind
	1	<1	Downwind
	<1	<1	Downwind
	<1	<1	Downwind
	<1	<1	Downwind
	<1	1	Downwind
Laying Windrows	42	8	Downwind
	4	<1	Downwind
	4	<1	Downwind
	2	1	Downwind
	1	2	Downwind
Turning Windrows	<1	<1	Downwind
	<1	<1	Downwind
	<1	<1	Downwind

highest value recorded was 33 CFU/m^3 downwind of the compost field, but the majority of samples were less than 10 CFU/m^3. Considering the background levels reported earlier for normal outdoor air and the high concentrations associated with certain conventional agricultural operations, the airborne concentrations of 1-60 CFU/m^3 observed at this windrow facility would appear not to present much, if any, risk.

Based on the two studies discussed here, it would appear that compost systems that do not use cellulosic amendments or bulking agents will not cause significant increases in airborne *Aspergillus* concentrations. Furthermore, available data indicate a reduction in the feed concentrations as a result of composting. High concentrations of *Aspergillus* are associated with the use of cellulosic bulking materials, particularly wood chips, required for the aerated static pile process. Apparently the substrate and other environmental conditions are suitable for growth of the fungus, resulting in elevated compost concentrations of *Aspergillus.*

Table 4-11. Concentrations of Airborne *Aspergillus* Spores from a Windrow Compost Field (Composting Mixture Consisted of Digested Sludge Blended with Recycled Compost) [59]

	Aspergillus Spores (CFU/m^3)	
Description of Sample Location	**Respirable**	**Nonrespirable**
Downwind from Single Windrow at Rest–Strong Wind	7	19
Center of Field–Equipment Working & Slight Wind	8	<1
	4	<1
Center of Field–No Equipment Working & Slight Wind	<1	1
	<1	<1
Downwind from Entire Field with Equipment Working–Slight Wind (3 samples)	<1	<1
	1	<1
	3	1
Downwind from Entire Field with No Equipment Working and		
Moderate Wind	<2	6
Moderate Wind	6	<2
Slight Wind	31	33
Upwind from Compost Field	<1	<1
	<1	<1
	<1	<1

KINETICS OF MICROBIAL GROWTH

The rate at which microbes consume substrate and grow is obviously of importance to all waste treatment processes including composting. Numerous models have been proposed to describe the rate of substrate use in solution. Here, the discussion will focus on the most commonly used of these models. The subject of composting kinetics will be discussed again in Chapter 10.

The effect of substrate concentration on the rate of substrate use by a microbe is illustrated in Figure 4-8. Suppose a series of test tubes are inoculated with the same mass of microbes. Suppose further that the test tubes have different concentrations of a substrate which can be used by the inoculated microbes. If all other substrates, such as oxygen, are plentiful, the rate of substrate use should plot in the form shown in Figure 4-8. Many mathematical equations can be used to describe the curve of Figure 4-8. Monod in 1942 proposed a hyperbolic relationship of the form

$$\frac{dS}{dt} = -\frac{k_m S X}{K_s + S} \tag{4-1}$$

where dS/dt = rate of substrate utilization, mass/volume-time
X = concentration of microbes, mass/volume
k_m = maximum utilization coefficient, maximum rate of substrate utilization at high substrate concentration, mass substrate/mass microbes-day
K_s = half-velocity coefficient, also referred to as the Michaelis-Menten coefficient, mass/volume
S = concentration of the rate limiting substrate, mass/volume

The form of the equation used by Monod is pleasing in that it is of similar form to equations developed by Henri (1902) and Michaelis and Menten (1913) for single enzyme-substrate systems. The negative sign is used to indicate that substrate concentration decreases as a result of microbial activity.

Use of Equation 4-1 assumes that mass transfer of substrate to the cell is not rate-limiting. In other words, kinetics are controlled only by the concentration of a limiting substrate S. It is common practice to measure substrate concentration in the immediate vicinity of the microbial cell. Thus, any subsequent mass transfer limitations across the cell wall or within the cellular protoplasm are included within the equation. Finally, the Monod model applies best to soluble substrates. Additional mass transfer limitations can be encountered with particulate substrates, a situation which is discussed in Chapter 10.

Recognizing these limitations, a physical interpretation of Equation 4-1 can be proposed. At high concentrations of substrate, the cellular enzyme

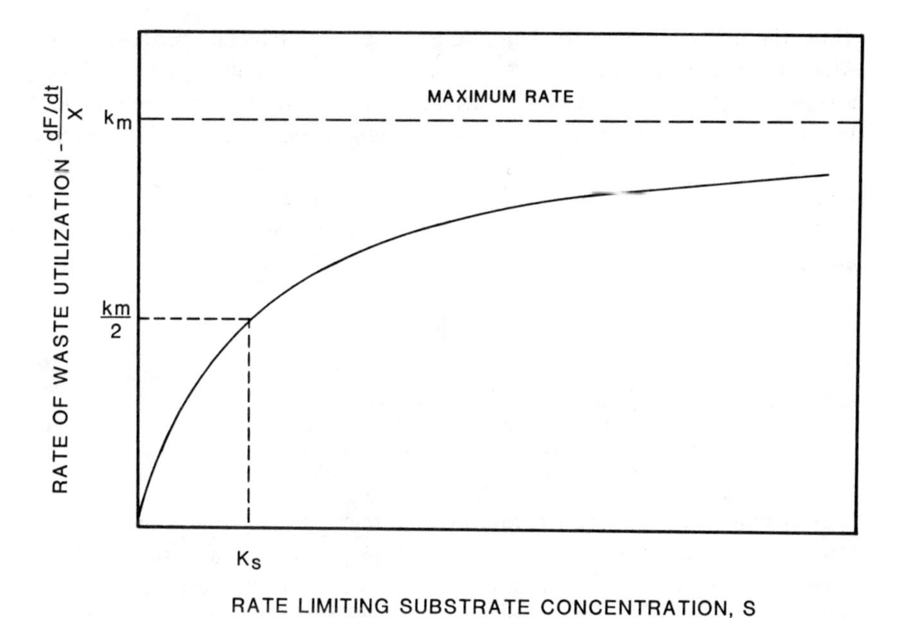

Figure 4-8. Effect of limiting substrate concentration on the rate of substrate use.

systems become saturated with substrate and process substrate as rapidly as possible. Further increase in substrate concentration causes no further increase in the rate of substrate use. In Equation 4-1, this corresponds to the condition where $S \gg K_S$ and the form of the equation reduces to:

$$\frac{dS/dt}{X} = -k_m \tag{4-2}$$

This is a zero-order reaction with regard to substrate concentration. At lower substrate concentrations, cellular machinery operates at reduced rates because of the limited supply of substrate. If $S \ll K_s$, Equation 4-1 reduces to:

$$\frac{dS/dt}{X} = -\frac{k_m}{K_s} S \tag{4-3}$$

which is first-order with respect to substrate concentration. Under the condition where $S = K_s$, Equation 4-1 reduces to

$$\frac{dS/dt}{X} = -\frac{k_m}{2} \tag{4-4}$$

Thus, the half-velocity coefficient corresponds to the substrate concentration where the rate of reaction is half the maximum as shown in Figure 4-8. Such a relationship has no real physical meaning in the model proposed by Monod, but results simply from the form of the hyperbolic equation. With single enzyme-substrate systems, however, the coefficient does have a physical interpretation, but one which will not be explored here.

The use of substrate can be related to microbial growth through an equation of the form:

$$\frac{dX}{dt} = Y_m\left(-\frac{dS}{dt}\right) - k_e X \tag{4-5}$$

where dX/dt = net growth rate of microbes, mass/volume-time
Y_m = growth yield coefficient, mass of microbes/mass of substrate
k_e = endogenous respiration coefficient, time^{-1} or mass of microbes respired/mass of microbes-time

It is assumed that the growth rate of new microbes is proportional to the rate of substrate utilization, with Y_m the proportionality coefficient. The endogenous respiration term is included to account for the observed fact that cellular mass decreases if no substrate is available. Microbes begin using stored reserves and eventually the essential protoplasm under such conditions. The rate of cell loss is generally proportional to the mass of organisms present in the solution.

Substituting Equation 4-1 into Equation 4-5 gives the combined growth equation:

$$\frac{dX}{dt} = Y_m \frac{k_m SX}{K_s + S} - k_e X$$

or

$$\frac{dX/dt}{X} = \frac{Y_m k_m S}{K_s + S} - k_e \tag{4-6}$$

The term $dX/dt/X$ is often referred to as the net specific growth rate μ. The product $Y_m k_m$ is termed the maximum net specific growth rate μ_m, achieved at high substrate concentrations ($S >> K_s$) and low endogenous respiration ($k_e \sim 0$). With these terms, Equation 4-6 becomes

$$\mu = \frac{\mu_m S}{K_s + S} - k_c \tag{4-7}$$

Equation 4-6 or 4-7 is the most common model used to describe microbial growth and substrate use in aqueous solutions. The model has been applied to a variety of biological processes used in sanitary and biochemical engineering practice, including the activated sludge process, oxidation towers and anaerobic digesters. Consideration of the model as applied to composting systems will be deferred to Chapter 10.

Referring to Equation 4-6, four kinetic coefficients, Y_m, k_m, K_s and k_e must be known for the particular substrate and microbe under consideration. Values for the coefficients have been determined for a variety of substrates under both aerobic and anaerobic conditions. As might be expected, the yield coefficient is strongly influenced by free energy yield of the energy reaction. In fact, a number of predictive models have been developed based on energy available to the microbes [60-63].

For aerobic heterotrophs, values of Y_m typically range from 0.25 to 0.5 g microbes/g substrate chemical oxygen demand (COD). It is common to base the yield coefficient on the COD of the substrate. As discussed in Chapter 3, substrate COD is proportional to the number of electrons transferred during aerobic metabolism. Furthermore, heat of combustion per electron transferred to a methane-type bond is reasonably constant at about 26.05 kcal/electron equivalent or, because O_2 has four such electrons, 104.2 kcal/mol O_2 (i.e., COD). For aerobic metabolism, therefore, COD turns out to be a measure of energy release. Thus, it is not surprising that yield coefficients for aerobic metabolism fall within a reasonably small range for a variety of substrates when based on COD units.

With anaerobic metabolism, yield coefficients are significantly reduced because of the lower energy yield from fermentation reactions. Values as low as 0.04 and as high as 0.2 g microbes/g substrate COD have been determined for fatty acid and carbohydrate substrates, respectively.

The maximum utilization coefficient k_m represents the maximum rate at which microbes will process substrate. Typical units for k_m are g substrate COD/g microbes-day. From the previous discussion, the substrate COD is related to the number of electrons transferred. Thus, k_m can be interpreted as an electron transfer rate. McCarty [64] analyzed a number of reported values for both heterotrophic and autotrophic metabolism and for aerobic, anoxic and anaerobic respiration. Values of k_m at 25°C appeared to vary between about 1 and 2 electron-mol/g microbe-day (an electron-mole is one Avogadro's number of electrons) which is equivalent to 8-16 g COD/g microbe-day. This strongly suggests that the rate of electron transport is the limiting factor at high substrate concentrations and that the rate is reasonably constant for a wide variety of metabolic types. Considering that the basic biochemical machinery is common to all organisms, this result is not surprising.

The rate of electron transport should be a function of temperature and should obey the Arrhenius relationship shown as Equation 3-20. McCarty

[64] reported values of E_a ranging from 9.8 to 18.2 kcal/mol (average 12.9) over the temperature range of 10-40°C. An E_a value of 12.25 kcal/mol corresponds to a doubling of the reaction rate for each 10°C rise in temperature. Temperature effects observed for microbial reaction rates are consistent with those observed for most chemical reactions.

Monod observed values of K_s on the order of 4 mg/l for glucose oxidation by dispersed aerobic growths. For pure substrates and with pure cultures of aerobic microbes, values are in the 20-mg COD/l range or lower. Higher values, usually in the range of 20-300 mg COD/l, have been determined with mixed cultures in activated sludge systems. This may be caused in part by added mass transport resistance into the sludge floc which would not exist in a dispersed growth system. Resulting concentration gradients into the floc would cause the apparent value of K_s to increase.

Measured K_s values for anaerobic processes are considerably higher than for aerobic metabolism. Reasons for this are not particularly clear. Nevertheless, O'Rourke [65] found values for fatty acid fermentation ranging from about 2000 to 5000 mg COD/l. Values for K_s also appeared to be a function of temperature. Over the temperature range 20-35°C, the value of K_s for a mixture of fatty acid intermediates approximated:

$$K_s = (1.8 \text{ g COD/l})\,(1.112^{35-T}) \tag{4-8}$$

where T = solution temperature (°C). The endogenous respiration coefficient is used to account for the observed fact that microbial mass in a batch reactor will decrease after the initial substrate is consumed. Measured values of k_e usually range from 0.02 to 0.15 day^{-1}. The meaning of the coefficient is perhaps clarified if the units are interpreted as grams of cells consumed per gram of cells per day. Endogenous respiration is the metabolic consumption of stored reserves and cellular protoplasm. Thus, the rate should be a function of temperature, although available data is too scattered to allow development of reasonable equations.

Equation 4-1 assumes that reaction rates are controlled by a single rate-limiting substrate. For most organisms a variety of substrates are involved in metabolism, any one or more of which can be potentially rate-limiting. For example, aerobic, organoheterotrophic bacteria must be supplied with both oxygen and organic matter as substrates. Depending on concentrations surrounding the cell, either substrate or both can exert a rate limitation. Such a condition has been mathematically modeled by modifying Equation 4-1 to the form:

$$\frac{dS}{dt} = -k_m X \left(\frac{S_1}{K_{S1} + S_1}\right)\left(\frac{S_2}{K_{S2} + S_2}\right) \cdots \left(\frac{S_n}{K_{Sn} + S_n}\right) \tag{4-9}$$

where S_n = concentration of the nth substrate which can be potentially rate limiting, and K_{sn} = half-velocity coefficient for the nth substrate. If the concentration of any substrate is much greater than the corresponding K_s value, the term in brackets will approach unity. If this condition exists for all but one substrate, Equation 4-9 reduces to Equation 4-1. It should be recognized that Equation 4-9 is not based on any fundamental analysis of cellular kinetics. Rather, it is a convenient mathematical form which has been applied usefully to situations involving multiple substrates.

In summary, rates of microbial growth and substrate use can be modeled mathematically using the Monod approach based on the rate-limiting substrate(s). The model as presented so far applies best to soluble substrates and dispersed growth of microbes. With particulate substrates and flocculated or fixed-film growths, additional mass transfer limitations must be considered. In composting systems the substrate is particulate by nature. To understand further the rate-limiting factors involved in composting, kinetics of cellular metabolism must be integrated with the rate of solubilization of particulate substrate and the subsequent mass transport of solubilized substrates and oxygen to the cell. These factors will be considered in Chapter 10.

SUMMARY

Composting is a biological process mediated by microbes belonging to the Protist kingdom, which includes bacteria, algae, fungi, protozoa and virus particles. Microbes can be classified into metabolic types based on the carbon and energy sources utilized by the cell. Autotrophs use carbon dioxide as a source of cell carbon, whereas heterotrophs use the carbon of organic molecules. Phototrophs obtain energy from light. Lithotrophs use the energy of inorganic chemical reactions while organotrophs use the energy of organic chemical reactions. Most bacteria and all fungi are organoheterotrophs using organic compounds both as a source of energy and for cell carbon. They are the microbes of importance in organic composting.

Cellular respiration can be aerobic, anoxic or anaerobic, depending on the electron acceptor used in the energy reactions. Oxygen and oxidized compounds of nitrogen and sulfur are electron acceptors for aerobic and anoxic respiration, respectively. Anaerobic fermentations are characterized by having the electron donor and acceptor originate from the same organic molecule.

Bacteria are the smallest living organisms, with representatives from all the metabolic and respiratory classifications. They are of procaryotic cell structure. Fungi are divided into molds and yeasts. Molds are usually strict aerobes while yeasts are of both aerobic and anaerobic types. Both molds and yeasts are of eucaryotic cell structure and are all organoheterotrophs.

The vast majority of microbes are harmless and even helpful to man. However, a small number can cause disease and are termed pathogenic. Of these, the enteric pathogens are indigenous to the intestines and can potentially be transmitted by sludges, sewage or water. Diseases caused by enteric pathogens include the following: poliomyelitis and infectious hepatitis caused by viral agents; cholera, typhoid and bacillary dysentery caused by becterial agents; amoebic dysentery and giardiasis caused by protozoan agents; and roundworm, tapeworm and fluke infestations caused by helminths.

A number of fungal organisms are termed opportunists and can cause infections in man if introduced into the body under appropriate conditions. Usually this is by inhalation of large quantities of spores. For the spores to cause infection, the host must usually be debilitated by other disease or in an otherwise weakened condition. *Aspergillus fumigatus* is the principal agent of aspergillosis and, unlike most other fungal opportunists, is thermotolerant and often isolated from composting material. Growth of *A. fumigatus* appears to be enhanced by use of cellulosic substrates such as wood chips in the aerated static pile process. However, no human infections as a result of composting operations have been reported.

Kinetics of microbial growth can be described in mathematical terms using the Monod model, which is similar in form to models used to describe a single enzyme-substrate complex. Rates of substrate use and microbial growth can both be described in terms of four kinetic coefficients: the yield coefficient Y_m; maximum rate of substrate utilization k_m; half-velocity coefficient K_s; and the endogeneous respiration coefficient k_e. The yield of organisms per unit of substrate used is a function of free energy available from the substrate. The maximum rate of substrate use is related to the electron transport rate within the cell and is reasonably constant for a wide variety of microbes.

CHAPTER 5

KINETICS OF HEAT INACTIVATION

INTRODUCTION

Processes used to control pathogens in sludge include aerobic and anaerobic digestion, heat drying, heat pasteurization, ionizing radiations and chemical treatment, usually with lime. To manage the disposal or reuse of sludges on land, the U.S. Environmental Protection Agency (EPA) has divided sludge processes into those which significantly reduce pathogens and those which cause yet a further reduction in pathogens [66]. Definition of these processes and typical operating parameters for each are shown in Table 5-1. Sewage sludge applied to the land surface or incorporated into the soil must be treated to significantly reduce pathogens. Public access to the site must be controlled for at least 12 months, and grazing by animals whose products are consumed by humans must be prevented for at least one month. Further public health protection is required if sludge is applied to land where crops for direct human consumption are grown less than 18 months after application. In these instances, sludge must be treated before application by a process to further reduce pathogens.

The above regulations are but one example of the types of management approaches being considered in various countries. It is reflective of the growing concern over sludge-borne pathogens and the desire to minimize any risk from use of sludge-based products. Although many approaches are possible, the one proposed by EPA recognizes that not all processes are equally effective in terms of pathogen destruction.

As discussed in Chapter 4, it is likely that both raw and digested sludges, whether dewatered by mechanical or air drying techniques, contain significant numbers of pathogens. Therefore, further pathogen destruction will probably be required if direct public contact with the material is expected as would usually be the case with compost. For a variety of reasons, heat

Table 5-1. Processes for Control of Pathogens in Sludge Based on EPA Interim Final Regulations 1979 [66]

A. Processes to Significantly Reduce Pathogens

Aerobic digestion: The process is conducted by agitating sludge with air or oxygen to maintain aerobic conditions at residence times ranging from 60 days at 15°C to 40 days at 20°C, with a volatile solids reduction of at least 38%.

Air drying: Liquid sludge is allowed to drain and/or dry on under-drained sand beds, or paved or unpaved basins in which the sludge is at a depth of nine inches. A minimum of three months is needed, two months of which temperatures average on a daily basis above 0°C.

Anaerobic digestion: The process is conducted in the absence of air at residence times ranging from 60 days at 20°C to 15 days at 35-55°C, with a volatile solids reduction of at least 38%.

Composting: Using the within-vessel, static aerated pile or windrow composting methods, the solid waste is maintained at minimum operating conditions of 40°C for 5 days. For four hours during this period the temperature exceeds 55°C.

Lime stabilization: Sufficient lime is added to produce a pH of 12 after two hours of contact.

Other methods: Other methods or operating conditions may be acceptable if pathogens and vector attraction of the waste (volatile solids) are reduced to an extent equivalent to the reduction achieved by any of the above methods.

B. Processes to Further Reduce Pathogens

Composting: Using the within-vessel composting method, the solid waste is maintained at operating conditions of 55°C or greater for three days. Using the static aerated pile composting method, the solid waste is maintained at operating conditions of 55°C or greater for three days.

Using the windrow composting method, the solid waste attains a temperature of 55°C or greater for at least 15 days during the composting period. Also, during the high temperature period, there will be a minimum of five turnings of the windrow.

Heat drying: Dewatered sludge cake is dried by direct or indirect contact with hot gases, and moisture content is reduced to 10% or lower. Sludge particles reach temperatures well in excess of 80°C, or the wet bulb temperature of the gas stream in contact with the sludge at the point where it leaves the dryer is in excess of 80°C.

Heat treatment: Liquid sludge is heated to temperatures of 180°C for 30 minutes.

Thermophilic aerobic digestion: Liquid sludge is agitated with air or oxygen to maintain aerobic conditions at residence times of 10 days at 55-60°C, with a volatile solids reduction of at least 38%.

Other methods: Other methods or operating conditions may be acceptable if pathogens and vector attraction of the waste (volatile solids) are reduced to an extent equivalent to the reduction achieved by any of the above methods.

Any of the processes listed below, if added to the processes described in Section A above, further reduce pathogens. Because the processes listed below, on their own, do not reduce the attraction of disease vectors, they are only add-on in nature.

Beta ray irradiation: Sludge is irradiated with beta rays from an accelerator at dosages of at least 1.0 megarad at room temperature (ca. 20°C).

Gamma ray irradiation: Sludge is irradiated with gamma rays from certain isotopes, such as ^{60}Cobalt and ^{137}Cesium, at dosages of at least 1.0 megarad at room temperature (ca. 20°C).

Pasteurization: Sludge is maintained for at least 30 minutes at a minimum temperature of 70°C.

Other methods: Other methods or operating conditions may be acceptable if pathogens are reduced to an extent equivalent to the reduction achieved by any of the above add-on methods.

pasteurization is currently the most widely accepted and commonly used method for providing additional pathogen destruction. Ionizing radiation, such as electron beams and gamma-ray sources, is currently being investigated in pilot-scale facilities [67]. Electron beams have been used primarily with liquid sludges, whereas gamma radiation may find its most promising application with dry or composted sludges. Either can provide improved quality control of the final product and may become significant sanitary engineering processes in the future.

Chemical disinfection of sludge is not widely practiced for pathogen control alone. However, both chlorine and lime have been added to sludge for conditioning purposes. Of these, lime is by far the most widely used. Both will reduce the microbial population in addition to their conditioning effects. However, Farrell [68] noted substantial survival of hookworm, amoebic cysts and *Ascaris* ova in a lime-treated sludge after 24 hr at pH 11.5. Because of the limited destruction expected with chemical treatment and the currently limited use of ionizing radiation, it would appear that pasteurization of sludge with heat will remain the predominant pathogen control method. Composting is one approach to supplying the heat required for pathogen destruction.

Heat death of a cell results in part from thermal inactivation of its enzymes. Enzymes that may be reversibly inactivated by mild heat are irreversibly inactivated by higher temperatures. If an enzyme denatures reversibly with temperature, at equilibrium a fraction of the enzyme will be in the active form, with the remainder in a denatured (inactive) form. Thermodynamic arguments indicate that the fraction in the active form decreases significantly over a narrow temperature range [31]. Without enzyme activity a cell cannot function and will die. Very few enzymes can withstand prolonged heat. Therefore, microbes can be expected to be very sensitive to thermal inactivation.

This chapter reviews the kinetics of heat inactivation, discusses the parameters of importance to heat pasteurization during composting and analyzes human health risks associated with various pathogen levels in the final compost. It should be noted, however, that heat inactivation is not the only method of pathogen destruction in a compost system. Organisms are also destroyed or controlled by competition with other microbes, antagonistic relationships and antibiotic or inhibiting substances produced by certain microbes [69]. Time is another factor that affects survival because of the natural die-off of pathogenic organisms in unsuitable environments. Nevertheless, temperature is the one factor which the engineer can measure and control during composting. For these reasons, regulatory agencies are prone to use temperature as a measure of pathogen destruction as illustrated in Table 5-1. Therefore, the discussion will center on kinetics of heat inactivation as applied to composting systems.

TEMPERATURE-TIME RELATIONSHIPS

Sterilization is the process of destroying all life forms. Bacterial spores and certain protozoan cysts are usually the most resistant microbial forms to temperature, consistent with earlier remarks on endospore formation as a survival mechanism. Cysts and spores, for example, resist exposure to dry heat of over 100°C. Dry sterilization, therefore, requires a temperature of about 180°C for 2-3 hr to assure complete destruction. Moist sterilization, on the other hand, requires only 15 min at 115°C (15 psig pressure in an autoclave or pressure cooker). Boiling at 100°C kills some cysts and spores, but induces germination in others. After a time lapse a second boiling is necessary to kill bacteria or protozoa which have emerged from the spores or cysts. Presumably, requirements for sterilization of a liquid sludge would be similar to those for moist sterilization, namely 115°C for 15 min.

The reason moist and dry heat differ in their inactivation effects is that enzyme denaturation is influenced by solvent concentration. The temperature needed to coagulate (denaturation followed by massive cross-linking of the denatured protein) egg protein albumin increases with decreasing moisture content as shown in Table 5-2. Resistance of bacterial endospores to heat is probably caused in part by their dehydrated condition.

Pasteurization implies heating to a specific temperature for a time sufficient to destroy pathogenic or undesirable organisms. In general, nonspore-forming bacteria and the vegetative cells of sporulating bacteria are destroyed in 5-10 min at temperatures of 60-70°C (moist heat). Data presented by Roedinger [70], shown in Table 5-3, suggest that pasteurization at 70°C for 30 min destroys pathogens found in sludge. Cysts of *Entamoeba histolytica* are reported to be destroyed at 50°C. A similar compilation of time-temperature relationships compiled by Gotaas [71], shown in Table 5-4, also indicates that *E. histolytica* is pasteurized at relatively low temperatures. Stern [72] reported that temperatures of 75°C for one hour destroyed enteric pathogens and reduced indicator organisms to less than 1000/100 ml.

Table 5-2. Effect of Hydration and Heat on Egg Albumin [48]

Water Content (%)	Approximate Coagulation Temperature (°C)
50	56
25	76
15	96
5	149
0	165

Table 5-3. Temperature and Time for Pathogen Destruction in Sludges after Roediger [70]

Microbe	Exposure Time (min) for Destruction at				
	50°C	55°C	60°C	65°C	70°C
Cysts of *Entamoeba histolytica*	5				
Eggs of *Ascaris lumbricoides*	60	7			
Brucella abortus		60		3	
Corynebacterium diptheriae		45			4
Salmonella typhi			30		4
Escherichi coli			60		5
Micrososccus pygogenes var. *aureus*					20
Mycobacterium tuberculosis var.					20
Viruses					25

Table 5-4. Temperature and Time for Pathogen Destruction in Sludges [71]

Microbe	Destruction Time-Temperature		Destruction Time-Temperature	
	Temperature (°C)	Time (min)	Temperature (°C)	Time (min)
Salmonella typhosa	55–60	30	60	20
Salmonella sp.	55	60	60	15–20
Shigella sp.	55	60		
Ent. histolytica cysts	45	Few	55	Few sec
Taenia	55	Few		
Trichinella spiralis larvae	55	Quickly	60	Instantly
Brucella abortis, Br. suis	62.5	3	55	60
Micrococcus pyogenes var. *aureus*	50	10		
Streptococcus pyogenes	54	10		
Mycobacterium tuberculosis var. *hominis*	66	15–20	67	Momentary
Corynebaclerium diptheriae	55	45		
Necator americanus	45	50		
Ascaris lumbricoides eggs	50	60		
Escherichia coli	55	60	60	15–20

At 70°C for 1 hr, pathogens were destroyed, although coliform indicator concentrations sometimes remained above 1000/100 ml.

For data presented in Tables 5-3 and 5-4 it is apparent that thermal inactivation is a time-temperature phenomena. A high temperature for a short period of time or a lower temperature for longer duration can be equally effective. Such time-temperature relationships for a variety of organisms are presented in Figures 5-1 through 5-5. Such data are far more informative than tabular summaries of the kind shown in Tables 5-3 and 5-4 because it allows development of mathematical expressions or models to represent the data.

Time-temperature conditions presented in Figure 5-1 for enteric bacteria and *Ascaris* generally support the inactivation conditions presented in Tables 5-3 and 5-4. It would appear that a temperature of 60°C held for 30 min provides about 6 log units reduction for both coliform and *Salmonella.* Fecal streptococcus appears to be somewhat more heat resistant, requiring about 67°C or a longer time for the same level of inactivation. A temperature of 60°C for 30 min should provide similar levels of reduction for the parasite *Ascaris lumbricoides,* considered to be one of the more heat resistant parasites.

Heat activation of virus is somewhat more complicated than inactivation of other classes of microbes. Time-temperature profiles developed by Ward et al. [73] for a poliovirus strain are presented in Figure 5-2 for liquid raw and digested sludge (about 5% solids) and a phosphate-buffered saline solution. Dramatically different responses to temperature are apparent in the data for raw and digested sludges. Raw sludge is quite protective of the viral particle, whereas protection is significantly reduced in digested sludge. Through a series of experiments, Ward et al. [74] concluded that raw sludge contains a protective substance whose activity was overwhelmed by a virucidal agent acquired during digestion. The latter was determined to be ammonia (NH_3) in the uncharged state. As would then be expected, virucidal activity is a function of pH, increasing at higher pH. Indeed, raw sludge that was protective at pH 7 became nonprotective at pH 9 because of the shift of ammonia from the charged to the uncharged form. Ammonia was also shown to be virucidal to strains of poliovirus, coxsackievirus and echovirus. Reovirus was found to be relatively insensitive to ammonia.

The effect of raw sludge solids content on heat inactivation of poliovirus is presented in Figure 5-3. Ward et al. [74] concluded that the protective agent identified previously was concentrated during the drying process because heat resistance was increased in the drier samples. Alternatively, it could be that resistance was increased because coagulating temperatures generally increase with solids content as shown in Table 5-2 for egg albumin. This is probably not the case, however, because later experiments using composted sludge showed that poliovirus is inactivated more rapidly in composted sludge then in raw sludge with the same moisture content (Figure 5-4). Apparently the protective agent is concentrated as raw sludge is dewatered but is removed or destroyed during composting.

Various organisms have at times been suggested as the most thermal tolerant of the enteric pathogens expected in sludge. Often mentioned candidates include cysts of the protozoan *E. histolytica* and eggs of *A. lumbricoides.* Enteric bacterial pathogens are not spore-forming and are usually considered less heat tolerant than protozoan cysts or helminth eggs. Certain vegetative bacterial cells, such as *Mycobacterium tuberculosis,* are reasonably heat resistant, however. Kruse [78] reported the data shown in Figure 5-6 comparing temperature sensitivity of human cysts with eggs of *Ascaris.* It was concluded that conditions which disinfect *Ascaris* would also destroy protozoan cysts, and that *Ascaris* would be an appropriate indicator for determining the degree of disinfection. Indeed, *Ascaris* has been accepted widely as an indicator organism because of its thermal-tolerant nature relative to other enteric pathogens.

MODELING HEAT INACTIVATION KINETICS

First-Order Decay Model

Referring to Figure 5-1, it is not uncommon to observe straight lines (or nearly so) through temperature survival data in semilog plots. Thus, inactivation kinetics are often modeled assuming first-order decay as follows:

$$\frac{dn}{dt} = -k_d n \tag{5-1}$$

where n = viable cell population and k_d = thermal inactivation coefficient. Equation 5-1 is often referred to as "Chick's law" after Harriet Chick [79] who reported on such exponential die-offs in 1908. However, some survival curves exhibit a nonexponential or aberrant nature such that Equation 5-1 is not applicable. Nonexponential curves will be discussed later.

If k_d is constant, integration of Equation 5-1 from an initial cell population n_o to a later population n_t at time t yields:

$$n_t = n_o e^{-k_d t} \tag{5-2}$$

Taking the log of both sides and rearranging

$$t = \left(ln \frac{n_o}{n_t} \right) \Big/ k_d$$

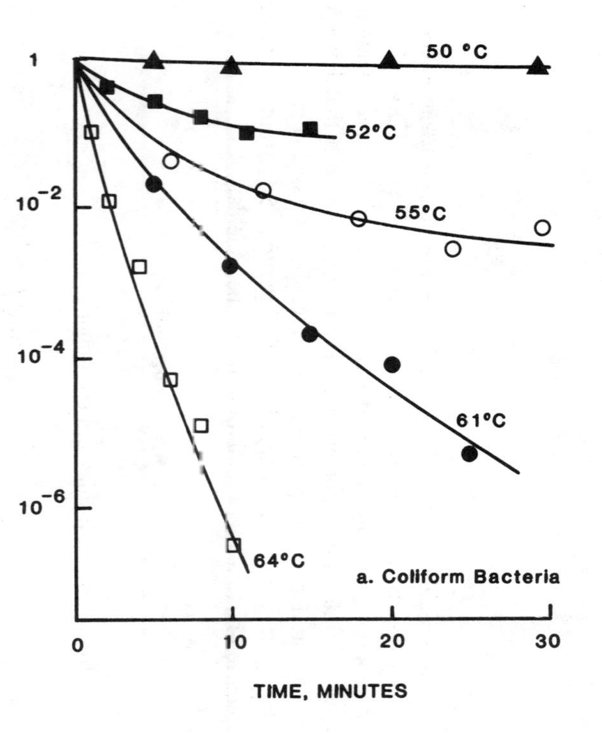

50 °C
52°C
55°C
61°C
64°C
a. Coliform Bacteria
SURVIVING FRACTION
1
10^{-2}
10^{-4}
10^{-6}
0
10
20
30
TIME, MINUTES

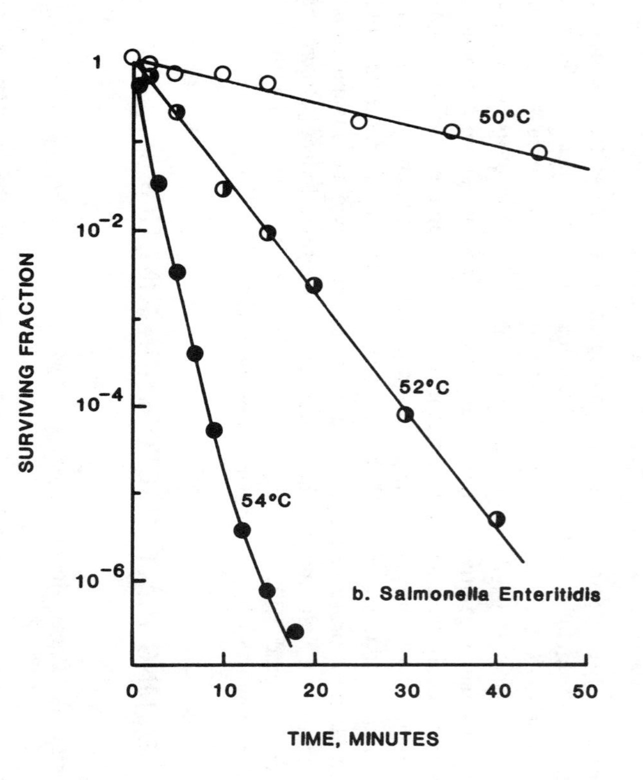

50°C
52°C
54°C
b. Salmonella Enteritidis
SURVIVING FRACTION
1
10^{-2}
10^{-4}
10^{-6}
0
10
20
30
40
50
TIME, MINUTES

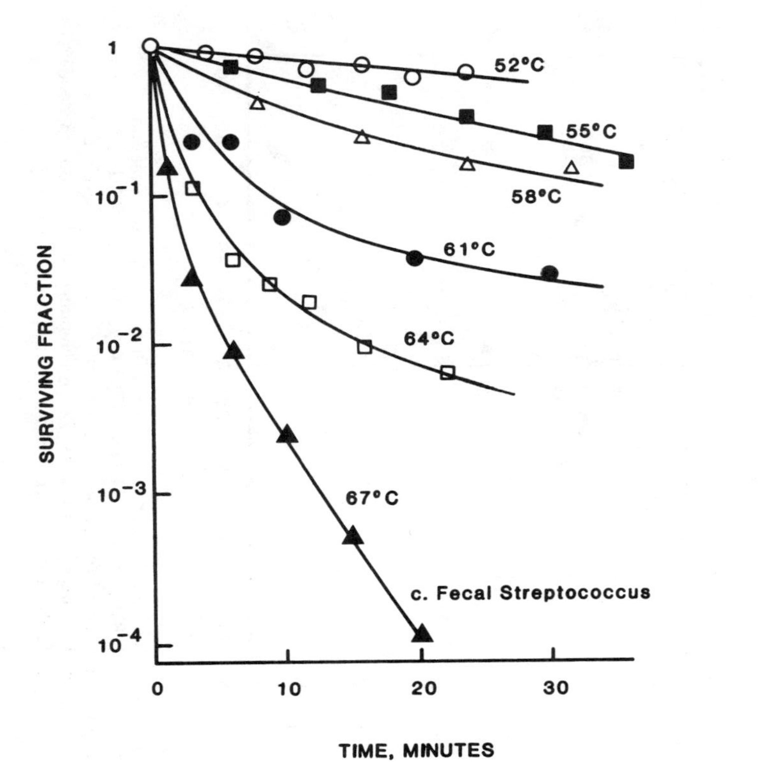

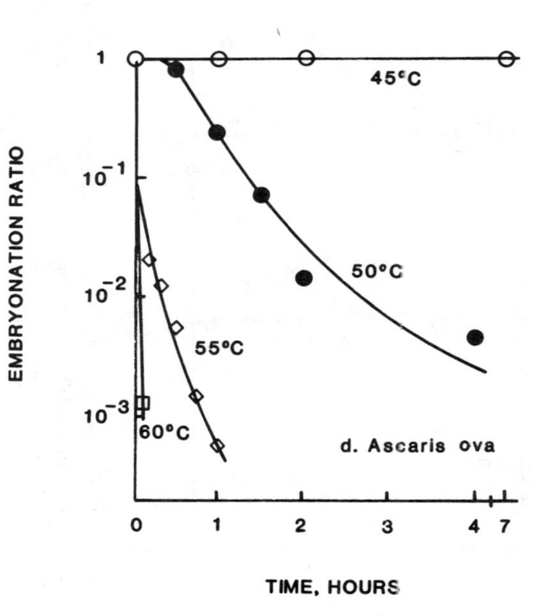

Figure 5-1. Heat inactivation of coliform bacteria, *Salmonella enteritidis*, fecal streptococcus and *Ascaris* ova in composting sludge (a, b, c after Ward and Brandon [75], d from Sandia Laboratories [76]).

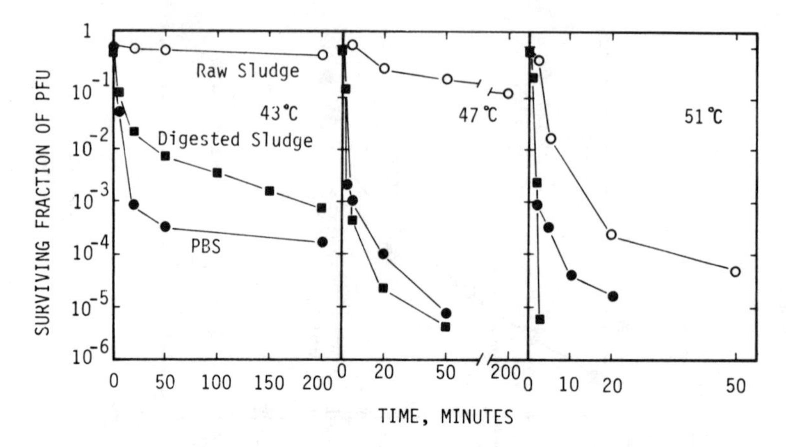

Figure 5-2. Effect of sludge on the rate of heat inactivation of poliovirus. After a tenfold dilution into buffer (PBS), raw sludge or anaerobically digested sludge, poliovirus strain CHAT was mixed for 15 min at room temperature and heated at the specified temperature. Samples were removed from the incubation bath at the times shown and immediately cooled in ice. After sonication in 0.1% SDS, each sample was directly assayed for total PFU on HeLa cells [73].

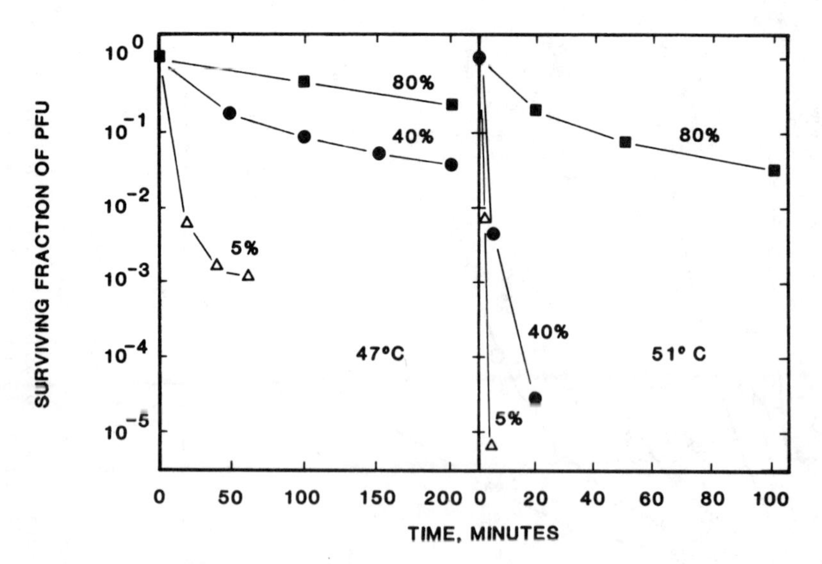

Figure 5-3 Rate of heat inactivation of poliovirus in raw sludge as a function of solids content of sludge. Poliovirus type 1 strain CHAT was heated at 47 and 51°C in raw sludge containing the percentages of solids specified. After exposure the solids contents of all samples were adjusted to 5% by blending in 0.1% sodium deodecyl sulfate; the loss of infectivity was determined by the plaque assay [76].

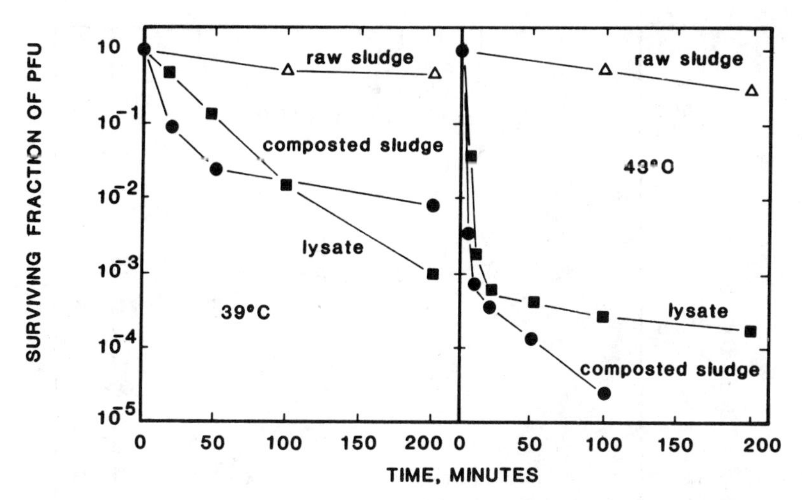

Figure 5-4. Comparative rates of heat inactivation of poliovirus samples in dewatered raw sludge, composted sludge and in lysate of poliovirus-infected cells. Poliovirus was blended with dried raw sludge or composted sludge whose final solids contents had been adjusted to 40%. The rates of viral inactivation by heat at 39 and 43°C were then determined and compared with that in a lysate of infected HeLa cells [76].

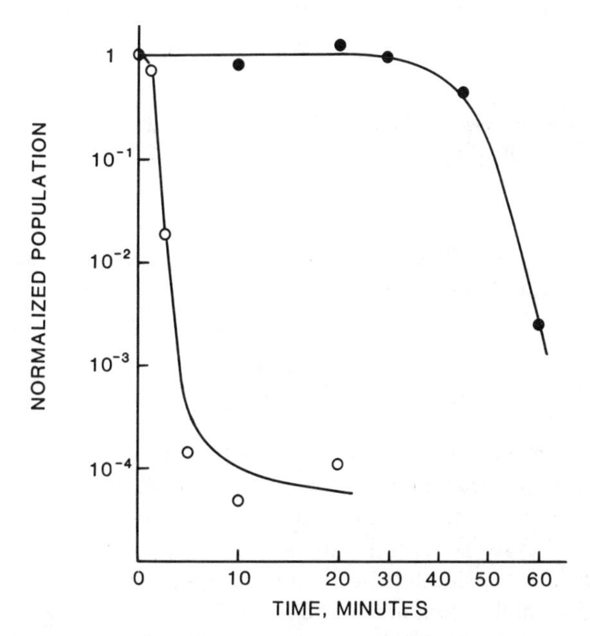

Figure 5-5. Heat inactivation (61°C) of *Aspergillus flavus conidia* in aqueous suspension (○) and in dried raw sludge (●) [77].

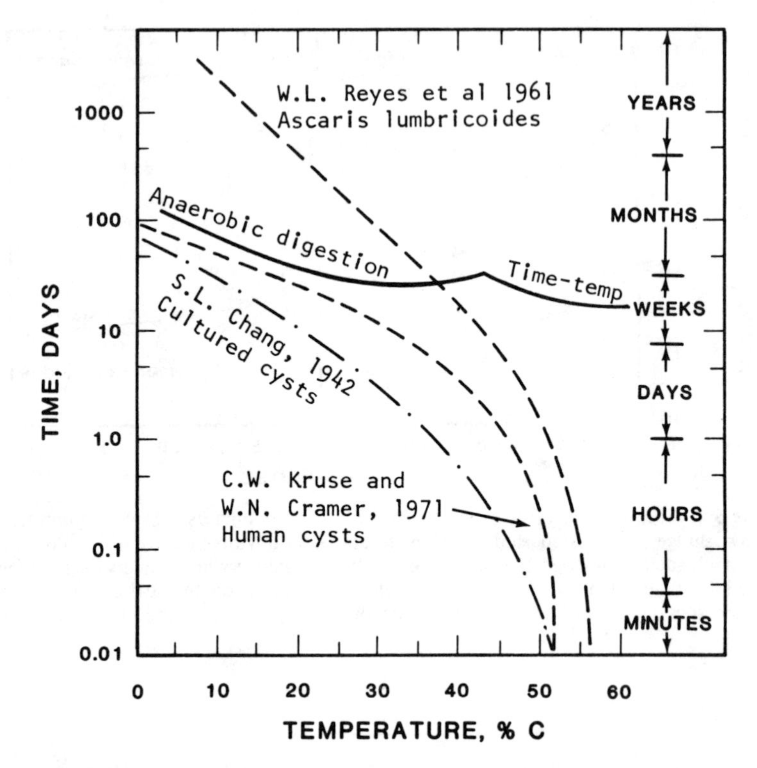

Figure 5-6. Die-away rate of cultured and human cysts of *Entamoeba histolytica* and ova of *Ascaris lumbricoides* var. *suum* under anaerobic digestion conditions giving 99.9% destruction [78].

Converting to base 10 logs and considering a one log reduction in cell concentration (i.e., a reduction of 90%),

$$t_{90} = D_r = \frac{2.303}{k_d} \tag{5-3}$$

The term D_r is the decimal reduction factor and is the time required to achieve a tenfold reduction in cell population. A number of D_r values obtained from the literature are presented in Table 5-5. Note that D_r for *E. histolytica* cysts is reported to be greater than that for *Ascaris* ova. This contradicts previously cited work by other authors and points out some of the variability seen in literature data.

The value of k_d is a function of temperature, the effect of which is most often modeled by the familiar Arrhenius form:

Table 5-5. Time Required for a Tenfold Population Reduction for Various Microbes [80][a]

Microbe	D_r (min)	
	55°C	60°C
Adenovirus, 12 NIAID	11	0.17
Poliovirus, Type 1	1.8	1.5
Ascaris Ova	–	1.3
Histolytica Cysts	44	25
Salmonella[b]	80	7.5
Bacteriophage f2	267	47

[a]See Burge et al. [80] for original cited references.
[b]Serotype Senftenburg 77W.

$$k_d = Ce^{-E_d/RT_k} \tag{5-4}$$

where T_k = temperature, °K. The range of inactivation energies E_d for many spores and vegetative cells is between 50 and 100 kcal/mol [31]. This implies that heat inactivation of microbes is much more sensitive to temperature than most chemical reactions. Logarithmic transformation of Equation 5-4 yields:

$$\log_{10} k_d = \log_{10} C - \frac{E_d}{R}\left(\frac{1}{T_k}\right) \tag{5-5}$$

Thus, a plot of the logarithm of k_d versus $1/T_k$ allows determination of the constant C and the inactivation energy E_d.

Example 5-1

Determine values for the thermal death coefficient k_d and the inactivation energy E_d for *A. lumbricoides* from data in Figure 5-1d.

Solution

1. Since the curves in Figure 5-1d are not straight lines, some approximation will be necessary to fit the first-order assumption. Fitting straight lines from the origin through the data as shown, the following table can be constructed:

Temperature (°C)	n_t/n_o	Time, t (min)	k_d (min^{-1})	$1/T_k$	T_c–50°C	D_r (min)
50	0.001	270	0.0256	30.96×10^{-4}	0	90.0
55	0.001	45	0.154	30.49×10^{-4}	5	15.0
60	0.001	7.5	0.921	30.03×10^{-4}	10	2.5

Values of k_d were determined from Equation 5-2 as follows:

$$ln \frac{n_o}{n_t} = k_d t$$

$$k_d = \frac{ln\, n_o/n_t}{t}$$

For 50°C

$$k_d = \frac{ln(1000)}{270} = 0.0256 \text{ min}^{-1}$$

2. Values of k_d are plotted as a function of $1/T_k$ and (T_c–50°C) in Figure 5-7. Note that a good straight-line fit is obtained, justifying use of the Arrhenius relationship. To determine E_d, values of k_d and $1/T_k$ from the curve in Figure 5-7 were applied to Equation 5-5 with the following results,

$$C = 1.81 \times 10^{49}$$

$$E_d = 75{,}200 \text{ cal/mol} = 75.2 \text{ kcal/mol}$$

Note that the value of E_d is consistent with the previously reported range of values for heat inactivation of microbes.

Substituting these values into Equation 5-4 and recalling that R = 1.99 cal/deg-mol

$$k_d = 1.81 \times 10^{49} e^{(3.78 \times 10^4)/T_k}$$

This equation is somewhat awkward because of the large exponents. Using alternative forms of the Arrhenius equation the following expression can be developed from the plot of k_d versus (T_c – 50):

$$k_d = 0.025e^{0.361(T_c-50)}$$

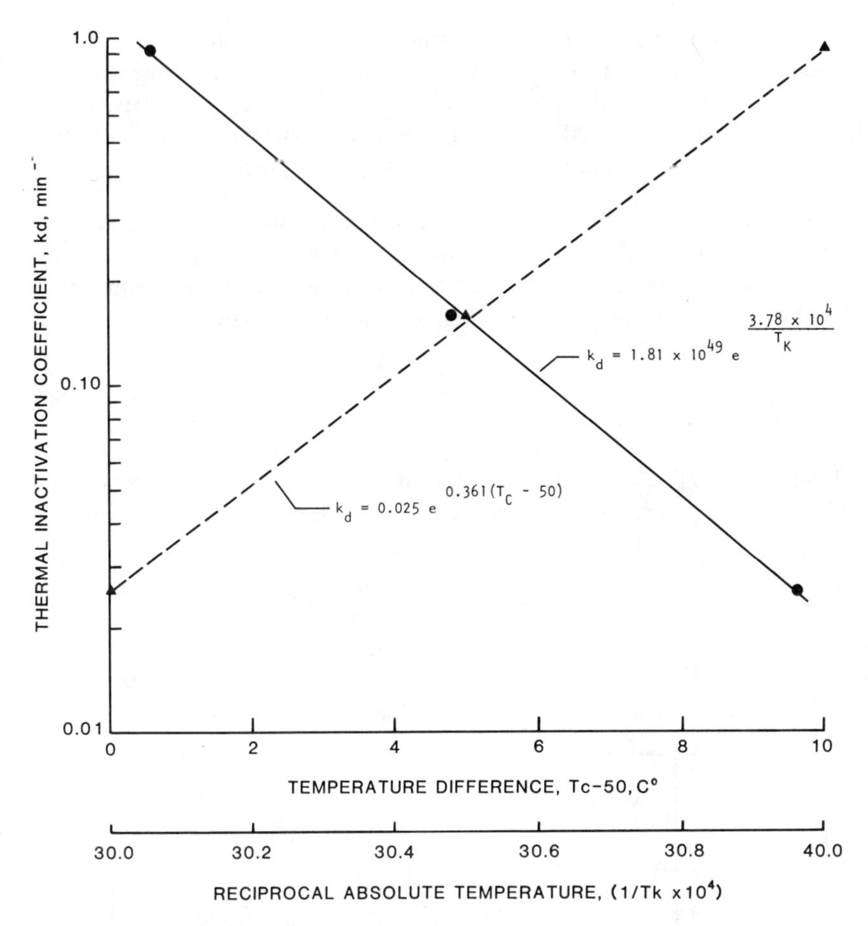

Figure 5-7. Thermal inactivation coefficient as a function of reciprocal temperature $1/T_K$ and temperature difference T_c-50. Data for *Ascaris* from Figure 5-1d.

3. The equation for k_d can be substituted into Equation 5-2 to give

$$ln_e \frac{n_o}{n_t} = 0.025te^{0.361(T_c-50)}$$

Converting to base 10 logs,

$$2.303 \log \frac{n_o}{n_t} = 0.025te^{0.361(T_c-50)}$$

Thus, for any combination of time t and temperature T_c, the log reduction can be calculated. Using this approach the curves in Figure 5-8 were calculated. Note that over 25 log reductions can be achieved with temperatures above 50°C. This is an astronomical number, indicating complete and total destruction for all practical purposes

In composting as well as in many batch sterilization processes, it is common for material to be heated, held at a relatively constant temperature for a time and then cooled. The kill resulting from such a time-temperature profile can be evaluated by first combining Equations 5-1 and 5-4:

$$\frac{dn}{dt} = -Cne^{-[E_d/RT_k(t)]}n \tag{5-6}$$

Temperature T_k is a function of time and is expressed as $T_k(t)$. Separating variables and integrating from initial conditions, $n = n_o$ at $t = 0$, to final conditions, $n = n_f$ at $t = t_f$:

$$ln\frac{n_o}{n_f} = \int_0^{t_f} Ce^{-[E_d/RT_k(t)]}dt \tag{5-7}$$

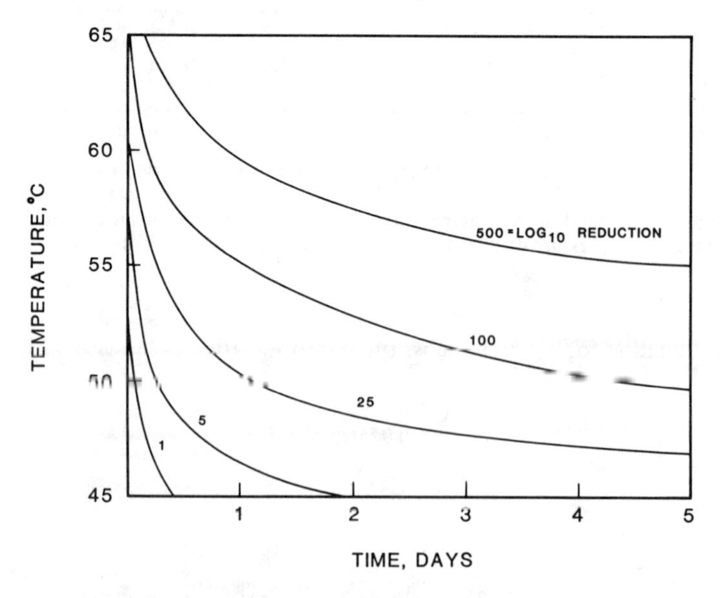

Figure 5-8. Time-temperature conditions to achieve stated $\log_{10}$ reductions for *Ascaris* ova using equations from Example 5-1.

Before the right side of Equation 5-7 can be integrated, the temperature profile must be known. Consider first a constant temperature profile, i.e., $T \neq f(t)$. Equation 5-7 then integrates directly to:

$$ln \frac{n_o}{n_f} = Ct_f e^{-(E_d/RT_k)} \tag{5-8}$$

where T_k = constant. Similar integrations have been performed for a variety of other time-temperature profiles including hyperbolic, exponential and linear increases. Bailey and Ollis [31] present further discussion of the analytic approach to solution. Regardless of the type of time-temperature profile, total kill is the sum of kills observed in the heating, constant (holding)-temperature and cooling periods. Thus,

$$ln \frac{n_o}{n_f}\bigg|_{Total} = ln \frac{n_o}{n_f}\bigg|_{heating} + ln \frac{n_o}{n_f}\bigg|_{holding} + ln \frac{n_o}{n_f}\bigg|_{cooling} \tag{5-9}$$

Actual time-temperature profiles observed in composting may not correspond to profiles for which analytical solutions are available. In such cases a more practical approach is to use graphical integration procedures to solve the right side of Equation 5-7. The graphical approach can be used with any profile and is not limited by our ability to describe the profile in formal mathematical terms. This method is illustrated in the following example.

Example 5-2

A time-temperature profile for a batch composting system is presented in Figure 5-9. Using a graphical integration approach determine the log reduction of *Ascaris* ova using equations developed in Example 5-1. Assume the minimum lethal temperature to be 45°C.

Solution

1. The time-temperature profile in Figure 5-9 is approximated by construction of a number of smaller elements as shown. Note that the first element begins at 45°C, the assumed minimum lethal temperature. The width of each element is Δt. Recall that Δt approaches dt as Δt becomes small. Average temperature for an element (indicated by points on the curve) is used to determine an average k_d for that element using the equation developed in Example 5-1:

$$k_d = 0.025e^{0.361(T_c-50)}$$

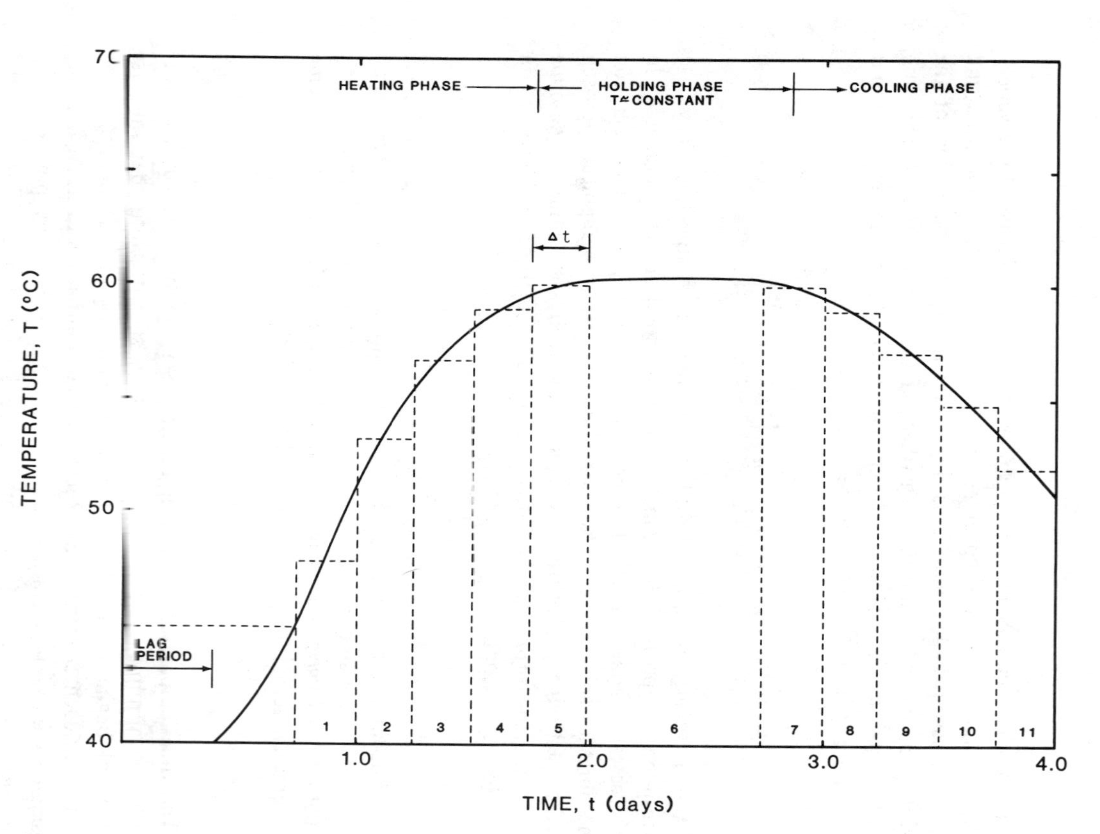

Figure 5-9. Idealized time-temperature profile for a batch composting system (Example 5-2).

Thermal inactivation for each element can then be determined from Equation 5-2 as:

$$ln \frac{n_0}{n_f} = k_d \Delta t$$

The following table was constructed for the elements in Figure 5-9:

Element No.	Δt (days)	Avg T_c (°C)	k_d (min^{-1})	$ln \frac{n_0}{n_f}$
1	0.25	47.7	0.011	4
2	0.25	53.2	0.079	29
3	0.25	56.7	0.281	101
4	0.25	59.0	0.644	232
5	0.25	60.0	0.924	333
6	0.75	60.3	1.030	1112
7	0.25	60.0	0.924	333
8	0.25	58.8	0.599	216
9	0.25	57.0	0.313	113
10	0.25	54.7	0.136	49
11	0.25	52.0	0.051	19

$$\sum ln \frac{n_0}{n_f} = 2541$$

2. The total $\log_{10}$ reduction is then determined as:

$$\log_{10} \frac{n_0}{n_f} = \frac{ln \frac{n_0}{n_f}}{2.303} = \frac{2541}{2.303} \sim 1100$$

This is 1100 $\log_{10}$ units of reduction! In other words:

$$n_f = n_0 10^{-1100} \sim 0$$

which is a very small number indeed. Data contained in Figure 5-1d are reasonably consistent with other data presented in Tables 5-3 and 5-4. Therefore, it must be concluded that essentially no *Ascaris* ova would survive the time-temperature profile of Figure 5-9. Even deviations from the first-order rate model as n becomes small are not likely to change this outcome.

Nonexponential Decay Curves

In some cases the assumption of first-order decay may be inappropriate to the data. Deviation from first-order decay is sometimes observed as the number of microbes decreases, as illustrated in Figure 5-10 for *Staphylococcus aureus.* Similar effects can be noticed in some of the data in Figure 5-1. In explaining these effects it has often been assumed that not all cells in the culture are equally vulnerable to the temperature conditions. More vulnerable cells are assumed to be killed first (higher k_d) leaving a small number of heat-resistant cells (lower k_d). Also, as the number of organisms becomes small, the assumption of a uniform population characterized by a single value of k_d begins to fail. Whether this explanation is correct or not, temperatures can be raised to the point where such fluctuations are minimized. Note that increasing the temperature from 53 to 57°C in Figure 5-10 results in a more linear plot with near complete inactivation of those microbes resistant at 53°C.

Time-temperature relationships observed during heat inactivation of microbes can be generalized by the forms shown in Figure 5-11. Curve A is the first-order or exponential decay form as previously discussed.

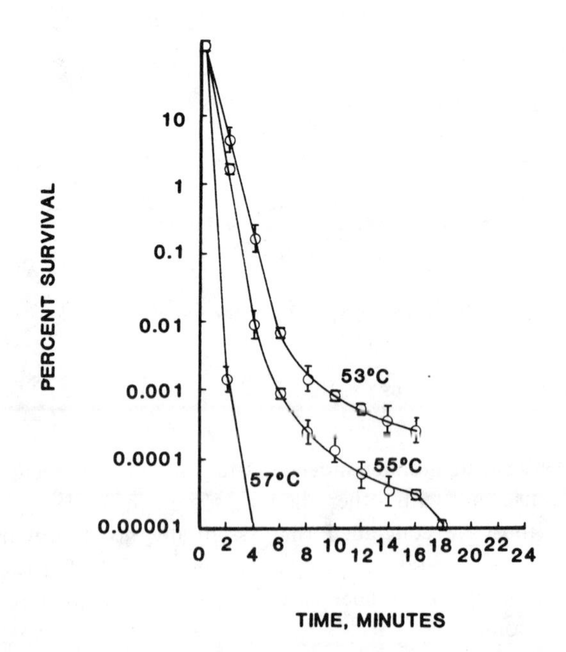

Figure 5-10. Heat destruction of *Staphylococcus aureus* in neutral buffer [48].

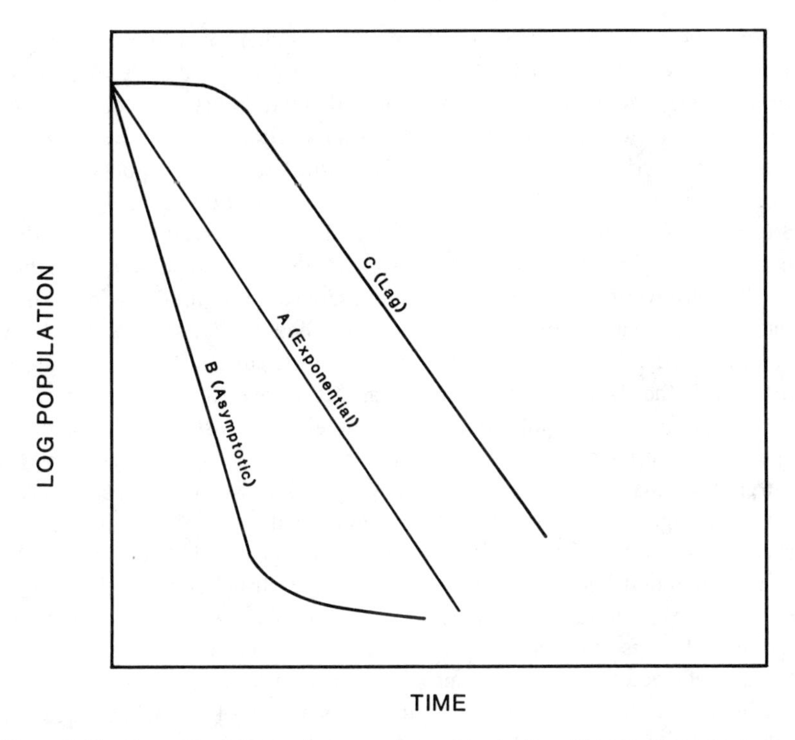

Figure 5-11. Generalized types of curves observed in studies of inactivation of microbes. (A) First-order or exponential decay; (B) first-order followed by asymptotic or retardant die-away; (C) initial lag or shoulder followed by first-order decay.

Curve B shows an initial phase of first-order decay followed by a subsequent decrease in the rate of inactivation. In some cases the curve can become almost asymptotic to the time axis. Curve C shows an initial lag period followed by first order decay. Sigmoid or S-shaped curves are possible by combination of curves B and C. Elements of these generalized curves can be seen in the time-temperature relationships already presented.

A disturbing question immediately comes to mind. If Curve B is applicable to some of the enteric pathogens, does this mean that there is a limit to expected inactivation at a particular temperature? Will a small but troublesome fraction of the population survive? By applying a first-order model have we been guilty of bending reality to fit a mathematically pleasing equation? To answer these questions the nature and cause of the asymptotic and lag portions of Curves B and C must be explored.

Several authors have advanced "multiple target" or "multiple hit" theories to explain the death of microbes as a result of heat, chemicals or ionizing radiation. In these theories a number of critical molecular sites are envisioned which can be inactivated by the disinfecting agent. For example, Moats [81] proposed that thermal injury and death could be explained by assuming that death results from inactivation of X_1 of a total of N critical sites per cell. The critical sites were thought not to be enzymes because death resulted from inactivation of only a small percentage of the total sites. Moats predicted that X_1 would vary depending on the recovery medium used for heating. This provided some explanation for the observation that some microbes are only injured by mild heat treatment, and can recover if grown on enriched media. Moat's model was able to fit the lag period in the type C curve but not the asymptotic portion of the B curve. The latter was attributed to the presence of a small heat-resistant population. Unfortunately, the asymptotic region is the one of most concern as previously discussed.

Wei and Chang [82] proposed an explanation which can account for both the lag and asymptotic phases. Microbial death in a disinfection process was assumed to result from random collision between molecules of a disinfectant and the microbe. Since the disinfectant is present in extremely large numbers compared to the microbes, a Poisson probability was used to describe the collision rate with a single microbe. Microbes were also allowed to exist in clumps of various sizes leading to a multi-Poisson distribution model. The model was tested using amoeba cysts of *Naegleria gruberi* and also applied to data on coliform and virus destruction.

The result was the realization that clumping of organisms is a major, if not the only, factor responsible for aberrant survival curves in disinfection studies. Exponential curves were obtained when cysts were treated to produce discrete particles.

The lag period of type C curves results from a large percentage of the population existing in clumps. The higher the percentage and the larger the clumps, the more prominent is the shoulder. To better understand this recall that plating techniques used to recover viable microbes cannot distinguish between a single microbe or a clump of microbes, where any from one to all of the microbes may be viable. Disinfection may proceed with first-order kinetics within a clump but as long as one organism remains viable, the cell count by traditional plating techniques will remain relatively unchanged. As the number of viable cells within each clump approaches unity, exponential rate kinetics again become evident.

The asymptotic curves of type B are caused by a large percentage of the population existing as discrete particles along with a number of clumps of unusually large size. The higher the percentage of singles, the more marked is the initial exponential drop. The situation with a sigmoid inactivation curve is shown in Figure 5-12. A large majority of small aggregates

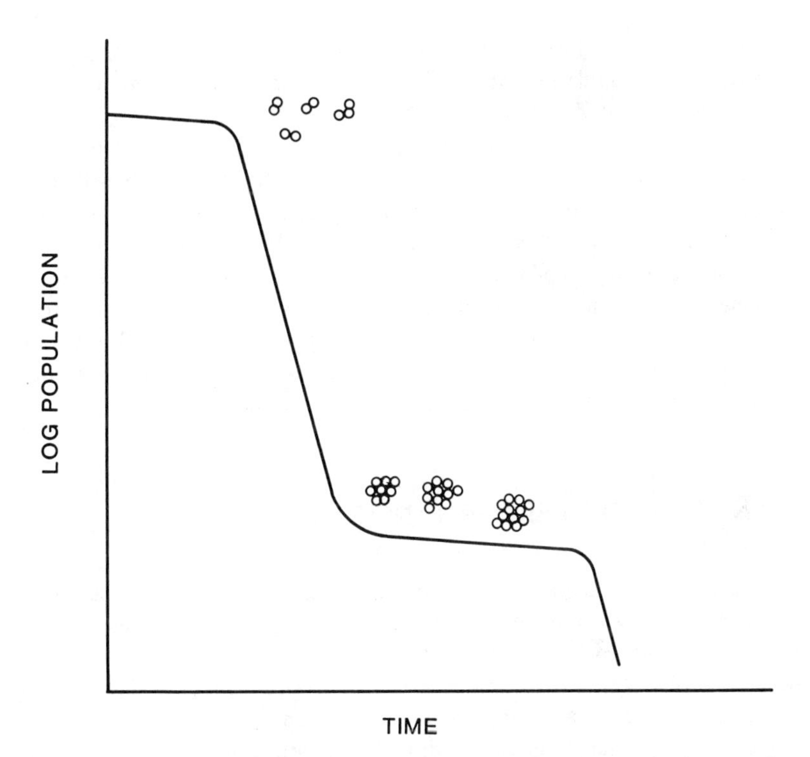

Figure 5-12. Sigmoid type inactivation curve with aggregate sizes of microorganisms producing the shoulder and plateau parts of the curve. On complete inactivation of large aggregates, the curve plunges to the time axis [80].

produce an initial shoulder and a small number of large aggregates produce a plateau that plunges to the time axis after all individual organisms are destroyed.

Wei and Chang also showed that if the asymptotic segment is located below the 5% survival level, the value of k_d for destruction of a single microbe can be computed from the exponential segment. If the asymptotic section is above the 10% survival level, survival of clumps will reduce the slope of the linear segment from that which would be observed if all organisms were singles. Similar limits apply to the type C curves. Within these limits, estimation of k_d and D_r values is relatively simple despite the aberrant nature of many curves.

An important result of this model is that the asymptotic segment will drop steeply once all organisms in the clumps have been destroyed. Furthermore, destruction of microbes within a clump appears to follow the same exponential decay coefficient observed for singles. As Burge et al.

pointed out, "this does much to erase from our minds the concept that a portion of a population will survive composting for some reason we have not been able to fathom" [80].

One final point will be made before leaving this subject. Other non-exponential curves have been observed, such as the sigmoid shaped, those which curve downward with constantly increasing slope (concave viewed from the origin), those which curve downward with a constantly decreasing slope (convex) and step-ladder–shaped curves. Each of these types can be explained by assuming clumps of various size and frequency of distribution. Estimation of k_d and D_r values for these types is mathematically more difficult. Nevertheless, it is reasonably certain that asymptotic sections do not represent heat-resistant portions of a population.

STANDARDS OF PERFORMANCE

Stochastic models have been applied to the problem of sterilization to statistically estimate numbers of viable organisms remaining after a certain temperature-time exposure. Assuming the usual case of a large number of initial microbes ($n_0 \gg 1$), application of first-order kinetics described by Equation 5-1, and assuming that each organism is independent and exposed to the same lethal condition, the probability of at least one organism surviving is given by:

$$1 - P_o(t) = 1 - e^{-n_t} \tag{5-10}$$

where n_t is determined from Equation 5-2. $P_o(t)$ is termed the extinction probability, in other words, the probability that all organisms are inactivated. It follows then that $1 - P_o(t)$ is the probability of at least one organism surviving the lethal condition.

A statistical approach of this type is often used in industries which require pasteurization or sterilization of materials. A considerable number of surviving organisms are acceptable in certain applications. For example, viable bacterial concentration in Grade A pasteurized milk must be less than 20,000 cells/ml [48]. Alternatively, a fermentation industry may require that feed substrates be assured of complete sterilization in 99 out of 100 batches. In this case, economic loss associated with a lost batch of substrate must be weighed against the cost of achieving a statistically higher probability of extinction $P_o(t)$. Perhaps the most severe requirements are found in the canning industry. A single surviving spore of *Clostridium botulinum* can lead to lethal concentrations of the exotoxin causing botulism poisoning. Therefore, assurance of virtually complete destruction is required. Design criteria often require that spore survival probability $1 - P_0(t)$ be reduced to 10^{-12} or less [31].

Requirements for pasteurization of milk are based on conditions necessary to inactivate the most heat-resistant pathogen expected to be transmitted by milk. For a long time this was considered to be *M. tuberculosis,* the causative agent of tuberculosis. *Coxiella burneti,* the causative agent of Q fever, was later discovered to be slightly more heat resistant, and time-temperature conditions for pasteurization were adjusted to account for this. Milk is presently batch-pasteurized at 63°C for 30 min (low temperature holding or LTH) or continuously pasteurized by heating to 71.7°C and holding for 15-30 sec followed by rapid cooling (high temperature short time, or HTST or flash).

A somewhat similar concept has been proposed [80] to assure satisfactory disinfection of enteric pathogens during composting. These authors have recommended use of bacteriophage f2 as a standard organism. From Table 5-5 it can be seen that bacteriophage f2 appears to be more heat tolerant than the enteric pathogens including viruses, bacteria, protozoan cysts and helminth ova. Therefore, conditions designed to achieve a certain log reduction in the standard organism would assure greater destruction of the enteric pathogens. Time-temperature conditions for various log reductions of bacteriophage f2 in tryptone yeast extract medium (TYE) are presented in Figure 5-13. Comparison with similar curves developed for *Ascaris* ova in Example 5-1 confirms that bacteriophage f2 is considerably more heat resistant. Some change in the curves of Figure 5-13 can be expected as more data are developed on survival of f2 in actual composting material. However, the more thermal-tolerant nature of f2 will probably remain unaltered.

Using Figure 5-13, one would predict that 25 log units reduction would be achieved at 55°C for about 3 days or 60°C for about 0.5 days. Reduction in enteric pathogens would be considerably greater because their D_r values are lower than that for bacteriophage f2. Using this approach, health officials can establish desired log reductions based on intended use of the final product and desired safety factors. Temperature-time conditions can then be established with assurance that enteric pathogens will be controlled to even safer levels. It is expected that this approach, using either bacteriophage f2 or other standard organisms, will find increasing application because of its straightforward and scientific basis and the well-established success of similar concepts used in pasteurization of milk and other products.

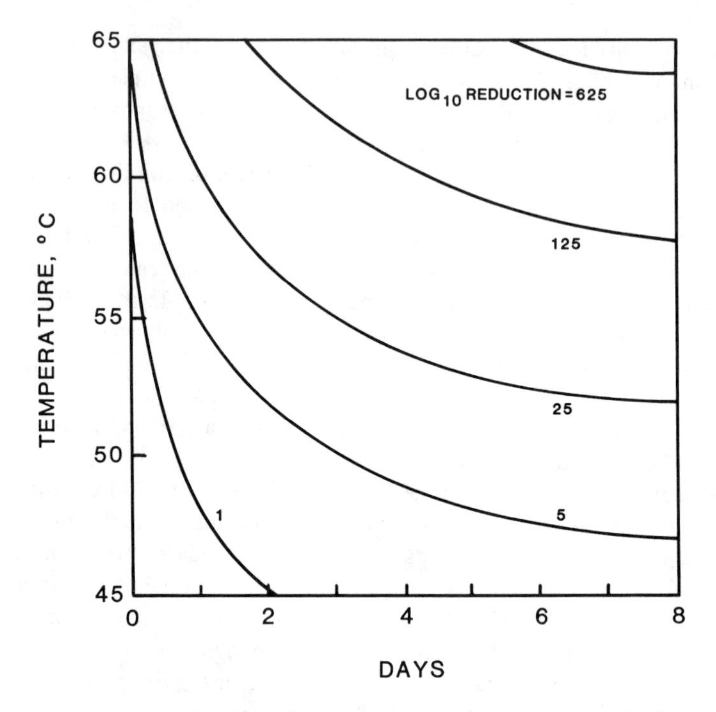

Figure 5-13. Curves showing the time by temperature regimes necessary for inactivation of a desired number of logs of f2 bacteriophage in tryptone yeast extract medium (TYE) [80].

LIMITATIONS ON MICROBIAL DESTRUCTION

Spore-Forming Bacteria

Rather dramatic reduction in response of *Ascaris* ova to a typical composting time-temperature profile was predicted in Example 5-2. However, other organisms may be more temperature resistant. Consider the data in Figure 5-14 for inactivation of *Bacillus subtilis* spores. Inactivation would not be expected until the temperature reached >100°C. The data of Figure 5-14 are typical for endospore-forming bacteria including types such as *Clostridium tetani* (tetanus) and *C. botulinum* (botulism). Thus, endospores formed by spore-forming bacteria would not be inactivated by composting. This statement does not extend to protozoan cysts and fungal spores which are much more susceptible to heat inactivation.

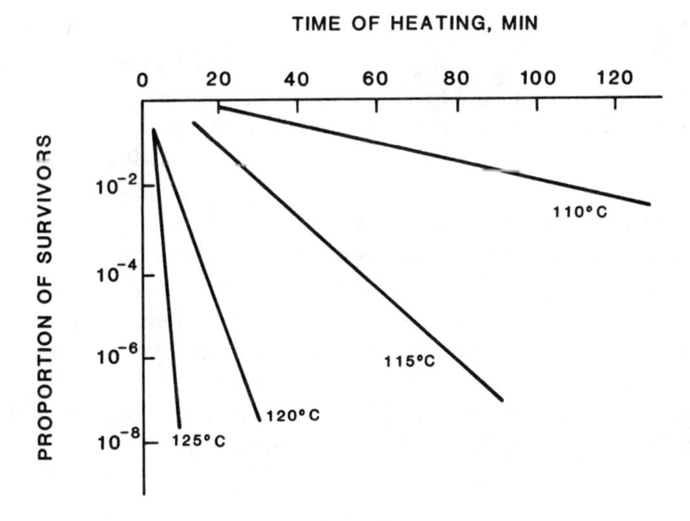

Figure 5-14. Thermal inactivation of endospores of *Bacillus subtilis.* Reprinted from Burton, H., and D. Jayne-Williams. "Sterlized Milk," in *Recent Advances in Food Science, Vol. 2, Processing*, J. Hawthorn and J. M. Leitch, Eds. (London: Butterworths & Co. Publishers Ltd., 1962). p. 107.

Fortunately, enteric bacterial pathogens are nonspore-forming. It is interesting to speculate on why this should be the case. One possibility is that the microbes found a suitable environment in the gut of man and never had to develop spore formation as a defense mechanism to assure their survival. Spore-forming pathogens may be found in sludge but probably not in levels much elevated from those naturally occurring in soils or other materials. *C. tetani,* for example, is a ubiquitous saprophytic organism commonly found in soil. The organism is dangerous to man only when introduced into wounds which become anaerobic, allowing growth of the microbe.

In summary, temperatures common to composting should be effective against enteric bacterial pathogens (as well as viral, protozoan and metazoan forms) but not endospores produced by bacterial spore formers. Therefore, care should be exercised in the use of compost just as one would exercise care in use of soil or other natural materials.

Clumping of Solids

Referring again to Example 5-2, the outcome is predicated on the assumption that all organisms experience the time-temperature profile to an equal extent. A number of factors could prevent full realization of this assumption. For one, large particles or balls may form in the compost. These may not receive adequate oxygen, which would significantly reduce heat buildup from within the particle itself. Heat transport from surrounding compost would then be necessary to assure adequate pathogen destruction with the particle. But does this occur at a sufficient rate, and for what size particles?

The situation can be examined analytically by assuming a spherical, homogeneous ball of material heated from the outside. This is a classical problem in heat transfer for which analytical solutions are well developed. The development of these solutions will not be explored here, but their results will be used. A wealth of solutions and other data for a variety of such problems is available in Carslaw and Jaeger [83].

Dimensionless temperature profiles within a sphere of radius R are shown in Figure 5-15 as a function of dimensionless time.

$$\text{dimensionless time} = \frac{kt}{\rho c_p R^2} \tag{5-11}$$

where k = thermal conductivity, cal/hr-cm²-°C/cm
t = time, hr
ρ = mass density, g/cm³
c_p = specific heat of the particle, cal/g-°C
R = radius of the spherical particle, cm

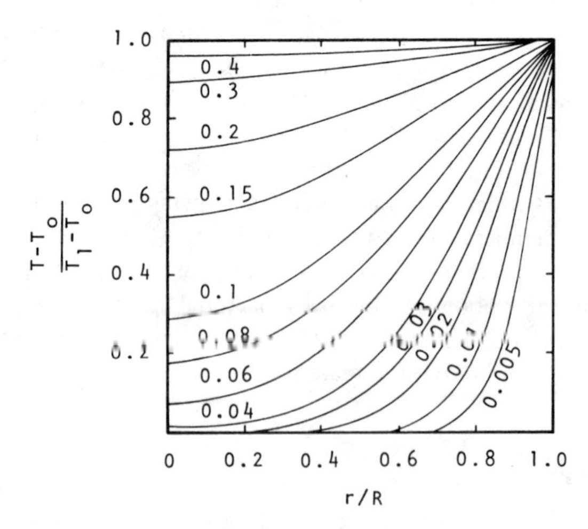

Figure 5-15. Temperature profiles within a sphere of radius R as a function of dimensionless time $kt/\rho c_p R^2$ (given as a parameter on the curves). Initially the sphere temperature is T_0 throughout, and the outer-surface temperature for $t > 0$ is T_1 [31].

For the solution shown in Figure 5-15, it is assumed that temperature is uniform throughout the sphere at $t = 0$ and is equal to T_0. The outer surface temperature is equal to T_1 for all $t > 0$. Thus, at $t > 0$ heat flows into the spherical particle until the temperature becomes uniform and equal to T_1 at $t = \infty$.

The thermal properties of compost are discussed at some length in Chapter 9. Here some reasonable assumptions will be made and further discussion will be delayed until Chapter 9. Assume a worst-case situation with a particle of high density and specific heat and low thermal conductivity. Because compost mixtures are commonly 50% moisture, let $\rho = 1$ g/cm^3 and $c_p = 1$ cal/g-°C. Approximate but typical values of thermal conductivity in cal/hr-cm^2-°C/cm for materials other than compost are: aluminum, 1800; stainless steel, 150; concrete, 15; water, about 5.6; water-saturated wood, 3; and corkboard, 0.40. Values between 2 and 4 cal/hr-cm^2-°C/cm have been measured for compost material, and a value of 3 will be assumed here.

The center of the sphere ($r = 0$) will heat most slowly. Therefore, if the center is adequately treated, the remainder of the particle should be as well. Let us consider a value of dimensionless temperature $(T - T_0)/(T_1 - T_0) = 0.9$ which will give an internal temperature close to the surrounding temperature T_1. From Figure 5-15, this corresponds to a dimensionless time of about 0.3 at $r/R = 0$. Thus:

$$\frac{kt}{\rho c_p R^2} = 0.3$$

Substituting assumed values for k, ρ and c_p, and rearranging:

$$t \simeq 0.1\ R^2 \tag{5-12}$$

Based on Equation 5-12 the data presented in Table 5-6 can be calculated readily.

Based on this analysis it would appear that the heating time for particles of 1-10 cm radius would be insignificant compared to detention times normally used in composting. If clumps larger than 20 cm radius are formed, however, heating times can be significant. This would then limit the potential kill of microbes to values less than that calculated using the approach in Example 5-2. Consideration should be given to breaking these clumps to assure full exposure of all organisms to the time-temperature profile.

Table 5-6. Estimated Heat Transfer Times into Spherical Compost Particles[a]

Particle Radius (cm)	Time to Reach $(T - T_0)/(T_1 - T_0) = 0.9$ (hr)
1	0.1
10	10
20	40
50	250
100	1000

[a]Assumed conditions: $\rho = 1$ g/cm³, $c_p = 1$ cal/g-°C, $k = 3$ cal/hr-cm²-°C/cm, $r/R = 0$.

Nonuniform Temperature Distributions

A second factor that may limit thermal inactivation during composting is the fact that uniform temperatures will not exist throughout the entire compost mass. Cold pockets or zones may allow pathogenic microbes to survive the composting conditions. When compost is not turned, such as in the aerated static pile process and certain reactor systems, the engineer must assure by proper design and operation that cold zones are not formed. In the windrow and many reactor systems in which compost is periodically agitated, the designer must assure a minimum statistical probability that all particles are exposed to the temperature conditions.

Compost Mixing and Turning

The case where compost is agitated at frequent intervals can be examined analytically provided a number of simplifying assumptions are made. Imagine the pile to be composed of two zones, each of which have uniform temperatures. Temperatures in the outer zone are sublethal and cause no organism destruction. Uniform lethal temperatures occur in the inner zone where thermal inactivation is described by Equation 5-2. The pile is turned at time intervals of Δt. Each time the pile is turned there is sufficient mixing energy to cause a random redistribution of material after each turning. Under these conditions, thermal inactivation can be described as:

$$n_t = n_o(f_l + f_h e^{-k_d \Delta t})^N \tag{5-13}$$

and

$$f_l + f_h = 1 \tag{5-14}$$

where n_t = number of organisms surviving
n_o = number of organisms initially present
f_l = fraction of composting material in the low-temperature, sublethal zone
f_h = fraction of composting material in the high-temperature zone
Δt = time interval between pile turnings
k_d = thermal death coefficient as defined by Equation 5-1
N = number of pile turnings

Equation 5-13 was solved for various assumed conditions with results shown in Figure 5-16. Two different values of f_l/f_h were evaluated. A value of $f_l/f_h = 1$ implies that half of the composting material is in the high-temperature zone at any time. For $f_l/f_h = 0.25$, 80% is in the high-temperature zone. Various values of $k_d\Delta t$ were assumed. Note that a value of 2.303 implies a one $\log_{10}$ reduction between each turning for material in the high-temperature zone. A value of ∞ for $k_d\Delta t$ implies that undesirable microbes in the high-temperature zone are completely inactivated between each turning. For the latter case Equation 5-13 reduces to:

$$n_t = n_o (f_l)^N \tag{5-15}$$

Equation 5-15 represents the maximum destruction achievable under the particular conditions of f_l and N.

The fraction of surviving organisms shown in Figure 5-16 is considerably greater than that which would be calculated by using procedures in Example 5-2. The latter example was based on a typical time-temperature profile using k_d values representative of the enteric pathogens and essentially complete destruction was predicted. This strongly suggests that average inactivation achieved during composting is determined as much by the ability to expose all material to the lethal temperature as by the time-temperature profile itself, provided, of course, that a reasonable value of $k_d\Delta t$ is maintained.

The amount of material in the low-temperature zone should be minimized to the extent possible. Suppose that a survival fraction of 10^{-6} is determined to be adequate. If $k_d\Delta t$ is 2.303, the desired inactivation could be achieved with 11 pile turnings at $f_l = 0.2$, increasing to 23 turnings at $f_l = 0.5$. In the windrow composting system, f_l can be minimized by constructing larger piles to reduce the surface/volume ratio.

The assumption of random redistribution of material during turning may not be reasonable in all cases. However, many commercial turning devices impart considerable energy to the pile during turning, and for these the assumption is probably reasonable. On the other hand, if a front-end loader is used, energy input is lower and the assumption is probably not valid. With a front-end loader, however, the operator can apply his knowledge as to the location of the low-temperature zone.

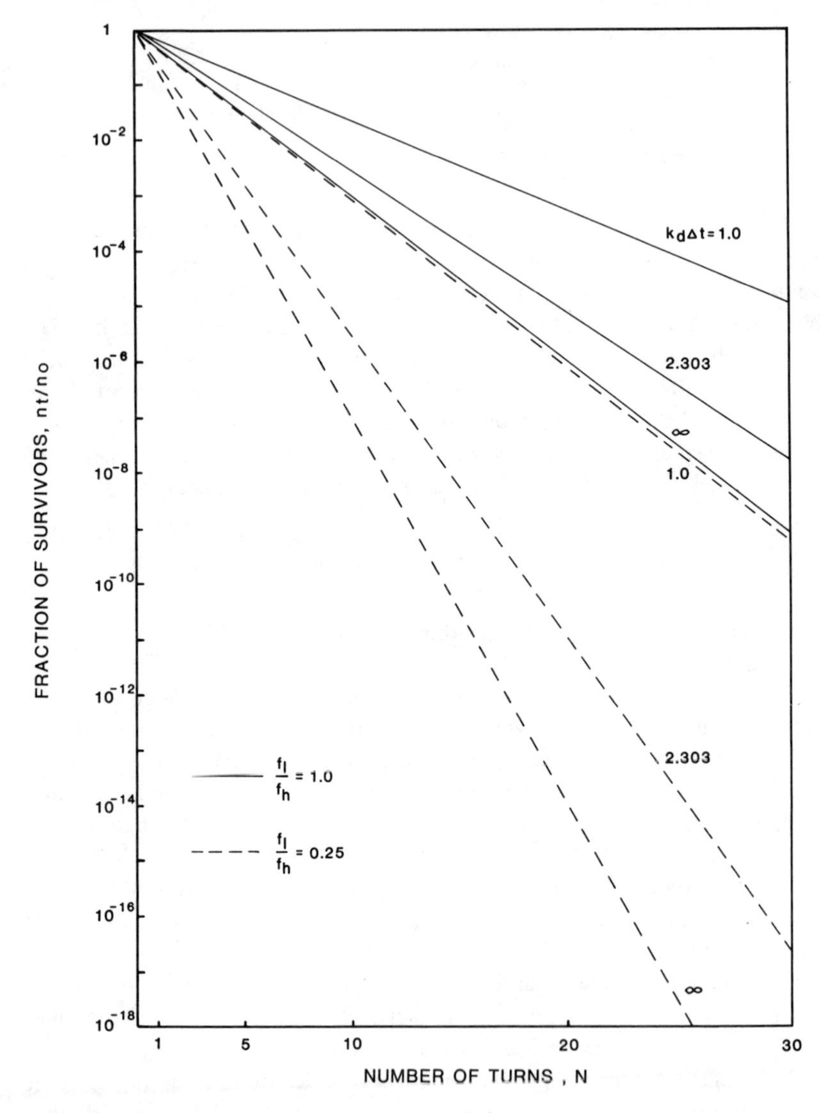

Figure 5-16. Microbe survival as a function of the number of turns with $k_d\Delta t$ and f_l/f_h as parameters.

He can then attempt to redistribute low-temperature material in the high-temperature zone for the next cycle. Thus, redistribution is purposely nonrandom. In other words, nonrandomized redistribution guided by information is used in one case compared to higher energy, randomized redistribution in the other. In either case, the basic conclusion remains that average thermal inactivation is determined in large measure by the ability to expose all material to the lethal temperature.

The model developed here assumes that all material is redistributed after each turning. Suppose a small amount of material near the edge of the pile escaped the turning action. If this material amounted to only 0.1% of the total composting mass and if no inactivation occurred within the material, average thermal inactivation would be limited to no more than 3 $\log_{10}$ units. Thus, it is most important that all material participate in the redistribution achieved during turning. If such care is exercised, inactivation levels can be achieved as determined in Figure 5-16.

At this point it would be advisable to mention that average thermal inactivation levels may be difficult to measure in practice. In the above example it was assumed that 0.1% of the material was not composted. If small grab samples of the final product were analyzed, acceptable destructions might be observed in as many as 999 out of 1000 samples. The one sample which finally did show high levels would probably be assumed to be in error. Thus, it is important to sample all portions of the final compost pile. Even so, it is still incumbent on the designer and operator to assure that all material is exposed to lethal temperatures.

Static Pile Systems

Static pile systems, in which compost is not turned, can be examined analytically provided temperature distributions within the piles are known. Burge et al. [80] examined 15 static piles at Beltsville, MD, which were composting a mixture of raw sludge and wood chips. Temperature readings were taken throughout the compost cycle from the pile center, lateral portions extending out from the center and the region designated as the toe just beneath the outer blanket at the lower edge of the pile. The latter has traditionally been the area of lowest temperature.

From the temperature history throughout the compost pile, confidence levels for achieving a particular time-temperature relationship can be established. Such confidence levels for the 15 piles tested by Burge et al. [80] are shown in Table 5-7. For example, a confidence level can be achieved of 99.9% that all material within the pile will attain 55°C for 9.4 days. Most material would see this temperature for a significantly longer period of time. Recall that 55°C for only three days is sufficient to achieve 25 $\log_{10}$ reductions in bacteriophage f2, which has been recommended as a standard test organism.

Mean temperature and standard deviations observed at the toes of the 15 test piles are shown in Figure 5-17. Significant reduction of bacteriophage f2 would be predicted not only on the basis of the mean temperature but

Table 5-7. Confidence Levels That All Material Will Obtain aTemperature Equal to or Greater Than a Particular Temperature for a Desired Number of Days[a]

Temperature (°C)	Confidence Levels (%)		
	95	99	99.9
	Days	Days	Days
⩾50	13.8	13.3	12.6
⩾55	10.6	10.1	9.4
⩾60	7.3	6.8	6.3
⩾65	4.3	3.9	3.4
⩾70	1.2	1.0	0.8

[a]Data were developed from analysis of temperatures taken from 15 piles during the composting of raw sewage sludge and wood chips by the aerated static pile method [80].

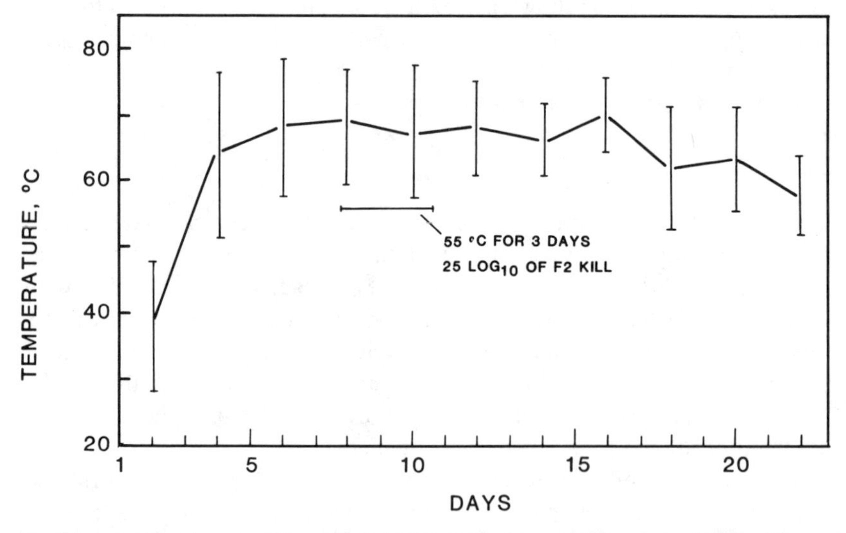

Figure 5-17. Mean temperature of the toe areas of 15 raw sludge compost piles. Vertical lines show values with ± the standard error and the horizontal line shows the time by temperature regime for 25 $\log_{10}$ reduction of f2 bacteriophage [80].

also the temperature obtained by subtracting the standard deviation from the mean.

It should be emphasized that confidence levels shown in Table 5-7 apply only to the 15 test piles examined by Burge et al. [80]. In general, such confidence levels will vary with the composting operation. Confidence levels for a particular operation would be a function of the substrate material (e.g., raw sludge, digested sludge, manure or refuse), the type of bulking agent used and the manner of operation of the piles. Thus, proficient design and operation are vitally important in establishing confidence that all material achieves an adequate time-temperature profile.

Bacterial Regrowth

Another factor which may limit thermal inactivation is that certain enteric bacteria can regrow in organic materials once temperatures are reduced to sublethal levels. This phenomenon has been observed with coliform, *Salmonella* and fecal streptococcus bacteria growing in liquid sludges and even composted material with moisture contents less than 40% [75,84]. The most dramatic regrowth is usually observed in sterilized material which is recontaminated by pathogenic bacteria as shown in Figures 5-18 and 5-19. In nonsterilized material such regrowth is restricted, if not eliminated, apparently by competition with the natural microbial flora. In practical applications regrowth should not pose overwhelming difficulties because natural flora are always present. Care must be exercised, however, to assure that all material is exposed to the time-temperature conditions, thus avoiding unnecessary reinoculation of the final product.

Selna and Smith [86] observed occasional regrowth of bacteria during windrow composting of digested sludge. Regrowth of coliform organisms was the most consistently observed, particularly if windrows were moist, as during winter months. Only sporadic regrowth of *Salmonella* was observed. In a supplementary series of experiments, samples of final compost were incubated at 30°C for 30 days under controlled laboratory conditions. In interpreting the results, regrowth was defined as a concentration increase of one or more logs to a final concentration greater than 10 most probable number (MPN)/g dry weight. Using this definition as a measure, regrowth of total coliform was observed in 25% of samples; fecal coliform in 8% of samples; and *Salmonella* in 8% of samples. The low moisture content of the final compost, estimated at 30-35%, along with the natural bacterial population may have inhibited bacterial regrowth.

With the exception of the bacteria mentioned above, enteric virus, protozoa and helminth pathogens are obligate parasites or require an intermediate host not present in sludge. Human viruses, for example, can reproduce only under suitable conditions in eucaryotic cells. Human viruses which survive one stage of sludge treatment do not regrow in the next. Thus, concern over regrowth is limited to a few bacterial species and is not of concern with other classes of pathogens.

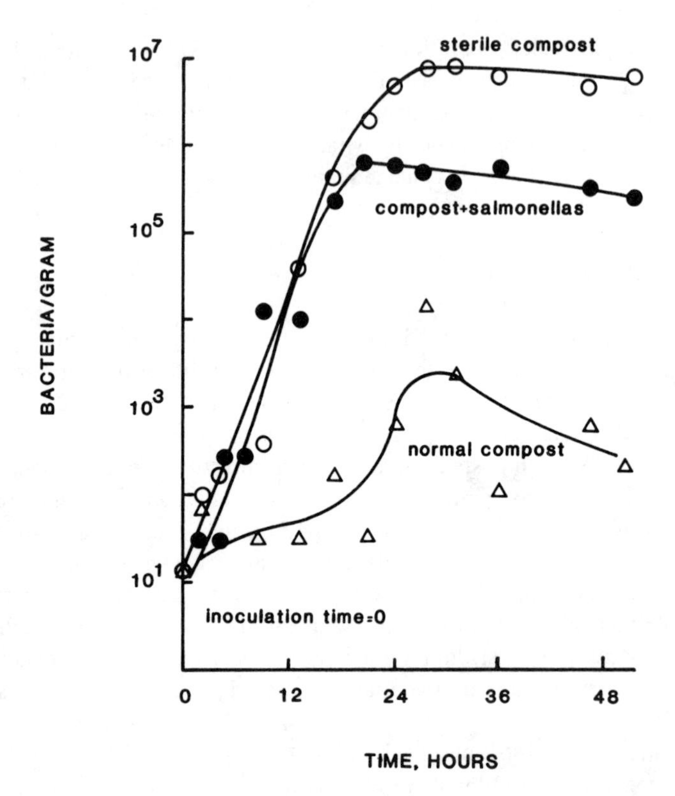

Figure 5-18. Growth of fecal streptococcus bacteria in normal composted sludge (Δ), in sterilized composted sludge (○) and in composted sludge saturated with *Salmonella* spp. (•). In the last case, following sterilization of the compost, *Salmonella* spp. were allowed to grow to approximately 10^8 bacteria/g [75].

FINAL PRODUCT STANDARDS

Product Quality in Practice

Thus far, time-temperature profiles, mathematical models of heat inactivation kinetics, proposed standards of performance using indicator organisms and potential limitations on microbial destruction have been discussed. Given all of the above, what microbial quality can be achieved in the final product in actual practice? To answer this, one must examine final product quality data developed from actual sludge composting installations.

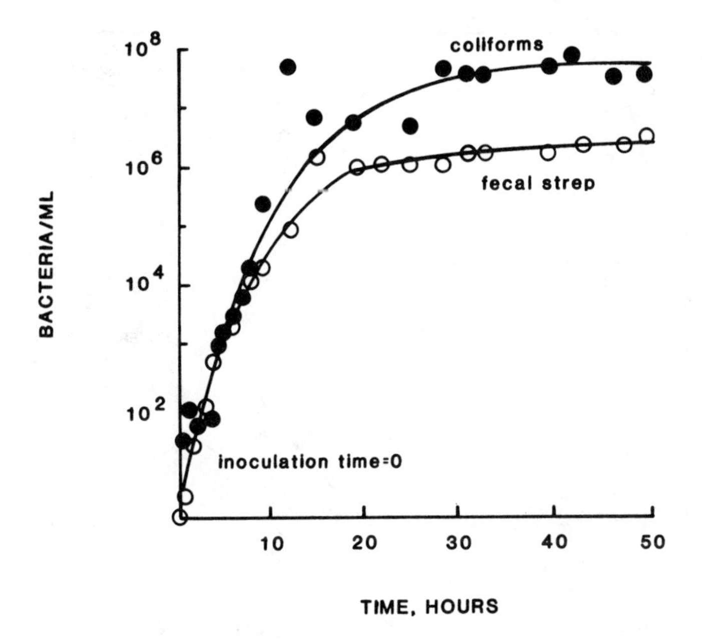

Figure 5-19. Growth of coliform (•) and fecal streptococcus (○) bacteria in sterile liquid digested sludge (35°C) [85].

To some extent the aerated static piles operated at Beltsville, MD, have already been discussed. Confidence levels shown in Table 5-7 should assure essentially complete destruction of enteric pathogens including virus, protozoa, helminths and obligate bacterial pathogens of enteric origin. The question of indicator bacteria, such as total and fecal coliform, and nonobligate bacterial pathogens such as *Salmonella* cannot be answered from the confidence levels in Table 5-7 because of the potential for regrowth.

Epstein et al. [13] reported fecal coliform reductions from initial concentrations of 10^6-10^7 cell/g solids to 0.03-30 cell/g during aerated static pile composting of both raw and digested sludges. A mixture of the two sludge types gave somewhat higher levels in the final product, although the reason for this was not clear because similar temperature profiles were observed. If detected at all, *Salmonella* was usually present at levels less than 1 cell/g solids.

Ettlich and Lewis [15] analyzed pathogen data from an aerated static pile system operated at Bangor, ME. Sufficient data were available to plot cumulative probability curves for sludge and final compost. Results of their analysis for total and fecal coliform are shown in Figures 5-20 and

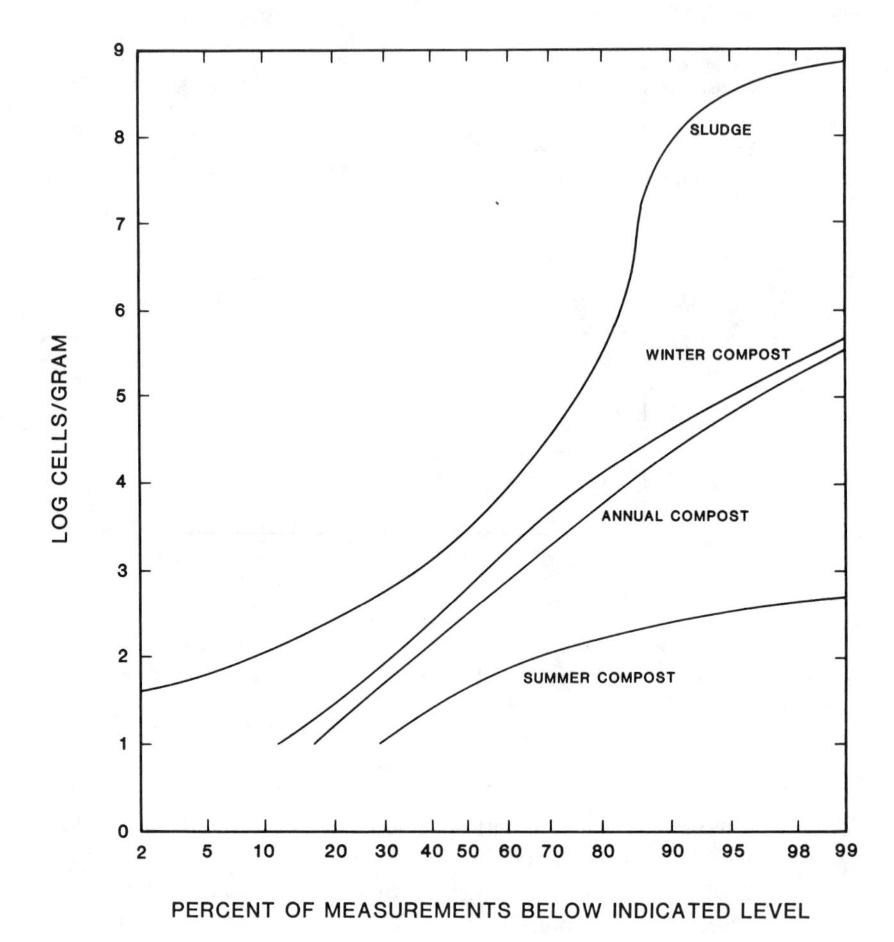

Figure 5-20. Statistical analysis of total coliform data from aerated static pile composting of raw sludge at Bangor, ME [15].

5-21. Levels of total and fecal coliform in winter compost were significantly higher than in summer compost. *Salmonella* were generally below detection limits and when detected were generally less than 10 cell/g. About 10% of winter compost samples were greater than 10 cell/g and about 2% reached levels of 10^3-10^4 cell/g solids. The occasional high levels of *Salmonella* may have been caused by operational difficulties encountered during start-up of the facility. The authors reported no detection of *Salmonella* in compost samples during the last year of the testing period.

The Los Angeles County Sanitation Districts have conducted extensive pathogen testing during windrow composting of digested sludge [86].

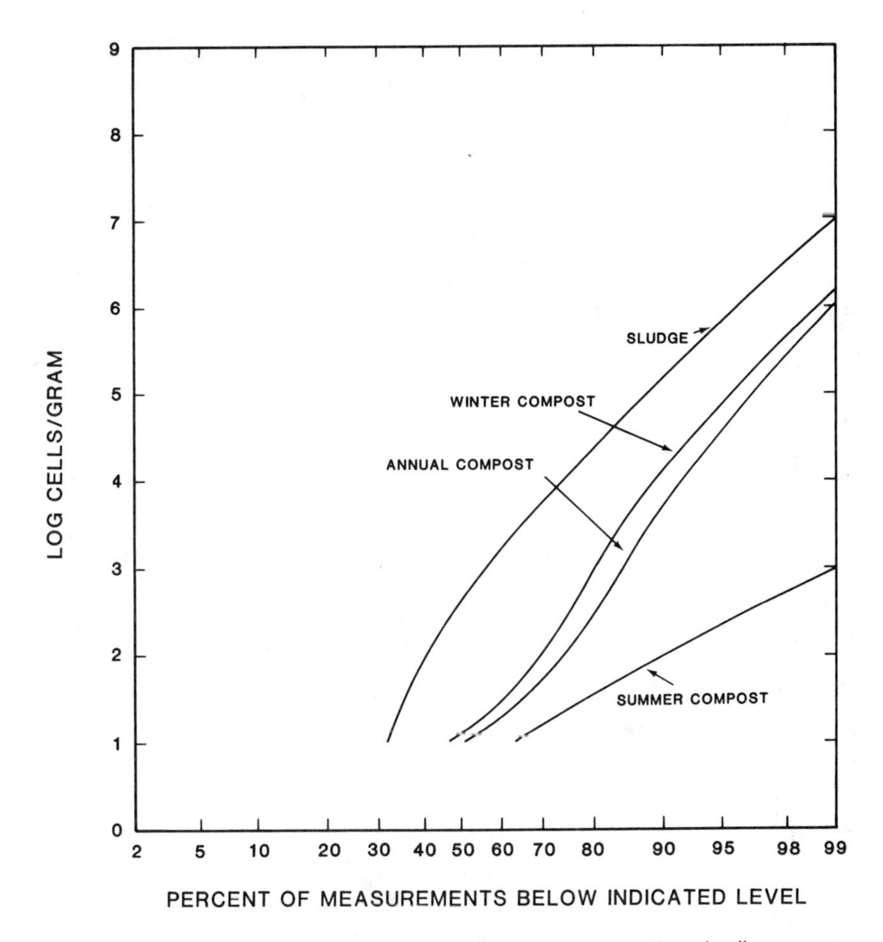

Figure 5-21. Statistical analysis of fecal coliform data from aerated static pile composting of raw sludge at Bangor, ME [15].

A one-year study was conducted to determine pathogen destruction during various weather conditions and modes of turning. A total of 19 windrows were extensively monitored during this period. Statistical analyses of their data are presented in Figures 5-22 and 5-23 for total coliform and *Salmonella.* Total coliform levels in digested sludge feed generally ranged from 10^7 to 10^9 cell/g solids. A statistical difference was observed between interior and exterior samples of the composting material. It was common to observe levels below 1 MPN/g solids in interior samples, but concentrations of 10^3-10^4 were occasionally observed in exterior samples. Further analysis of their data indicated that almost all of the higher concentrations were

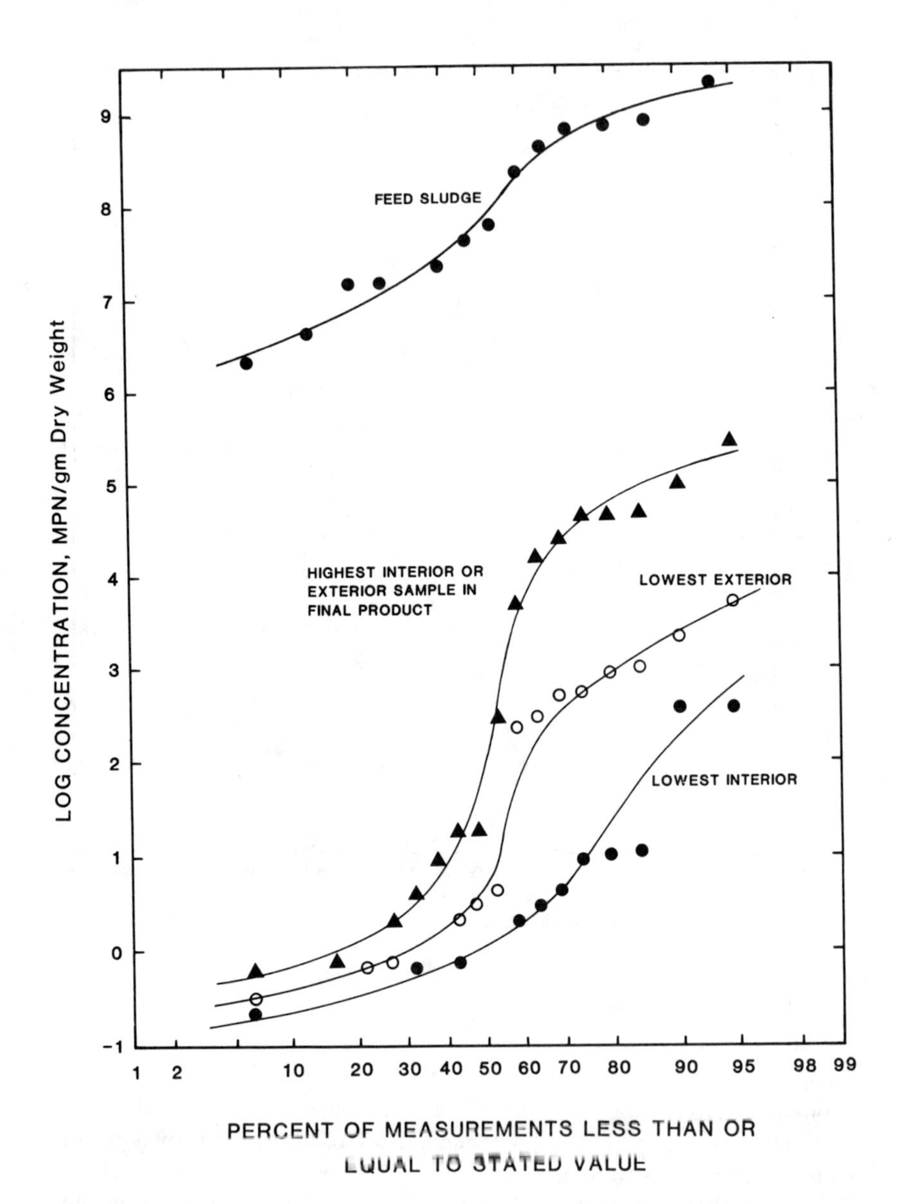

Figure 5-22. Statistical evaluation of total coliform data during a one-year study of windrow composting of digested sludge blended with recycled compost. A total of 19 windrows were examined under various modes of turning. Lowest interior and exterior concentrations recorded during each compost period are shown along with the highest sample recorded in the final product. Raw data from Selna and Smith [86].

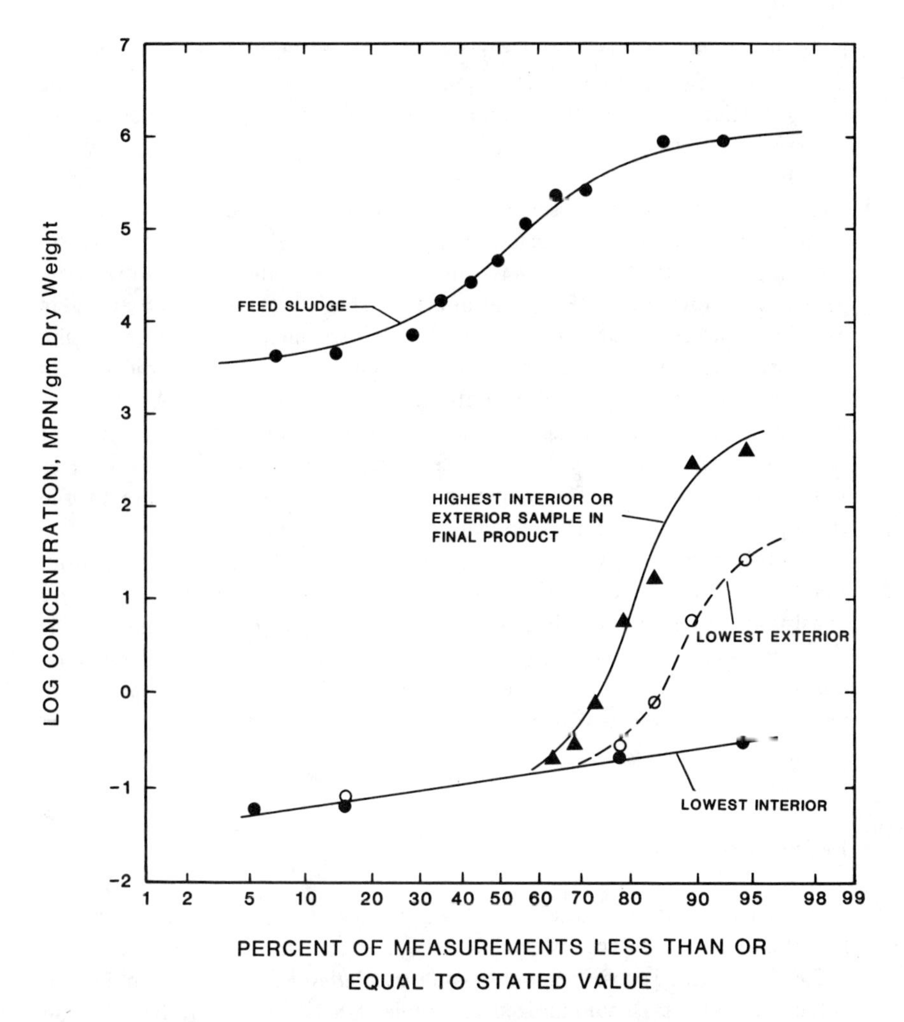

Figure 5-23. Statistical evaluation of *Salmonella* data during a one-year study of windrow composting of digested sludge blended with recycled compost. A total of 19 windrows were examined under various modes of turning. Lowest interior and exterior concentrations recorded during each compost period are shown along with the highest sample recorded in the final product. Raw data from Selna and Smith [86].

associated with winter operations. In this particular facility, the windrows are not covered and frequently became excessively wet during winter. Mechanical difficulties with turning equipment also were noted during the winter period. It is likely that the rising tails on the frequency curves could be significantly lowered by better winter operations. Note also that the difference between interior and exterior samples is much reduced in the lower portions of the curves which correspond to nonwinter conditions. A frequency distribution of the final product quality is also shown in Figure 5-22. Occasional coliform regrowth was responsible for the higher levels, and again this was observed mainly during winter operation.

Salmonella levels in digested sludge feed ranged from about 10^4 to 10^6 MPN/g solids. Lowest levels observed in the compost windrows were generally below 0.3 MPN/g, although higher levels were observed in exterior samples about 20% of the time. These higher levels were, again, associated with the winter period and could probably be avoided with improved system design and operation. Some regrowth was observed but was considerably less frequent than with total coliform. A typical plot of total coliform and *Salmonella* concentrations as a function of time is shown in Figure 5-24 for nonwinter operation. The temperature profile corresponding to this data was shown previously in Figure 2-11.

Human parasitic ova were monitored during the same test period. Viable *A. lumbricoides, Trichuris trichiura* and hookworm ova were consistently present only during the first 7-10 days of composting. Viable ova were isolated in only 3 final compost samples and only 8 of 140 samples collected after more than 10 days of composting. This same general pattern was true for viruses, which were consistently not isolated in final compost samples.

Health Risks

Actual health risk associated with various levels of pathogens in compost is difficult to assess accurately but estimates can be made. Hornick et al. [87,88] published dose response data for *Salmonella typhosa* on human volunteers which is summarized in Table 5-8. In these studies, human volunteers were inoculated with various doses of *S. typhosa* and the frequency of illness was noted. *S. typhosa* tends to produce more severe symptoms than other *Salmonella* and is generally credited with greater virulence [89]. Data reported on other species and strains of *Salmonella* are presented in Table 5-9. It is obvious from these data that large (10^4-10^9) numbers of *Salmonella* must be introduced to the body to achieve a high risk of illness. If we assumed that 1 g of final compost solids was ingested by a human, the risk of illness would be very low even at *Salmonella* concentrations of 10^3 MPN/g.

Mechales et al. [89] conducted extensive and sophisticated statistical analyses of the data of Hornick and others. They concluded that infectivity

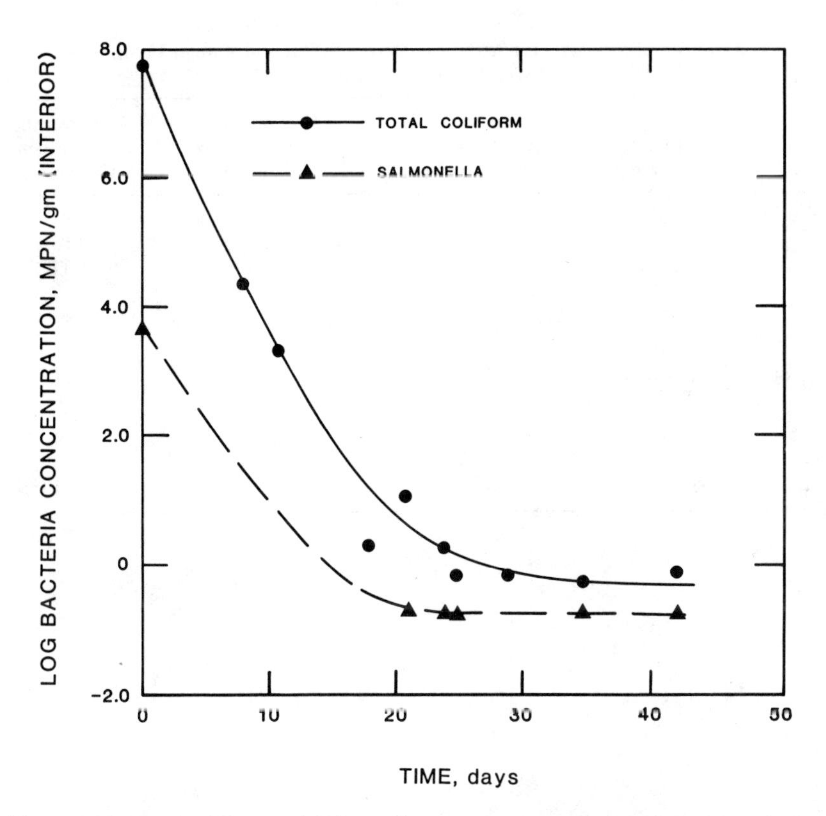

Figure 5-24. Total coliform and *Salmonella* concentrations as a function of time during windrow composting of digested sludge. Data correspond to the temperature profile and other conditions of Figure 2-11 and are representative of nonwinter conditions. Windrows were turned once per day [9].

of *S. typhosa* was similar to other *Salmonella* species and that the entire *Salmonella* group could be handled as a single factor. Although details of their analysis cannot be presented here, they were able to develop estimates of disease risk as a function of *Salmonella* concentration in recreational water. Partial results of their analysis are shown in Figure 5-25. It was assumed that 10 ml of water would be imbibed by a recreationist either by direct swallowing or nasal inhalation and subsequent swallowing. Thus, the concentration can be converted to a total ingested dose. This allows extrapolation of the results to the ingestion of compost. The authors were quick to point out that their analysis was based on limited data (even though the literature search was exhaustive) and that Figure 5-25 must be interpreted accordingly.

Table 5-8. *Salmonella typhosa* Dose-Response Data on Human Volunteers (After Hornick et al. [87,88])

Test	*S. typhosa* Strain	Challenge Dose Concentration	Number of Volunteers Who Became Ill	Total Number of Volunteers Challenged
a	Quailes	10^3	0	14
		10^5	32	116
		10^7	16	32
		10^8	8	9
		10^9	40	42
b	Quailes	10^7	16	30
	Zermatt	10^7	6	11
	TY2V	10^7	2	6
c	Quailes	10^5	28	104
		10^7	15	30
		10^9	4	4

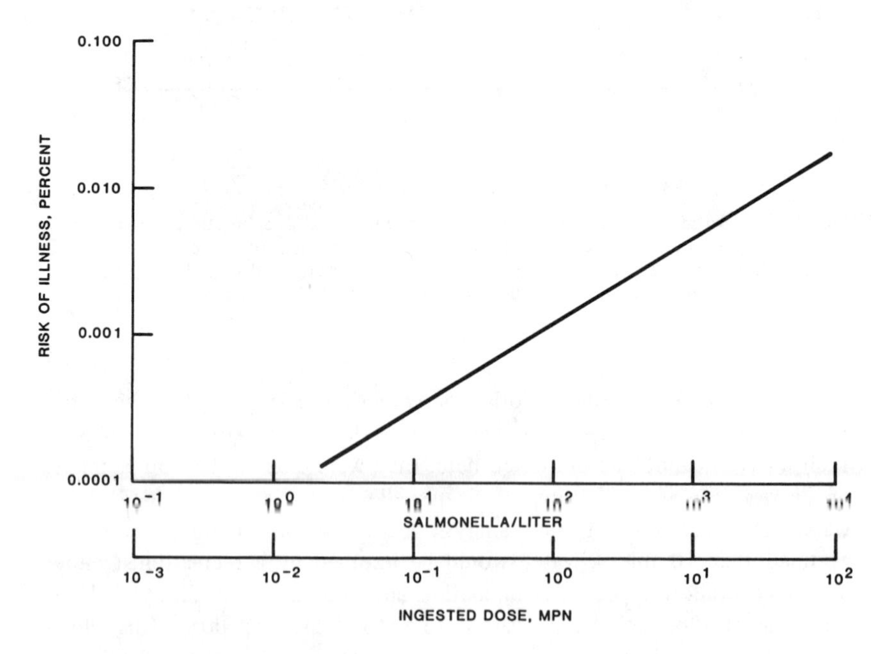

Figure 5-25. A risk-dosage diagram for *Salmonella* developed by Mechalas et al. [89]. Original work was for a water recreationist who was assumed to imbibe 10 ml of water. This assumption was used to convert the concentration axis to a total dosage to allow extrapolation of the results to inhalation or ingestion of compost.

Table 5-9. Dose of Various Species and Strains of *Salmonella* That Caused Disease in Human Volunteers (From McCullough and Eisele [90-93])

Salmonella Species/Strain	Dose at Which 50% or More Respond[a]
S. meleagridis 1	50,000,000
S. meleagridis 11	41,000,000
S. meleagridis 111	10,000,000
S. anatum 1	860,000
S. anatum 11	67,000,000
S. anatum 111	4,700,000
S. newport	1,350,000
S. derby	15,000,000
S. bareilly	1,700,000
S. pullorum 1	1,795,000,000
S. pullorum 11	163,000,000
S. pullorum 111	1,295,000,000
S. pullorum IV	1,280,000,000

[a]Develop clinical disease.

Recognizing the limitations, let us assume that a health risk of 0.001% or less is desired on ingestion or inhalation of 1 g of compost solids. In other words, a risk of ill effect in 1 person out of 100,000 who each ingest 1 g. There is really no statistical basis for the assumption of 1 g ingested, but it does seem to be a conservative estimate. From Figure 5-25 this level of risk corresponds to ingestion of about 0.80 MPN. Because we are considering 1 g of compost, the corresponding concentration would be 0.8 MPN/g solids. At a level of risk of 1 in 10,000 the concentration can be increased to 30 MPN/g. This analysis indicates a relatively low risk of Salmonellosis from ingestion or inhalation at concentrations below 10 and even perhaps 100 MPN/g, which is readily achievable during composting.

Mechalas et al. [89] also investigated risks associated with ingestion of virus. Here the infective dosages seem to be markedly different than for bacterial agents. Considerable scatter is evident in literature data, but there is evidence that as little as 1 plaque forming unit (PFU) can cause infection, as shown in the data of Table 5-10.

Considerable variation in viral concentrations in raw sludges can be expected, depending largely on the incidence rate of viral disease in the population. Various sources indicate a range of about 20-700 PFU/100 ml of raw sewage. If the virus particles are associated with particulate solids, a range of about 600-25,000 PFU/g dry raw sludge solids could be expected. Another rule of thumb suggests about 15 virus particles per 10^6 coliform organisms [89]. Assuming 10^8 coliform/g of sludge, a viral

Table 5-10. Infection of Human Volunteers with Attenuated Poliovirus I (Adapted from Mechalas et al. [89] from Original Data by Koprowski [94])

Dose (PFU)	Number Infected/Number Fed	Per Cent Infected
0.2	0/2	0.0
2.0	2/3	66.7
20.0	4/4	100.0
200.0	4/4	100.0

concentration of 1500 PFU/g can be estimated. This is consistent with the range estimated above. Obviously, the sludge treatment train must provide many logs reduction in virus to achieve an acceptable compost product.

Mechalas et al. [89] analyzed data from viral infectivity studies which listed 812 volunteer exposures. Statistical analysis yielded the dose-risk diagram for virus shown in Figure 5-26. It is obviously difficult to extrapolate dose-response data of the type shown in Table 5-10 to the low levels of risk shown in Figure 5-26. Nevertheless, it does allow estimation of the approximate level of risk associated with a given dosage.

Let us assume a desired level of risk of 1:10,000 on ingestion of 1 g of compost, similar to the risk level previously assumed for *Salmonella*. From Figure 5-26 the risk corresponds to a concentration of about 7 PFU/l. Since it was assumed that 10 ml were imbibed, the viral dosage would be 0.07 PFU. Considering 1 g of compost solids, the corresponding concentration would be 0.07 PFU/g (dry wt). In other words, if 1 g of compost at an average concentration of 0.07 PFU/g were administered to 10,000 individuals, about one case of infection would be expected. Obviously, with an average concentration of only 0.07 PFU/g, most of our 10,000 "volunteers" would not have ingested a single virus in their sample. This means that to achieve the same level of risk between virus and *Salmonella*, the virus concentration must be reduced essentially to zero. Fortunately, human virus cannot reproduce itself extracellularly and re-growth in compost is not a concern. Available data indicate that virus can be reliably reduced to below detection limits, at least with present isolation and identification methods. Reasonable detection limits for virus are currently about <0.1 infective virus/g.

The situation with protozoan and helminth disease agents would be expected to be similar to that for virus. Ingestion of one viable ovum or cyst is likely to produce a significant risk of infection. Therefore, viable helminth ova and protozoan cysts should be below detection limits in the final compost. Reasonable detection limits for viable *Ascaris* ova are currently about <0.5 viable ova/g. Again, this level appears to be achievable in practice.

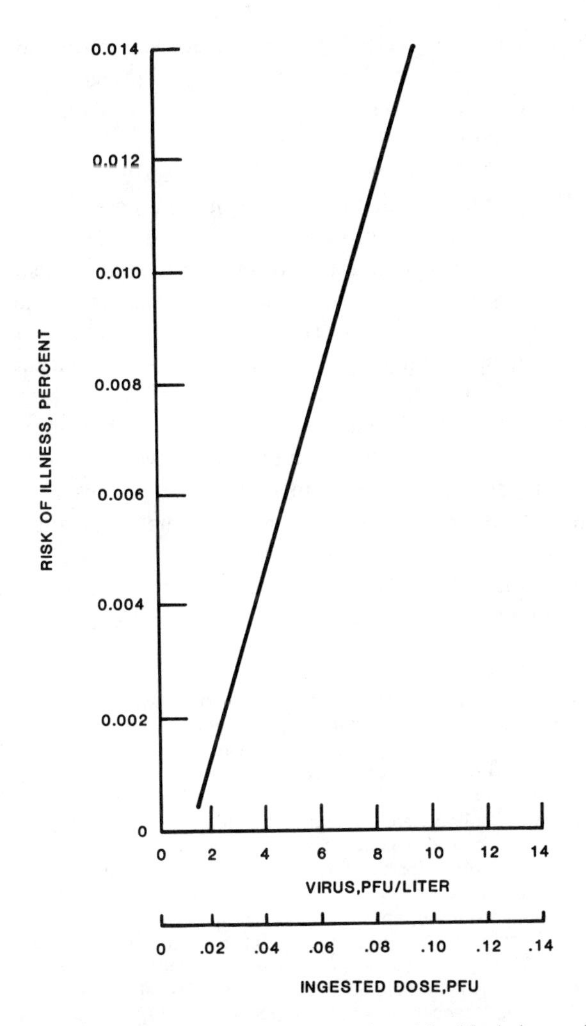

Figure 5-26. A risk-dosage diagram for virus developed by Mechalas et al. [89] for the case of a water recreationist who was assumed to imbibe 10 ml of water. This assumption was used to convert the concentration axis to a total dosage to allow extrapolation of the results to inhalation or ingestion of compost. The data indicate a relatively high level of risk from ingestion of small numbers of virus.

Suggested Product Standards

Establishment of microbial standards on the final compost is an approach which regulatory agencies can use to assure an acceptably low risk to users of the material. However, setting numerical standards is at best a difficult task. The level of risk associated with proposed use of the material must be considered along with the type of pathogens likely to be present, levels of reduction which can actually be achieved in practice and the statistical nature of final product quality. In the previous analysis the relative risk of infection from direct ingestion of compost was considered. The number of people who would purposely ingest compost would hopefully be small. Ingestion would more likely result from inhalation of dusts, accidental hand to mouth contact or adherence of compost to crops which are to be eaten raw. Thus, in a practical sense the levels of risk are probably less than calculated above because the probability of a person actually ingesting compost must also be considered.

Establishment of standards must also recognize the fact that there is no such thing as "risk-free" actions. Recovery of any resource, such as energy or compost from waste products, will always involve some risk. Unfortunately, much recent environmental legislation has not recognized this fact and efforts toward resource recovery have often been delayed because of it. For example, residents near proposed composting sites often want guarantees that there will be no odors, dusts, noise or risk of any pathogens. Of course, it is impossible to guarantee the negative.

Speaking of the concept of "risk-free," Akin et al. [95] wrote:

> The concept of "acceptable risk" rather than "risk free" must be embraced as the only realistic approach for grappling with environmental health questions. It is a nebulous concept but a required one, nonetheless. However, it would be absurd to attempt to quantify the number of cases of any serious disease within a population that would be acceptable. The term "acceptable risk" has practical definition only in an economic and political sense. When a situation becomes of such little health concern that its investigation and control cannot demand sufficient priority for funding (either from the absence of documented evidence of a hazard or lack of public interest and pressure), then that risk by definition has become acceptable. To a health scientist the acceptable level is achieved only when exhaustive study with the most sensitive tools at his command has failed to demonstrate disease transmission. This will usually be determined through the application of the skills of epidemiology. When disease transmission through a single source e.g., land application of waste, occurs below the background disease transmission from all other sources, then the limits of epidemiological discernment have been reached and the health scientist is obligated to consider the risk to be at an acceptable level. Unfortunately perhaps, the economic and political forces will normally have accepted the level of risk before exhaustive epidemiological studies could be performed.

Comar [210] suggested the following guidelines in dealing with risks: (1) eliminate any risk that carries no benefit or is easily avoided; (2) eliminate any large risk (about 1 in 10,000 per year or greater) that does not carry clearly overriding benefits; (3) ignore for the time being any small risk (about 1 in 100,000 per year or less) that does not fall into category 1; and (4) actively study risks falling between these limits, with the view that the risk of taking any proposed action should be weighed against the risk of not taking that action. Comar was quick to point out that establishing such risk levels is an oversimplification of a very complex technical, social and political problem. However, the approach should promote understanding about how to deal with risk in the real world, focus attention on actions that can effectively improve or safeguard health, and avoid squandering resources attempting to reduce small risks while leaving large ones unattended.

With the above comments in mind, the following standards are proposed as being achievable with good management practices, economically reasonable, and representing an acceptable low level of risk to the user (all in the author's opinion of course).

1. *Virus:* No infective viruses detected by an acceptable laboratory method with a minimum detection limit of 0.1 infectious unit (IU)/g solids.
2. *Ascaria Ova:* No viable *Ascaris* ova detected by an acceptable laboratory method with a minimum detection limit of 0.5 viable ova/g solids. *Ascaris* will be considered representative of all parasites, i.e., helminth ova and protozoan cysts.
3. *Salmonella:* Median of all samples to be less than 1 MPN/g solids. Not more than 10% of samples to exceed 10 MPN/g solids. No sample >100 MPN/g solids.
4. *Total Coliform:* Although the coliform group is not generally considered pathogenic, their destruction is indicative of good composting practice. Median of all samples to be <10 MPN/g solids. Not >20% of samples to exceed 1000 MPN/g. No sample to exceed 10^4 MPN/g.

Along with these standards, statistical assurance must be provided that all material is exposed to compost temperatures. Samples must also be collected from all regions of the windrow, pile or reactor to assure that material does not escape the composting.

SUMMARY

Heat inactivation of microbes is a function of both temperature and length of exposure. A high temperature for a short period of time or a lower temperature for longer duration can be equally effective. Heat inactivation kinetics are often modeled assuming first-order decay. The inactivation coefficient k_d is a function of temperature, which can be described by the Arrhenius relationship. The inactivation energy E_a for many spores and vegetative cells is between 50 and 100 kcal/mol. This means that heat inactivation is strongly influenced by temperature, and a change in temperature of a few degrees Celsius can cause significant changes in the rate of heat inactivation. If the value of k_d is known as a function of temperature for a particular microbe, thermal inactivation resulting from a particular time-temperature profile can be estimated.

Based on known time-temperature relationships, heat inactivation of enteric pathogens should be readily accomplished with the conditions common to composting. Temperatures of 55-60°C for a day or two should be sufficient to kill essentially all pathogenic virus, bacteria, protozoa (including cysts) and helminth ova to acceptably low levels. Endospores produced by spore-forming bacteria would not be inactivated under these conditions. However, enteric pathogens are nonspore-forming.

A number of factors can reduce the pathogen inactivation calculated by the time-temperature conditions in the composting process. These include: (1) clumping or balling of solids, which can isolate material from the temperature effects; (2) nonuniform temperature distribution, which can allow pathogens to survive in colder regions; (3) short-circuiting of the feed substrate; and (4) bacterial regrowth which has been observed with coliforms, *Salmonella* and fecal streptococcus. When compost is not agitated, such as in the aerated static pile system, it is important that uniform airflow and temperature be achieved throughout the pile and that excessive clumping of solids be avoided. When compost is agitated, such as in the windrow and many reactor systems, it is important that turning frequency be sufficient to assure a minimum probability of a microbe escaping the high temperature zone. It is also important that short-circuiting be reduced to the point that all solids are exposed to high temperatures for a minimum period of time.

Analysis of available data on final product quality indicates that high product standards can be achieved with good management practices. Viruses and *Ascaris* ova should be reduced to below detection limits. *Salmonella* and total coliforms should normally be reduced to levels below 1 and 10 MPN/g solids, respectively. Attainment of these standards should assure a very low risk of disease infection to users of the material.

CHAPTER 6

MOISTURE AND VOLATILE SOLIDS CONTROL

INTRODUCTION

Decomposition of organic matter depends on the presence of moisture to support microbial activity. As Golueke [44] pointed out, the theoretical ideal moisture content would be one that approaches 100%, because under such conditions, biological decomposition would occur in the absence of any moisture limitation. Because of technical and economic reasons, practical moisture contents must be less than 100%. If the compost is to be placed in windrows, static piles or a reactor system, the question arises as to the maximum moisture content to begin the process. General ranges of moisture contents found suitable for various wastes are listed in Table 6-1. The values shown are related to the structural strength of the composting material. Fibrous or bulky material such as straw or wood chips can absorb relatively large quantities of water and still maintain their structural integrity and porosity. For example, McGauhey and Gotaas [96] were able to compost mixtures of vegetable trimmings and straw that had initial moisture contents as great as 85%, but 76% moisture was too great when paper was used instead of straw.

Many composting materials, such as municipal refuse and many agricultural residues, begin the composting process in a relatively dry form. Even animal manures are often field-dried before composting. Most composting systems have been developed to process such dry materials and allowance is usually made for moisture addition as composting proceeds. Furthermore, the fibrous and bulky nature of such materials allows absorption of relatively large quantities of water.

Sludges and other wet organic wastes such as manure differ on both accounts. Municipal sludge seldom exceeds 30% solids, and sludge is not a fibrous material capable of supporting such high moisture contents. If dewatered cake were composted alone, nearly constant mechanical agitation

Table 6-1. Maximum Recommended Moisture Contents for Various Composting Materials (After Golueke [44])

Type of Waste	Moisture Content (% of total weight)
Theoretical	100
Straw[a]	75-85
Wood (Sawdust, Small Chips)	75-90
Rice Hulls[a]	75-85
Municipal Refuse	55-65
Manures	55-65
Digested or Raw Sludge	55-60
"Wet" Wastes (Lawn Clippings, Garbage, etc.)	50-55

[a]Serves as a moisture absorbent and source of carbonaceous material. Requires addition of nitrogenous material to lower C/N ratio to a proper level.

would be required to provide aeration because of the lack of porosity or free airspace (FAS) in which to store oxygen.

Maintenance of proper moisture levels becomes a matter of balancing numerous competing forces. Moisture levels must be high enough to assure adequate rates of biological stabilization, yet not so high that void spaces are eliminated, thus reducing the rate of oxygen transfer and in turn the rate of biological activity. In addition to this, there is the desire to produce a reasonably dry final product, one that can be stockpiled and transported economically for subsequent reuse. It may not be possible to optimize all of these factors at the same time. Tradeoffs will usually be necessary. Relationships between these competing factors will be explored further during discussion of process dynamics in Chapters 11-14. Discussion here will focus on methods available to control moisture content in the starting mixture, and the closely related subject of volatile solids (VS) control.

MOISTURE CONTROL

The importance of proper moisture control was highlighted in work conducted by Senn [21] on the composting of dairy manure. Composting was conducted in 2.4-m-deep bins equipped with a forced aeration system mounted in the bin floor (see Figure 2-20). The influence of moisture content on subsequent temperature development is shown in Figure 6-1. At 66% moisture the temperature rose to about 55°C but no higher. The bin was unloaded, dry material was added and the mix was reloaded at 61% moisture. Temperature rose rapidly to >75°C. In a parallel test on material

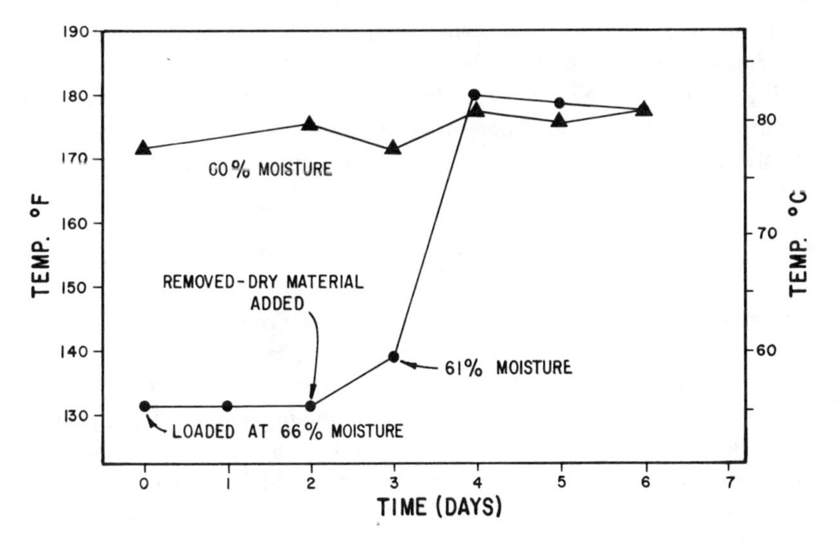

Figure 6-1. Effect of initial moisture content on temperatures developed during composting of dairy manure in deep bins [21].

loaded at 60% moisture, the temperature quickly rose above 75°C and remained for several days. Elevated temperatures were reached in six hours, accounting for the high temperature indicated at time zero. Apparently, excessive moisture can result in packing and reduced void space, which prevent proper air movement throughout the composting material. The 60% moisture content which Senn found to be adequate for dairy manure in deep bins may not be applicable to sewage sludge. Sludge is less fibrous than manure and may require a drier starting mixture when piled in similar depths. Indeed, data presented in Chapter 2 suggest a maximum moisture content of about 50% if sludge is composting in a similar bin system.

A number of approaches are available to overcome the problem of high moisture content in the feed substrate. First, dry, previously composted material can be recycled to adjust moisture content of the starting mixture. Second, dry amendments, such as sawdust or ground refuse, can be added to the feed sludge either with or without compost product recycle. Third, bulking agents, such as wood chips, can be added to the sludge to maintain structural integrity and porosity of the pile. Fourth, constant agitation could be provided by mechanical equipment. Shell and Boyd [97] successfully used this approach on a bench-scale, but even then found that compost recycle was advantageous. Finally, dewatered cake could be air- or heat-dried to decrease the moisture content before composting.

At this point the discussion will center on use of recycled compost and amendments and on using supplemental drying before composting. Use of bulking agents will be described in detail in Chapter 7. In general, the quantity of bulking agent is determined as much by the need for structural support and porosity as by requirements for moisture control.

Moisture Control With Compost Recycle

A mass balance diagram for a generalized sludge composting system is presented in Figure 6-2. The diagram is a generalized model for performing mass balances and is applicable to windrow, aerated static pile and reactor systems. Inputs of sludge cake, amendment, bulking agent and compost are shown. This is not meant to imply that recycled compost, bulking agent and amendment would all be used at the same time. Whether one or all are used will depend on the particular system being analyzed.

Nomenclature in Figure 6-2 and subsequent discussion is as follows:

X_c = total wet weight of dewatered sludge cake produced per day
X_p = total wet weight of compost product produced per day
X_r = total wet weight of compost product recycled per day
X_a = total wet weight of organic amendment, other than sludge cake or compost recycle, added to mixture per day

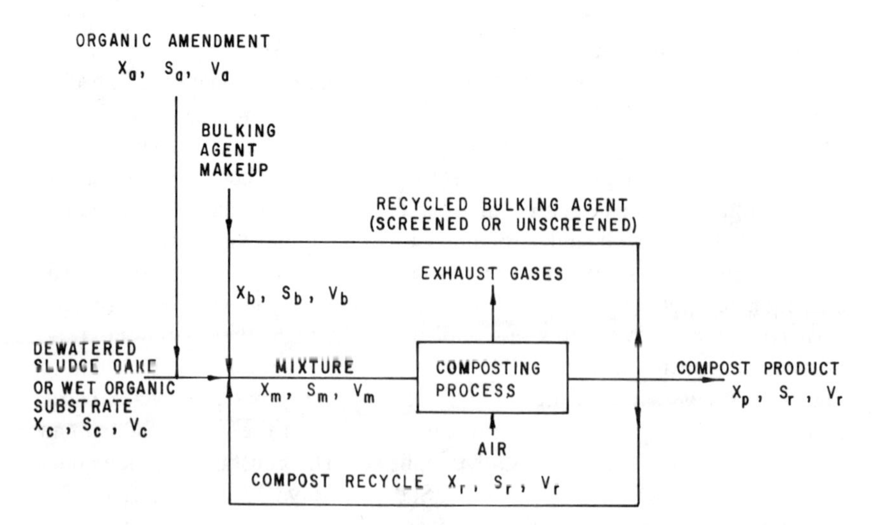

Figure 6-2. Generalized mass balance diagram for sludge composting showing inputs of dewatered sludge cake, compost product recycle, organic amendment and bulking agent.

X_b = total wet weight of bulking agent added to mixture per day
X_m = total wet weight of mixed material entering the compost process per day
S_c = fractional solids content of dewatered sludge cake
S_r = fractional solids content of compost product and compost recycle
S_a = fractional solids content of amendment
S_b = fractional solids content of bulking agent
S_m = fractional solids content of mixture before composting
V_c = volatile solids content of sludge cake, fraction of dry solids
V_r = volatile solids content of compost product and recycle, fraction of dry solids
V_a = volatile solids content of amendment, fraction of dry solids
V_b = volatile solids content of bulking agent, fraction of dry solids
V_m = volatile solids content of mixture, fraction of dry solids

The quantity of compost to be recycled in a process can be determined from the mass balance shown in Figure 6-2. Total dry solids produced per day is S_cX_c. Similarly, dry solids in recycled compost product is S_rX_r. Assuming no amendment or bulking agent addition, a mass balance on total wet solids yields:

$$X_c + X_r = X_m \tag{6-1}$$

Similarly, a mass balance on dry solids gives:

$$S_cX_c + S_rX_r = S_mX_m \tag{6-2}$$

Substituting Equation 6-1 into 6-2:

$$S_cX_c + S_rX_r = S_m(X_c + X_r) \tag{6-3}$$

Let R_w be defined as the recycle ratio, based on total wet weight of compost product recycled to total wet weight of dewatered sludge cake. Then:

$$R_w = X_r/X_c \tag{6-4}$$

Substituting into Equation 6-3 and rearranging yields:

$$R_w = \frac{S_m - S_c}{S_r - S_m} \tag{6-5}$$

Let R_d be defined as the recycle ratio, based on dry weight of compost product recycled to dry weight of the dewatered sludge cake. Then;

$$R_d = \frac{S_rX_r}{S_cX_c} \tag{6-6}$$

Substituting the expression for R_d into Equation 6-3 and rearranging yields:

$$R_d = \frac{\left(\frac{S_m}{S_c} - 1\right)}{\left(1 - \frac{S_m}{S_r}\right)} \tag{6-7}$$

Equations 6-5 or 6-7 can be used to calculate the required recycle ratio, either on a dry- or total-weight basis, as a function of the sludge cake solids S_c, compost recycle solids S_r and desired solids content in the mixture S_m.

Example 6-1

Ten dry ton/day of digested dewatered sludge is to be windrow-composted using recycled compost to dry the initial composting mixture. Desired mixture solids is 40% and recycled compost is 70% solids. Calculate the required recycle ratio, both wet- and dry-basis, if the dewatered cake is 30% solids. What is the total weight of material to be processed daily?

Solution

1. The dry weight recycle ratio R_d is determined from Equation 6-7:

$$R_d = \left(\frac{0.40}{0.30} - 1\right) \Big/ \left(1 - \frac{0.40}{0.70}\right) = 0.777$$

2. From Equation 6-5, the weight recycle ratio R_w is:

$$R_w = \frac{0.40 - 0.30}{0.70 - 0.40} = 0.333$$

3. The total weight to be processed can be determined from either R_d or R_w as follows:

$$\text{total weight} = \text{sludge weight} + \text{recycle weight}$$

$$\text{Based on } R_d\text{: total weight} = \frac{10}{0.30} + \frac{10(0.777)}{0.70} = 44.4 \text{ wet ton/day}$$

$$\text{Based on } R_w\text{: total weight} = \frac{10}{0.30} + \frac{10(0.333)}{0.30} = 44.4 \text{ wet ton/day}$$

Equations 6-5 and 6-7 were solved for various dewatered cake and compost product solids contents, assuming a desired mixture solids of 40% (S_m =

0.4). Resultant total and dry weight recycle ratios are presented in Figures 6-3 and 6-4, respectively.

In Figure 6-3, R_w has a value of zero at a cake solids of 40%, reflecting the fact that a 40% solid mixture was assumed to be necessary to begin the process. As dewatered cake solids decrease below 40%, R_w increases, depending on the solids content of the recycled product. Obviously, the drier the recycled product, the lower the required recycle ratio. Equation 6-5 can be solved for the case where $S_c = 0$, in other words, no solids in the dewatered cake (i.e., pure water). Clearly, this is a mathematical anomaly which results from the necessity of assuming a solids content for the recycle product. Both

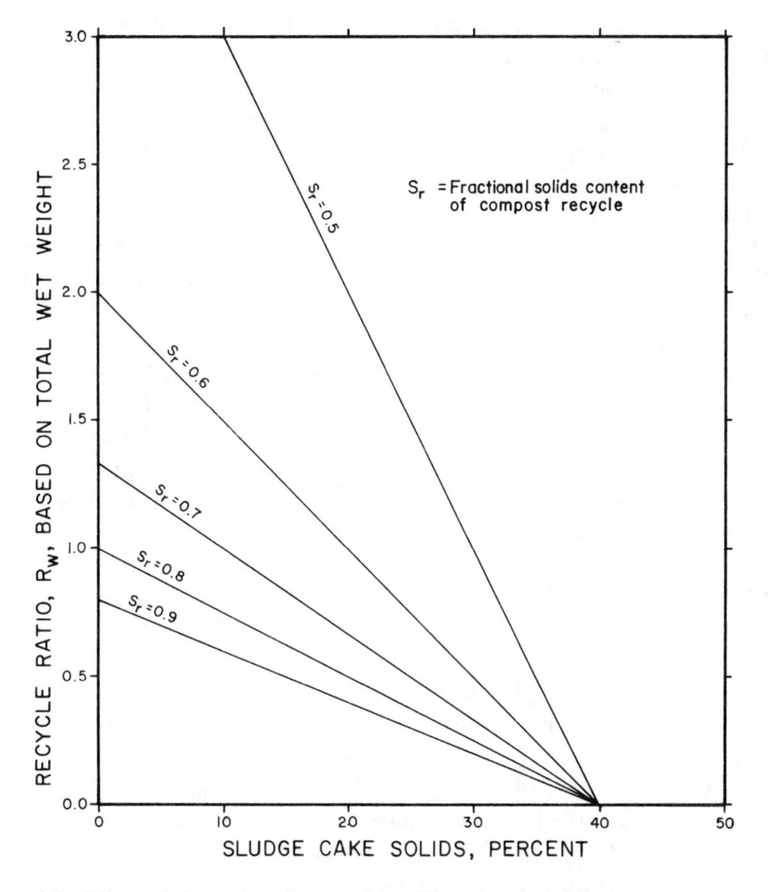

Figure 6-3. Effect of dewatered sludge cake solids content on the wet weight recycle ratio needed to achieve a 40% mixture solids content [98,99].

Equations 6-5 and 6-7 are mathematically valid over the entire range of solids contents from $0 \leqslant S_c \leqslant S_m$. However, the practical range of application is probably limited to dewatered cake solids greater than ~10%. Furthermore, valid application of both equations depends on the ability to achieve the assumed compost product solids, a subject of discussion in later chapters.

The total quantity of mixed material to be handled each day depends largely on the initial dewatered cake solids. Assuming a dewatered cake solids of 30% and a compost product of 70% solids, an R_w of 0.33 is determined from Figure 6-3. Assuming 100 dry ton/day of sludge, X_c would be 333 wet ton/day, with an X_r of 110 wet ton/day, giving a total weight of 443 wet ton/day to be handled. If dewatered cake solids were reduced to 20%, R_w would increase to 0.67, X_c to 500 wet ton/day, X_r to 335 wet ton/day, and the total quantity of mixed material to 835 wet ton/day. Thus, reduction in dewatered cake solids from 30 to 20% would result in nearly doubling the total weight of mixed material to be handled each day.

The recycle ratio based on dry weight R_d is presented in Figure 6-4. As S_c approaches zero, Equation 6-7 tends toward infinity and the boundary problems associated with Equation 6-5 are not encountered. Based on Figure 6-4 the quantity of recycled compost on a dry-weight basis can, in certain cases, exceed the quantity of dewatered cake solids on a dry-weight basis. Assuming a dewatered cake of 30% solids and a compost product of 70% solids, R_d calculated from Equation 6-7 is 0.78. In other words, 0.78 g of compost product (dry basis) would be recycled for each 1.0 g of input dewatered cake (dry basis) to achieve 40% solids in the initial mixed material. If dewatered cake solids are reduced to 20%, R_d increases to 2.34. Thus, in the example presented here, a decrease in cake solids from 30 to 20% would increase the total dry weight in the mixed material from 1.78 to 3.34 g, and would increase the percentage of recycled compost (dry basis) in the mixture from 44 to 70%. Also, as previously mentioned, the total wet weight of mixed material in process each day would nearly double. This is not meant to imply that sludge cake of 20% solids cannot be composted. It merely points out the importance of initial cake solids in determining both the quantity of material to be handled each day and the relative percentage of compost product in the mixture.

Equation 6-5 was again solved, in this case to determine the effect of mixture solids content on the recycle ratio. Dewatered sludge cake was assumed to be constant at 25% solids with recycle compost solids ranging from 50 to 90%. Results are presented in Figure 6-5. Mixture solids content has a significant effect on the required recycle ratio, particularly at lower solids contents in the recycled compost. As recycle compost becomes drier, the effect of mixture solids becomes less pronounced. Assuming recycle compost at 70% solids, R_w increases from 0.5 to 1.25 as S_m increases from 40

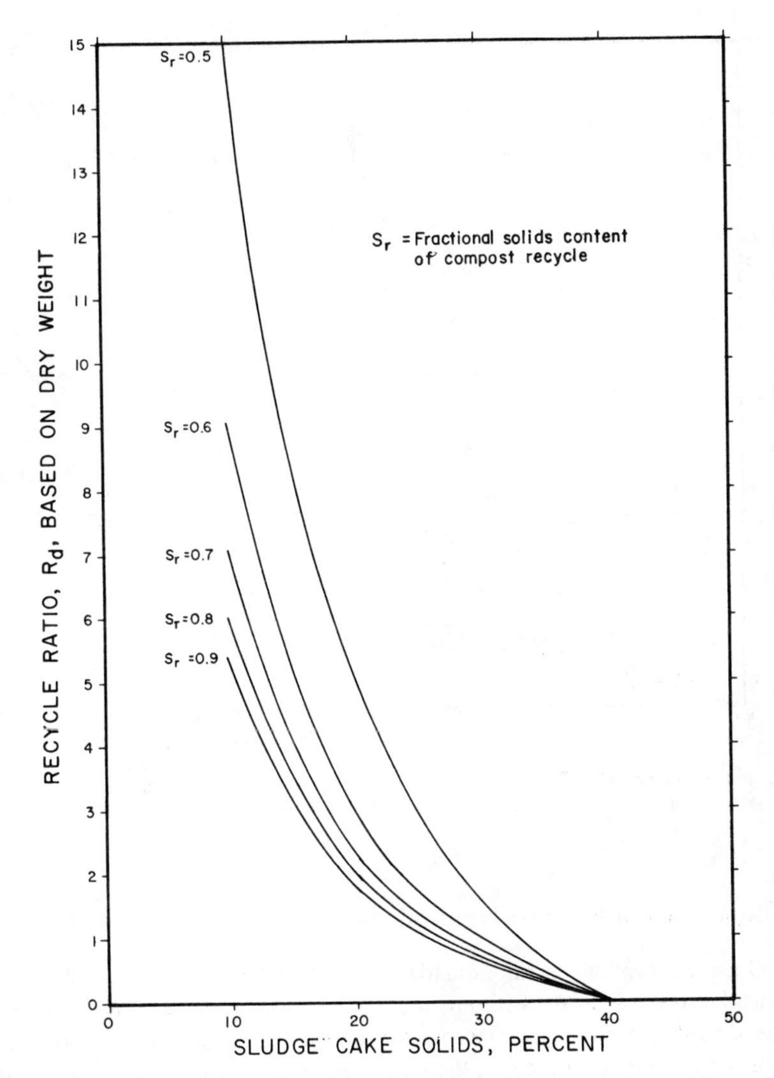

Figure 6-4. Effect of dewatered sludge cake solids content on the dry weight recycle ratio needed to achieve a 40% mixture solids content [98, 99].

to 50% solids. In practice it would be desirable to minimize the mixture solids S_m to reduce the weight and volume of material to be processed daily. Factors which determine the minimum mixture solids include the need for structural integrity and for porosity or FAS to promote aeration. These factors will be discussed in greater detail in Chapter 7.

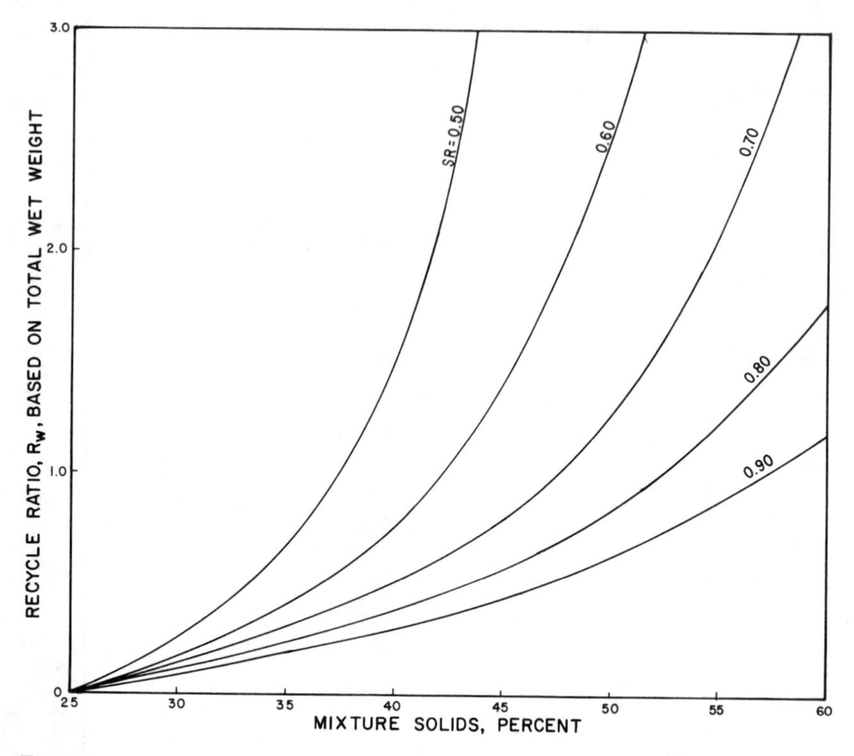

Figure 6-5. Effect of mixture solids content on the wet weight recycle ratio for various recycle product solids contents. Dewatered cake solids S_c assumed to be 25%.

Moisture Control By Amendment Addition

Organic amendments can be added to a composting mixture for moisture control either with or without use of compost recycle. Obviously, dry amendments are more advantageous for moisture control. If an amendment is used with no compost recycle, the mass balance depicted in Figure 6-2 can be solved by substituting X_a for X_r in Equations 6-1 and 6-2. If an amendment is added along with recycled compost, the solids mass balance as given by the latter equations cannot be uniquely solved since there are three unknowns (X_r, X_a and X_m) and only two equations. An approach to solution of problems where both amendment and recycled compost product are used is presented in Example 6-2.

Use of an organic amendment for moisture control without compost recycle can result in consumption of considerable quantities of amendment. This can be expensive, depending on local availability of suitable material.

Furthermore, the quantity of final product will be increased compared with systems which rely all or in part on compost recycle for moisture control. Whether this is an advantage or disadvantage depends on the expected market for the final product.

A concept of using compost recycle for the task of moisture control and amendment addition for VS control is described later in this chapter. The quantity of amendment required for VS control is significantly reduced compared to the case where amendment is used for moisture control without compost recycle.

Example 6-2

Both recycled compost and an organic amendment are added to a sludge cake for moisture control. The organic amendment used is sawdust with a solids content of 70%. Dewatered cake and recycled compost are 25 and 60% solids, respectively. Sludge, compost and amendment are blended in a 1:0.5:0.5 ratio by wet weight. Determine the mixture solids content.

Solution

1. For each unit weight of sludge added to the mixture the corresponding solids and water in the mixture are:

$$\text{solids} = 0.25 + (0.5)(0.60) + (0.5)(0.70) = 0.9$$

$$\text{water} = 0.75 + (0.5)(0.40) + (0.5)(0.30) = 1.1$$

Therefore, total weight of the mixture is 2.0 g/g sludge cake feed. The solids content of the mixture is:

$$S_m = \frac{0.9}{1.1 + 0.9} = 0.45$$

2. If compost recycle is not used, the quantity of amendment to achieve the same mixture solids can be determined from Equation 6-5:

$$R_w = \frac{S_m - S_c}{S_r - S_m}$$

Because no compost recycle is used, S_a can be substituted for S_r:

$$R_w = \frac{X_a}{X_c} = \frac{S_m - S_c}{S_a - S_m} = \frac{0.45 - 0.25}{0.70 - 0.45} = 0.80$$

Therefore, the blend of cake and amendment would be 1:0.8 on a wet-weight basis. Note that this is a considerably greater quantity of amendment than required in part 1 to achieve the same mixture solids.

Moisture Control by Drying

In lieu of using compost recycle or amendments to improve the moisture balance, sludge cake could be dried further to reduce the moisture content before composting. Use of an air-drying technique has been used at a compost facility in southern California to reduce the moisture content of dewatered digested sludge before windrow composting [100]. Sludge cake at about 20% solids is spread to a depth of about 0.3 m. Material is then turned approximately daily to enhance surface drying. Little if any composting occurs as a result of the turning because of very limited aeration. Except in wet-weather months, the cake will dry to about 40-50% solids in about two weeks. At this point it is formed into large windrows and composted. Composting temperatures as high as 70°C have been recorded [100], so the technique is apparently effective. Previously dried compost and organic amendments, such as paunch, are used in lieu of air drying during wet weather.

The Los Angeles County Sanitation Districts experimented with the concept of air drying before windrow composting of digested sludge to remove excess moisture and reduce the need for compost recycle [17]. In these experiments windrows were formed using cake/compost recycle mixtures, while others were composed entirely of sludge cake containing no recycled compost. Sludge cake averaged 25% solids in both cases. The compost cycle was conducted during summer months without any precipitation so that air drying occurred under optimum conditions.

Windrows composed entirely of sludge cake were initially too wet to maintain adequate void space for proper aeration. As a result, little biological activity was noted until about day 20, when temperatures began to increase. By this time surface evaporation had removed sufficient water to give a 40-45% solids material. By about day 25, temperatures in the windrow exceeded 65°C. By contrast, windrows using compost recycle for moisture control exhibited significant temperature elevations by about day 10. Peak temperatures were highest in the windrows without compost recycle, probably the result of the higher VS content in the starting material.

The tradeoffs between using compost recycle or air drying for moisture control are probably as follows. Compost recycle allows formation of a porous windrow at the onset of the composting cycle. As a result, aeration in the early stages is more effective, and elevated temperatures are produced

Figure 6-6. Not only must the weight of various mixture components be considered but also the homogeneity achieved in the mixture. The above photographs show a windrow system using digested sludge blended with recycled compost. Correct proportions are generally achieved on an average basis, but the mixture is not homogeneous as evidenced by the wet and dry zones. Areas where windrows have slumped consist of dewatered cake with very little recycle material. Even though windrows are turned frequently, little mixing occurs along the length of the windrow and wet zones are clearly evidenced in windrows 1–2 weeks old. Mechanical metering and mixing can be used to achieve a more homogeneous mixture which greatly facilitates development of uniform compost temperatures. The above situation has since been improved and a more uniform starting mixture is achieved.

earlier in the cycle. The starting windrow is drier and structurally more stable. Odor control may also be more effective because of improved aeration. A disadvantage is the lowered volatility of the mixture, particularly with low feed cake solids. On the other hand, air drying should maintain a higher initial volatility because no compost recycle is used. As a result, the thermodynamic balance is improved (see Chapter 9), and peak temperatures may be higher. A disadvantage is that air drying is subject to the vagaries of weather. Even if the drying area is covered, low evaporation rates may preclude rapid drying. Also air drying of raw sludge would appear to be impractical because of the large potential for odor production.

To estimate the time required for evaporative drying consider a layer of wet cake of depth d_c and unit surface area. The depth of water which must be evaporated from the unit surface can be expressed as:

$$d_w = d_c p \frac{\gamma_c}{\gamma_w} \tag{6-8}$$

where d_w = depth of water to be evaporated, cm
d_c = depth of sludge cake, cm
p = weight of water which must be evaporated to reach the desired cake solids per weight of original sludge cake, g water/g cake
γ_c = bulk weight of sludge cake, g/cm³
γ_w = bulk weight of water, g/cm³

The time required to achieve the desired cake solids can then be determined as:

$$t_a = \frac{p d_c \gamma_c}{(E - P)\gamma_w} \quad (E > P) \tag{6-9}$$

where E = evaporation rate, cm/day
P = precipitation rate, cm/day
t_a = time required for air drying, days

The area required for drying can then be determined knowing the daily weight of sludge cake.

$$\text{area} = \frac{\text{volume/day}}{\text{depth}} (t_a)$$

$$\text{area} = \frac{X_c}{\gamma_c d_c} \frac{p d_c \gamma_c}{(E - P)\gamma_w} = \frac{X_c p}{(E - P)\gamma_w} \quad (E > P) \tag{6-10}$$

Should precipitation exceed the evaporation rate during any period, no drying will result. Thus, Equations 6-8 to 6-10 apply to periods of the year when evaporation exceeds precipitation or to situations where the drying area is covered to prevent rainfall from reaching the drying cake.

The most readily available information on evaporation rates will usually be pan evaporation data recorded in the vicinity of the proposed facility. However, evaporation rates from sludge may not necessarily be equal to the pan evaporation rate. This would obviously be true if a dry crust were allowed to form, restricting passage of water to the surface. It is assumed that sludge would be turned frequently to reduce the likelihood of such a rate limitation. Even so, evaporation rates per unit of surface area are likely to be less than pan evaporation rates, particularly as the cake moisture content decreases. On the other hand, surface area of cake per unit of ground surface is probably greater than unity because of the irregular surface of the dry cake. In the absence of actual field data and for purposes of estimation, pan evaporation rates can be assumed over the ground surface area occupied by sludge.

Example 6-3

One hundred fifty dry ton/day of digested, dewatered cake is to be air dried from 25 to 40% solids by placing the cake in a 0.3-m layer and turning daily to prevent formation of a hard, dry surface which would impede evaporation. Available weather data indicate a pan evaporation rate of 0.6 cm/day during summer months with essentially no precipitation. Estimate the time and land requirements for air drying under these conditions. Assume cake bulk weight γ_c to be 1.065 g/cm^3 and that for water as 1.00 g/cm^3.

Solution

1. The weight of water to be evaporated p can be determined as:

$$p = \frac{\text{initial water} - \text{final water}}{\text{total initial weight}}$$

$$p = \left[\left(\frac{1}{0.25} - 1\right) - \left(\frac{1}{0.40} - 1\right)\right] \Big/ (1/0.25) = 0.375 \text{ g } H_2O/\text{g cake}$$

2. Determine the time required from Equation 6-9, assuming the evaporation rate from the cake to be equal to the measured pan evaporation rate.

$$t_a = \frac{0.375\ (30.0)\ 1.065}{0.6(1.00)} = 20.0 \text{ days}$$

3. Determine the land required from Equation 6-10.

$$\text{area} = \frac{\left(\frac{150}{0.25}\right)10^6 \ (0.375)}{(0.6)(1.00)} = 3.75 \times 10^8 \text{ cm}^2 = 3.75 \text{ hectares (9.26 acres)}$$

Equations 6-8 to 6-10 represent a rather simplified approach to analysis because both evaporation and precipitation are assumed to be constant over the time of the estimate. However, evaporation and precipitation rates vary throughout the year. As a result, both the time required for drying and the land area will vary dynamically throughout the year. Simulation models have been developed to predict the drying time and area requirements in response to changing weather conditions [101]. Development of such models is beyond the present scope and is left to the interested reader.

Many parts of the world have relatively long periods of dry weather followed by predictable periods of wet weather. Many compost systems experience their most difficult operating problems during wet weather. This is particularly true of sludge composting by the open windrow system. Advantage can be taken of the cyclic weather pattern in such a case. Part or all of the sludge can be air dried during dry months and then stockpiled. Air drying would continue until the sludge reaches 70-90% solids. This is sufficient to reduce rates of biological activity to very low levels and serves two purposes: (1) the potential for odor production in the stockpile will be reduced; and (2) organics will be conserved for use during wet weather when the thermodynamic balance is most stressed. During wet months, air-dried sludge from the stockpile would be blended with wet cake to achieve a suitable starting mixture, e.g., 40-50% solids. If no organic decomposition has occurred in the stockpile, this will be equivalent thermodynamically to producing sludge with 40-50% solids from dewatering (see Chapter 9). In other words, composting of dewatered cake amended with air-dried sludge will have a thermodynamic advantage, and significant drying should be possible even with moderate water additions from rainfall (see Chapter 14). Note that air-dried sludge that has been stockpiled should not be reused until it has been composted to control pathogens or until tests show that desiccation and detention time have provided sufficient pathogen destruction.

Another approach to increasing the solids content of dewatered cake would be to heat dry either all or a portion of the cake before composting. Consider a dewatered cake at 25% solids. If the cake is heat dried to 40% solids before composting, about 2.5 g of water must be removed per g of cake solids. Further drying of the cake to 70% solids would require removal of only 1.1 g additional water. It would appear that essentially complete drying may be obtained with only modest additional investment over that required to dry to 40% solids. Thus, heat drying the wet cake for moisture control in composting may not be a practical alternative.

A more practical application of heat drying might be to dry the compost product and/or recycled compost. This would aid in assuring a proper moisture content in the final product and would not be subject to adverse weather conditions. If compost product were dried from 50 to 70% solids, only about 0.57 g water need be removed per g of compost solids, significantly less than that calculated for drying of wet cake. Moisture removal that can be achieved during composting is determined by the thermodynamic balance of the system, which is discussed in Chapter 9. For the present purposes, it can be stated the compost process can remove significant quantities of water using the heat of biological decomposition. Heat drying can then be used to "polish" the composted product to the desired moisture content. This is an attractive concept because energy requirements for heat drying can be significantly reduced without placing the total drying burden on the compost process or on the uncertainties of natural open-air drying. The reader should refer back to the mass balance of Figure 2-19 for a practical application of this approach.

VS CONTROL

The aerobic composting process results in the generation of carbon dioxide, water and heat. The loss of weight from conversion of VS to gases and the evaporation of moisture substantially reduces the weight and volume of the composted product. Understanding the relationship of VS to moisture content in the composting process is essential.

Control of VS in the composting process is intimately related to control of moisture content. Heat is released during composting by organic decomposition. It is this heat which causes the temperature elevations observed during composting and the associated pathogen destruction. If excessive moisture is present, temperature elevation will be less for a given quantity of heat released. On the other hand, low moisture can decrease the rate of microbial activity and thus reduce the rate of heat evolution. Moisture contents below 45-50% can become rate-limiting, and bacterial metabolism generally ceases below 10-15% moisture [44].

The quantity of VS in a sludge cake/compost recycle mixture can be determined from a mass balance on VS. Referring to Figure 6-2, VS in dewatered sludge cake can be expressed as $V_cS_cX_c$. Performing a mass balance on cake and compost components of the mixture:

$$V_cS_cX_c + V_rS_rX_r = V_mS_mX_m$$

Substituting Equations 6-1 and 6-5, rearranging and solving for V_m:

$$V_m = \frac{V_c S_c + V_r S_r R_w}{S_m(1 + R_w)} \tag{6-11}$$

A similar equation can be developed based on the dry weight ratio R_d:

$$V_m = \frac{V_c + V_r R_d}{1 + R_d} \tag{6-12}$$

Digested Sludge

The effect of cake solids on mixture volatility was calculated by Equation 6-12. Volatility of cake and compost product solids was assumed to be 0.5 and 0.35, respectively, corresponding to approximate values observed in composting of digested sludge. (The term "volatility" as used here refers to the VS fraction of the total solids.) It was further assumed that sufficient compost product would be recycled to adjust the mixture solids content to 40%. Results and other assumptions used in the analysis are presented in Figure 6-7. Under the specific conditions of this example, mixture volatility is determined by dewatered cake solids and to a lesser extent by solids content of recycled compost product. Obviously, as cake solids decrease the amount of recycled compost must increase to achieve the same solids content in the mixture. Mixture volatility will also decrease. This leads to concern over the actual quantity of degradable organics in the mixture, a subject of later discussion.

One very effective method of increasing mixture volatility is to increase dewatered cake solids, as is evident by study of Figure 6-7. The most straightforward method of accomplishing this is to employ dewatering methods capable of producing drier cakes. However, this is not always practical. Another approach might be to air or heat dry dewatered cake before composting.

Raw Sludge

Another method of increasing mixture volatility is to compost raw rather than digested sludge solids. Again using Equation 6-12, the effect of cake solids on mixture volatility was calculated assuming a raw sludge volatile solids content of 0.70. Compost product was again assigned a volatility of 0.35, the rationale being that the quantity of nondegradable

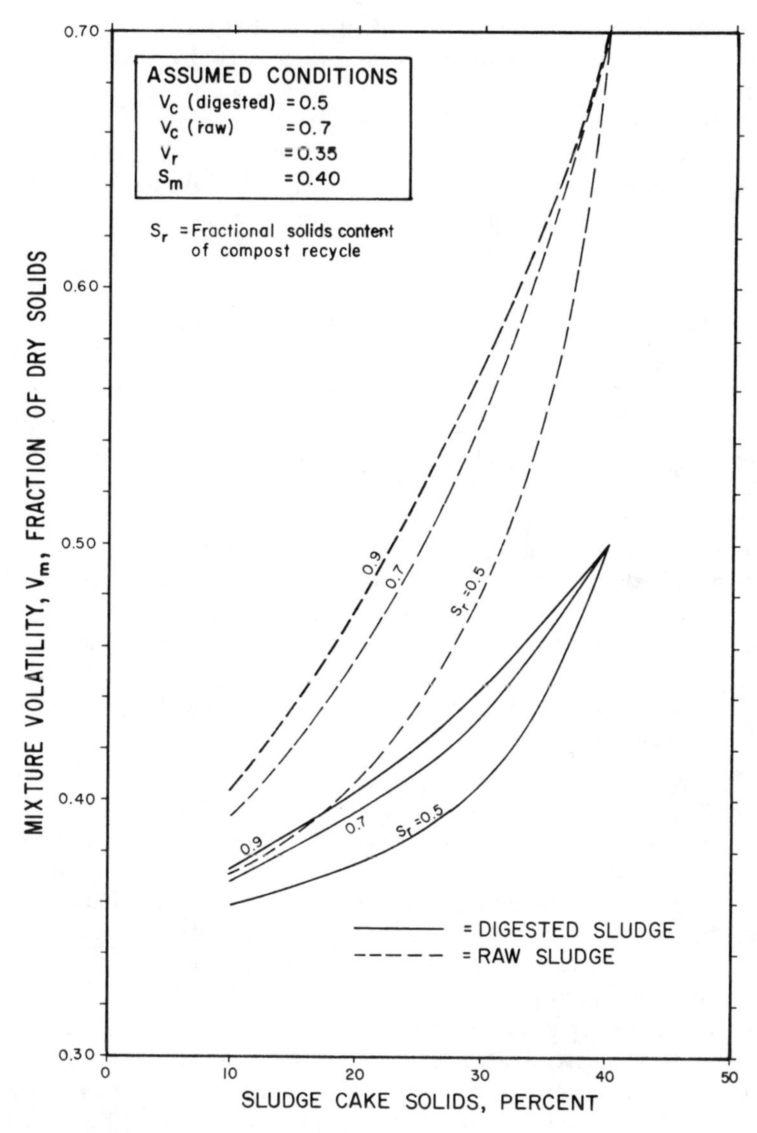

Figure 6-7. Effect of sludge cake solids content on volatility of mixtures of sludge cake and recycled compost [98, 99].

organics, and thus the final volatility, would be about the same whether the starting material was raw or digested. Available but limited data on composted raw sludge seem to justify this assumption [11]. Results of the analysis are presented in Figure 6-7.

As was noted for digested sludge, mixture volatility is largely determined by dewatered cake solids content and to a lesser extent by recycled compost solids content. For an S_r of 0.7 and a cake solids of 20%, mixture volatility would be about 0.455 for raw sludge compared to 0.395 for digested. It is interesting to note that the effect of compost solids content was quite pronounced as S_r increased from 0.5 to 0.7. Beyond an S_r of 0.7, however, the effect on mixture volatility is greatly diminished.

Addition Of Organic Amendments

Another approach to controlling mixture volatility is to add a supplemental source of degradable organics, termed an amendment, to the mixture. This concept is certainly not new. The practice of adding sludge as an amendment to composting refuse is well established. In the majority of such cases, however, the sludge served as a source of moisture and nitrogen and, on a dry-weight basis, represented a small fraction of the refuse. To an agency charged with the responsibility for sludge disposal, such a situation would probably be unattractive. Large quantities of refuse or other material would have to be amended with the sludge, which would greatly increase the quantity of final product to be marketed or disposed.

In the aerated static pile system, use of a degradable bulking agent such as wood chips amounts to the addition of amendment even though it is not the primary function of the bulking agent. Volatilities of new and recycled wood chips have been reported [11] to exceed 98 and 80%, respectively. When chips are blended with raw sludge, mixture volatilities have exceeded 75% [11]. Given these relatively high volatilities, the need for additional amendment beyond the bulking agent would appear to be unlikely. Therefore, the discussion on volatility control will again focus on windrow and reactor systems where bulking agents are not usually added.

Considering windrow and reactor systems, a compromise between sludge-only composting and the addition of large quantities of amendment is possible. If compost is recycled for moisture control, relatively small quantities of amendment can be added for volatility control. The major function of the amendment would be to control mixture volatility, with the bulk of moisture control provided by dry compost recycle.

The quantity of amendment needed to adjust mixture volatility can be determined from the mass balance diagram shown in Figure 6-2. For a given situation, solids content and volatile fraction of cake, compost product and amendment would be known or assumed along with daily mass of sludge cake to be composted (X_c). Desired mixture solids content and volatility would also be assumed. This leaves the daily mass of recycled compost (X_r), amendment (X_a) and mixture (X_m) as unknowns. Since there are three unknowns, three equations are required for solution. These were developed by performing mass balances on total wet weight, dry solids and VS, in accordance with the mass balance diagram presented in Figure 6-2. Based on the mass balance equations, the following solutions can be obtained.

$$X_a = \frac{X_c\left[S_r\left(\frac{S_c}{S_m} - 1\right)(V_r - V_m) + S_c\left(\frac{S_r}{S_m} - 1\right)(V_m - V_c)\right]}{S_r\left(1 - \frac{S_a}{S_m}\right)(V_r - V_m) - S_a\left(\frac{S_r}{S_m} - 1\right)(V_m - V_a)} \quad (6\text{-}13)$$

$$X_r = \frac{S_cX_c(V_m - V_c) + S_aX_a(V_m - V_a)}{S_r(V_r - V_m)} \quad (6\text{-}14)$$

$$X_m = X_a + X_r + X_c \quad (6\text{-}15)$$

The procedure for any problem would be to assume all input variables and consecutively solve Equations 6-13, 6-14 and 6-15 for X_a, X_r and X_m, respectively.

Example 6-4

One hundred dry ton/day of digested dewatered sludge is to be windrow-composted. Cake solids and volatility are 25 and 50%, respectively. It is desired to maintain a mixture solids of 45% with a VS content of 55%. Compost is expected to be 60% solids, with 40% volatile content. Sawdust is available for use as an organic amendment and is estimated to be 85% solids and 80% volatile. Determine the quantities of recycled compost and amendment to achieve the desired mixture characteristics. What would be the mixture volatility if only recycled compost were used?

Solution

1. The quantity of amendment can be determined from Equation 6-13:

$$X_a = \frac{\frac{100}{0.25}\left[0.60\left(\frac{0.25}{0.40} - 1\right)(0.40 - 0.55) + 0.25\left(\frac{0.60}{0.45} - 1\right)(0.55 - 0.50)\right]}{0.60\left(1 - \frac{0.85}{0.45}\right)(0.40 - 0.55) - 0.85\left(\frac{0.60}{0.45} - 1\right)(0.55 - 0.80)}$$

$X_a = 117$ wet ton/day

2. The required quantity of recycle is then given by Equation 6-14:

$$X_r = \frac{100(0.55 - 0.50) + 0.85(117)(0.55 - 0.80)}{0.60(0.40 - 0.55)}$$

$$X_r = 221 \text{ wet ton/day}$$

3. The total quantity of mixed material to be handled daily is:

$$X_m = \frac{100}{0.25} + 117 + 221 = 738 \text{ wet ton/day}$$

4. If no amendment is added, the required recycle ratio R_w to achieve a mixture solids of 45% can be determined from Equation 6-5:

$$R_w = \frac{0.45 - 0.25}{0.60 - 0.45} = 1.33$$

The mixture volatility is then given by Equation 6-11:

$$V_m = \frac{(0.50)(0.25) + (0.40)(0.60)(1.33)}{0.45(1 + 1.33)} = 0.424$$

5. For the recycle determined in part 4, the total mixed material would be:

$$X_m = \frac{100}{0.25} + \frac{100}{0.25}(1.33) = 932 \text{ wet ton/day}$$

6. Note that under conditions of this problem total weight to be processed daily is actually less with amendment addition than without and the volatility is significantly greater.

The most suitable amendment for composting with sludge is one that is both dry and highly degradable. Dry material aids moisture control while degradability improves the energy balance for composting.

For purposes of illustration, suppose an amendment is available with a solids content of 90% ($S_a = 0.9$) and volatility of 90% ($V_a = 0.9$). Further assume that volatility of cake and compost product are 0.5 and 0.35, respectively, and that it is desired to maintain a mixture solids of 0.4 and volatility of 0.5. Equations 6-13 to 6-15 were solved consecutively under these conditions for an input sludge quantity of 100 dry ton/day and various cake solids. Results based on total wet weight are presented in Figure 6-8 and on the basis of dry weight in Figure 6-9.

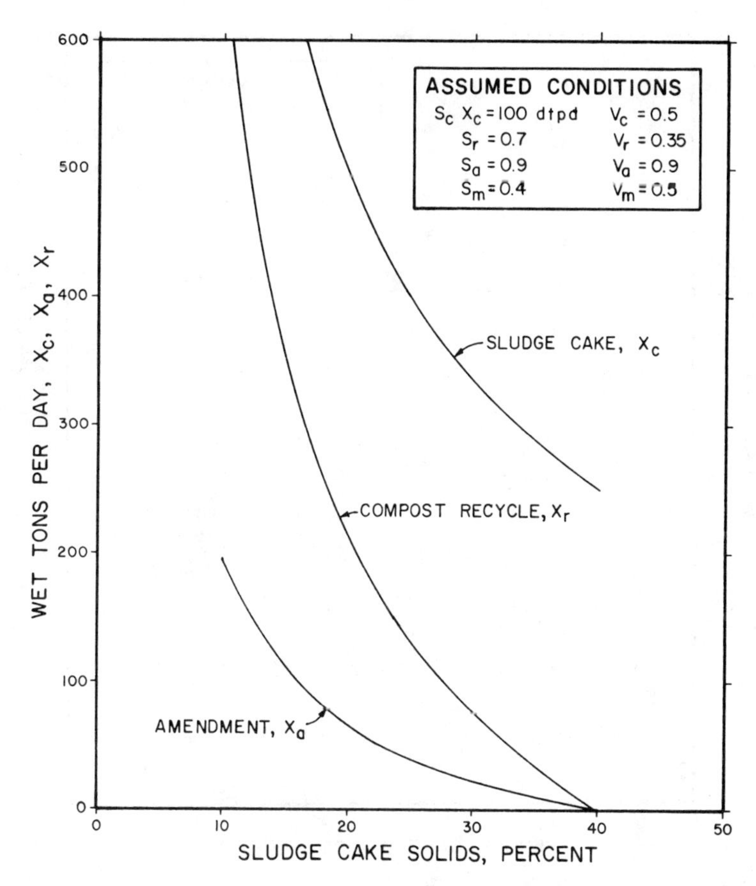

Figure 6-8. Total weight of amendment and recycled compost to achieve a mixture solids content of 40% and volatility of 50% as a function of sludge cake solids. Based on 100 dry ton/day of dewatered digested sludge [98].

On a total-weight basis, required quantities of amendment and compost product are considerably less than the total weight of sludge cake processed daily. At 20% cake solids, total quantity of amendment, recycled compost, and cake would be 65, 214 and 500 wet ton/day, respectively, giving a total weight of 779 wet ton/day. Achieving the same mixture solids without amendment would require about 335 wet ton/day of recycled product for the same 500 wet ton/day of cake (from Figure 6-3). Volatility of the latter mixture would be only 39.5% (from Figure 6-7). The total weight with amendment added would be 779 wet ton/day compared to 835 wet ton/day without amendment. These results are consistent with those of Example 6-4.

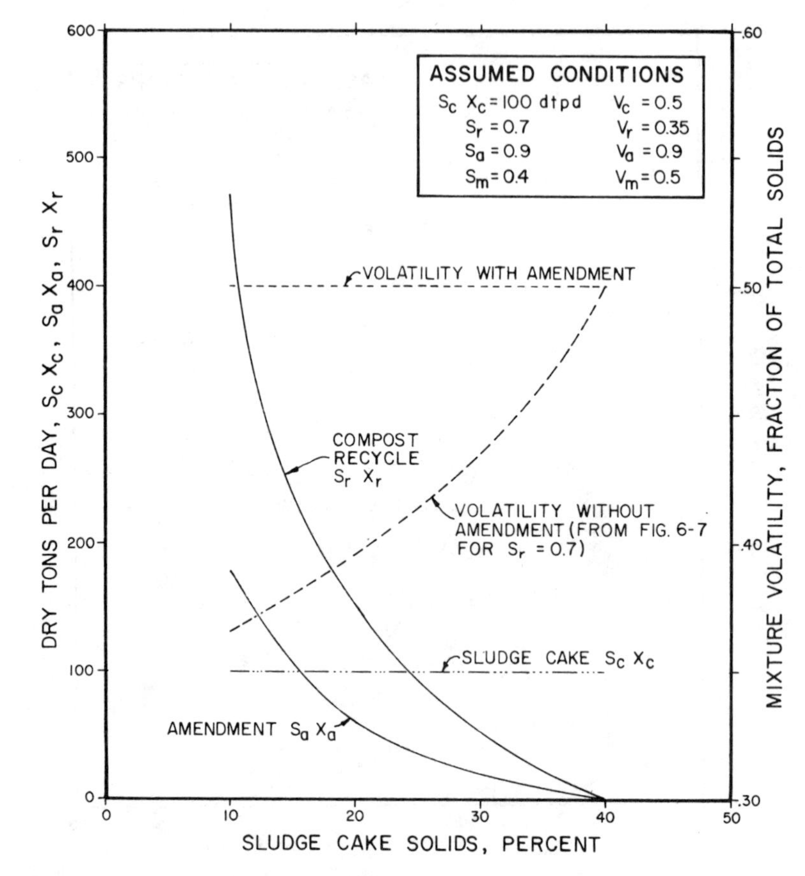

Figure 6-9. Dry weight of amendment and recycled compost to achieve a mixture solids content of 40% and volatility of 50% as a function of sludge cake solids. Based on 100 dry ton/day of dewatered digested sludge [98].

Corresponding results based on dry weight are shown in Figure 6-9. At about 15% cake solids, the dry weight of the amendment would equal that of the sludge cake, while the dry weight of recycled compost would be about 2.6 times that of the sludge cake. At increasing sludge solids the quantity of amendment and recycled compost decrease rapidly.

If no compost were recycled for moisture control, the quantity of amendment required would increase dramatically. Assuming cake and amendment solids of 20 and 90%, respectively, about 200 wet ton/day of amendment would be needed for each 500 wet ton/day of sludge to achieve a mixture solids of 40%. This is significantly greater than the 65-wet ton/day of amendment calculated above when compost was recycled. Therefore, reliance on compost product recycle for moisture control and amendment addition for volatility control as needed would appear to be an attractive way of reducing the material to be handled and the quantity of amendment required.

Equations 6-13 and 6-14 were again solved to determine the amendment requirements as a function of mixture volatility at a fixed cake solids of 20% ($S_c = 0.2$). Results are presented in Figure 6-10. It was assumed that sufficient

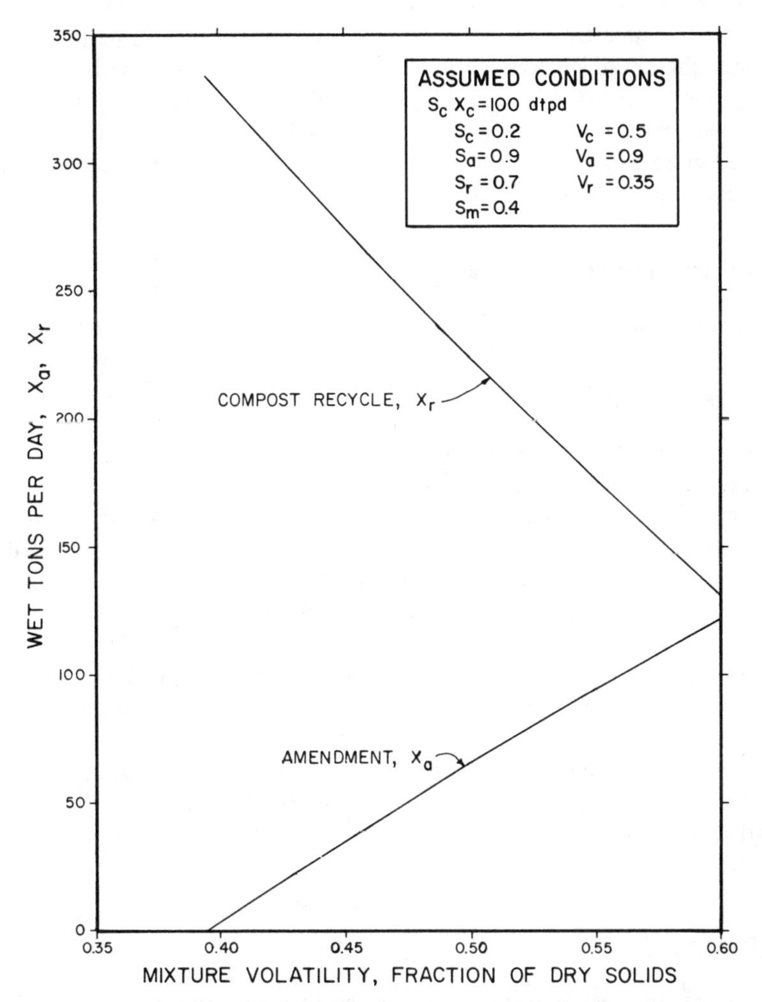

Figure 6-10. Total weight of amendment and recycled compost as a function of mixture volatility assuming a dewatered cake solids of 20% [98].

compost would be recycled to adjust the mixture solids content S_m to 0.40. All other assumptions were similar to those previously described. Required amendment is nearly linear over the range of volatilities examined. Requirements for recycled product decrease as mixture volatility increases. However, even at 60% mixture volatility, the amendment requirement is less than that of recycled compost.

The reader is cautioned to view the above results carefully. The amendment was assumed to be dry with a low inert content. In a practical application, available amendment may not be as optimal as that chosen for this example. In such a case, Equations 6-13 to 6-15 can be used to estimate material requirements under conditions appropriate to the situation. Nevertheless, the example above illustrates that volatility control can be exercised by proper addition of a suitable amendment. Furthermore, by relying on compost product for the task of moisture control, the quantity of amendment can be minimized which, in turn, reduces the quantity of final product remaining for disposal or reuse.

Predicting Final Product Volatility

Application of Equations 6-11 to 6-14 requires knowledge of the final product volatility that would also equal the volatility of recycled product V_r. In the previous examples, values for V_r were assumed based on available field and experimental data. However, it should be recognized that final volatility of any compost will depend on the feed materials used and their respective degradabilities. If the volatility and degradability of each feed component is known, the final product volatility can be calculated. An algorithm for predicting product volatility based on feed characteristics is presented in Chapter 11 as part of the development of simulation models. If compost recycle is used in the initial mixture, an iterative procedure is required for solution.

Information on degradability of various feed materials under composting conditions is limited. Furthermore, degradability would be expected to vary depending on the extent to which optimum conditions for biological decomposition are maintained. Assuming a final volatility based on available information and experience is a reasonable approach and will yield answers sufficiently accurate for many purposes, including those of the present chapter.

SUMMARY

Composting wet organic substrates, particularly sewage sludges, differs from composting many other materials because the feed substrate is still mostly water. Maintaining a proper moisture balance is a critical factor in design of sludge composting facilities. The ability to aerate the compost and the thermodynamic balance of the system are influenced by proper moisture control.

Because of the high moisture content, sludge cake generally lacks structural strength to maintain a properly shaped windrow or pile and the friability or porosity necessary to assure aerobic conditions in the composting material. Five techniques have been described to deal with the problem of high moisture content:

1. recycle of dry, previously composted material;
2. addition of dry amendments either with or without compost product recycle;
3. use of bulking agents such as wood chips;
4. drying the wet cake before composting to reduce the moisture content; and
5. use of nearly constant agitation to achieve aeration with or without the above techniques.

Recycle of compost product has been used in the windrow and many reactor compost systems. In some cases it is the only conditioning material added to the feed substrate. Bulking agents are usually used in the aerated static pile process. Constant agitation has been successful only in bench-scale applications and is not considered a feasible alternative for full-scale practice.

Recycle of compost can be used to adjust the mixture moisture content to a suitable range, increase mixture friability and, depending on its degradability, aid in achieving a proper energy balance. Compost recycle should be adjusted to maintain an initial mixture of at least 40% solids (60% moisture). The latter value has been found to be generally suitable for the windrow system, but higher mixture solids contents may be necessary in other systems. As sludge solids decrease, the quantity of recycled compost to achieve a 40% solids mixture is increased. Increased recycle of material will, in turn, reduce the mixture volatility and increase the total quantity of mixed material. Reduction in dewatered cake solids from 30 to 20% results in nearly doubling the total weight of mixed material to be handled each day.

Addition of a degradable amendment of high VS content is effective in increasing mixture volatility. If compost is recycled for moisture control, the quantity of amendment required for volatility control is significantly reduced compared to the case where amendment is used for moisture control without compost recycle. In some cases, total wet weight to be processed daily can be less using compost recycle with amendment addition compared to use of compost recycle alone. Another method of increasing mixture volatility is to compost raw rather than digested sludge.

The solids content produced during dewatering is probably the single most important variable in determining the successful composting of sludge.

Moisture and volatile solids control, and the energy budget for the system are largely influenced by this parameter. Implementation of any composting system should be coordinated with design of the sludge dewatering process so as to obtain cake solids with sufficient dryness to reduce the problem of moisture and volatile solids control. This conclusion is valid regardless of the type of composting system, whether windrow, aerated static pile or reactor.

CHAPTER 7

CONCEPTS OF POROSITY AND FREE AIRSPACE

INTRODUCTION

The purpose of this chapter is to explore the relationships between bulk weight, porosity and free airspace (FAS). These factors are important in maintaining aerobic conditions within a composting matrix. Porosity and FAS are related to the moisture content of the feed substrate. Approaches to controlling moisture were discussed in Chapter 6. When combined with the contents of Chapter 7, the reader should have a good understanding of the physical factors essential to aerobic composting.

The composting matrix can be considered to be a network of solid particles that contains voids and interstices of varying size. Voids between particles are filled with air, water or a mixture of air and water. When voids are completely filled with water the sludge has little bearing strength and behaves as a plastic. Oxygen transfer is greatly restricted and aerobic composting becomes impractical in the absence of constant agitation. As the voids become filled with air, oxygen transfer becomes possible and aerobic composting can be initiated.

A schematic representation of the composting matrix considered as a three-phase system of solids, water and gas is shown in Figure 7-1. The figure must be understood as a schematic representation because it is evident that all void and solid volumes cannot be segregated as shown. However, the sketch will greatly facilitate understanding relationships between terms given in this chapter. As shown in Figure 7-1, total volume v_t consists of two essential parts, the volume of solid matter v_s and the volume of voids v_v. The volume of voids is further distinguished into water volume v_w and gas volume v_g.

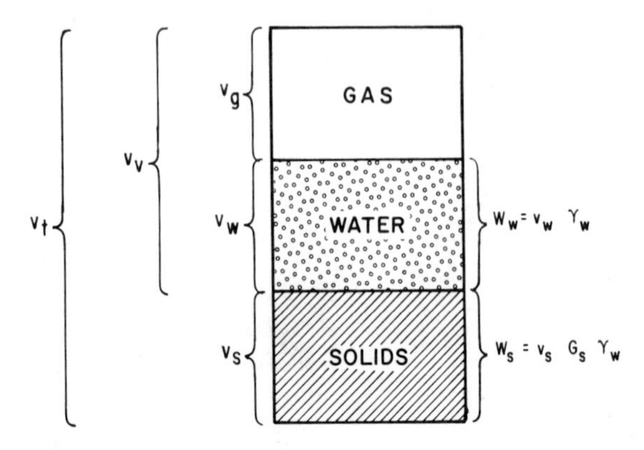

Figure 7-1. Diagrammatic representation of the compost matrix as a three-phase system of solids, water and gas.

WEIGHT RELATIONSHIPS

The specific gravity of a solid is a function of the specific gravities of component parts making up the solid. If specific gravities of organic (volatile) and ash components are known, the specific gravity of the solids as a whole can be calculated as:

$$\frac{1}{G_s} = \frac{V_s}{G_v} + \frac{(1 - V_s)}{G_f} \tag{7-1}$$

where V_s = volatile fraction of the total solids
G_s = specific gravity of the total solids
G_v = specific gravity of the volatile fraction
G_f = specific gravity of the fixed or ash fraction

Normally, specific gravity of the volatile solids (VS) is about 1.0 and that of fixed solids about 2.5.

The total volume of a composting mixture equals the sum of the volumes of water, solids and gas contained in the mixture. Assuming the gas volume to be zero, total volume becomes:

$$v_t = \frac{W_s}{G_s \gamma_w} + \frac{W_w}{\gamma_w} = \frac{W_s}{G_s \gamma_w} + \frac{W_s(1 - S_c)}{S_c \gamma_w}$$

$$v_t = \frac{W_s}{\gamma_w}\left(\frac{1}{G_s} + \frac{1 - S_c}{S_c}\right) \quad (v_g = 0) \tag{7-2}$$

where v_t = total volume of the solids and water
W_s = weight of dry solids
W_w = weight of water
γ_w = unit weight of water

Let γ_c = unit bulk weight = total wet weight per unit volume, and γ_c (dry) = unit dry weight = dry weight per unit volume. Based on Equation 7-2, unit bulk and dry weights can be calculated as:

$$\gamma_c = \frac{W_s}{S_c v_t} = \frac{W_s}{S_c \frac{W_s}{\gamma_w}\left(\frac{1}{G_s} + \frac{1 - S_c}{S_c}\right)}$$

$$\gamma_c = \frac{\gamma_w}{\left(\frac{S_c}{G_s} + 1 - S_c\right)} \quad (v_g = 0) \tag{7-3}$$

$$\gamma_c\,(\text{dry}) = S_c \gamma_c \tag{7-4}$$

Equation 7-3 is valid as long as pore spaces within the material are completely filled with water. At some point, however, continued moisture removal will leave FAS that results in actual unit weights less than that calculated by Equation 7-3.

Available data on unit bulk weight of sludge and sludge compost are limited. Information from the literature and the author's own data are presented in Figure 7-2. The theoretical sludge bulk weight which would result if there were no air voids is also shown. The latter was calculated from Equation 7-3 using the procedure of Example 7-1. Although there is considerable scatter in the data, unit bulk weight appears to decrease with increasing solids content. Furthermore, measured values begin to deviate from the theoretical at a solids content between about 30 and 40%. Thus, FAS becomes measurable above about 40% solids. The latter is a recommended minimum solids content in the starting mixture for all sludge composting systems, with the exception of those that use bulking agents, which will be discussed later.

Corresponding values of unit dry weight are shown in Figure 7-3. Unit dry weight is reasonably constant beyond solids contents of 35-50%. Within this range, it is likely that moisture evaporates from the solids causing little net change in the total volume.

Scatter in the data of Figures 7-2 and 7-3 is probably caused in part by different levels of compaction and consolidation which are, of course, a function of methods used in handling the material. Compaction was minimized during collection of data shown in Figures 7-2 and 7-3. It should be noted that relatively high unit weights are possible if FAS is removed by

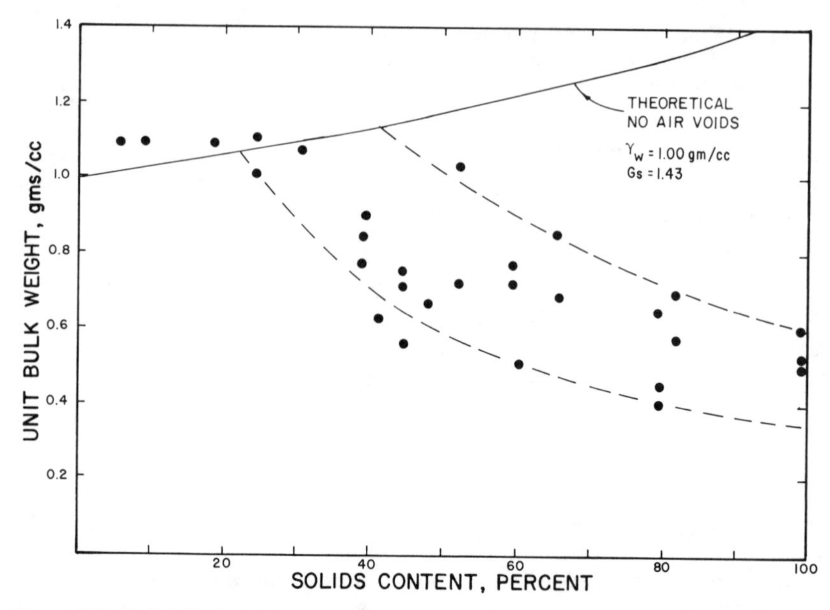

Figure 7-2. Unit bulk weight of sludge and compost as a function of solids content. Data were derived from sludge-only systems and do not include effects of amendment or bulking agent additions [102].

compaction. Volume reductions have often been observed during windrow composting caused in part by consolidation of material. One of the functions of compost agitation or turning is to decrease the unit bulk weight and minimize effects of compaction.

Because of the scatter in available information on unit bulk and dry weights, it would be advisable to measure these constituents on the actual substrate in question. Furthermore, amendments and bulking agents can significantly affect the bulk weight and, to the extent possible, should be included in the analysis. For example, new wood chips used at Beltsville are reported to have a bulk weight of only 0.30 g/cm^3 (18.5 lb/ft^3). The weight of the final compost would certainly be reduced in proportion to the quantity of wood fiber in the final product.

As stated above, FAS should begin to occur at a solids content of about 40%. However, this value should be expected to vary depending on the depth of material in the pile or reactor. The overburden pressure and, hence, the tendency to consolidate are functions of depth of material. Higher mixture solids contents may be necessary in such cases. Indeed, a 50% starting solids

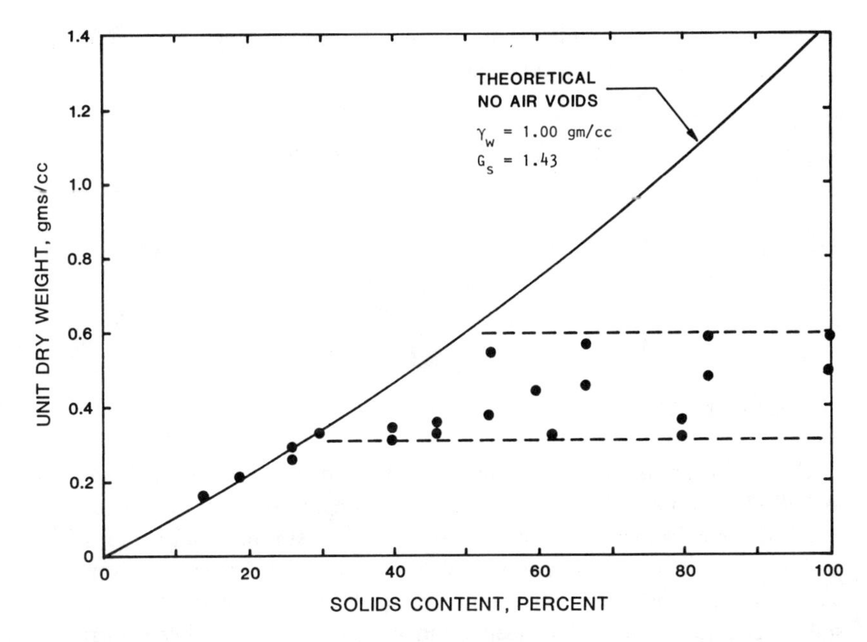

Figure 7-3. Unit dry weight of sludge and compost as a function of solids content. Data were derived from sludge-only systems and do not include effects of amendment or bulking agent additions.

content was recommended during composting of sludge/recycled compost mixtures in 3-m-deep bins (see Chapter 2).

Example 7-1

A digested sludge with a volatility of 0.50 is dewatered to 30% cake solids. Estimate the unit bulk and dry weights of the dewatered cake assuming no measurable gas volume in the cake. Assume the specific gravities of volatile and ash fractions to be 1.0 and 2.5, respectively.

Solution

1. The specific gravity of the sludge solids is given by Equation 7-1:

$$\frac{1}{G_S} = \frac{0.5}{1.0} + \frac{(1-0.5)}{2.5}$$

$$G_S = 1.43$$

2. Unit bulk weight can then be determined from Equation 7-3, assuming the unit weight of water as 1.00 g/cm³.

$$\gamma_c = \frac{1.00}{\left(\frac{0.3}{1.43} + 1 - 0.3\right)} = 1.099 \text{ g/cm}^3$$

3. Using Equation 7-4, the unit dry weight is

$$\gamma_c(\text{dry}) = S_c\gamma_c = (0.30)(1.099) = 0.33 \text{ g/cm}^3$$

POROSITY AND FAS

One might argue that the volume of voids shown in Figure 7-1 should not include the water volume v_w. After all, if the substrate is saturated with water it seems somewhat incorrect to say that it still has void volume. The problem is that the concept of void volume, and related ideas such as porosity, void ratio and degree of saturation, have their origin in the science of soil mechanics. Both the concepts and nomenclature have been borrowed and applied to composting systems. To change nomenclature at this point would be difficult and of dubious value to the student who must still contend with terms commonly used in the composting literature. Instead, let us define terms as clearly as possible and overlook minor problems of nomenclature.

Volume ratios commonly used in composting are the porosity and FAS. Porosity n of a composting mass is defined as the ratio of void volume to total volume.

$$n = \frac{v_v}{v_t} = \text{porosity} \tag{7-5}$$

$$n = \frac{v_t - v_s}{v_t} = 1 - \frac{v_s}{v_t} \tag{7-6}$$

Considering a unit total volume:

$$n = 1 - \frac{\gamma_n S_m}{G_m \gamma_w} \tag{7-7}$$

where γ_m = unit bulk weight of the mixed material to be composed, wet weight per unit volume
S_m = fractional solids content of the mixture
G_m = specific gravity of mixture solids.

FAS f is defined as the ratio of gas volume to total volume:

$$f = \frac{v_g}{v_t} \tag{7-8}$$

$$f = \frac{v_t - v_s - v_w}{v_t} \tag{7-9}$$

Again consider a unit total volume:

$$f = 1 - \frac{\gamma_m S_m}{G_m \gamma_w} - \frac{\gamma_m(1 - S_m)}{\gamma_w} \tag{7-10}$$

The FAS of a composting mixture is important in determining the quantity and movement of air through the mixture. As discussed in Chapter 6, a range of moisture contents have been reported as optimum for different composting materials. In general, materials of a more fibrous and friable nature with lower bulk weights can maintain higher moisture contents during composting. It is felt that the optimum moisture content for a particular material is related to maintenance of a certain minimum FAS. Thus, different materials can hold different moisture levels while still maintaining the same FAS. From Equation 7-10, it is advantageous to maintain the bulk weight at a minimum because this will allow a higher moisture content in the composting matrix. Recall that the term "optimum moisture" represents a tradeoff between the moisture requirements of the microbes and their simultaneous need for an adequate oxygen supply.

The effect of moisture content on FAS is shown in Figure 7-4 for various composting mixtures. Reported optimum moisture contents tend to fall in the range of FAS between 30 and 35%. Jeris and Regan [103] further examined the effect of FAS on oxygen consumption rates of mixed refuse samples. Approximately 67% moisture and 30% FAS were found to be optimum conditions as shown in Figure 7-5. About 95% of the maximum oxygen consumption rate was maintained when FAS was between about 20 and 35%. Working with garbage and sludge mixtures, Schultz [104] concluded that a minimum of about 30% FAS should be maintained.

The minimum FAS is probably influenced to some extent by method of aeration. With constant tumbling or turning, a minimum FAS should be required because of repeated exposure of new surfaces to the oxygen. However, constant agitation is not characteristic of most compost systems. Windrow, aerated static pile and most reactor systems will require FAS more in keeping with previously quoted results. It would seem, therefore, that a

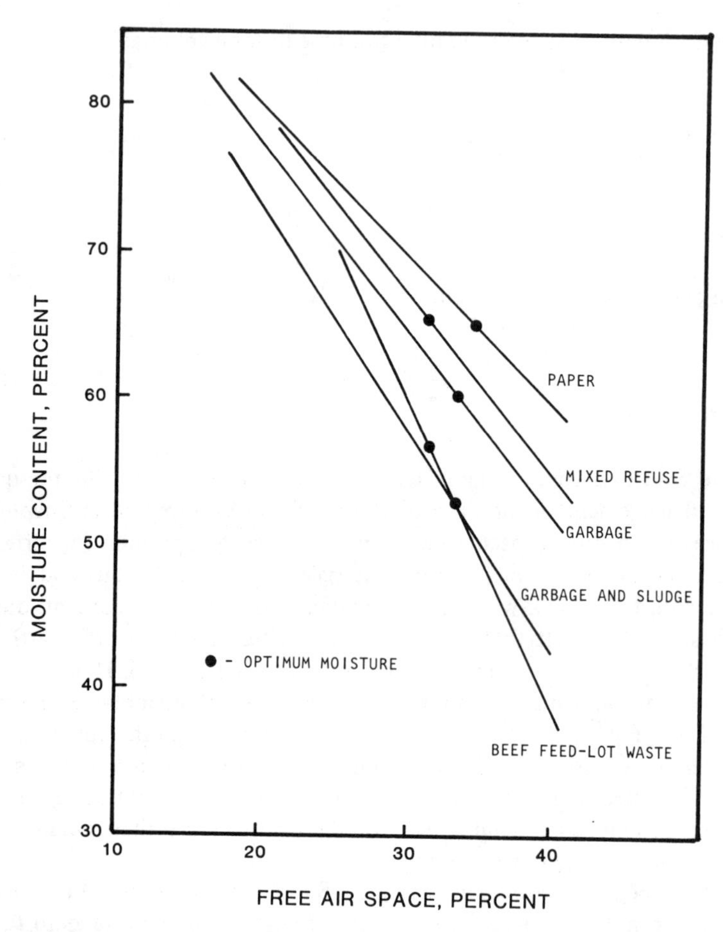

Figure 7-4. FAS as a function of moisture content for various feed materials. Reported optimum moisture contents tend to fall within a reasonably narrow range of FAS [103].

minimum FAS of about 30% should be maintained for a wide variety of composting materials and composting systems.

Example 7-2

A digested sludge is dewatered to 30% solids and blended with recycled compost to give a mixture solids of 45%. Assuming a mixture volatility of 0.50 and a unit bulk weight of 0.85 g/cm^3, calculate the porosity and FAS of the mixed material.

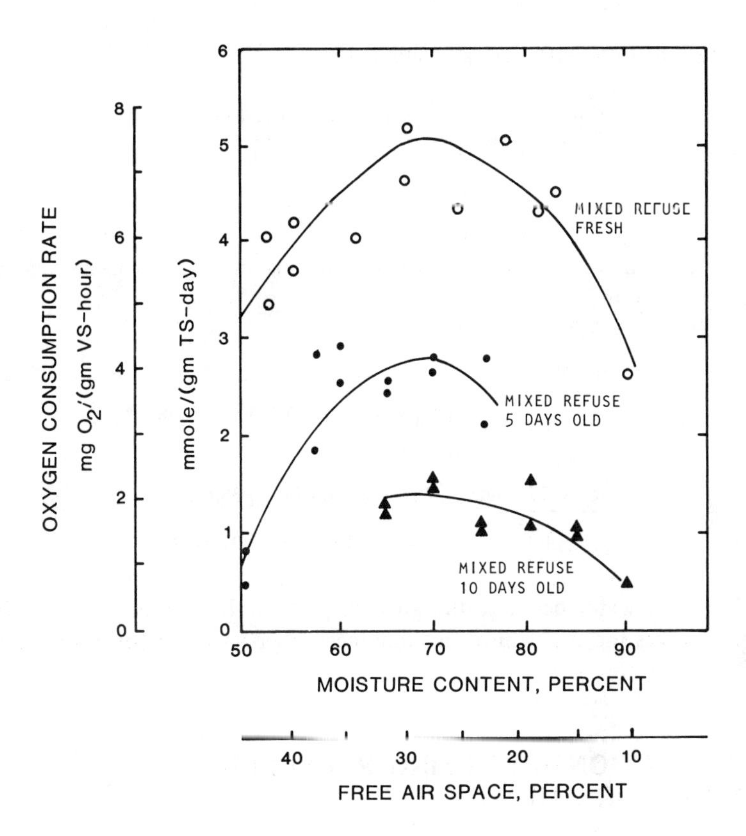

Figure 7-5. Effects of moisture and FAS on the oxygen consumption rate of mixed refuse samples. Data were developed from batch Warburg respirometer runs. Incubation temperatures were not specified [103].

Solution

1. From Example 7-1, specific gravity of the solids is:

$$G_S = 1.43$$

2. From Equation 7-7, the porosity is calculated as:

$$n = 1 - \frac{0.85\,(0.45)}{1.43\,(1.00)}$$

$$n = 0.73$$

3. Using Equation 7-10, FAS is calculated as:

$$f = 1 - \frac{0.85\,(0.45)}{1.43\,(1.00)} - \frac{0.85\,(1 - 0.45)}{1.00}$$

$$f = 0.27$$

4. Volume and weight relationships for the mixture solids can be schematically illustrated as:

$v_g = 0.27$ gas

$v_W = 0.46$ water $\begin{cases} W_W = 0.85\ (1 - 0.45) = 0.46\ \text{g/cm}^3 \\ v_W = 0.46 \end{cases}$

$v_S = 0.27$ solid $\begin{cases} W_S = 0.85\ (0.45) = 0.38\ \text{g/cm}^3 \\ v_S = 0.38/(1.43)(1.00) = 0.27 \end{cases}$

$v_t = 1.00$

Note that water occupies the greatest part of the mixture volume. This emphasizes the problem of maintaining proper moisture control with wet feed substrates.

FAS RELATIONSHIPS USING RECYCLED COMPOST OR AMENDMENTS

Using Equation 7-10, FAS was calculated for various conditions of unit bulk weight and solids content. The specific gravity of the solids was calculated from Equation 7-1 assuming a VS fraction of 0.50. Results are presented in Figure 7-6. It is evident that FAS is strongly influenced by both bulk weight and solids content of the composting material. For sludges with solids contents less than about 30%, the unit bulk weight is probably that calculated through use of Equations 7-2 and 7-3, corresponding to zero FAS. Obviously, this is an undesirable situation as regards aeration and explains the need for addition of a bulking agent or compost recycle to increase the FAS.

If compost is recycled to give an initial mixture of 40% solids, a decrease in bulk weight from that calculated by Equation 7-2 and 7-3 can be expected. Using average values from Figure 7-2, a bulk weight of 0.96 g/cm^3 (60 lb/ft^3) would not be unreasonable at $S_m = 0.40$. Assuming a volatility of 0.50, the resultant FAS from Figure 7-6 would be about 0.15. As moisture evaporates from the compost, the volume and weight of water will decrease and the bulk weight should decrease. Assuming a final compost solids content of 70%

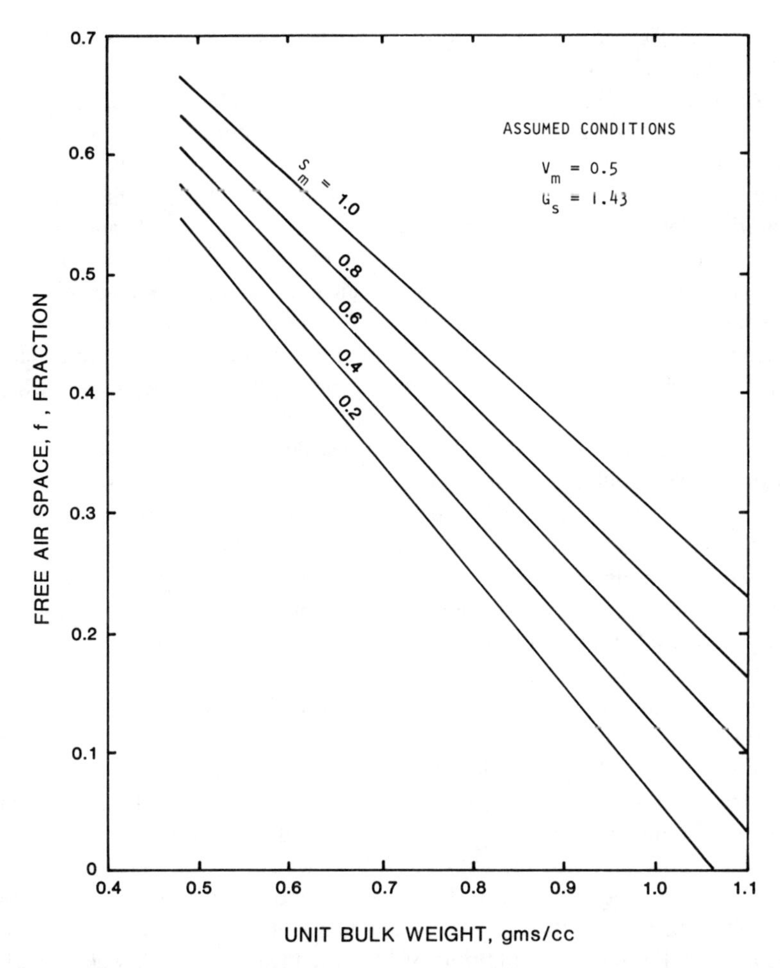

Figure 7-6. Effect of unit bulk weight and mixture solids content on FAS. Based on a VS fraction of 0.50 and solids specific gravity of 1.43.

and bulk weight of 0.48 g/cm^3 (30 lb/ft^3), the final FAS from Figure 7-6 would be about 0.62. Assuming a range in the bulk weight of compost product from about 0.48 to 0.72 g/cm^3 and solids contents from 60 to 80%, the likely range in final FAS would be from about 0.40 to 0.65. This remarkable increase in FAS as composting proceeds is caused almost entirely by moisture evaporation. Obviously, if drying is limited or compaction of the sludge occurs during composting the increase in FAS will not be as dramatic as that calculated above. The loss of VS and resultant increase in specific gravity of the remaining solids was not considered in the above analysis. However, this would serve to further increase values of FAS over those estimated above.

A number of conclusions can be drawn from the above example. Dewatered sludge cake can be assumed to have essentially no FAS. This is particularly true if centrifuge dewatering devices are employed; these characteristically produce a cake with little structural integrity. Filtration dewatering devices, such as belt presses, typically produce a "pancake-thin" cake which may maintain some airspace much as when pancakes are randomly stacked on a plate. Little or no airspace would be expected within the individual pancake pieces, however. Furthermore, if mechanical turning is employed the structure will likely be disturbed allowing the cake to collapse on itself. Without FAS, oxygen cannot be stored in void spaces, which means a constant need for exposure of the sludge by turning or mixing. Indeed, Shell and Boyd [97] composted raw dewatered sludge with no compost recycle or bulking agent and demonstrated the need for constant mixing by paddle mixers along with forced aeration to maintain aerobic conditions.

If compost is recycled for moisture control, only limited airspace can be expected in the initial mixture. Therefore, aeration and oxygen transfer will likely be most difficult in the early stages of composting. Some control over the initial FAS is possible by varying the amount of compost product recycle and, thus, the moisture content of the starting material.

As moisture evaporates from the compost the bulk weight should decrease while the FAS increases. Airspaces as high as 0.5–0.6 should not be uncommon in final compost product at solids contents of $\geq$70%. The increase in airspace means that maintenance of aerobic conditions in the composting substrate should become easier as composting and drying proceed.

Finally, dry amendments with a low bulk weight can be very effective in decreasing the average mixture bulk weight. From Figure 7-6, any decrease in bulk weight will result in a corresponding increase in FAS. For example, Houser [19] found that addition of 3% shredded paper (dry-weight basis) to a mixture of sludge and recycled compost resulted in a decrease in mixture bulk weight from about 0.67 to 0.51 g/cm^3 (see Chapter 2). It would appear that even small amendment additions can have a rather dramatic impact, depending, of course, on the characteristics of the amendment.

Example 7-3

A digested sludge with a volatility of 0.50 is dewatered to 30% cake solids. It is to be blended with recycled compost that is 70% solids with a bulk weight of 0.65 g/cm^3 and a volatility of 0.35. Estimate the FAS of the mixed material if the recycle ratio is adjusted to provide 40% solids in the mixture. What if recycle is increased to give a 50% solids mixture?

Solution

1. From Example 7-1 the bulk weight of the sludge cake γ_c is about 1.099 g/cm³.
2. The required recycle ratios to achieve a 40% mixture can be determined from Equations 6-5 and 6-7:

$$R_W = \frac{S_m - S_c}{S_r - S_m} = \frac{0.40 - 0.30}{0.70 - 0.40} = 0.333$$

$$R_d = \left(\frac{S_m}{S_c} - 1\right) \Big/ \left(1 - \frac{S_m}{S_R}\right) = \left(\frac{0.40}{0.30} - 1\right) \Big/ \left(1 - \frac{0.40}{0.70}\right) = 0.778$$

3. The bulk weight of mixed material can be estimated as the weighted average of the two components:

$$\gamma_m = \frac{1.099\,(1.0) + 0.65\,(0.333)}{(1.0 + 0.333)} = 0.987 \text{ g/cm}^3$$

4. From Equation 7-1 the specific gravity of solids in the recycled compost is:

$$\frac{1}{G_s} = \frac{0.35}{1.0} + \frac{(1 - 0.35)}{2.5}$$

$$G_s = 1.64$$

In Example 7-1 the specific gravity of sludge solids was calculated to be 1.43.

5. The specific gravity of solids in the mixture can be estimated as the weighted average of components on a dry weight basis:

$$G_m = \frac{1.43\,(1.0) + 1.64\,(0.778)}{1.0 + 0.778} = 1.52$$

6. From Equation 7-10 the FAS of the mixture is given by:

$$f = 1 - \frac{0.987\,(0.40)}{1.52\,(1.00)} - \frac{0.987\,(1 - 0.40)}{1.00} = 0.148$$

7. Repeating the above calculations for a mixture solids of 50%:

$$R_w = \frac{0.50 - 0.30}{0.70 - 0.50} = 1.00$$

$$R_d = \left(\frac{0.50}{0.30} - 1\right) \Big/ \left(1 - \frac{0.50}{0.70}\right) = 2.333$$

$$\gamma_m = \frac{1.099\,(1.0) + 0.65\,(1.00)}{1.0 + 1.0} = 0.875\ \text{g/cm}^3$$

$$G_m = \frac{1.43\,(1.0) + 1.64\,(2.333)}{1.0 + 2.333} = 1.58$$

$$f = 1 - \frac{0.875\,(0.50)}{1.58\,(1.00)} - \frac{0.875\,(1 - 0.50)}{1.00} = 0.286$$

8. Note that in this example the FAS is about doubled by increasing the mixture solids content from 40 to 50%. Note also that the required weight of recycled compost is tripled. Thus the tradeoffs in design become evident. Increasing mixture solids will increase the FAS and help assure proper aeration. However, it will also increase the required recycle ratio, the daily weight of material to be handled and the land requirements. Striking a proper balance is the art of successful composting.

FAS RELATIONSHIPS USING BULKING AGENTS

Development of a Conceptual Model

From the previous section, it was observed that the FAS of dewatered cake can reasonably be assumed to be zero. The function of a bulking agent is to provide structural support for the dewatered sludge, to provide FAS within the voids between particles and to increase the size of pore spaces and allow easier air movement through the mixture. A bulking agent can be viewed as a three-dimensional matrix of solid particles capable of self-support. The void volume and size of the pore spaces is determined by the shape and size of bulking particles. Conceptually, sludge can be viewed as occupying part of the void volume in the bulking agent, as shown schematically in Figure 7-7.

If the sludge cake is assumed to occupy the void spaces, limits must exist on the ratio of bulking agent to sludge. If too little bulking agent is added the individual bulking particles will not be in contact with each other. Instead they will be immersed in the sludge and no practical increase in FAS or pore size will result. On the other hand, addition of bulking agent beyond that

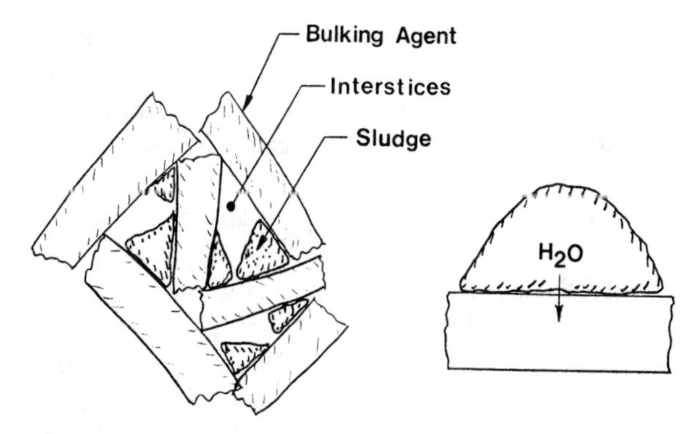

Figure 7-7. Schematic illustration of sludge/bulking agent mixture showing water absorption from wet cake into bulking particles.

required to assure adequate FAS will increase the quantity of material to be handled daily, increase the consumption of bulking agent (assuming it to be degradable) and result in greater land requirements and higher costs.

As previously discussed, numerous bulking agents have been used, including wood chips, straw, pelleted refuse, shredded tires, rice hulls, peanut shells and other materials. Most of these bulking agents are composed of cellulosic material and are degradable to some extent under composting conditions. Decomposition will use up a portion of the bulking agent, and size reduction will allow an additional fraction eventually to pass through the final screening process with the composted sludge. Thus, continual makeup of bulking agent is necessary to balance that which is degraded and that which becomes part of the final product. Certain bulking agents, such as shredded tires and plastic materials, are extremely resistant to microbial decomposition and are probably not affected by the composting process. Hence, they should act as conservative substances. Assuming that bulking agent can be screened efficiently from the compost mixture and reused, little or no makeup would be required in such a case.

Bulking agents can also be classified as to their moisture-absorbing characteristics. Most natural cellulosic materials are porous and capable of significant moisture absorption. Other, nonporous materials, such as plastics, for all practical purposes can be assumed to be nonabsorptive. Similarly, a porous bulking material that is saturated with water is not capable of further absorption unless it is dried before reuse.

The moisture-absorptive capacity of the bulking agent is important in determining the required quantity of such material. Assume the bulking agent to be dry and porous before mixture with the sludge cake. Once mixed, moisture will be drawn from the sludge cake into the bulking agent. Thus, a volume of water will penetrate the bulking agent, leaving an equivalently greater void volume remaining. If the bulking agent is nonporous or water-saturated, a greater quantity of bulking agent will be required to produce the same void volume.

The quantity of moisture that can be absorbed by a bulking agent such as wood chips is considerable. Assume 1 m^3 of dry wood chips with a bulk weight of 288 kg/m^3 (18 lb/ft^3) is saturated with water to a final moisture content of 60%. In such a case, the 288 kg of dry chips would absorb 434 kg of water, i.e., (288/0.4) − 288, or 0.43 m^3. Thus, a volume of moisture greater than 40% of the volume of wood chips can be absorbed. Considering that the porosity of randomly piled wood chips is probably on the order of 40%, the importance of moisture absorption can easily be recognized.

Mathematics of the Conceptual Model

A mass balance diagram for the addition of bulking agent to sludge cake is presented in Figure 7-8. The required ratio of bulking agent to sludge cake can be determined from analysis of the mass balance diagram provided certain assumptions are made. First, assume that sludge occupies the interstices of the mixture and does not increase the total volume of the mixture. Few field data are available to verify this assumption. However, because most dewatered sludge cakes are semifluid in nature, the assumption should be reasonable provided the cake and bulking agent are well mixed and the cake is not too dry. Additional comments on this assumption will be made after development of the model. Second, sufficient bulking agent must be added so that contacts between the bulking particles provide the structural support for the mixture. In general, this condition will be satisfied as long as the

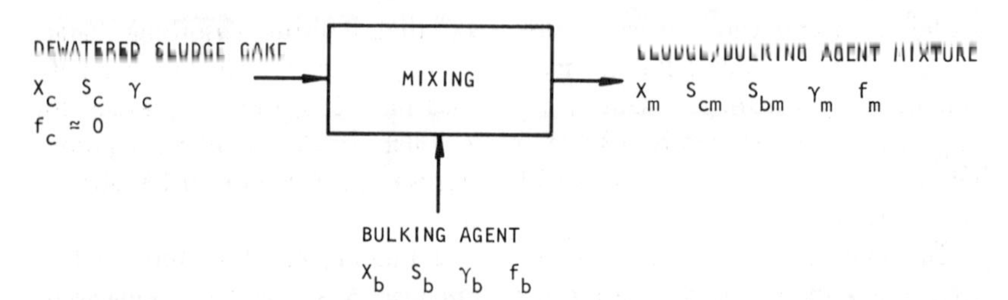

Figure 7-8. Mass balance diagram for bulking agent addition to sludge cake.

cake/bulking agent mixture remains porous. If insufficient bulking agent is added, bulking particles will be suspended in the sludge cake, and both sludge and bulking agent will contribute to the volume of mixture. Third, assume that moisture absorption by the bulking agent is limited to a maximum moisture content (i.e., minimum solids content). Fourth, moisture release from the sludge cake is limited to a maximum solids content. Fifth, no other amendments or recycled compost are added to the mixture. Sixth, assume individual bulking particles to be solid with no internal FAS. Free airspace that might exist in the cellular matrix of a wood chip or other porous particle is of little value to sludge in the void space between particles, at least in terms of oxygen transfer.

Moisture relationships with absorption limited by the bulking agent are shown in Figure 7-9 and the case for moisture absorption limited by sludge

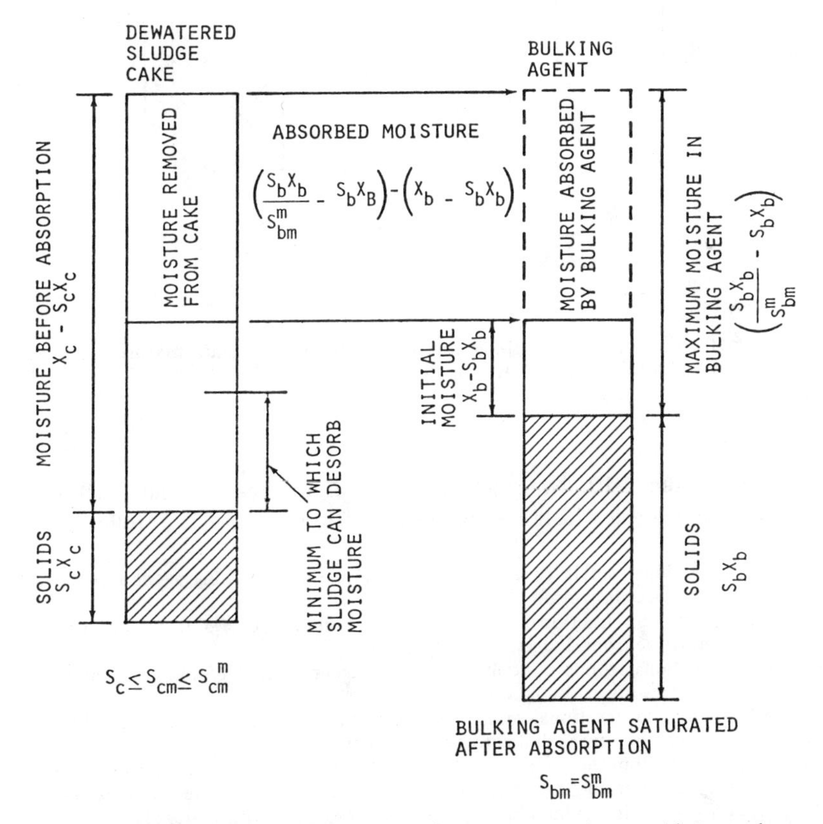

Figure 7-9. Moisture relationships for a sludge/bulking agent mixture with water absorption limited by the bulking agent [102].

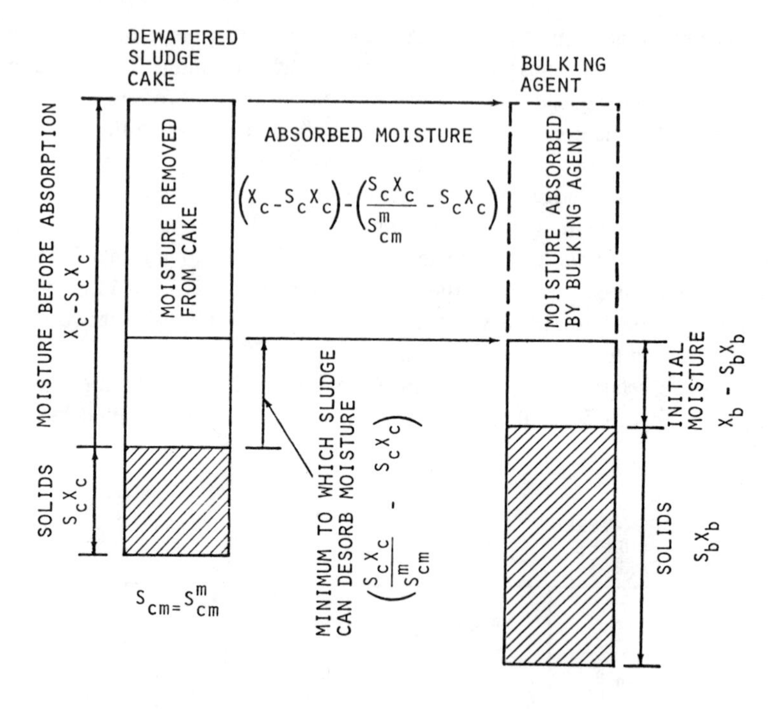

Figure 7-10. Moisture relationships for a sludge/bulking agent mixture with water absorption limited by the sludge cake [102].

cake in Figure 7-10. Nomenclature used in Figures 7-9 and 7-10 and in subsequent analysis which has not been previously defined is as follows:

S_{cm} = fractional solids content of sludge in a sludge/bulking agent mixture after moisture absorption

S_{cm}^m = maximum fractional solids content of sludge achievable by absorption of moisture from sludge to bulking agent

S_{bm} = fractional solids content of bulking agent in sludge/bulking agent mixture after moisture absorption

S_{bm}^m = minimum fractional solids content of bulking agent achievable by absorption of moisture from sludge to bulking agent

f_b = FAS within the interstices of a bulking agent before sludge addition

f_c = FAS within the interstices of a sludge cake, usually assumed to be zero

f_m = FAS within the interstices of a mixture of composting materials

γ_b = unit bulk weight of bulking agent, wet weight per volume

For the case where moisture transfer is limited by absorption into the bulking agent, the following equation describes the volume of sludge cake in the void spaces after moisture absorption:

(initial sludge volume) − (absorbed moisture volume) = (mixture volume)

− (bulking particle volume) − (mixture FAS) (7-11)

Substituting terms from Figure 7-9:

$$\frac{X_c}{\gamma_c} - \frac{1}{\gamma_w}\left[\left(\frac{S_b X_b}{S_{bm}^m} - S_b X_b\right) - (X_b - S_b X_b)\right]$$

$$= \frac{X_m}{\gamma_m} - \frac{X_b}{\gamma_b}(1 - f_b) - \frac{X_m}{\gamma_m} f_m \qquad (7\text{-}12)$$

If the sludge does not add to the total mixture volume but occupies only the interstices:

volume of bulking agent = volume of mixture

$$\frac{X_b}{\gamma_b} = \frac{X_m}{\gamma_m} \qquad (7\text{-}13)$$

Let the volume of bulking agent to sludge cake be defined as the volumetric mixing ratio M_{bc} as follows:

$$M_{bc} = \text{volumetric mixing ratio} = \frac{X_b/\gamma_b}{X_c/\gamma_c} \qquad (7\text{-}14)$$

Substituting Equations 7-13 and 7-14 into Equation 7-12 and rearranging:

$$\frac{1}{M_{bc}} = \frac{\gamma_b}{\gamma_w}\left(\frac{S_b}{S_{bm}^m} - 1\right) + (f_b - f_m) \qquad (7\text{-}15)$$

Equation 7-15 can be used to estimate the volumetric mixing ratio when the quantity of moisture removed from the sludge cake is limited by that which the bulking agent can absorb. One difficulty with use of Equation 7-15 is that for a porous bulking agent the bulk weight γ_b is a function of the moisture content and, hence, the solids content S_b. Because both terms appear in Equation 7-15 the relationship between the two must be known or determined. This subject will be discussed shortly.

It was assumed that moisture absorption is limited by the bulking agent. Therefore, the solids content of the bulking agent in the mixture S_{bm} must equal the minimum solids content S_{bm}^m. Sludge cake solids, on the other hand, will increase to a level between the initial cake solids S_c and the maximum cake solids S_{cm}^m. The actual value of S_{cm} can be determined as:

$$S_{cm} = \frac{\text{wt solids}}{\text{wt solids} + \text{wt initial } H_2O - \text{wt } H_2O \text{ absorbed}}$$

Substituting terms from Figure 7-9:

$$S_{cm} = \frac{S_cX_c}{S_cX_c + \left[(X_c - S_cX_c) - \left(\frac{X_bS_b}{S_{bm}^m} - X_bS_b - (X_b - X_bS_b)\right)\right]}$$

Rearranging:

$$S_{cm} = \frac{S_c}{1 + M_{bc}\frac{\gamma_b}{\gamma_c}\left(1 - \frac{S_b}{S_{bm}^m}\right)} \qquad S_c \leqslant S_{cm} \leqslant S_{cm}^m \tag{7-16}$$

As previously discussed, sludge cake cannot give up unlimited quantities of water. Thus, a situation can arise where moisture transfer is limited by the amount that the sludge cake is capable of yielding, a situation illustrated in Figure 7-10. Substituting appropriate terms from Figure 7-10 into Equation 7-11 yields:

$$\frac{X_c}{\gamma_c} - \frac{1}{\gamma_w}\left[(X_c - S_cX_c) - \left(\frac{S_cX_c}{S_{cm}^m} - S_cX_c\right)\right] = \frac{X_m}{\gamma_m} - \frac{X_b}{\gamma_b}(1 - f_b) - \frac{X_m}{\gamma_m}f_m \tag{7-17}$$

Rearranging and substituting Equations 7-13 and 7-14 gives:

$$M_{bc} = \frac{1 - \frac{\gamma_c}{\gamma_w}\left(1 - \frac{S_c}{S_{cm}^m}\right)}{f_b - f_m} \tag{7-18}$$

If moisture absorption is limited by the sludge cake, the solids content of the bulking agent S_{bm} will decrease to a value less than the initial solids content S_b, but greater than the minimum solids content with complete moisture absorption S_{bm}^m. The value of S_{bm} can be estimated as:

$$S_{bm} = \frac{\text{wt bulking agent solids}}{\text{wt solids} + \text{wt initial } H_2O + \text{wt } H_2O \text{ absorbed}}$$

Substituting terms from Figure 7-10:

$$S_{bm} = \frac{S_b X_b}{S_b X_b + \left[(X_b - S_b X_b) + (X_c - S_c X_c) - \left(\frac{S_c X_c}{S_{cm}^m} - S_c X_c\right)\right]}$$

Rearranging:

$$S_{bm} = \frac{S_b}{1 + \frac{1}{M_{bc}} \frac{\gamma_c}{\gamma_b}\left(1 - \frac{S_c}{S_{cm}^m}\right)} \qquad S_{bm}^m \leqslant S_{bm} \leqslant S_b \qquad (7\text{-}19)$$

Under these conditions S_{cm} will equal the maximum cake solids S_{cm}^m.

Since either case may be limiting, the procedure for calculation of the required M_{bc} for a given situation is as follows:

1. Calculate M_{bc} and the corresponding S_{cm} from Equations 7-15 and 7-16, respectively.
2. Calculate M_{bc} and the corresponding S_{bm} from Equations 7-18 and 7-19, respectively.
3. Determine the required mixing ratio as the greater of the two calculated values. If M_{bc} calculated by Equation 7-15 is greater, moisture transfer is limited by the amount which can be absorbed by the bulking agent and, therefore, $S_{bm} = S_{bm}^m$. If M_{bc} calculated by Equation 7-18 is greater, absorption is determined by the amount released from the sludge cake and, therefore, $S_{cm} = S_{cm}^m$. Equations 7-15 and 7-18 will calculate to the same M_{bc} if $S_{cm} = S_{cm}^m$ and $S_{bm} = S_{bm}^m$.

Model Results

Sample calculations were performed for various values of dewatered cake solids and bulking agent moisture content. The bulking agent was assumed to be wood chips with a porosity of 0.40, capable of absorbing water to a maximum 60% moisture, $S_{bm}^m = 0.40$. Available data suggested a bulk weight of 0.288 g/cm^3 for oven-dried chips. Bulk weight at other moisture contents was estimated as:

$$\gamma_b = \frac{0.288 \text{ g/cm}^3}{S_b} \qquad (7\text{-}20)$$

Sludge cake was assumed to release moisture to a maximum solids content of 40%, S^m_{cm} = 0.40. Bulk weight of sludge cake was determined from Equations 7-3 assuming a specific gravity of 1.43 for the solid fraction and 1.00 for water. A summary of the method of calculation is presented in Table 7-1. Results of the complete analysis are presented in Figure 7-11.

Analysis of published data for the aerated pile process indicates a FAS of ~20% in the mixed material. Referring to Figure 7-11, therefore, consider

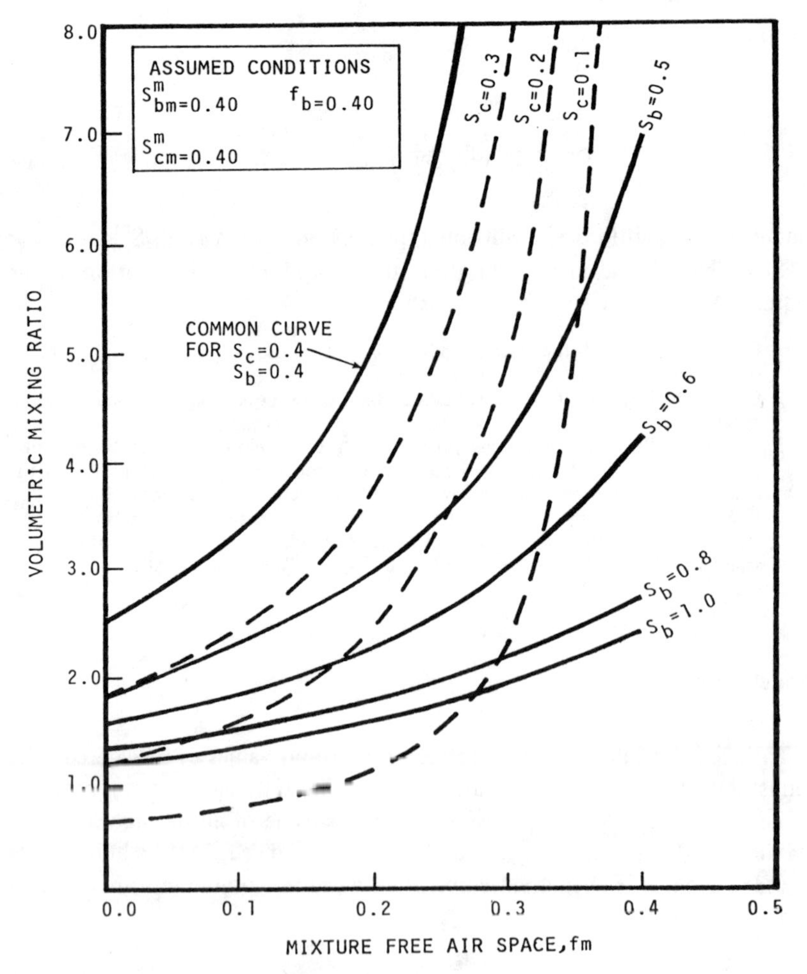

Figure 7-11. Volumetric mixing ratio for bulking agent/sludge mixtures as a function of mixture FAS for various values of dewatered cake solids S_c and bulking agent solids S_b [102].

Table 7-1. Summary Calculations to Determine the Volumetric Mixing Ratio M_{bc}

S_c[a] Assumed	S_b[a] Assumed	γ_c[b] (g/cm³)	$\gamma_b = \frac{0.288}{S_b}$ (g/cm³)	f_m Assumed	M_{bc}[c] Equation 7-15	M_{bc}[c] Equation 7-18	S_{cm} Equation 7-16	S_{bm} Equation 7-19
0.20	0.50	1.064	0.576	0.0	1.84	1.17	0.266	0.400[d]
				0.1	2.25	1.56	0.288	0.400[d]
				0.2	2.91	2.34	0.330	0.400[d]
				0.3	4.09	4.68	0.400[e]	0.418
				0.4	6.93	∞	0.400[e]	0.500
0.20	0.60	1.064	0.480	0.0	1.56	1.17	0.309	0.400[d]
				0.1	1.85	1.56	0.344	0.400[d]
				0.2	2.27	2.34	0.400[e]	0.407
				0.3	2.94	4.68	0.400[e]	0.485
				0.4	4.16	∞	0.400[e]	0.600

[a]Calculations repeated for various values of S_c and S_b to construct curves shown in Figure 7-11.
[b]γ_c calculated from Equations 7-1 and 7-3 assuming $G_v = 1.0$, $G_f = 2.5$, $V_c = 0.50$, $\gamma_w = 1.00$.
[c]Greater of the two calculated values represents the required M_{bc} value.
[d]$S_{bm} = S_{bm}^m$ because moisture absorption is limited by the bulking agent.
[e]$S_{cm} = S_{cm}^m$ because moisture absorption is limited by the sludge cake.

a mixture f_m of 0.20. Reading upward, the first intercepted curve is for a cake solids of 10%. All of the solids curves, which represent different bulking agent solids contents, lie above the line for $S_c = 0.10$. Under these conditions, moisture transfer will be limited by the absorptive capacity of the bulking agent (solid lines). If $S_b = 0.80$, for example, the required M_{bc} will be about 1.8, increasing to $M_{bc} = 2.9$ if $S_b = 0.50$. Thus, the influence of bulking agent moisture content in determining the required M_{bc} is evident. If sludge solids increase to 20%, M_{bc} will be determined by moisture release from the sludge cake if S_b is greater than 0.60. In other words, the dashed curve for $S_c =$ 0.20 lies above the solid curves for S_b equal to or greater than 0.60 at $f_m =$ 0.20. In this case drying the bulking agent beyond 60% solids would not decrease the required M_{bc} of about 2.3. If S_b is less than 0.60, M_{bc} will be determined from the solid curves. Considering an S_b of 0.40, for example, M_{bc} would be about 5.0, decreasing to about 2.9 at $S_b = 0.50$.

As seen above, moisture content of the bulking agent can have a significant influence on the total quantity of mixed material to be processed each day. Drying the bulking agent before recycling should be considered as a way to reduce the required M_{bc}. However, as illustrated above, drying beyond an S_b of 0.60 with a sludge cake of 20% solids will no longer reduce M_{bc} because moisture removal from the sludge becomes limiting at about that point.

Another interesting observation can be made from interpretation of Figure 7-11. Consider an f_m of 0.20, an S_b of 0.60 or greater and cake solids of 20 and 30%. Under these conditions moisture absorption is limited by the sludge cake. At 20% cake solids M_{bc} is 2.34, increasing to 3.63 at 30% cake solids. If sludge production is assumed to be constant at 10 dry ton/day, about 50 wet ton/day would be produced at $S_c = 0.20$ and 33.3 wet ton/day at $S_c = 0.30$. These wet weights correspond to volumes of about 42.7 and 27.5 m^3/day, respectively. Multiplying by the respective M_{bc} values, the quantity of bulking agent is determined to be about 100 m^3/day in either case. This rather surprising result is caused by the fact that with S_b greater than 0.60, moisture absorption is limited by the sludge cake. Thus, the additional water associated with the wetter cake would be absorbed by the bulking agent, leaving the same quantity of solids and water in the void spaces as with the drier cake. Of course, the bulking agent will have a higher moisture content with the wetter cake and will require more drying before recycling to assure that S_b is greater than 0.60. Nevertheless, if moisture absorption is limited by the sludge cake (dashed lines in Figure 7-11) and if f_m is constant, the quantity of bulking agent is reasonably independent of sludge cake solids.

In the above example, if the 20% sludge cake produced a wetter bulking agent which could only be dried to $S_b = 0.50$, from Figure 7-11 the required M_{bc} would increase to 2.90. Moisture absorption would no longer be limited by the sludge cake. Thus, the required volume of bulking agent would increase, in this case to about 124 m^3/day.

It must be remembered that the curves in Figure 7-11 are based on assumed values for a number of variables. Although the values assumed appear to be reasonable, variations should be anticipated, depending on conditions specific to a particular problem. The examples discussed above were intended primarily to illustrate the factors involved in determining the volumetric mixing ratio. If better input data are available for a specific case, Equations 7-15 to 7-19 should be used to develop curves similar to those presented in Figure 7-11.

Even though assumed values have been used in the previous analysis, it is interesting to note that the general range of M_{bc} values is in good agreement with values reported from various field experiments on aerated pile composting. With wood chips as a bulking agent and cake solids of about 20%, volumetric mixing ratios of 2.0–3.0 have been reported [11-13]. Based on the model results in Figure 7-11, M_{bc} values will likely be between 1.0 and 3.0 if the bulking agent is capable of some moisture absorption and dewatered cake solids are less than 40%. The general agreement is encouraging and suggests that the model is a reasonable representation of the physical events that occur when using bulking agents for control of FAS.

Example 7-4

A wood chip bulking agent is to be used to provide airspace for composting 200 dry ton/day of raw dewatered sludge of 25% solids and volatility of 70%. The bulking agent has a porosity of 35% and a moisture content of 40%. The minimum FAS to be maintained in the mixture is 10%. It is expected that the sludge can further dewater to a maximum of 45% solids (S_{cm}^m = 0.45) and the bulking agent absorb water to a maximum of 65% moisture (S_{bm}^m = 0.35). Determine the required volumetric mixing ratio, the moisture contents of the cake and bulking agent in the mixture and the total volume and weight of mixed material to be processed daily.

Solution

1. Specific gravity of sludge solids from Equation 7-1 is:

$$\frac{1}{G_s} = \frac{V_s}{G_v} + \frac{(1 - V_s)}{G_f} = \frac{0.7}{1.0} + \frac{(1 - 0.7)}{2.5}$$

$$G_s = 1.22$$

2. Bulk weight of the sludge solids from Equation 7-3 is:

$$\gamma_c = \frac{1.00}{\left(\frac{0.25}{1.22} + 1.0 - 0.25\right)}$$

$$\gamma_c = 1.047$$

3. Estimate the bulk weight of wood chips from Equation 7-20 as:

$$\gamma_b = \frac{0.288}{0.6} = 0.48 \text{ g/cm}^3$$

4. Calculation of M_{bc} by Equation 7-15:

$$M_{bc} = \frac{0.48}{1.00}\left(\frac{0.60}{0.35} - 1\right) + (0.35 - 0.10)$$

$$M_{bc} = 1.69$$

5. Calculation of M_{bc} by Equation 7-18:

$$M_{bc} = \frac{1 - \frac{1.047}{1.00}\left(1 - \frac{0.25}{0.45}\right)}{(0.35 - 0.10)} = 2.14$$

Therefore, the required M_{bc} is 2.14 and moisture absorption is limited by the sludge cake. Therefore:

$$S_{cm} = S_{cm}^{m} = 0.45$$

6. S_{bm} is then given by Equation 7-19:

$$S_{bm} = \frac{0.60}{1 + \frac{1}{2.14}\frac{1.047}{0.48}\left(1 - \frac{0.25}{0.45}\right)} = 0.413$$

Note that $S_{bm}^{m} \leqslant S_{bm} \leqslant S_b$

$$0.35 \leqslant 0.413 \leqslant 0.60$$

7. Calculation of the total daily volume of mixed material:

$$\text{volume of sludge cake} = 100 \text{ dry ton/day} \times \frac{1}{0.25} \times \frac{\text{m}^3}{1047 \text{ kg}} = 382 \text{ m}^3\text{/day}$$

$$\text{volume of bulking agent} = 2.14\ (382) = 818 \text{ m}^3\text{/day}$$

$$\text{total volume} = 382 + 818 = 1200 \text{ m}^3\text{/day}$$

8. Calculation of the total daily weight of mixed material:

$$\text{wt sludge cake} = 100 \text{ dry ton/day} \times \frac{1}{0.25} = 400 \text{ wet ton/day}$$

$$\text{wt bulking agent} = 818\ \text{m}^3/\text{day}\left(\frac{480\ \text{kg}}{\text{m}^3}\right)\frac{1\ \text{ton}}{1000\ \text{kg}} = 393 \text{ wet ton/day}$$

$$\text{total wt} = 400 + 393 = 793 \text{ wet ton/day}$$

Example 7-5

If a nonporous bulking agent is substituted for the wood chips in Example 7-4, calculate the required volumetric mixing ratio. As in Example 7-4. assume that the bulking agent has a porosity of 35% and that a final mixture FAS of 10% is desired.

Solution

1. No further moisture is assumed to be absorbed by the bulking agent and:

$$S_{bm}^{m} = S_{bm} = S_b$$

$$S_{cm} = S_c$$

2. Calculation of M_{bc} from Equation 7-15:

$$\frac{1}{M_{bc}} = \frac{\gamma_b}{\gamma_w}\left(\frac{S_b}{S_{bm}^{m}} - 1\right) + (f_b - f_m) = f_b - f_m$$

$$\frac{1}{M_{bc}} = (0.35 - 0.10)$$

$$M_{bc} = 4.00$$

3. M_{bc} calculated from Equation 7-18 will be 2.14 as in Example 7-4. Therefore, the required M_{bc} is 4.00, as calculated above.
4. Note that use of a nonporous bulking agent, or one that is already saturated with water, would nearly double the required mixing ratio over that calculated in Example 7-4. This would in turn significantly increase the daily volume of material to be handled and the land area requirements for composting.

Example 7-6

Data [11] indicate the following mass flow of materials for a mixture to be composted by the aerated static pile technique. Estimate the FAS of the mixture.

	Bulk Weight kg/m³	Wet Ton/Day	Percent Solids
Raw Sludge	1068	600	20.0
New Chips	297	58	67.1
Recycled Chips	415	444	62.8
Mixed Blend	726	1102	40.0

Solution

1. Calculate the average solids content and bulk weight of wood chips and sludge:

$$S_b = \frac{58\,(0.671) + 444\,(0.628)}{58 + 444} = 0.633$$

$$\gamma_b = \frac{58\,(297) + 444\,(415)}{58 + 444} = 401 \text{ kg/m}^3$$

$$\gamma_c = 1068 \text{ kg/m}^3$$

2. Calculate the volumetric mixing ratio as

$$M_{bc} = \frac{\text{volume bulking agent}}{\text{volume sludge}} = \frac{\dfrac{444\,(1000)}{415} + \dfrac{58\,(1000)}{297}}{\dfrac{600\,(1000)}{1068}}$$

$$M_{bc} = 2.25$$

3. Assume S_{bm}^{m} to be 0.40, S_{cm}^{m} as 0.40 and f_b as 0.40. First consider the case where moisture transfer is limited by that which the bulking agent can absorb and estimate f_m from Equation 7-15:

$$\frac{1}{2.25} = \frac{0.401}{1.00}\left(\frac{0.633}{0.40} - 1\right) + (0.40 - f_m)$$

$$f_m = 0.189$$

The corresponding S_{cm} is determined from Equation 7-16:

$$S_{cm} = \frac{0.20}{1 + 2.25 \dfrac{0.401}{1.068}\left(1 - \dfrac{0.633}{0.40}\right)} = 0.394$$

$$S_{cm} = 0.394 < 0.40; S_{cm} < S_{cm}^{m}$$

4. Now consider the case where moisture transfer is limited by that which can be released by the sludge cake. Estimate f_m from Equation 7-18:

$$2.25 = \frac{1 - \dfrac{1.068}{1.00}\left(1 - \dfrac{0.20}{0.40}\right)}{0.40 - f_m}$$

$$f_m = 0.193$$

The corresponding value of S_{bm} is determined from Equation 7-19:

$$S_{bm} = \frac{0.633}{1 + \dfrac{1}{2.25}\dfrac{1.068}{0.401}\left(1 - \dfrac{0.20}{0.40}\right)}$$

$$S_{bm} = 0.397 < 0.40; S_{bm} < S_{bm}^{m}$$

5. The condition $S_{bm} < S_{bm}^{m}$ cannot exist, so the conditions as determined in part 3 are correct and:

$$f_m = 0.189$$

$$S_{cm} = 0.394$$

$$S_{bm} = 0.40$$

6. Note that when a bulking agent such as wood chips is used, it is probably not necessary to maintain as high a value of FAS as when using compost recycle or other materials such as sawdust or rice hulls. The reason for this is the larger size of void spaces produced by the bulking particles. It will be shown in Chapter 8 that the dimensions of the void spaces are important in establishing proper airflow and in aiding natural ventilation.

The Effect of Mixture Volume

A key assumption in the previous analysis is that sludge occupies only interstices of the mixture and does not add to total mixture volume. From a theoretical standpoint, this should be a reasonable assumption at low cake solids where the sludge is still pliable and fluid. As case solids increase, however, the structural strength also increases and sludge becomes capable of partially supporting its own weight. From Figure 7-2, it was noted that FAS would begin to occur as cake solids approached about 40%. Therefore, it is likely that as cake solids increase the sludge will begin to add to total mixture volume.

One attempt to determine the effect of cake solids on the volumetric mixing ratio is illustrated in Figure 7-12 for blends of wood chips and dewatered sludge. Unfortunately, FAS was not measured at the different test points in these studies. Some of the variation in volumetric mixing ratio may have been caused by differences in FAS. Therefore, the relationship shown in Figure 7-12 should be viewed with some caution until more data become available.

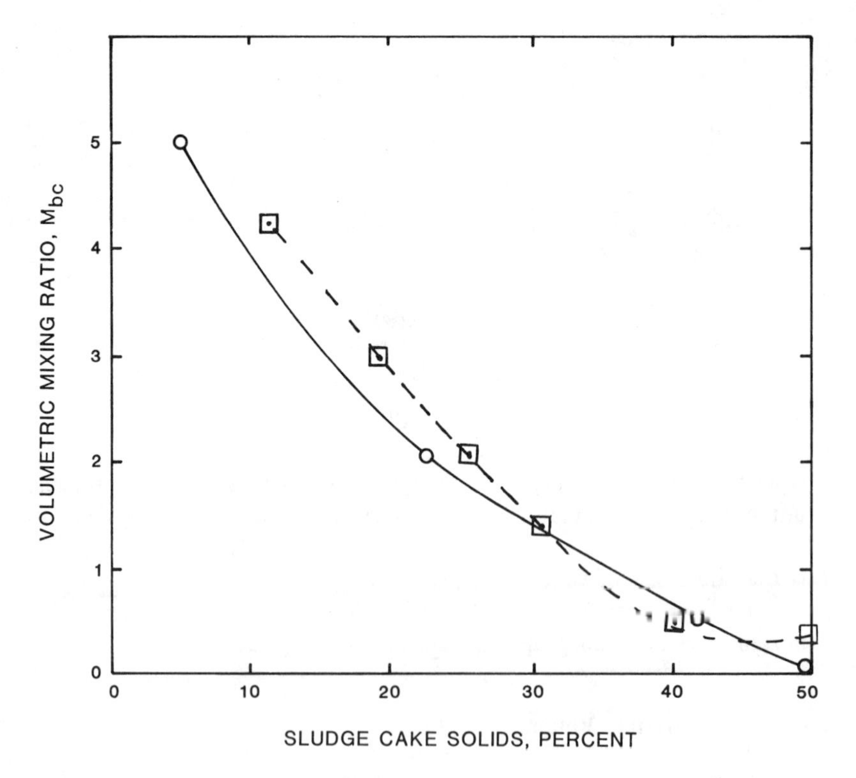

Figure 7-12. Effect of sludge cake solids on the ratio of woodchips to sludge by volume (volumetric mixing ratio). Solid line represents Willson's [14] data; dashed line represents that of Epstein [211]. FAS was not measured in these tests, so values may have varied at the different test points.

The potential increase in volume after mixing is also evident in the material balance of Table 2-3. New and recycled chips are shown to have a total volume of 1149 m^3/day compared to a mixed blend of 1377 m^3/day, an increase of about 20%.

Let M_{mb} be defined as the volume ratio of mixed materials to bulking agent:

$$M_{mb} = \frac{\text{volume mixture}}{\text{volume bulking agent}} = \frac{X_m/\gamma_m}{X_b/\gamma_b} \tag{7-21}$$

$$M_{mb} = f(S_c, M_{bc})$$

M_{mb} would be 1.0 if mixture volume equalled the bulking agent volume, which was the assumption of the previous analysis. The value of M_{mb} should be a function of sludge cake solids, as previously discussed, and also a function of the volumetric mixing ratio, M_{bc}. Unfortunately, the relationship between these variables has not been determined experimentally, and additional work is needed in this area.

Mathematical relationships developed previously can be extended to the case where $M_{mb} > 1$. For the case where moisture transfer is limited by absorption into the bulking agent, Equations 7-11 and 7-12 are still applicable. Substituting Eqations 7-14 and 7-21 into Equation 7-12 and rearranging:

$$\frac{1}{M_{bc}} = \frac{\gamma_b}{\gamma_w}\left(\frac{S_b}{S_{bm}^m} - 1\right) + M_{mb}(1 - f_m) - (1 - f_b) \tag{7-22}$$

With moisture transfer limited by the bulking agent, S_{bm} will equal S_{bm}^m, and S_{cm} can be determined from Equation 7-16. Note that if $M_{mb} = 1$, Equation 7-22 reduces to Equation 7-15.

For the case where moisture transfer is limited by the sludge cake, Equations 7-11 and 7-17 are applicable. Substituting Equations 7-14 and 7-21 into Equation 7-17 and rearranging:

$$M_{bc} = \frac{1 - \dfrac{\gamma_c}{\gamma_w}\left(1 - \dfrac{S_c}{S_{cm}^m}\right)}{M_{mb}(1 - f_m) - (1 - f_b)} \tag{7-23}$$

With moisture transfer limited by the sludge cake, S_{cm} will equal S_{cm}^m, and S_{bm} can be determined from Equation 7-19. If $M_{mb} = 1$, Equation 7-23 reduces to Equation 7-18.

The relationship between M_{mb}, S_c and M_{bc} can be defined better only if more experimental data are developed. For the present it will be necessary to assume a value for M_{mb} based on the limited information available. This approach is illustrated in the following example.

Example 7-7

Using the material flow data of Example 7-6 and allowing for the increase in mixture volume, estimate the FAS in the mixture. Assume S_{bm}^{m} to be 0.40, S_{cm}^{m} as 0.40 and f_b as 0.40.

Solution

1. From Example 7-6 the following values apply:

$$\gamma_c = 1068 \text{ kg/m}^3 = 1.068 \text{ g/cm}^3$$

$$\gamma_b = 401 \text{ kg/m}^3 = 0.401 \text{ g/cm}^3$$

$$\gamma_w = 1.00 \text{ g/cm}^3$$

$$S_c = 0.20$$

$$S_b = 0.633$$

$$M_{bc} = 2.25$$

2. From the material flow data the value of M_{mb} can be determined as follows:

$$\text{volume of mixture} = \frac{1102\,(1000)}{726} = 1520 \text{ m}^3/\text{day}$$

$$\text{volume of bulking agent} = \frac{58\,(1000)}{297} + \frac{444\,(1000)}{415} = 1265 \text{ m}^3/\text{day}$$

$$M_{mb} = 1520/1265 = 1.20$$

3. As determined in Example 7-6, moisture transfer will be limited by the bulking agent and f_m can be determined from Equation 7-22:

$$\frac{1}{2.25} = \frac{0.401}{1.00}\left(\frac{0.633}{0.40} - 1\right) + 1.20(1 - f_m) - (1 - 0.40)$$

$$f_m = 0.324$$

The increase in FAS is considerable compared to that calculated in Example 7-6, and points to the importance of including the effect of M_{mb}.

4. To maintain an $f_m = 0.189$ (as calculated in Example 7-6) the volumetric mixing ratio can be reduced. If M_{mb} is constant at 1.20, the value of M_{bc} to maintain f_m at 0.189 can be calculated from Equation 7-22.

$$\frac{1}{M_{bc}} = \frac{0.401}{1.00}\left(\frac{0.633}{0.40} - 1\right) + 1.20(1 - 0.189) - (1 - 0.40)$$

$$M_{bc} = 1.65$$

A reduction in the volumetric mixing ratio means a reduction in the total quantity of material to be handled daily. It would appear that increasing cake solids from dewatering will increase the value of M_{mb}. This in turn will reduce not only the wet weight of cake to be handled but also the quantity of bulking agent.

SUMMARY

The specific gravity of sludge solids is a function of the specific gravities of the organic and ash components making up the solids. If the components are known the average specific gravity of the total solids can be calculated. The unit bulk weight of sludge can then be determined for any solids content assuming porosity within the cake to be zero.

Based upon available data for composting sludges, unit bulk weight begins to decrease from the theoretical at solids contents between about 30 and 40%, and continues to decrease with increasing solids content. This indicates that a sludge or composting mixture begins to assume a measurable FAS somewhere between about 30 and 40% solids, which increases with further drying of the compost. Barring compaction or consolidation during composting, the unit dry weight appears to remain relatively constant.

Because most dewatered sludges are less than 30% solids, essentially no void volume will exist in the dewatered sludge cake. This will significantly limit any oxygen storage within the voids and would require nearly constant mixing to achieve satisfactory oxygenation during composting. Conditioning agents, such as compost product and organic amendments, and bulking particles, such as wood chips, can be blended with dewatered cake to increase the mixture FAS.

FAS is important in maintaining aerobic conditions within a composting material. There is considerable evidence that the optimum moisture content for a particular material is related to maintenance of a certain minimum FAS. Reported optimum moisture contents tend to fall in the range of FAS between about 20 and 35%.

When compost is used as a conditioning agent, initial mixture FAS is probably ~0.10–0.15 if mixture solids are about 40%. As composting proceeds, moisture evaporation will result in increased void volume. FAS of the final product may be as high as 0.40–0.65. Therefore, aeration and oxygen transfer will likely be most difficult in the early stages of composting, becoming easier as composting proceeds and the mixture dries. Some control over FAS is possible by varying the amount of compost product recycled and, thus, the moisture content of the starting mixture.

The function of a bulking agent, such as wood chips, is to provide structural support for the dewatered sludge, assume adequate FAS and increase the size of pore spaces to allow easier air movement through the mixture. Conceptually, sludge can be viewed as occupying void spaces between bulking particles. Sufficient bulking agent must be added to assure that sludge volume does not exceed the available void volume. Use of a dry bulking agent is advantageous because moisture will be absorbed from the sludge, thus increasing the available FAS. Use of a nonporous bulking agent, or one that is already saturated with moisture, will increase the required quantity of bulking agent because no moisture can be absorbed from the sludge. In some cases twice the volume of bulking agent may be required, depending on whether sludge moisture is absorbed. This can have significant effect on the daily volume of material to be handled and the land requirements for composting.

CHAPTER 8

AERATION REQUIREMENTS AND MECHANISMS

INTRODUCTION

Oxygen must be provided to any composting material to supply demands imposed by organic decomposition. This same oxygen or air will be heated by the composting mass, pick up moisture from it and thus dry the remaining material. Drying can occur as part of oxygenation and is an important benefit to be achieved during composting. Obviously, aeration and drying are related because both depend on supplying air to the compost. However, the quantity of air required to accomplish both purposes may be significantly different and it may be difficult in some cases to accomplish both purposes simultaneously.

Another function of aeration is to remove heat from the composting system. Depending on characteristics of the feed substrate, composting temperatures can reach such elevated levels that biological activity is actually impeded. The aeration rate can be used to control the rate of heat removal and thereby adjust the system temperature. This aspect of aeration will be discussed further in Chapters 12-14. For the present, attention will be focused on aeration requirements for organic decomposition and moisture removal.

OXYGEN REQUIREMENTS FOR ORGANIC DECOMPOSITION

The stoichiometric oxygen requirement can be determined from the chemical composition of the organic solids and the extent of degradation during composting. Representative chemical compositions of various organics and organic mixtures are presented in Table 8-1. Assuming an average composition for sludge organics of $C_{10}H_{19}O_3N$, the stoichiometric oxygen requirement can be determined as:

$$C_{10}H_{19}O_3N + 12.5O_2 \rightarrow 10CO_2 + 8H_2O + NH_3 \tag{8-1}$$

Ammonia formed as a result of organic decomposition will probably volatilize from the compost because conditions of elevated temperature and pH >7 are conducive to ammonia volatilization. Thus, oxygen demands imposed by nitrification generally need not be considered.

Based on Equation 8-1, approximately 2.0 g O_2 will be required per g of organic oxidized. This value will vary from a low of about 1.0 g O_2/g organic for highly oxygenated wastes such as cellulose to a high of about 4.0 g/g for saturated hydrocarbons (refer to Table 3-2). On a practical basis, however, the values for domestic sludge should not vary significantly from that calculated by Equation 8-1.

The fraction of organics that will actually degrade during composting must be determined by laboratory or field measurements. Let k_c, k_r, k_a and k_b be termed "degradability coefficients" defined as follows:

k_c = fraction of the sludge cake volatile solids (VS) degradable under composting conditions
k_r = fraction of the compost recycle VS degradable under composting conditions
k_a = fraction of the amendment VS degradable under composting conditions
k_b = fraction of the bulking agent VS degradable under composting conditions.

Note that the coefficients are not defined in terms of total degradable organics but the degradable fraction to be expected under composting conditions. Given enough time all organics could be expected to degrade, but this is not practical under normal conditions. Therefore, a functional definition was developed.

Limited data are available to determine values for the degradability coefficients. Reduction in volatility from about 50-55 to about 35-40% has been reported during windrow composting of primary digested sludge [9,105]. Based on these values, the degradability coefficient k_c for digested cake can be estimated as ranging from 0.33 to 0.56, with an average value of about 0.45. In other words, about 33-56% of the organics in digested primary sludge cake should degrade during composting. This will depend also on conditions maintained during composting and digestion.

Raw sludge will have a volatility of about 70-75%. It is reasonable to assume that volatility after composting would be similar to that for digested, composted sludge, about 0.35-0.40 [11]. Based on this expected change in volatility, the degradability coefficient for raw cake can be calculated as ranging from 0.72 to 0.82. By way of comparison, about 50% organic destruction is observed during anaerobic digestion of raw sludge. Because further decomposition is observed when digested sludge is composted, the estimated range in the degradability coefficient for raw sludge appears reasonable.

Almost no data are available to assess the degradability of compost product. The degree of stabilization achieved during composting depends largely on proper operation of the system, composting and curing times, and a number of other site-specific parameters. As long as organics are present in compost product, continued degradation should occur, but at greatly reduced rates. Therefore, even with a well-stabilized compost it is likely that k_r will be greater than zero.

A well-stabilized sample of digested, windrow-composted sludge was rewetted to a moisture content of 60% and incubated under controlled aerobic conditions at 49°C (120°F). After 18 days, the volatility decreased from 0.325 to 0.29, corresponding to a degradability coefficient k_r of 0.15 for the organic fraction. Although considerably more data are desirable, the value of 0.15 determined above probably indicates the general range of values to be expected.

For the case of digested sludge with a volatility of 50% and a degradability coefficient of 0.5, about 0.25 g will decompose for each g of sludge solids. Stoichiometrically, about 0.50 g of oxygen would be consumed in the oxidation, based on Equation 8-1. This is equivalent to about 2.18 g of air or

Table 8-1. General Chemical Compositions of Various Organic Materials

Waste Component	Typical Chemical Composition	Reference
Carbohydrate	$(C_6H_{10}O_5)_X$	
Protein	$C_{16}H_{24}O_5N_4$	
Fat & Oil	$C_{50}H_{90}O_6$	
Sludge		
Primary	$C_{22}H_{39}O_{10}N$	
Combined	$C_{10}H_{19}O_3N$	145
Refuse (Total Organic Fraction)	$C_{64}H_{104}O_{37}N$	67
	$C_{99}H_{148}O_{59}N$	37
Wood	$C_{295}H_{420}O_{186}N$	37
Grass	$C_{23}H_{38}O_{17}N$	37
Garbage	$C_{16}H_{27}O_8N$	37
Bacteria	$C_5H_7O_2N$	
Fungi	$C_{10}H_{17}O_6N$	

1.7 liters of air/g sludge solids. These values would approximately double under conditions appropriate to raw sludge.

In any practical composting system it will be necessary to supply an excess of air over the stoichiometric requirement to assure fully aerobic conditions. For purposes of this text the excess air ratio (EAR) has been defined as the ratio of actual air supplied to that stoichiometrically required.

Example 8-1

A waste is to be composted with a chemical composition for the volatile fraction of $C_5H_7O_2N$. The volatility of the waste is 80% with a degradability coefficient estimated at 0.60. Calculate the weight and volume of air required for biological oxidation per g of original waste.

Solution

1. The balanced stoichiometric equation for oxidation of the organic fraction is:

$$\underset{113}{\overset{1.0\,g}{C_5H_7O_2N}} + \underset{5(32)}{\overset{x}{5O_2}} \rightarrow 5CO_2 + 2H_2O + NH_3$$

$$x = \frac{5(32)}{113}(1.0) = 1.42 \text{ g } O_2/\text{g organic oxidized}$$

2. Because the original feed is 80% volatile with a degradability of 60%, the oxygen required based on the original feed is:

$$1.42(0.8)(0.6) = 0.68 \text{ g } O_2/\text{g waste}$$

3. Air at 25°C and 1 atm pressure has a specific weight of 1.20 g/l (0.075 lb/ft^3) and contains 23.2% oxygen by weight. The theoretical air volume and weight are determined as:

$$\frac{0.68}{(1.20)(0.232)} = 2.44 \text{ liter/g waste}$$

$$\frac{0.68}{0.232} = 2.93 \text{ g air/g waste}$$

Organic amendments are often added to composting materials for a variety of reasons. Amendments added to sludge are usually used to increase the content of degradable material, and thus improve the thermodynamic balance, or to increase the mixture free airspace (FAS). In many cases such amendments are derived from plant materials, such as wood, sawdust, garden debris, agricultural wastes and refuse, or animal materials such as chicken, horse and steer manure, and paunch. Manures and paunch are composed largely of undigested plant material. Thus, the degradability of plant materials is of central importance when considering amendment additions.

Cellulose and hemicellulose are the major structural molecules used by plants. Each is a polymeric material composed primarily of simple sugar subunits, glucose in the case of cellulose, and glucose, xylose, galactose and others in the case of hemicellulose. Cellulose comprises more than 50% of the total organic carbon (TOC) in the biosphere. Wood is often 45% cellulose and cotton nearly 95% cellulose. The cellulose contained in wood and other sources contains lignin and other binders in a very complex structure. Lignin is a complex, three-dimensional polymer which retards decomposition of the cellulose. The retardation is thought to be strictly physical because the presence of lignin between the cellulose fibrils decreases the surface area accessible to microorganisms [106].

The degradability of various cellulosic amendments during composting will depend in large measure on the previous processing. A fairly high degree of decomposition is observed in paper products which have been kraft-processed to remove lignin. This would include most paper and cardboard products, except newsprint. Golueke [107] reported that about 90% of the cellulose in kraft pulp was destroyed during digestion at 37°C for 30 days. Newsprint is mechanically pulped and would be expected to have a digestibility somewhere between that of kraft pulp and native wood. Mechanical pulping does not remove lignin, but disrupts the structure, providing increased surface area and accessibility to microbial attack. Golueke [107] noted that cellulose in newsprint was 50% decomposed in anaerobic digestion at 37°C for 30 days, whereas native wood showed little, if any, degradation of the cellulose.

The composition and degradability of common organic amendments is presented in Tables 8-2 and 8-3. It should be noted that many of the

Table 8-2. Compositions and Degradability of Municipal Refuse and Its Components [106]

Item	Percent of Total Dry Weight Organics	Degradability (%)
Cellulose, Kraft	40	90
Cellulose, Mechanical Pulp	15	50
Hemicellulose	10	70
Other Sugars	10	70
Lignin	10	0
Lipids	8	50
Protein	4	50
Other (Plastics, etc.)	3	0

Table 8-3. Degradability of Certain Organic Amendments

Amendment	Degradability (%)	Reference
Refuse (Total Organic Fraction)	43–54	106–108
Garden Debris	66	184
Chicken Manure	68	184
Steer Manure	28	184
Garbage	66	184

degradability measurements were made under anaerobic conditions and some differences may be expected in aerobic composting environments. However, the general range of values should be applicable.

Another approach to estimating oxygen requirements during aerobic composting is based on the composition of feed substrate and final product. This approach is useful if bench-scale composting studies can be conducted to estimate final product composition. Degradability of the feed substrate is thereby included in the analysis. Rich [109] suggested the following stoichiometric equation:

$$C_aH_bO_cN_d + 0.5(ny + 2s + r - c)O_2 \rightarrow nC_wH_xO_yN_z + sCO_2 + rH_2O + (d - nz)NH_3 \tag{8-2}$$

where $r = 0.5[b - nx - 3(d - nz)]$
$s = a - nw$

The terms $C_aH_bO_cN_d$ and $C_wH_xO_yN_z$ represent the compositions of feed substrate and final product, respectively. An elemental analysis is required for evaluation of the subscripts in these terms. Analysis of compost produced from digested sludge amended with recycled product indicated an approximate elemental composition of $C_{25}H_{42}O_{10}N$ [110]. Use of this approach is illustrated in the following two examples adapted from Rich [109].

Example 8-2

Bench scale tests of aerobic composting were conducted on a feed substrate of composition $C_{31}H_{50}O_{26}N$. Tests indicated that 1000 kg of feed organic would be reduced to only 200 kg at the completion of composting. Final product composition was determined to be $C_{11}H_{14}O_4N$. Determine the stoichiometric oxygen requirement per 1000 kg of feed.

Solution

1. Moles of organics entering the process = 1000/852 = 1.173

2. Moles of organics leaving the process per mole entering process = n = 200/(1.173)(224) = 0.76
3. Because $a = 31, w = 11$

$$b = 50, \quad x = 14$$
$$c = 26, \quad y = 4$$
$$d = 1, \quad z = 1$$
$$r = 0.5[50 - 0.76(14) - 3(1 - 0.76 \times 1)] = 19.33$$
$$s = 31 - 0.76(11) = 22.64$$

4. From Equation 8-2, the quantity of oxygen required by the process is:

$$W = 0.5[0.76(4) + 2(22.64) + 19.33 - 26]\ (1.173)(32) = 783 \text{ kg}$$

5. Checking with a materials balance:

In: organic material	1000
oxygen	783
	1783 kg
Out: organic material	200
carbon dioxide 1.173(22.64)(44)	1170
water 1.173(19.33)(18)	408
ammonia [1 – 0.76(1)](1.173)(17)	5
	1783 kg

Example 8-3

How much energy is released as heat in the composting process described in Example 8-2?

Solution

1. The percentage elemental compositions of the organics entering and leaving the process are:

In: $\%C = 31 \times 12/852 = 43.6$
$\%H = 50 \times 1/852 = 5.9$
$\%O = 26 \times 16/852 = 48.9$
Out: $\%C = 11 \times 12/224 = 59.0$
$\%H = 14 \times 1/224 = 6.3$
$\%O = 4 \times 16/224 = 28.6$

2. From Equation 3-26, the R-values of the organics entering and leaving the process are:

$$\text{In } R = \frac{100[2.66(43.6) + 7.94(5.9) - 48.9]}{398.9} = 28.6$$

$$\text{Out } R = \frac{100[2.66(59.0) + 7.94(6.3) - 28.6]}{398.9} = 44.8$$

3. From Equation 3-27, the unit heats of combustion of the organics entering and leaving the process are:

$$\text{In } h = 127(28.6) + 400 = 4030 \text{ cal/g}$$
$$\text{Out } h = 127(44.8) + 400 = 6100 \text{ cal/g}$$

4. Total energy released as heat is:

$$\begin{aligned} \text{Energy in:} &\quad 1000 \times 1000 \times 4030 = 4030 \times 10^6 \text{ cal} \\ \text{Energy out:} &\quad 200 \times 1000 \times 6100 = \underline{1220 \times 10^6 \text{ cal}} \\ &\quad 2810 \times 10^6 \text{ cal} \end{aligned}$$

AIR REQUIREMENTS FOR MOISTURE REMOVAL

Determination of the air requirement for drying requires further analysis of the composting process. Referring to the mass balance diagram shown in Figure 6-2, and neglecting any organic amendment or bulking agent addition, the quantity of water to be evaporated daily is given by:

$$\Delta H_2O = (X_c - S_cX_c) - (X_p - S_rX_p) \quad (8\text{-}3)$$

Considering the quantity of ash to be conservative, a mass balance on the inorganic fraction yields:

$$(1 - V_c)S_cX_c = (1 - V_r)S_rX_p \quad (8\text{-}4)$$

Solving Equation 8-4 for X_p, substituting into Equation 8-3, and rearranging gives:

$$\frac{\Delta H_2O}{S_cX_c} = \left(\frac{1 - S_c}{S_c}\right) - \left(\frac{1 - V_c}{1 - V_r}\right)\left(\frac{1 - S_r}{S_r}\right) \quad (8\text{-}5)$$

Equation 8-5 was solved for various cake and compost solids contents. Results and other assumptions used in the calculations are presented in Figure 8-1. Conditions appropriate to a digested sludge were used to obtain the results shown in Figure 8-1. However, similar analysis for the case of a raw sludge results in very little change from the values indicated. As would be expected, the quantity of moisture to be evaporated is determined primarily by the input solids at cake solids below about 30%. Above 30%, both feed solids and the final compost solids are important in determining the quantity of moisture to be evaporated. Rates of biological activity begin to decrease

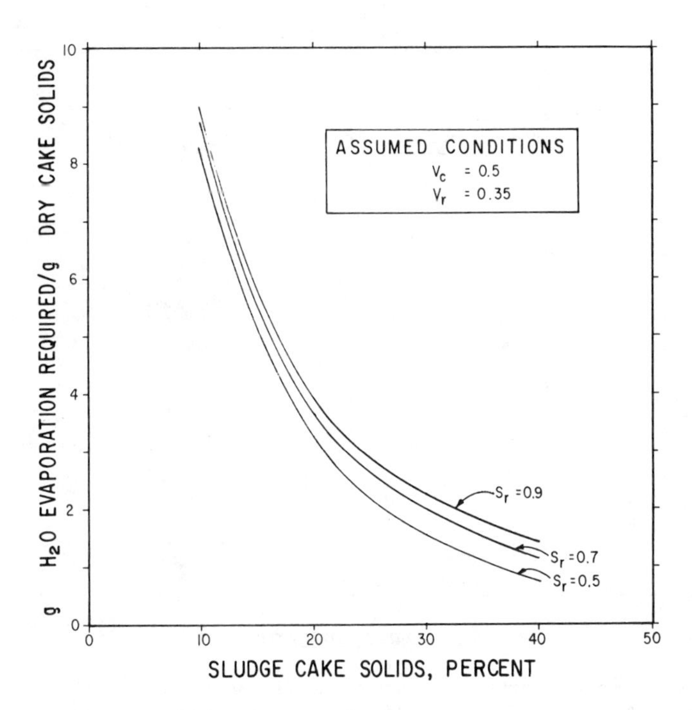

Figure 8-1. Effect of sludge cake solids and final compost product solids on required moisture evaporation [99].

below moisture levels of about 40-50% [44]. Therefore, drying beyond a moisture level of 30-40% with heat supplied by composting may not be practical.

The quantity of water vapor which can be carried in saturated air at different temperatures is shown in Figure 8-2. The curve was constructed from standard psychrometric charts and steam tables [34] for a total atmospheric pressure of 760 mm Hg. The quantity of moisture in saturated air increases exponentially with increasing air temperature. Air that leaves a wet composting material can be assumed to be saturated, or nearly so. Furthermore, the exit air should be at about the same temperature as the composting material. If temperatures common to composting are obtained, considerable moisture will be removed with the exit air. Also, if the temperature difference between inlet and exit air is greater than about 25°C, relative humidity of the inlet air will have a minor effect in determining the overall moisture removal. This means that drying can occur even in climates with high natural humidity.

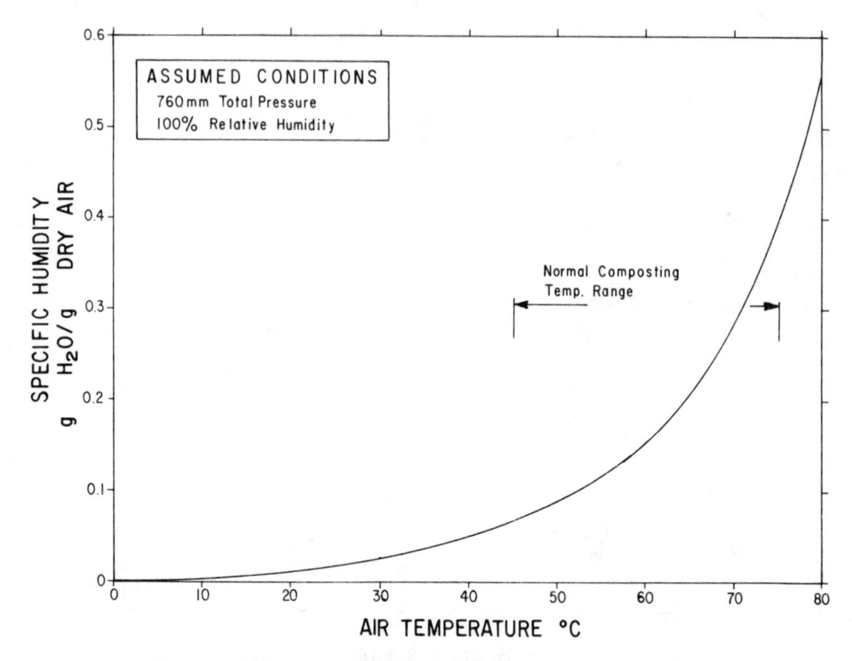

Figure 8-2. Specific humidity as a function of air temperature under conditions of 100% relative humidity and 760 mm Hg total pressure. See Chapter 11 for the mathematics of this relationship.

Example 8-4

A sewage sludge is to be composted and dried to a final moisture content of 30%. The dewatered cake is 30% solids with a volatility of 0.50. Compost product will be used as a conditioning agent and is expected to have a volatility of 35%. Determine the weight and volume of required air if exit and ambient air temperatures are 60 and 20°C, respectively.

Solution

1. Determine the required moisture evaporation from Equation 8-5:

$$\frac{\Delta H_2O}{S_C X_C} = \left(\frac{1-0.30}{0.30}\right) - \left(\frac{1-0.50}{1-0.35}\right)\left(\frac{1-0.70}{0.70}\right)$$

$$= 2.00 \text{ g } H_2O/\text{g dry cake solids}$$

2. Determine moisture-carrying capacity of exit air. From Figure 8-2 the specific humidity of air at 20 and 60°C is 0.015 and 0.152 g H_2O/g dry air,

respectively. Assuming exit and ambient air to be saturated, moisture removed from the composting sludge would be:

$$0.152 - 0.015 = 0.137 \text{ g } H_2O/\text{g dry inlet air}$$

Note the relatively small moisture contribution of the inlet air if the temperature difference is about 20°C or greater.

3. Weight of required air:

$$\frac{2.00}{0.137} = 14.6 \text{ g dry air/g dry cake solids}$$

4. From Example 8-1 the specific weight of air at standard temperature and pressure (STP) is 1.20 g/1. Therefore:

$$\frac{14.6}{1.2} = 12.2 \text{ liters dry air/g dry cake solids}$$

Note that the air required for drying is significantly greater than that for biological oxidation as calculated in Example 8-1.

Based on data contained in Figures 8-1 and 8-2, the quantity of air required for moisture removal was calculated as a function of sludge cake solids. Inlet air was assumed to be 20°C and 100% relative humidity. Exit air temperature from the composting material was assumed to be 40°C or greater. Thus, inlet air conditions are not critical to the problem. A final compost moisture content of 30% was assumed. Results and other assumptions used in the analysis are presented in Figure 8-3. The range of stoichiometric air requirements for biological oxidation of digested and raw sludge is also shown. The air requirement for drying is significantly greater than the air requirement for biological oxidation and is influenced largely by the cake solids content and exit air temperature. Only at cake solids approaching 30-40% and an exit air temperature of 70°C do the air requirements for drying and biological oxidation become reasonably equivalent. At cake solids of 20%, the air requirement for drying can be as much as 10-30 times that for biological oxidation. Obviously, if too much air is supplied the composting material will be cooled, exit air temperatures will drop, and air requirements for drying will increase even further over those calculated above. Achieving both drying and composting requires control between energy release as a result of biological oxidation and heat demands for drying, a subject of discussion in Chapter 9.

Data presented in Figure 8-3 are particularly significant for those composting systems in which air contact is controlled, for example, the

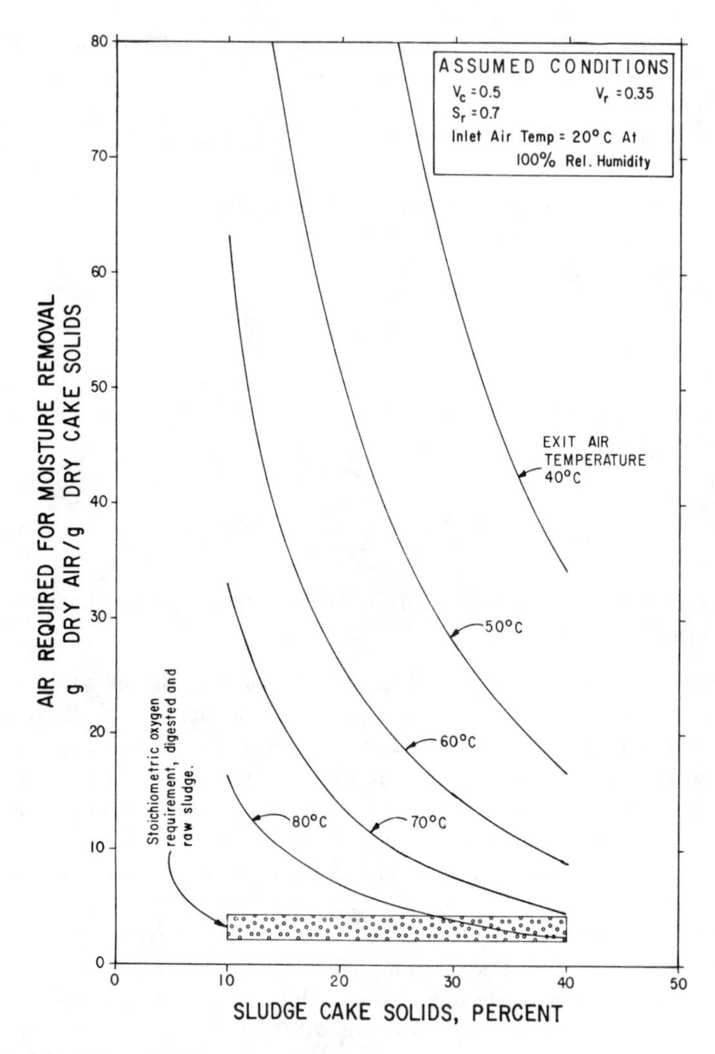

Figure 8-3. Effect of sludge cake solids and exit air temperature on air requirements for moisture removal to final compost solids content of 70% [99].

aerated static pile and many mechanical systems. A measure of operational control is provided that is not enjoyed by systems where air contact is uncontrolled. If air supply is limited to the stoichiometric requirement, only limited drying will occur. Air supply to the aerated pile system is often regulated by monitoring the oxygen content of air in the pile interior. In such a case, oxygen supply may be only slightly greater than stoichiometric.

Even though only limited drying is possible with stoichiometric oxygen supply, process modifications to achieve better drying are not precluded. Air supply is the key variable that controls the rate of moisture and heat removal from the composting system. Relationships between these parameters are quite complex and a more complete analysis must await development of simulation models and the discussion of process dynamics in Chapters 11 to 14. However, a number of approaches to improve moisture removal are possible. As an example, air supply could be maintained at stoichiometric levels throughout the composting cycle; this will result in only limited moisture removal. After stabilization the material could be spread out into windrows or thin layers and turned at regular intervals to encourage surface drying. The potential for odors should be reduced greatly because of the stable nature of the drying compost. However, climatic conditions will have a large bearing on the ability to surface dry in this manner. The latter method has been used at Beltsville with general success [11,13].

Alternatively, the air supply could be limited to the stoichiometric requirement in the early stages of composting, thus encouraging maximum temperature elevations and pathogen destruction. The air supply could then be increased to effect better moisture removal. Obviously, the increased air supply will increase the rate of heat removal from the pile, resulting in lower exit air temperature. Whether sufficient energy is available for the purposes of drying will depend on the thermodynamic balance established within the system (see Chapter 9).

As indicated in Figure 8-3, the quantity of air required for moisture removal increases dramatically as the temperature difference between inlet and outlet air is reduced. A number of compost systems employ secondary reactors for compost drying which use either ambient air or heated air. Thus, the task of drying is separated from the composting process. Estimation of the air quantity required for such a drying approach is illustrated in the following example.

Example 8-5

Compost is produced from an enclosed reactor system at a solids content of 50%. The compost is to be dried to 90% solids for reuse as a conditioning agent. Ambient air temperature is 20°C with a specific humidity at saturation of 0.015 g H_2O/g dry air and relative humidity of 75%. Estimate the air requirements for drying if ambient air is used. What if air is preheated to 60°C (specific humidity = 0.152 g H_2O/g dry air)?

Solution

1. The analysis will assume that inlet and outlet air are at the same temperature and that exit air is saturated. Water to be removed per gram of

solids can be determined as:

$$\left(\frac{1}{0.5} - 1\right) - \left(\frac{1}{0.9} - 1\right) = 0.889 \text{ g } H_2O/\text{g solids}$$

2. For the case using ambient air at 20°C, the theoretical air requirement is estimated as:

$$0.889 \frac{\text{g } H_2O}{\text{g solids}} \times \frac{1 \text{ g dry air}}{0.015 \text{ g } H_2O} \times \frac{1}{(1.0 - 0.75)} \times \frac{10^6\text{g}}{\text{ton}} \times \frac{22.4 \text{ liters}}{29 \text{ g air}} \times \frac{1 \text{ m}^3}{1000 \text{ liters}} =$$

$$183{,}000 \text{ m}^3 \text{ dry air/dry ton compost}$$

3. If air is preheated to 60°C the moisture removal capacity is:

$$0.152 - (1.0 - 0.75)(0.015) = 0.148 \text{ g } H_2O/\text{g dry air}$$

The air requirement is then:

$$0.889 \times \frac{1}{0.148} \times 10^6 \times \frac{22.4}{29} \times \frac{1}{1000} = 4640 \text{ m}^3 \text{ dry air/dry ton}$$

4. A number of simplifying assumptions are involved in the above analysis which should be noted. First, evaporative cooling will occur in the dryer, and the exit air temperature may be reduced from the inlet temperature. This would increase air requirements over that calculated above. Second, adjustment of specific humidity by direct multiplication with the relative humidity is not strictly correct as will be shown later in Chapter 11. The error introduced in the present example is quite small, however.

5. Drying with ambient air requires use of very large air volumes. The volume can be reduced significantly if some preheat is used. Waste heat sources are often available for such purposes. Air preheat by heat exchange with hot exhaust gases from composting is one possibility.

MECHANISMS OF AERATION

For the aerated static pile and many reactor compost systems, the mechanisms for supplying oxygen are fairly obvious. For example, air exchange in the aerated pile process is provided by a forced ventilation system using blowers to pull or push air through the pile. Most reactor systems employ either forced aeration and/or near constant tumbling and agitation of solids to expose new surfaces to the air. In other systems, particularly the windrow process, mechanisms of aeration are not nearly so obvious. On first impulse one is tempted to assume that windrow turning is the primary method for oxygen replenishment within the FAS. As will be seen this does not appear to be the principal mechanism.

The purpose of this section is to explore the factors responsible for supplying the oxygen required in composting. Particular attention will be given to the windrow process as its aerating mechanisms are the least defined of the sludge composting processes.

Exchange Volumes

For any composting mixture the air volume required to supply the stoichiometric oxygen demand can be estimated. The calculation requires a knowledge of the type of sludge, amendments and bulking agents making up the mixture, and their composition and expected degradability during composting. Once the total air volume is estimated the number of exchange volumes can be calculated provided the FAS in the mixture is known. The number of exchange volumes is defined as the number of times the mixture FAS must be renewed with air to supply the stoichiometric oxygen demand. Knowledge of the number of exchanges required can provide insight into the responsible aerating mechanisms.

The required number of exchange volumes were estimated for a digested sludge under various conditions of mixture unit bulk weight and cake solids S_c. In all cases it was assumed that compost product was recycled to achieve a 40% solids mixture. Results of the analysis are presented in Figure 8-4, using the same assumptions and method of calculation shown in Example 8-6 below.

Example 8-6

A digested, dewatered sludge at 20% solids and 50% volatility is blended with recycled compost at 70% solids and 35% volatility to achieve a mixture solids of 40%. If unit weight of the mixture is 0.64 g/cm^3, estimate the number of volume exchanges required to satisfy the oxygen demand during composting.

Solution

1. Determine the required recycle ratio R_d from Equation 6-7:

$$R_d = \left(\frac{0.40}{0.20} - 1\right) \Big/ \left(1 - \frac{0.40}{0.70}\right) = 2.33$$

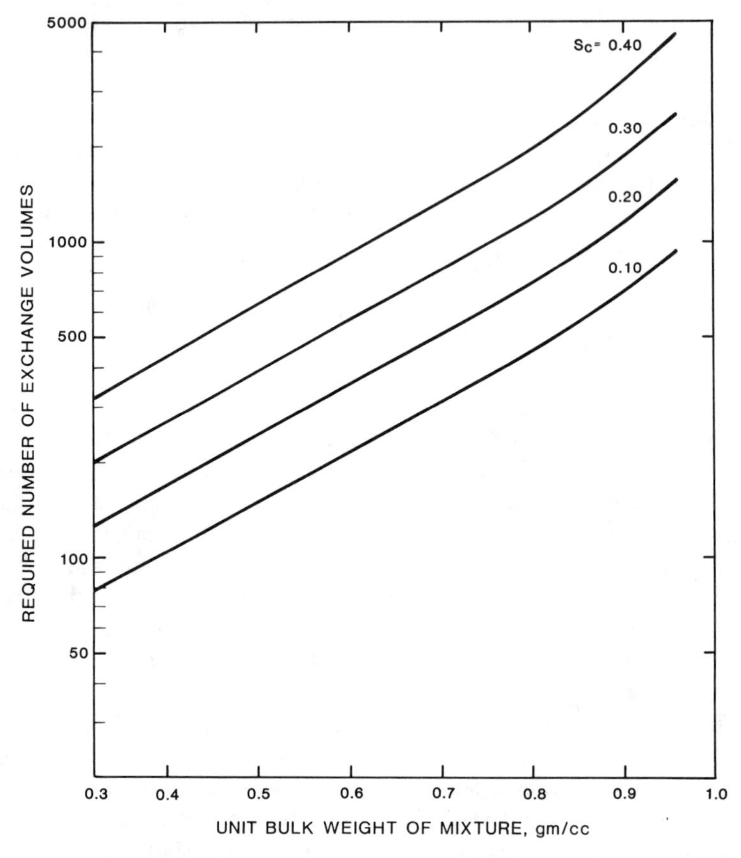

Figure 8-4. Effect of mixture bulk weight and sludge cake solids on the required number of exchange volumes.

2. The specific gravities of solids in the cake and compost recycle can be calculated from Equation 7-1:

$$\text{Cake} \quad \frac{1}{G_S} = \frac{0.30}{1.0} + \frac{(1 - 0.30)}{2.5} \Rightarrow G_S = 1.43$$

$$\text{Compost} \quad \frac{1}{G_S} = \frac{0.35}{1.0} + \frac{(1 - 0.35)}{2.5} \Rightarrow G_S = 1.64$$

The specific gravity of solids in the mixture can be calculated as the weighted average of the two components:

$$\text{Mixture} \quad \frac{1}{G_S} = \frac{1.43(1.0) + 1.64(2.33)}{1.0 + 2.33} = 1.58$$

3. Determine the mixture FAS from Equation 7-10:

$$f_m = 1 - \frac{0.64(0.40)}{1.58(1.00)} - \frac{0.64(1 - 0.40)}{1.00} = 0.453$$

4. The air requirements for composting can be estimated from the stoichiometry of Equation 8-1. Assume a degradability of 0.5 for the digested cake and 0.10 for recycle compost:

$$\text{air for cake} \quad \frac{2.00\ gO_2}{g\ \text{organic}} \frac{(0.5)(0.5)}{(0.23)} = 2.17\ \text{g air/g cake solid}$$

$$\text{air for compost} \quad \frac{2.00(0.35)(0.10)}{(0.23)} = 0.304\ \text{g air/g compost solids}$$

5. The total air required for 1 liter of mixture can be estimated as:

$$\text{air for cake} \quad (10^3)\ \frac{0.64\ g}{cm^3}\ (0.40)\left(\frac{1.00}{1.00 + 2.33}\right)\ 2.17 = 166.7\ g/l$$

$$\text{air for compost} \quad (10^3)(0.64)(0.40)\left(\frac{2.33}{1.00 + 2.33}\right)\ 0.304 = 54.5\ g/l$$

Therefore, the total air requirement is 166.7 + 54.5 = 221.2 g/l of mixed material.

6. The required number of void volume exchanges can then be estimated assuming air density at 1.20 g/l:

$$\frac{221.2}{(0.453)1.20} = 407\ \text{exchanges}$$

Note that this corresponds to the value presented in Figure 8-4.

Based on Figure 8-4, required exchange volumes can vary over nearly two orders of magnitude within the range of variables assumed. The number of exchanges increases with increasing unit bulk weight since the denser mixtures contain both less FAS and more decomposable organics per unit volume. Thus, unit weight is a critical parameter in determining the number of exchanges. The reader should note that Figure 8-4 uses a semilog scale. For example, with $S_c = 0.20$ the number of exchanges increases from about 230 to 740 as the bulk weight increases from 0.48 to 0.80 g/cm^3. If bulk weight continues to increase beyond the 0.95 g/cm^3 shown in Figure 8-4, the FAS will eventually reach zero and the number of exchanges will become infinite. This explains the upward trend to curves shown in Figure 8-4 at higher bulk weights. Obviously, if a composting mixture really had zero FAS, no air could be stored within the composting matrix and nearly constant agitation would be necessary to continually expose the mixture to the air.

The number of exchanges is also shown to be a function of dewatered cake solids. Because recycle of compost product was used to achieve a 40% solids mixture, increased cake solids decreases the amount of compost product in the mixture. Degradability of compost product was assumed to be significantly less than that of dewatered cake. Therefore, at higher cake solids the quantity of degradable organics per unit volume is greater, which results in a larger number of required air exchanges.

In the early stages of composting, the bulk weight of cake/compost recycle mixtures is probably in the range of 0.7-0.95 g/cm^3. Required exchange volumes would range from about 300 to 4500 under these conditions. As composting proceeds, moisture evaporation will result in an increased FAS and bulk weight will likely be reduced to about 0.5-0.7 g/cm^3. Even so, at least 150 exchange volumes would be necessary and perhaps as many as 1500.

Considering the case of windrow composting, suppose it is assumed that each turning of the windrow completely renews air contained in the voids and that no other aerating mechanisms are at work. With these assumptions, the number of windrow turnings must equal the number of exchange volumes required. Therein lies the problem. The number of turnings normally used in windrow composting is nowhere near the exchange volumes estimated in Figure 8-4. Furthermore, no combination of reasonable values for the variables used in the analysis seems likely to reduce significantly the required number of exchange volumes.

Available data on the windrow composting system operated by the Los Angeles County Sanitation Districts [9,105] indicate that daily turning of the windrow for composting periods ranging from 20 to 50 days is the most common operational mode. More frequent turning may be used in the first few days of composting. This operational procedure was developed while windrow composting a digested sludge of 30-35% solids, using recycled compost to adjust the initial mixture to 40% solids. With a reported mixture bulk weight of 0.9 g/cm^3, and referring to Figure 8-4, nearly 2000 exchange volumes would be required.

Smith [111] and Iacoboni et al. [112] observed that oxygen concentrations in the interiors of composting sludge windrows often decreased to undetectable levels within a few minutes to several hours after turning. A profile of oxygen content with depth into such a windrow is shown in Figure 8-5. The data were recorded 5 hr after turning. Based on the windrow dimensions it was estimated that about 50% of the windrow volume was

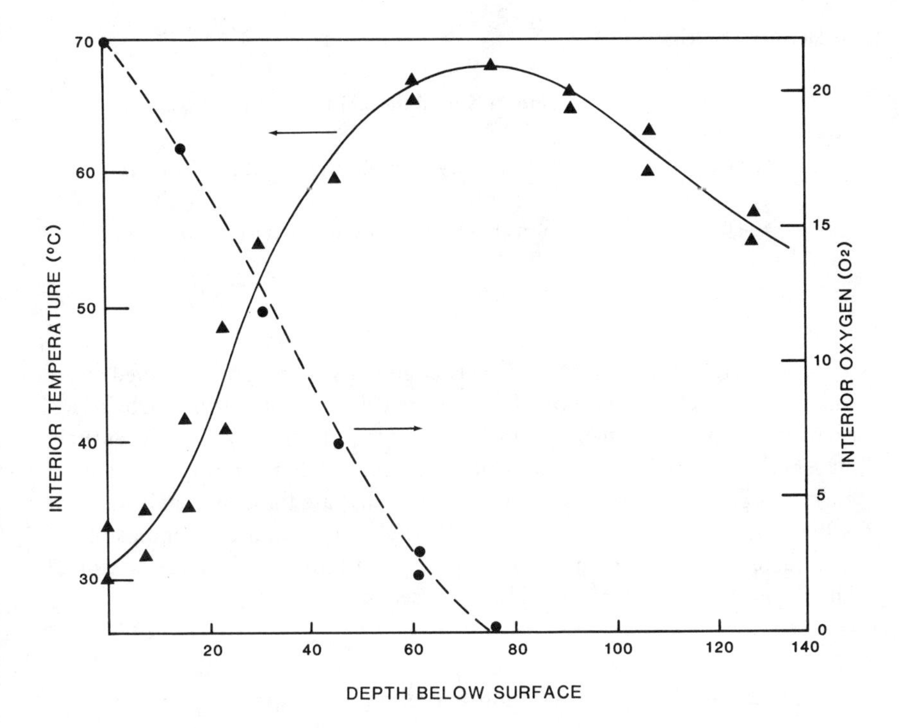

Figure 8-5. Profiles of oxygen concentration (%) and temperature with depth (cm) into a windrow composed of digested sludge cake and recycled compost. Measurements were made 5 hr after windrow turning [112].

aerobic, while the rest was either slightly aerobic or anaerobic. Maier et al. [113] observed a similar phenomenon in windrows composed of finely ground refuse (particle size <1.25 cm) and reported oxygen depletion in the bottom interior 4-5 hr after turning. These observations would seem reasonable in light of the analysis of required exchange volumes.

In discussing the turning frequency required for windrow composting of ground refuse, Golueke [44] indicated that the windrow need be turned every other day for a total of only 4-5 turns. By the fourth or fifth turn the material should be so advanced toward stabilization as to require no further turning. More frequent turning would be beneficial if the material were very wet or compacted. Based on experience gained during windrow composting studies at the University of California, McGauhey and Gotaas [96] and later Golueke [114] recommended the following turning schedule as adequate for municipal refuse (2.5-5 cm probable particle size):

Moisture Content %	
>70	Turn daily until the moisture content is reduced to <70%.
60-70	Turn at 2-day intervals. Number of required turns is about five.
40-60	Turn at 3-day intervals. Approximate number of turns is four.
<40	Add moisture.

The question that remains is how such minimal turning as described above can result in the aerobic conditions desired during composting. Obviously, exchanging air contained in the FAS by windrow turning is entirely inadequate to supply the air quantities required. Other aeration mechanisms must be operating in the windrow process. What are these mechanisms, what is the function of windrow turning and why should windrow composting of such materials as refuse appear to require less turning than sludge mixtures? These questions are addressed in the next section.

Aerating Mechanisms In The Windrow Process

In the time interval between turnings there are two ways oxygen can be supplied through pore spaces into the interior of the windrow: (1) by molecular diffusion and (2) by mass movement of air through the pores in response to an energy gradient. Molecular diffusion results from constant and random collisions between molecules of a fluid. As a result of such collisions, there is a tendency for molecules to move from a zone of high concentration to a zone of lower concentration. If oxygen is depleted within the windrow, there will be a net movement of oxygen from the surrounding air into the windrow as a result of molecular diffusion. Similarly, carbon dioxide and water vapor will diffuse from the interior of the windrow where it is produced to the outside air where the concentration is lower. The problem with molecular diffusion, however, is that the process is extremely slow compared to rates at which oxygen is required for composting. Shell [115] and Snell [116] studied the diffusion of oxygen through composting refuse and concluded that diffusion alone supplied only a small percentage of the maximum oxygen needs during composting. Furthermore, the rate of diffusion decreased as moisture content of the compost increased, a definite problem with sludge mixtures. Only when refuse was placed in thin (5- to 10-cm) layers, did molecular diffusion become significant. Even in this case the rate of oxygen diffusion was estimated to be <5% of the maximum rate of oxygen demand.

Because molecular diffusion is not a practical transport mechanism, it would appear that only a mass flow of air in response to an energy gradient can supply the required oxygen. Remember, forced ventilation or aeration by means of mechanical blowers or fans is not considered here. The question, then, is "What forces are operating within a composting windrow or pile to produce a mass flow of air?"

Referring to Figures 2-8, 2-9 and 2-29, it is common to observe steam continually issuing from the top of windrows even though they may only be turned daily. Obviously, this implies a mass flow of water vapor from the pile interior. The movement is likely in response to the high temperature of gases within the windrow, analogous to movement of hot gases up a chimney.

The density of dry and saturated air as a function of temperature is shown in Figure 8-6. As temperature increases, the density of dry air decreases. However, the degree of saturation can also affect the density, particularly at higher temperatures, because the saturation vapor pressure increases exponentially with temperature (Figure 8-2). The effect of water vapor on density is explained by the fact that the molecular weight of water is considerably less than that of the oxygen and nitrogen it displaces. Carbon dioxide produced from organic decomposition would have the opposite effect because its molecular weight is greater than either oxygen or nitrogen.

The density difference between warm moist air within the interior of the windrow or pile and colder, less moist ambient air produces an upward buoyant force which in turn induces a natural ventilation of the windrow. A schematic illustration of the process is presented in Figure 8-7. The concept of natural draft is easily understood. However, parameters have not yet been described here to determine the rate of ventilation and determine whether the air supplied is sufficient to satisfy oxygen demands.

A Model For Natural Ventilation

To determine if natural ventilation can supply sufficient oxygen for composting, one would ideally construct an experiment to verify the hypothesis. Unfortunately, such information on composting sludge mixtures is virtually nonexistent. Therefore, recourse must be made to a theoretical model. Although this is less satisfying than experimental data, it will permit the examination of relevant variables that influence rates of natural ventilation.

To simplify the physical system for mathematical analysis the pore spaces are envisioned as cylindrical tubes running between the composting

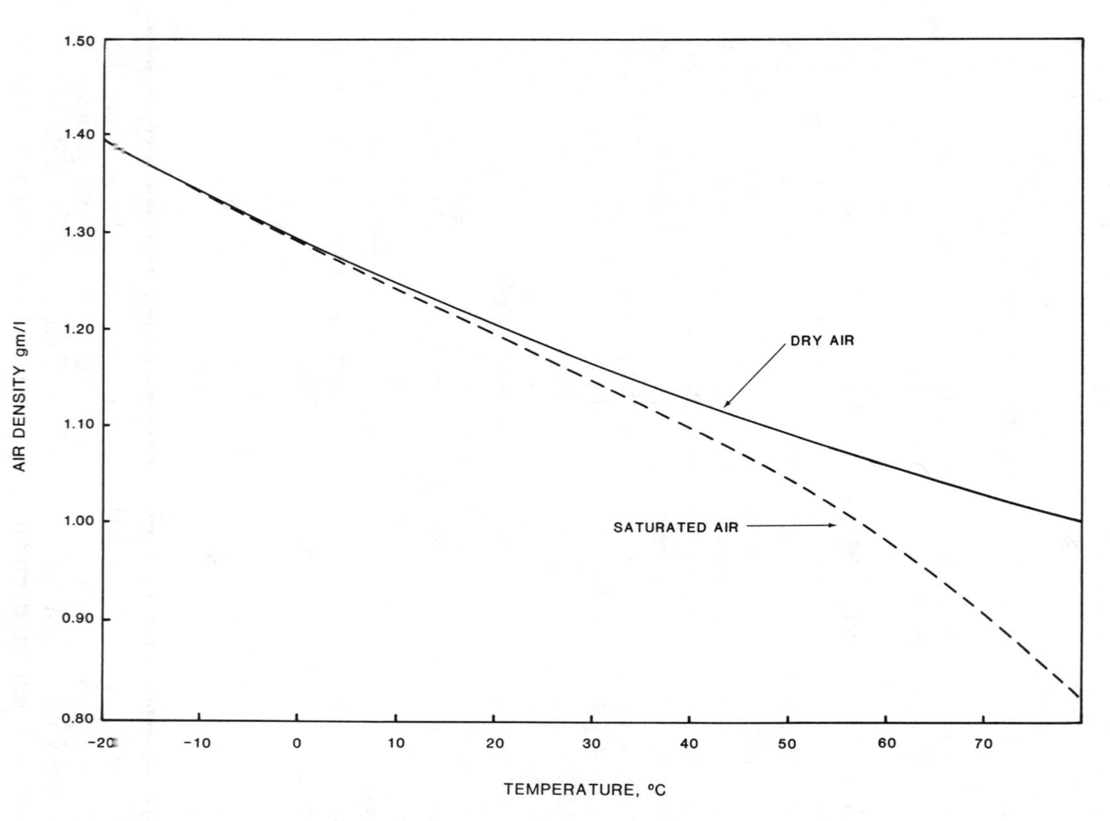

Figure 8-[illegible]. Density of dry and saturated air as a function of air temperature (dry bulb) at 760 mm Hg pressure.

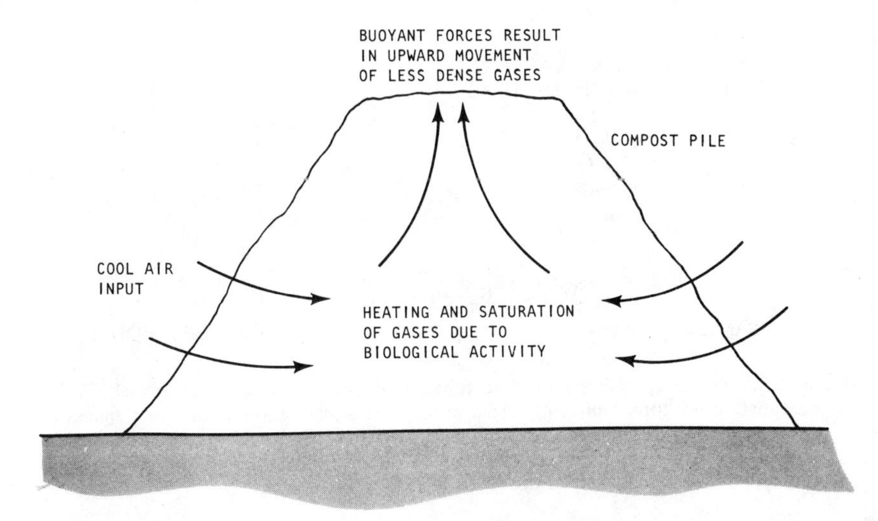

Figure 8-7. Schematic illustration of the natural draft and ventilation induced by buoyant forces acting on hot, moist air produced during active composting.

particles as shown in Figure 8-8. Composting particles are assumed to be discrete, spherical particles of radius r. Void spaces are assumed to be free of water so that porosity and FAS are equivalent. Average pore size, total FAS and total surface area are retained in the envisioned cylindrical tubes. Size of the tubes is a function of the diameter of particles in the composting mixture and the porosity or free air space. For a cylindrical tube the volume is:

$$V = \pi R^2 L \tag{8-6}$$

and the surface area:

$$S = 2\pi RL \tag{8-7}$$

The tube radius is thus:

$$R = \frac{2V}{S} \tag{8-8}$$

where V = total volume of pore tubes which equals the void volume in the composting mixture
R = radius of pore tube
r = radius of composting particles
S = surface area of pore tubes which equals the surface area of particles in the composting mixture
L = length of pore tubes which equals the height of the composting mixture

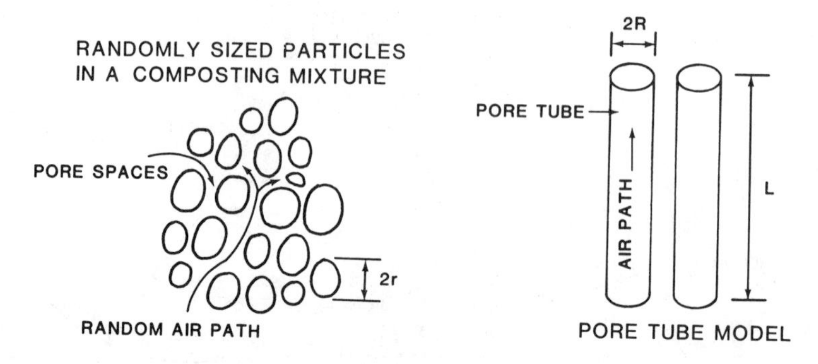

Figure 8-8. Schematic diagram of pore tubes used to represent pore spaces within the composting mixture. Uniformly sized pore tubes replace the random pore spaces in the actual composting mixture but retain the same total void volume and surface area. Composting solids are assumed to be discrete particles and the void volume (porosity) is assumed to equal the FAS.

The volume in Equation 8-8 corresponds to the total void volume in the composting mixture, while the surface area corresponds to the total surface area of particles. The surface area of a single particle was determined by assuming it to be a sphere. The total number of particles in a given volume was then estimated from the volume of a single particle and the total mixture volume minus the FAS. From the void volume and total surface area, the average pore radius was determined from Equation 8-8.

The velocity of gas flow within a pore tube as a result of gas heating can be determined from analysis of the forces acting on the pore tube. The pressure potential from the density difference between ambient air gases within the pore tube can be expressed as:

$$\text{buoyancy} = \frac{\Delta\gamma}{\gamma} L = \frac{\Delta\rho(g)}{\gamma} L \qquad (8\text{-}9)$$

Head losses which result from fluid movement will balance the forces of buoyancy. These losses are related to both friction and exit velocity. With small flowrates the exit velocities can be neglected. Frictional losses can be estimated from the Hagen-Poiseuille law for laminar flow in tubes.

$$\text{friction head loss} = h_L = \frac{8\mu}{\gamma} \frac{L}{R^2} \nu \qquad (8\text{-}10)$$

where γ = specific weight of fluid
ν = average fluid velocity in the pore tube
ρ = mass density of fluid
μ = fluid viscosity

Equating Equations 8-9 and 8-10 and simplifying:

$$\nu = \frac{R^2 \Delta \rho g}{8\mu} \tag{8-11}$$

If porosity and average particle size within a composting mixture are assumed, the radius and number of pore tubes can be determined. If the specific weight difference is then assumed, the velocity of gas flow within each tube can be determined from Equation 8-11. The flowrate within each tube multiplied by the number of pore tubes yields the estimate of the natural ventilation rate. Additional information on the method of calculation can be found in Appendix III.

This procedure was repeated for various combinations of FAS and particle size with results presented in Figure 8-9. A density difference of 0.22 g/l was assumed between gases in the pore tube and ambient air. From Figure 8-6 this corresponds to a temperature difference of about 40°C. It can be seen that the natural ventilation rate is a function of both the FAS and the size of particles in the composting mixture. Two composting mixtures may have equivalent void volumes but the ventilation rate may be significantly different due to particle size differences. For a given FAS in Figure 8-9, an increase in particle size by a factor of 10 will increase the ventilation rate by a factor of 100.

The influence of particle size on natural ventilation has implications for those systems which use bulking agents such as wood chips. Not only does the bulking agent increase the mixture FAS, it also increases average void size and thus promotes natural ventilation in two ways. However, most systems which use bulking agents also employ forced aeration and thus do not use the naturally high ventilation rates which should result.

A range of probable stoichiometric air rate requirements for composting mixtures is also shown in Figure 8-9. These rates were calculated assuming a nominal 20-day composting time, excess air ratios from 1 to 3, pile heights between 1 and 3 m, and degradability coefficients discussed previously. If the particle size is <0.01 cm, it is unlikely that natural ventilation can satisfy the oxygen demands. Increasing the particle size to 0.1 cm, however, significantly increases the likelihood that natural ventilation can keep pace with the rate of oxygen demand. As moisture leaves the composting material the effective FAS should increase, thus increasing the ventilation rate. Therefore, ventilation may be inadequate in early stages of composting but increase to more adequate levels as drying occurs. For a particle size of 0.1 cm, note the large increase in ventilation rate as FAS increases from 0.20 to 0.60. Such an increase in FAS is not unlikely during windrow composting.

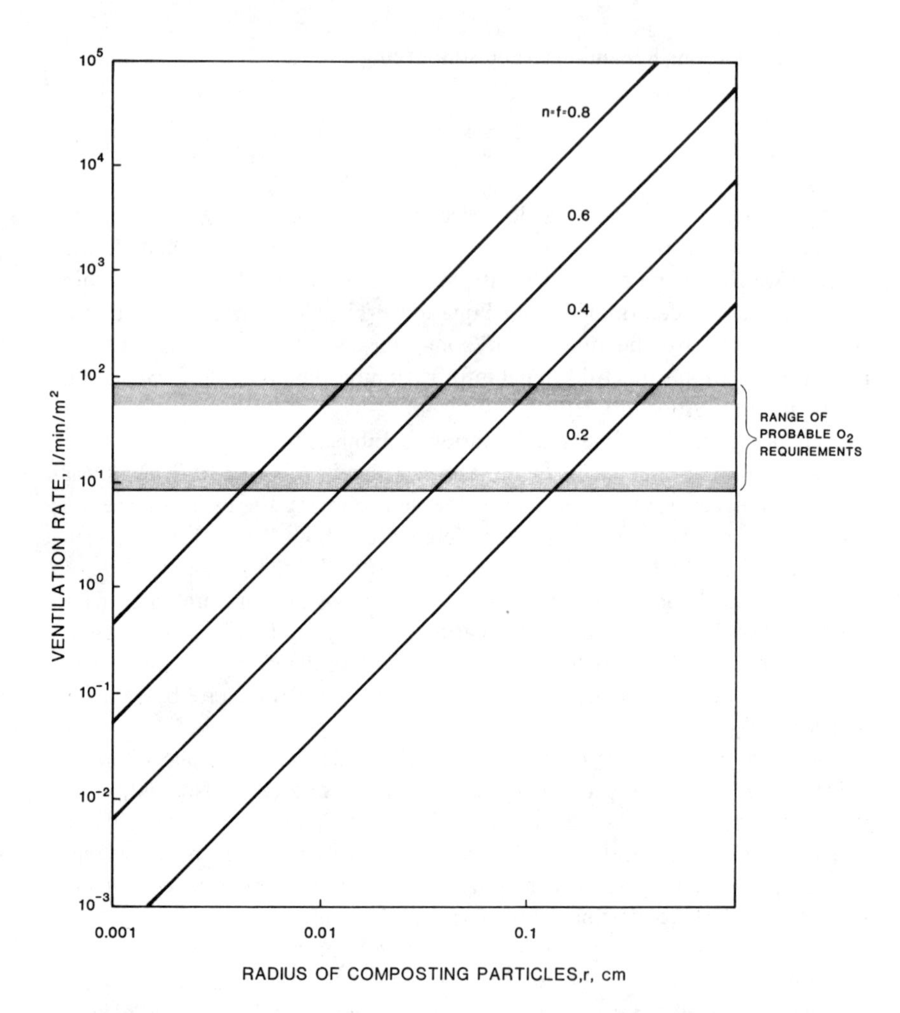

Figure 8-9. Natural ventilation rates predicted by the pore tube model as a function of particle size and mixture porosity. See Appendix III for the method of calculation. The range of probable oxygen requirements corresponds to a range of EAR from 1 to 3 and pile heights from 1 to 3 m.

Iacoboni et al. [17] conducted a limited study of particle sizes in a digested sludge/compost recycle mixture undergoing windrow composting. Particle size distributions are presented in Figure 8-10. It is interesting to note that on average about 80% of the particles by weight were greater than 0.10 cm in radius. The authors also reported that moisture and VS did not differ significantly among the various particle sizes.

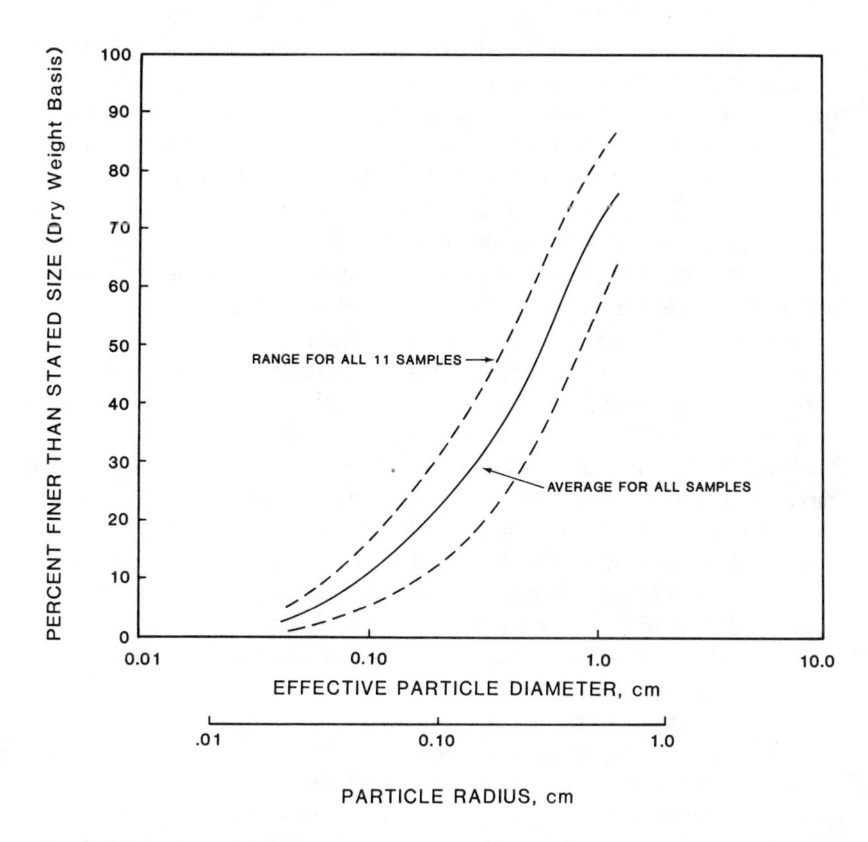

Figure 8-10. Particle size distribution observed during windrow composting of digested sludge blended with recycled compost. Solids content of samples ranged from 62 to 78% and volatile solids content from 25 to 41% [17].

At this point it seems good to reiterate that the pore tube model is at best a much simplified version of an actual composting system. The model should not be expected to yield actual values for the rate of ventilation. Instead it focuses attention on important variables affecting natural ventilation and probably can yield "order of magnitude" estimates. Since the purpose here is to explore the mechanisms of aeration, the model is a useful tool. However, accurate prediction of ventilation rates must await development of better experimental data.

As previously discussed, mechanical turning in the windrow system appears to be entirely inadequate to supply the air quantities required. It is difficult to prove conclusively that natural ventilation is the primary aerating mechanism for windrow composting. However, based on results of the pore tube model, observation of actual composting systems and available data on particle size distributions, it would seem that natural ventilation, caused by the density difference between the windrow interior and outside ambient air, is the driving force for most oxygen transfer in a conventional (no forced aeration) windrow system. In this regard, Maier et al. [113] observed that aeration of windrows composed of finely ground refuse was improved when the windrow was placed on elevated racks so that air could permeate from below and produce a natural upward draft. The natural draft was reported to successfully aerate windrows placed about 1.2 m deep.

With regard to aeration, the function of daily or weekly turning of the windrow must be to assure that adequate FAS is maintained. Ground refuse, for example, is noted for its tendency to consolidate during composting. If left unchecked, FAS may decrease to the point where ventilation becomes inadequate. Periodic turning would decrease the unit bulk weight of the mixture, assuring the highest possible ventilation rate for the particular particle sizes in the mixture. With windrow composting of sludge, excessive moisture is always a problem and initial mixture FAS is relatively low when compared to other composting materials such as ground refuse and straw. Daily and sometimes more frequent turning would maintain a minimum bulk weight and promote moisture evaporation which further increases the FAS.

Another method to increase the rate of ventilation is to increase particle sizes within the compost mixture. With a particle size in the range of 0.1-1.0 cm there is a good likelihood that natural ventilation can match the rate of oxygen demand. Pelleting a sludge/compost recycle mixture to produce a particle size within this range may be an effective way of increasing ventilation. Addition of suitable amendments such as straw, rice hulls or even sawdust is another approach. The latter materials would affect both the average particle size and unit bulk weight.

Based on Equation 8-11, the velocity within any pore tube is independent of the height L. This means that for a given material the ventilation rate should be independent of pile height, all other factors being constant. As the height increases, the rate of natural ventilation will remain relatively constant while the rate of oxygen consumption per unit of area will increase. Thus, there is a limit to the height of pile beyond which adequate oxygen cannot be supplied by natural ventilation.

SUMMARY

Air is required during aerobic composting for a number of purposes, including supplying oxygen to support biological activity, removing moisture from the composting mass and removing heat to prevent excessively high temperatures. Oxygen must be provided to any composting mixture to supply demands imposed by organic decomposition. The stoichiometric oxygen demand will depend on chemical composition of the organics. For municipal sludge, a value of about 2.0 g oxygen/g organic oxidized appears reasonable.

Air supplied to a composting mixture will be heated by the composting mass. The quantity of moisture contained in saturated air increases exponentially with increasing temperature. Thus, heated gases in the composting mixture can transport significant quantities of moisture from the compost even if ambient air is saturated or has a high relative humidity.

The air requirement for drying is significantly greater than the requirement for biological oxidation, and is influenced largely by dewatered cake solids and exit air temperature. Only at cake solids in the range of 30-40% and exit air temperature of 70°C do the air requirements for drying and biological oxidation become reasonably equivalent. At cake solids of 20%, the air requirement for drying can be as much as 10-30 times that for biological oxidation. The differing air requirements allow control over the extent of drying in those systems in which air supply is controlled, for example, the aerated static pile and many reactor systems. Control over the extent of drying will have a significant influence on the thermodynamic balance achieved during composting.

Forced ventilation by means of blowers or fans is the major aerating mechanism in the aerated static pile and many reactor systems. In the windrow composting system, however, the mechanism of aeration is not as obvious. Required air volumes to supply the stoichiometric oxygen requirement are significantly greater than can be accounted for by periodic mechanical turning, even if practiced on a daily or more frequent basis.

A model was developed which suggests that natural ventilation is the most significant aerating mechanism in the windrow process. Natural ventilation occurs as a result of the density difference between warm, moist gases contained within the windrow and cooler, less moist, ambient air. The rate of ventilation is a function of the density difference, FAS and particle size of the composting mixture and is enhanced by increasing both the FAS and particle size.

With regard to aeration, periodic mechanical turning probably increases the mixture FAS, assuring the highest possible ventilation rate for the particular particle sizes in the mixture. Even though periodic turning by itself does not transfer significant amounts of oxygen, maintaining a high FAS is important because it strongly influences the rate of natural ventilation.

CHAPTER 9

THERMODYNAMICS OF COMPOSTING

INTRODUCTION

The first law of thermodynamics states that energy can neither be created nor destroyed. Thus, the energy which flows into a system must be fully accounted for either as energy stored within the system or energy which flows out of the system. In a composting process the major energy input is that contained in organic molecules of the feed material. As these molecules are broken down by biological activity, energy is either transformed into new organic molecules within the microorganisms or is released as heat into the surroundings. Thus, it is energy release from organic decomposition which drives the composting process, causes temperature elevation and produces the drying that is so desirable in sludge composting. Indeed, it is the capture of a portion of this energy which causes microorganisms to decompose the organics in the first place.

This chapter will examine material and energy balances accompanying the compost process. At the risk of revealing the ending of the chapter, it can be stated that composting can be divided into two distinct thermodynamic regions: one in which sufficient energy is available for both composting and evaporative drying, and another in which energy is sufficient only for composting with limited drying.

MECHANISMS OF HEAT TRANSFER

Three distinct mechanisms of heat or energy transfer can be described: conduction of heat, convection and radiation. Conductive transfer may be defined as the transfer of heat between two points caused by a temperature difference but without any mass movement between the points. Conduction of heat within a compost particle or from one particle in contact with another are examples. One example of conductive transfer, in Chapter 5, was the time required to heat the interior of a spherical particle when heated from the outside.

The quantitative law of heat conduction was formulated in 1822 by Fourier as a generalization of his experimental investigations. Fourier's law states:

$$\frac{dq}{dt} = -kdA \frac{dT}{dx} \tag{9-1}$$

Fourier's equation describes the amount of heat dq that passes through a plane of area dA in time dt, in response to a temperature gradient dT/dx. The proportionality constant in the equation (k) is termed the thermal conductivity and usually is expressed in units of cal/hr-cm^2-°C/cm. Thermal conductivity of a substance depends on the state of the substance (solid, liquid or gas) and for a given state will vary somewhat with temperature. Typical values of thermal conductivity for a number of substances were presented in Chapter 5.

Thermal diffusivity is related to thermal conductivity by:

$$a = \frac{k}{\rho c_p} \tag{9-2}$$

where a = thermal diffusivity, cm^2/hr
ρ = mass density, g/cm^3
c_p = specific heat, cal/g-°C

Like thermal conductivity, thermal diffusivity is a property of the substance. The term $k/\rho c_p$ appears in many heat conduction problems and is related to the diffusion coefficient used in mass transfer problems.

Solutions to Equation 9-1 are beyond the scope of this book and the interested reader is referred to specialized textbooks devoted to heat transfer. However, many useful estimates can be made for the case of one-dimensional steady-state conditions. Integrating Equation 9-1:

$$q = kA \frac{\Delta T}{\Delta x} \tag{9-3}$$

where q = the rate of heat transfer
k = the average thermal conductivity (assumed constant)
Δx = the length of the flow path

Convective heat transfer results from movement of mass between a high temperature zone, where heat is accepted, to a zone of lower temperature, where heat is released. Convective heat transfer occurs as a result of fluid motion in contrast to conductive heat transfer which occurs entirely by means of intermolecular energy transfers. Movement of air within a bed of composting particles is an example of convective heat transfer. Another is the evaporation of water, its transport and condensation in another part of the compost bed.

In many cases both conduction and convection operate to transfer heat. A case in point is heat transfer from the surface of a compost pile or from the walls of a compost reactor. Mass movement of ambient air across the pile will result in convective heat loss. In response to the resulting temperature difference, heat will be conducted from the interior of the pile or across the walls of the reactor. Steady state heat transfer in such cases is often modeled by an equation of the form:

$$q = UA(T_1 - T_2) \tag{9-4}$$

where U = overall heat transfer coefficient which includes effects of both conductive and convective heat transfer, cal/hr-cm^2-°C
A = area perpendicular to direction of heat transfer, cm^2
T_1, T_2 = temperatures at points 1 and 2, °C

Estimates of overall heat transfer coefficients cannot be made entirely on the basis of theoretical arguments. Empirical relationships based on experimental data are usually required. For example, the value of U for heat loss from the surface of a compost pile would likely be a function of wind speed, relative humidity and perhaps other factors. Even so, once the surface has cooled, further heat loss is probably limited by transport of heat (either by conduction or convection) from the interior of the pile.

A final form of heat transfer can occur as a result of radiant or electromagnetic energy exchange between two bodies of unequal temperature. Radiant energy can be transmitted in a vacuum and does not depend on direct physical contact or the movement of any fluid between the bodies. Radiant energy is always being exchanged, but a net exchange in one direction results only if there is a temperature difference between the two bodies. At temperatures below 300°C practically all radiant energy is in the infrared region of the spectrum. The reader is probably familiar with infrared satellite photographs which highlight temperature differences between objects. If composting is successful, its temperature should be greater than that of the surroundings and some radiative losses can be expected from the surface.

Radiant energy transfer between two bodies A and B is described by the Stefan-Boltzmann Law:

$$q = \sigma A(T_{a_1}{}^4 - T_{a_2}{}^4)F_a F_e \qquad (9\text{-}5)$$

where σ = Stefan-Boltzmann constant, 4.87×10^{-8} kcal/hr-m²-°K⁴
F_a = configurational factor to account for the relative position and geometry of the bodies
F_e = emissivity factor to account for nonblack-body radiation
T_{a_1}, T_{a_2} = absolute temperature of bodies A and B

Ideal radiators emit radiant energy at a rate proportional to the fourth power of the absolute temperature. As a result radiant energy losses become more significant at higher temperatures.

Example 9-1

Determine the heat transfer rate from 1 m² of surface at a temperature of 60°C radiating to the ambient environment at 20°C. Assume an emissivity factor of 0.9 and configurational factor of 1.0. Compare this to the conductive transport of heat through a 30-cm layer of compost with a uniform temperature gradient from 60 to 20°C. Assume a k of 4.0 cal/cm²-hr-°C/cm.

Solution

1. Estimate the radiant energy transfer from the Stefan-Boltzmann equation:

$$q = 4.87 \times 10^{-8}(1)[(273 + 60)^4 - (273 + 20)^4](0.9)(1.0)$$
$$q = 216 \text{ kcal/hr}$$

2. Using Equation 9-3 the conductive heat rate per m² of surface area can be estimated as:

$$q = 4.0\,(10^4 \text{ cm}^2)(60 - 20)/30 \text{ cm} = 5.3 \times 10^4 \text{ cal/hr}$$
$$q = 53 \text{ kcal/hr}$$

3. Using Equation 9-4, estimate a U value corresponding to the conductive heat loss rate in part 2:

$$q = UA(T_1 - T_2)$$
$$53 = U(1 \text{ m}^2)(60 - 20)$$
$$U = 1.3 \text{ kcal/m}^2\text{-hr-°C}$$

4. The estimated value of U is consistent with values used for calculating losses from insulated, concrete anaerobic digesters. The latter typically range from about 0.5 to 1.5 kcal/m²-hr-°C.

When a warm compost surface is exposed to cooler ambient air, such as after turning a windrow or pile, a rapid surface temperature drop should result from radiative and convective losses with the outside ambient air. Once the outer surface is cooled, further heat loss should be limited by conductive transport from the pile interior. Convective transport from air moving through the pile also occurs, but can be accounted for from the temperature and quantity of output gases.

THERMAL PROPERTIES OF COMPOST

Before proceeding with a thermodynamic analysis of composting, it is clear that the thermal properties of compost must be examined. Unfortunately, few if any tests have been reported on sludge compost. However, Mears et al. [117] have conducted some excellent work on compost produced from swine wastes. Their results should be reasonably applicable to sludge-based compost as well. Thermal properties of importance to the analysis include the specific heat (at constant pressure) and thermal conductivity.

Before proceeding to the results obtained by Mears et al., it is interesting to examine their laboratory approach. A calorimetric technique was used for determination of specific heat. A wide-mouth Thermos was used as a calorimeter and was tested with a fluid of known specific heat to determine the fluid equivalent of the thermos as a function of depth. Water was the fluid of choice in these studies. Material to be tested was placed in the calorimeter and a known quantity of hot water was added. The slurry was mixed for 5 min before taking final temperature measurements. Mixing of compost and water at the same temperature did not result in any measurable rise in temperature, indicating that any heat released in the wetting of compost was insignificant. This was not true for oven-dried samples, however, which did exhibit a significant heat of wetting. Therefore, if one were to measure the specific heat of very dry material by the method of mixtures, the heat of wetting would have to be considered.

Thermal conductivity was measured by placing the sample in a long, thin-walled cylinder constructed of material with a high thermal conductivity, in this case aluminum. Insulating plugs were placed in the ends of the cylinder to restrict heat transfer in the axial direction. The cylinder with its sample was then placed in a hot water bath. Subsequent heat transfer can be described as occurring in the radial direction into a cylinder of infinite length. Solutions to this problem have been developed with results similar to those shown in Figure 5-15 for the case of heat transfer into a sphere. By measuring temperature differences between the center of the cylinder of compost and the hot water bath vs time, thermal conductivity can be calculated. One problem with this technique is the fact that convective currents can arise from differential temperatures, and hence densities, of gases in the void spaces between particles. This is a problem encountered in measuring thermal conductivity of any fluid and the effects of convective heat transfer cannot be entirely avoided. Apparently, these effects were minimized to a sufficient degree in the test equipment. Measurements of thermal conductivity of known materials which contained void spaces, such as sand, in the same device were within 4% of accepted values.

Using these techniques, Mears et al. determined the thermal properties of composting material composed of swine waste with the addition of 5% by weight straw (10% by volume). The swine waste consisted of feces, uneaten feed, bones, plastics, paper and glass. Large pieces of bone, plastics, etc., were excluded from the analysis, but usually totaled only 2-10% of the sample weight. A windrow composting technique was used and representative samples collected at weekly intervals. Results are presented in Figure 9-1 and Table 9-1.

In all cases the specific heat and thermal conductivity varied linearly with moisture content. Extrapolation of the curves to 100% moisture content gave results very close to the actual specific heat of water, 1.0 cal/g-°C,

Table 9-1. Specific Heat and Thermal Conductivity of Swine Compost [117]

Sample No.	Age of Windrow (days)	Zero Moisture Intercept I	Slope of Curve S	100% Moisture Intercept I + 100 S
		Specific Heat (cal/g-°C)[a]		
1	0	0.0550	0.00940	0.9951
2	7	0.0699	0.00921	0.9909
3	14	0.0771	0.00922	0.9991
4	20	0.0834	0.00906	0.9894
5	30	0.1289	0.00834	0.9629
6	35	0.1551	0.00813	0.9831
		Thermal Conductivity (cal/hr-cm²-°C/cm)[b]		
1	0	0.875	0.0498	5.854
2	7	1.086	0.0473	5.816
3	14	1.323	0.0440	5.726
4	20	1.341	0.0403	5.374
5	30	1.854	0.0383	5.682
6	35	2.071	0.0341	5.476

[a]Specific heat of water = 1.00 cal/g-°C.
[b]Thermal conductivity of water = 5.62 cal/(hr-cm²-°C/cm).

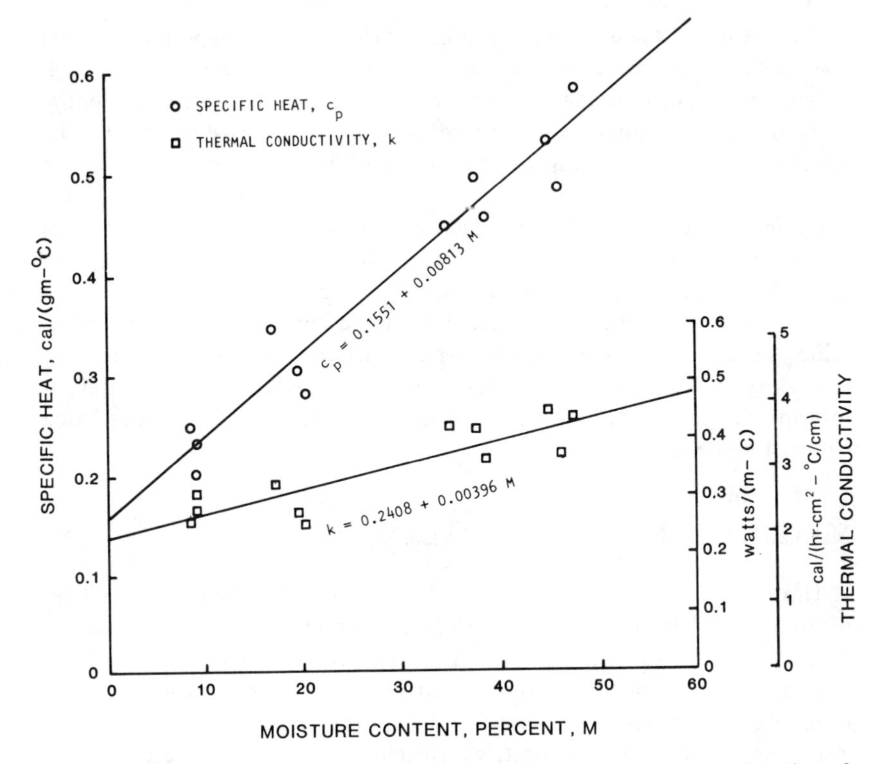

Figure 9-1. Specific heat c_p and thermal conductivity k of compost as a function of moisture content. Compost was produced from swine waste blended with about 5% straw by weight. Data shown correspond to Sample No. 6 from Table 9-1 [117].

and thermal conductivity, 5.62 cal/cm^2-hr-°C/cm. This means that in a thermodynamic analysis the water and solid fractions can be treated separately. The water portion of compost material can be assumed to have the thermal properties of water. The solid fraction can be assumed to have thermal properties equal to the extrapolated value at zero moisture. It also means that a compost sample can be tested at a single moisture content and the entire relationship established by using this point and the value for 100% water.

From data in Table 9-1, it is apparent that both specific heat and thermal conductivity of the solid fraction increase with compost age. This is probably caused by an increase in the proportion of inorganic material (ash) as a result of organic decomposition. Specific heat and thermal conductivity of inorganic components are generally greater than values found for the organic fraction.

As a result of these studies it can be concluded that compost will have a low specific heat that increases with increasing moisture content. Although a function of moisture, thermal conductivity is relatively low over the entire range of moisture contents. Therefore, a large compost pile will tend to be self-insulating, and heat losses by conduction should generally be small. In the case of sludge composting, values of specific heat and thermal conductivity for the mixture will likely decrease during composting since the loss of moisture should overshadow effects of increased ash content. Finally, thermal properties are likely a function of the particular compost material in question. Given the basic organic nature of most composts, however, values should not be significantly different from those reported here. The laboratory techniques used to measure thermal properties are very straightforward and it is hoped that the approach will be applied to sludge based compost in the future.

MATERIAL AND ENERGY BALANCES

Heat liberated from the decomposition of organics increases the temperature of solids and liquid in the composting mixture. As drying occurs heat will be used to evaporate the water. Because the compost is at a higher temperature than the surroundings, heat loss will occur from exposed surfaces of the compost. This loss will be mitigated to some extent by the insulating effect of the compost, which limits conduction of heat from the pile or windrow interior. Loss will also occur as windrows are turned for aeration. In the aerated pile system, energy will be continually expended to heat air mechanically drawn into the pile. Under equilibrium conditions, compost temperature will rise to a point where energy inputs are balanced by outputs. However, maximum obtainable temperatures are limited to about 75-80°C, because rates of biological activity, and hence heat evolution, begin to decrease above about 55°C [44].

To illustrate the method of analysis, a materials and energy balance to compost and dry 1 g of digested solids to final compost product is presented in Figure 9-2. A 20% solids cake was assumed and compost recycle was adjusted to give an initial mixture of 40% solids. Energy requirements to raise both solids and water to a temperature of 60°C were estimated along with requirements for evaporation of water contained in the cake and that produced during aerobic decomposition. The latter was estimated from Equation 8-1 to be about 0.18 H_2O/g original solids, assuming a digested cake volatility of 0.50 and degradability coefficient k_c of 0.50. Because of organic decomposition, about 0.75 g of cake solids would remain after composting for each gram of digested feed solids. It was assumed that no degradation of recycled compost or bulking agent would occur ($k_r = 0$).

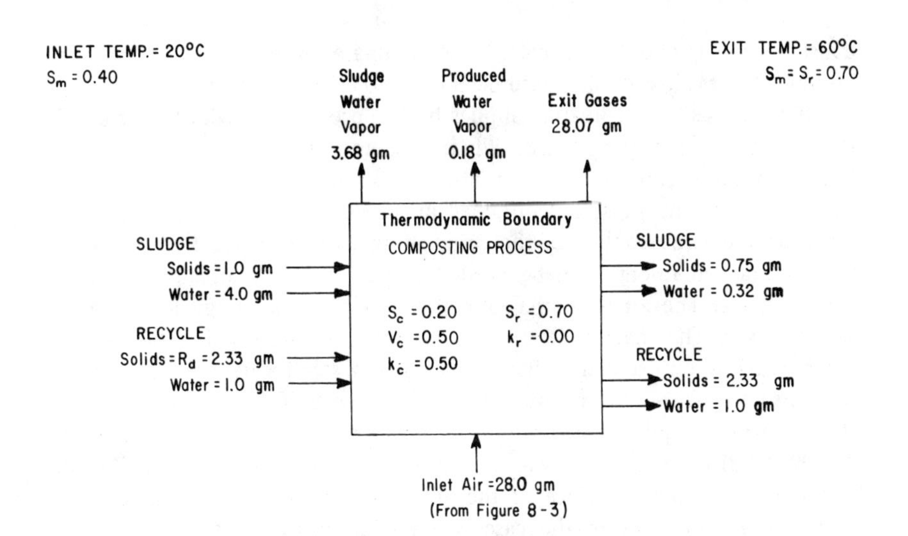

	Calories[a] / gm Sludge Solids
HEAT TO SOLIDS, q_s	
$q_s = mc_p\Delta T$ = 3.33 gm (0.25 cal/gm°C)(60-20°C)	30
HEAT TO WATER, q_w	
q_w = 5.0 gm (1.0 cal/gm °C)(60-20°C)	200
EVAPORATIVE BURDEN, q_v	
q_v = 540 cal/gm (3.86gm)	2080
HEAT TO AIR, q_a	
q_a = 28.0gm(0.25 cal/gm °C)(60-20°C)	280
TOTAL ENERGY REQUIRED[b]	2590

a. All values rounded to nearest 10
b. Excludes losses to surroundings

Figure 9-2. Materials and energy balance for composting and drying [99].

Energy required for air heating will depend on whether the air supply is regulated by requirements for aeration or drying. For the example shown in Figure 9-2, it was assumed that air supply was sufficient for drying. From Figure 8-3, for an exit air temperature of 60°C, the air requirement is about 28 g air/g cake solids. Specific heat of air in the temperature range under consideration is about 0.25 cal/g-°C. Assuming a temperature increase from 20 to 60°C, heat lost to air would be about 280 cal/g sludge solids.

This analysis for air heating applies best to processes in which air is drawn through the composting mass. Windrow composting is more difficult to analyze because composting material is thrown into the air during mechanical turning. Warm, moist air contained in the voids is replaced by cooler ambient air and some cooling of the composting solids will occur accompanied by evaporation. Energy will be needed to heat the new air and reheat the cooled solids. The analysis in Figure 9-2 also neglects energy losses to the surroundings. The magnitude of such losses is obviously a function of the composting system and may be significant in systems where the compost is not insulated from the surroundings. Offsetting this, surface drying from the windrow or pile, which would occur when atmospheric conditions of <100% relative humidity prevail, was also neglected. Both phenomena would affect only the outer surface of the pile. However, surface drying could be particularly important in the case where windrows are turned frequently, thus exposing much of the material to surface drying. Further consideration of these factors will be presented in Chapter 11.

Adding the energy demands calculated in Figure 9-2, a 20% digested sludge cake requires about 2590 cal/g of cake solids for composting and drying. The evaporative burden represents over 75% of the total energy required in the composting and drying process. The latter value has been found to be relatively constant over the range of cake solids from 10 to 40%.

Because aeration demands for drying will usually exceed those for biological oxidation, the energy budget was recalculated for the case of composting with limited drying (aeration regulated by stoichiometric needs). The stoichiometric air requirement for the case of digested sludge was estimated from Equation 8-1 to be 2.18 g air/g solids. It was assumed that three times this value would be supplied (excess air ratio (EAR) = 3). Moisture removal was estimated from Figure 8-2 to be about 0.137 g H_2O/g air for inlet and exit air temperatures of 20 and 60°C, respectively. Results of the calculations are presented in Figure 9-3.

About 0.89 g of water would be removed with the air while about 0.18 g of water would be produced from aerobic decomposition. This leaves about 4.20 g of moisture in the 3.08 g of solids remaining after composting, giving a solids content of about 42%, which is only slightly greater than the 40% assumed in the original mixture. Since it was assumed that recycled compost or bulking agent was 70% solids, considerable air drying would be required after composting.

Total energy demand for the case of composting with limited drying is about 780 cal/g solids, which is about 30% of the requirement for both composting and drying. This is caused almost entirely by reduction of the

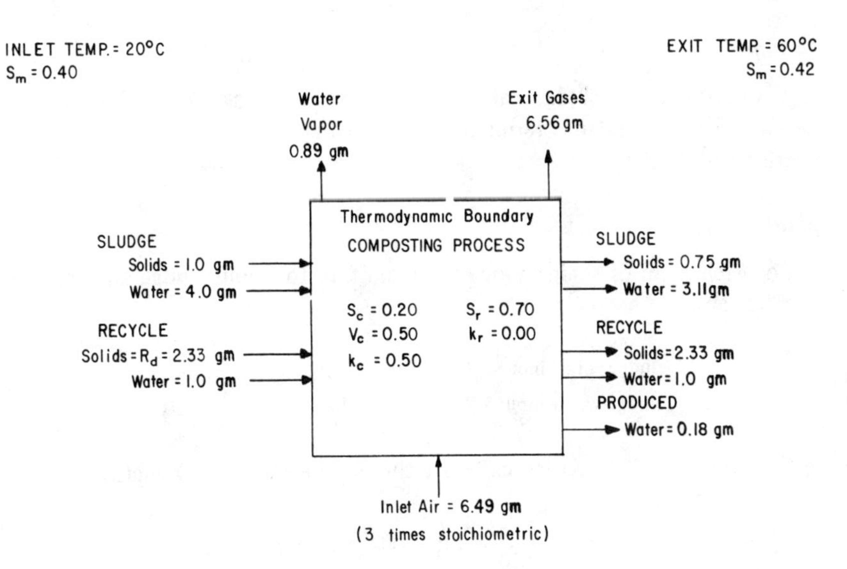

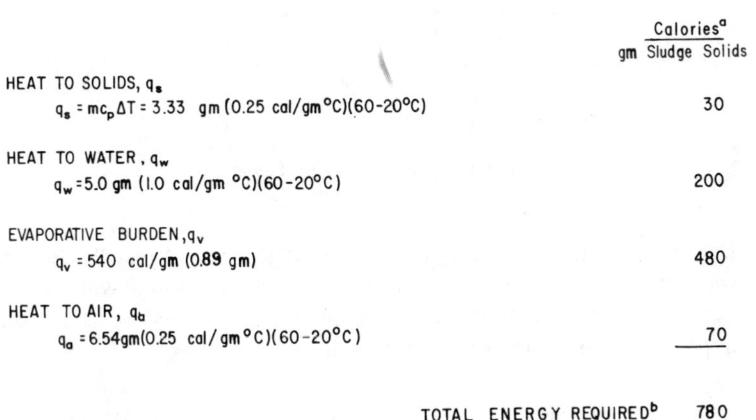

Figure 9-3. Materials and energy balance for composting with limited drying [99].

evaporative burden when air supply is limited by the stoichiometric requirement. Even so, the heat required for evaporation still represents over 60% of the total energy used. Because energy budgets for the two cases are so markedly different, each case will be examined in greater detail.

Example 9-2

Given the input sludge cake and other variables defined in Figure 9-2, develop the mass balance terms presented in Figure 9-2 for the case of composting and drying.

Solution

1. For each gram of sludge solids the mass flux for input sludge and recycle solids is:

$$\text{sludge solids input} = 1 \text{ g (given in problem statements)}$$
$$\text{sludge water input} = 1/0.20 - 1 = 4 \text{ g}$$

To achieve a 40% mixture, calculate the required R_d from Equation 6-7:

$$R_d = \left(\frac{0.40}{0.20} - 1\right) \bigg/ \left(1 - \frac{0.40}{0.70}\right) = 2.333$$

Therefore:

$$\text{recycle solids input} = 2.333 \text{ g}$$
$$\text{recycle water input} = \frac{2.333}{0.70} - 2.333 = 1.00 \text{ g}$$

2. Determine solids and water in composted sludge and recycle. Because k_r was assumed to be zero for this problem, recycle solids will not be lost during composting. Thus, recycle solids and water will be as determined in part 1. Sludge solids after composting will be:

$$\text{sludge solids output} = 1.0 - 1.0\,(V_c)(k_c) = 1.0 - 1.0\,(0.5)(0.5) = 0.75 \text{ g}$$

$$\text{sludge water output} = \frac{0.75}{0.70} - 0.75 = 0.32 \text{ g}$$

3. Determine moisture production from organic decomposition. Assume an organic composition of $C_{10}H_{19}O_3N$:

$$\underset{201}{\overset{0.25}{C_{10}H_{19}O_3N}} + \underset{12.5\,(32)}{12.5\,O_2} \rightarrow \underset{10\,(44)}{10\,CO_2} + \underset{17}{NH_3} + \underset{8\,(18)}{8\,H_2O}$$

$$\frac{0.25\,(8)(18)}{201} = 0.18 \text{ g } H_2O \text{ produced}$$

Total water lost in vapor will equal the sludge water lost plus the produced water, or:

$$(4.00 - 0.32) + 0.18 = 3.68 + 0.18 = 3.86 \text{ g}$$

4. Air required for moisture removal can be determined from Figure 8-3 or by calculation given the inlet and exit air temperatures. Assume the following conditions:

inlet air sat. at 20°C carries 0.0148 g H_2O/g dry air
outlet air sat. at 60°C carries 0.152 g H_2O/g dry air

Therefore, 0.152–0.0148 = 0.1372 g water/g dry air can be removed from the composting material. Total air requirement is, then:

$$\frac{0.386 \text{ g } H_2O}{\text{g solid}} \times \frac{1}{0.1372} = 28 \text{ g air/g cake solids}$$

5. The quantity of exit gases other than water vapor will equal the inlet air, less oxygen consumed, plus the carbon dioxide and ammonia formed:

$$\text{exit gases} = 28.0 - \frac{0.25\,(12.5)(32)}{201} + \frac{0.25(10)(44)}{201} + \frac{0.25(17)}{201}$$

$$= 28.07 \text{ g}$$

Example 9-3

Develop the mass balance terms shown in Figure 9-3 for the case of composting with limited drying.

Solution

1. The mass of water and solids in sludge and recycle components and the mass of produced water are calculated as in Example 9-2.
2. The inlet air required is determined from the organic feed composition. Assuming $C_{10}H_{19}O_3N$, and referring to part 3 of Example 9-2:

$$O_2 \text{ required} = (3)\,\frac{0.25\,(12.5)(32)}{201} = 1.49 \text{ g } O_2\text{/g cake solids}$$

Because air is 23% oxygen by weight:

$$\text{air required} = 1.49/0.23 = 6.49 \text{ g air/g cake solids}$$

3. Referring to part 4 of Example 9-2, moisture removal can be determined as:

$$6.49\,(0.1372) = 0.89 \text{ g water}$$

4. Solids content of the mixture after composting is calculated as:

$$S_m = \frac{0.75 + 2.33}{0.75 + (4.0 - 0.89) + \dfrac{2.33}{0.7} + 0.18} = 0.42$$

COMPOSTING WITH DRYING

The energy requirement for composting and drying 20% digested sludge cake to 70% product solids was estimated in Figure 9-2 to be about 2590 cal based on 5 g of water in the initial mixture, or 518 cal/g water. Heat released during organic decomposition can be estimated at about 5500 cal/g of organics oxidized to carbon dioxide and water. Therefore, a water-to-degradable organic ratio (W) of about 10.6 g H_2O/g degradable organic can be estimated (i.e., 5500/516). Defining **W** on the basis of water content is a rational approach because evaporation represents most of the energy demand in the composting and drying system.

Considering the assumptions involved, a value rounded to 10 g H_2O/g degradable organics can be used to judge the thermodynamic characteristics of the composting process. Because water evaporation is the major energy use, and as long as moisture in the dewatered cake is the major water input, the factor should apply equally to windrow, aerated pile or reactor systems. If W is $\leqslant 10$, sufficient energy should be available for temperature elevation and water evaporation even if no surface drying results from climatic conditions. If the compost operation is properly managed and excessive moisture additions from rainfall are prevented, both temperature elevation and drying should occur during composting. If W is >10, composting alone will not provide sufficient energy, and lower temperatures or less drying can be expected unless surface drying can be used to carry off excessive moisture.

Some surface drying would generally occur in all but the wettest climates or the wettest months of the year. To maximize surface drying, sludge could be composted and spread out to dry and remove remaining water. Alternatively, the sludge could be spread into thin layers to dry and lower W before composting. The latter technique is used at a compost facility in southern California to reduce the moisture content of dewatered digested sludge before windrow composting [100]. However, spreading for air drying would be practical only in dry climates or sheltered facilities in wet climates. These and other methods of moisture control have been discussed in Chapter 6. Obviously, it would be desirable to maintain a W <10 in the initial dewatered cake to avoid reliance on surface drying. Alternatively, a degradable amendment could be added before composting to reduce W to <10.

To determine W for a particular sludge, consider the weight of water per unit weight of degradable organic in the mixed material. Referring to Figure 6-2, weight of water in a mixture of sludge cake, compost product and amendment is given by:

$$\text{weight water} = (X_c - S_cX_c) + (X_r - S_rX_r) + (X_a - S_aX_a) \tag{9-6}$$

The quantity of degradable organics in the mixture would be given by:

$$\text{degradable organics} = k_cV_cS_cX_c + k_rV_rS_rX_r + k_aV_aS_aX_a \tag{9-7}$$

where k_c, k_r and k_a are degradability coefficients as previously defined. Combining Equations 9-6 and 9-7:

$$W = \frac{\text{weight water}}{\text{weight degradable organics}}$$

$$= \frac{(X_c - S_cX_c) + (X_r - S_rX_r) + (X_a - S_aX_a)}{k_cV_cS_cX_c + k_rV_rS_rX_r + k_aV_aS_aX_a} \tag{9-8}$$

Assuming that no amendment is added and recalling that $R_w = X_r/X_c$:

$$W = \frac{(1 - S_c) + R_w(1 - S_r)}{k_cV_cS_c + k_rV_rS_rR_w} \tag{9-9}$$

The corresponding equation based on dry weight recycle R_d is:

$$W = \frac{\left(\frac{1}{S_c} - 1\right) + R_d\left(\frac{1}{S_r} - 1\right)}{k_cV_c + k_rV_rR_d} \tag{9-10}$$

If amendment or bulking agent is added to the mixture, the W ratio can be determined most easily from directly solving Equation 9-8.

The rationale for use of total infeed water in Equations 9-8 to 9-10, instead of the amount actually evaporated, is that most of the infeed water is evaporated and the calculations are thus simplified. It should be remembered that W is intended only as a tool or rule of thumb to judge the thermodynamic characteristics of the composting process. Thus, some compromise in technical accuracy is justified to give a tool which can be applied easily.

W is not a substitute for a complete energy balance of the type shown in Figure 9-2.

Equation 9-9 was solved for the cases of raw and digested sludge cake. Cake degradability coefficients of 0.5 and 0.7 were assumed for digested and raw sludge, respectively. The assumed k_r values ranged from 0.0 to 0.25. Results and other assumptions used in the calculations are presented in Figure 9-4. Values of W are strongly determined by the cake solids content, particularly if recycled product is assigned a degradability coefficient of zero. The influence of cake solids diminishes as the value of k_r increases. Also, W values for raw sludge are considerably lower than for digested sludge, reflecting the higher degradability of raw sludge. These results are logical and consistent with previous discussions.

For $k_r = 0$, W nearly doubles as cake solids decrease from 30 to 20%. It appears that digested sludge would be amenable to composting and drying

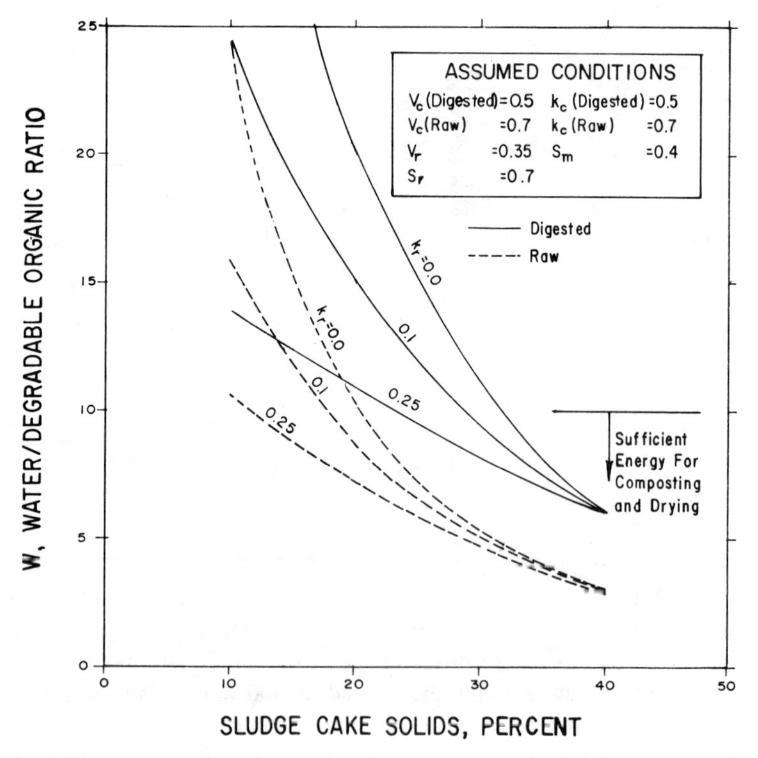

Figure 9-4. Effect of cake solids content on the ratio of water to degradable organics for raw and digested sludges [98, 99].

in most climates if cake solids of 30% or greater were obtained during dewatering. W would be about 10 or less regardless of the degradability coefficient for compost recycle k_r. Recycle ratios R_d and R_w decrease with increasing cake solids; therefore, k_r has less influence on W at higher cake solids. At 20% digested cake solids, however, k_r has a strong influence because of the increased quantity of recycled material. Assuming a k_r of 0.10 as a conservative estimate, W would be about 15, which would probably preclude composting with drying except perhaps in arid climates. Below 15-20% cake solids, composting and drying of digested sludge alone would appear to be impractical.

Achieving an adequate water/degradable organic ratio with raw sludge is an easier matter because of the higher volatility compared with digested sludge. From Figure 9-4, a W of about 10 could be maintained even at cake solids of only 20%. At 15% cake solids, composting with some drying would still be practical if k_r values exceeded about 0.10.

Addition of a degradable bulking agent in the aerated pile system can significantly increase the total quantity of degradable organics in the mixed material because of the large quantity of bulking agent normally used. Until development of better information, it is probably reasonable to assume a k_r of about 0.25 when a degradable bulking agent such as wood chips is used. A k_r of 0.0 would apply to the case where a bulking agent of low degradability is used, such as shredded rubber tires.

Example 9-4

Using the assumed conditions given in Figure 9-4, calculate W for both a raw and digested sludge cake at 20% solids. Assume the recycled compost degradability coefficient to be 0.10.

Solution

1. The required recycle ratio to achieve a 40% solids mixture, $S_m = 0.40$, is determined from Equation 6-5:

$$R_w = \frac{0.40 - 0.20}{0.70 - 0.40} = 0.667$$

2. Using Equation 9-9 for the case of raw sludge:

$$W = \frac{(1 - 0.20) + 0.667\,(1 - 0.70)}{(0.70)(0.70)(0.20) + (0.10)(0.35)(0.70)(0.667)} = 8.7$$

3. Using Equation 9-9 for the case of digested sludge:

$$W = \frac{(1 - 0.20) + 0.667\ (1 - 0.70)}{(0.50)(0.50)(0.20) + (0.10)(0.35)(0.70)(0.667)} = 15.1$$

Example 9-5

Determine W for the mixture of cake, compost and amendment determined in Example 6-4. Assume the degradability coefficients for cake, compost and amendment to be 0.5, 0.1 and 0.4, respectively.

Solution

1. Using Equation 9-8 and the values of X_c, X_r and X_m from Example 6-4:

$$W = \frac{(400 - 100) + [221 - 0.60\ (221)] + [117 - 0.85\ (117)]}{0.50\ (0.50)(0.25)(400) + 0.10\ (0.40)(0.60)(221) + 0.40\ (0.80)(0.85)(117)}$$

$$W = 6.53$$

2. Note that considerable control over the W ratio is possible by addition of suitable amendments.

COMPOSTING WITH LIMITED DRYING

If W is >10, composting can still be practiced, provided the air supply is regulated to limit evaporation and, hence, drying. To investigate this concept, calculations similar to those shown in Figure 9-3 were performed for various sludge cake solids, exit air temperatures and EAR. Results are presented in Figure 9-5 for the case of digested sludge and in Figure 9-6 for raw sludge. The quantity of available energy was estimated by assuming a heat release of 5500 cal/g of organics oxidized and the values of V_c and k_c indicated in Figures 9-5 and 9-6. No energy contribution from recycled compost or added bulking agent was assumed ($k_r = 0$).

Referring to Figure 9-5 for the case of digested sludge, all curves below the available energy line are possible situations. Sufficient energy for both composting and drying to 70% solids is available only at digested cake solids above about 30%. This is consistent with results shown in Figure 9-4 for $k_r = 0.0$. If air supply is regulated by the stoichiometric requirement, however, evaporation can be reduced, thus reducing the total energy requirement. Cake solids as low as 10% have sufficient energy for composting if drying is

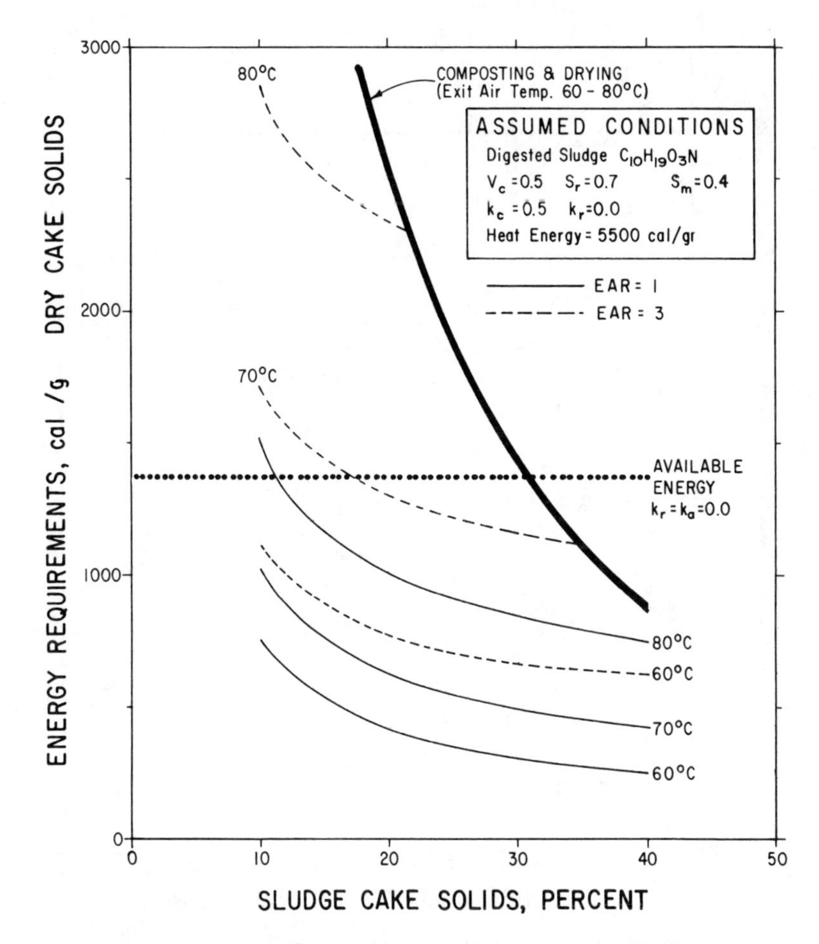

Figure 9-5. Effect of cake solids content on composting energy requirements for digested sludge under various conditions of exit air temperature and EAR [99].

reduced. This is extremely important because it is often difficult to produce 30% cake solids with conventional mechanical dewatering equipment.

The energy curve for composting and drying is steeply sloped because the evaporative burden increases significantly with decreasing cake solids. By contrast, energy curves for composting without drying have a relatively flat slope which increases slightly as cake solids approach 10%. This is because, for a given exit air temperature and EAR, the evaporative burden is constant and not a function of cake solids. As cake solids decrease, however,

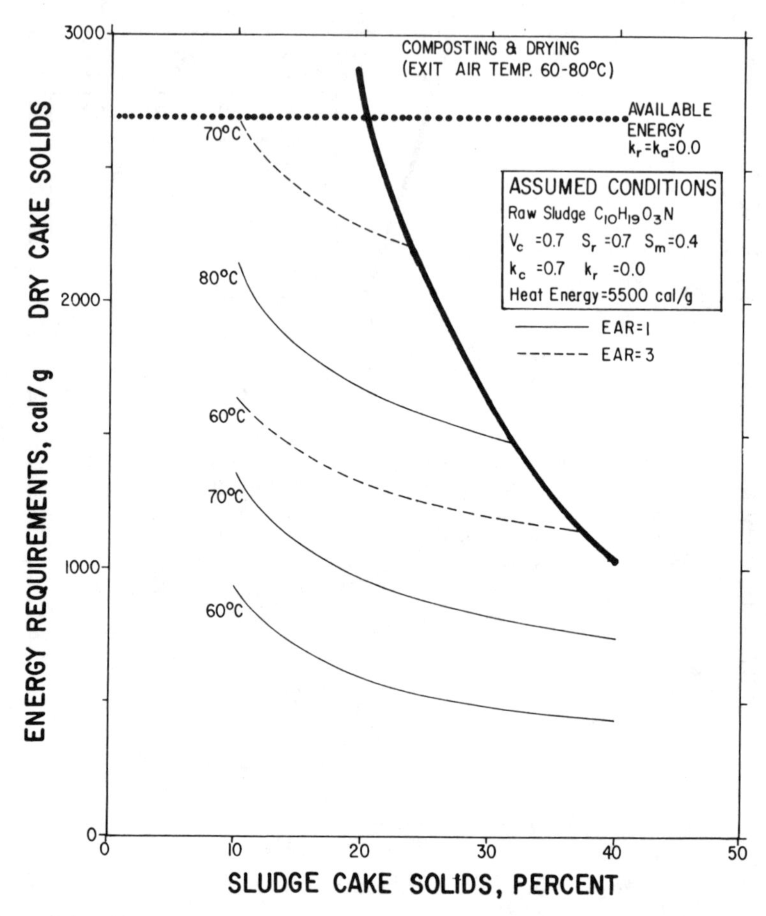

Figure 9-6. Effect of cake solids content on composting energy requirements for raw sludge under various conditions of exit air temperature and EAR [99].

the total quantities of water and compost recycle per weight of cake solids increase, thus increasing the energy required to heat these materials.

The energy requirement for composting with limited drying is strongly influenced by the EAR. From Figure 9-5, for example, at 20% digested cake solids, the energy requirement to produce 60°C exit air at an EAR of 3.0 is about 800 cal/g cake solids. If the EAR is reduced to 1.0, an exit air temperature of 80°C can be achieved with only a slightly greater energy requirement of 1000 cal/g. Therefore, air supplies should be controlled closely to ensure that excessive amounts of air are not provided, which would result in lower exit air temperatures.

Energy losses to surroundings were not included in the previous energy budgets. Therefore, the difference between the energy requirement curves and the available energy line represents a factor of safety. Referring to Figure 9-5 at 20% digested cake solids, exit air temperature of 70°C and EAR of 1.0, about 600 cal/g of cake solids are required, whereas nearly 1400 cal/g will likely be available from organic oxidation. The difference of 800 cal/g represents the heat which can be lost to the surroundings without reducing the exit air temperature below the assumed 70°C. With proper insulation, actual losses to surroundings should be less than this value.

Similar comments can be made for the case of raw sludge composting as shown in Figure 9-6. Available energy is greater for the case of raw sludge and, as a result, cake solids of 20% should have sufficient energy for both composting and drying. Again, this is consistent with results shown in Figure 9-4 for $k_r = 0.0$. Energy requirements for composting with limited drying are significantly reduced and it would appear that cake solids as low as 10% can be composted if air supplies are regulated by the stoichiometric oxygen requirement.

Mixture solids after composting were calculated for each of the conditions examined in Figures 9-5 and 9-6, using the same method of calculation as shown in Figure 9-3 and Example 9-3. In all cases the mixture before composting was assumed to be 40% solids. Results and other assumptions are presented in Figures 9-7 and 9-8 for the cases of digested and raw sludge, respectively. Final mixture solids content is a function of the initial sludge cake solids, exit air temperature and EAR. At an exit air temperature of 60°C and an EAR of 1.0, mixture solids can actually decrease during composting from loss of solids by oxidation coupled with very limited evaporation. Maximum obtainable solids content is a function of the available energy, which in turn is a function of initial cake solids. At about 30% digested cake solids, drying to a 70% solids mixture is possible, which is consistent with Figures 9-4 and 9-5. Similarly, drying to a final mixture solids of 70% is possible at raw sludge solids greater than about 20%. Drying beyond 70% solids was not considered because of the reduced biological activity caused by moisture limitation.

Figures 9-5 to 9-8 show what is thermodynamically achievable under various conditions of feed sludge type, cake solids, EAR and exit air temperature. The figures are useful because they define thermodynamic limits and constraints within which all compost systems must operate. However, the reader should recall that thermodynamics and kinetics are not synonymous. Kinetic rates must not be limited, so that what is achievable thermodynamically can indeed be realized. The integration of thermodynamics with kinetics will be developed in the discussion of process dynamics in Chapters 11 to 14.

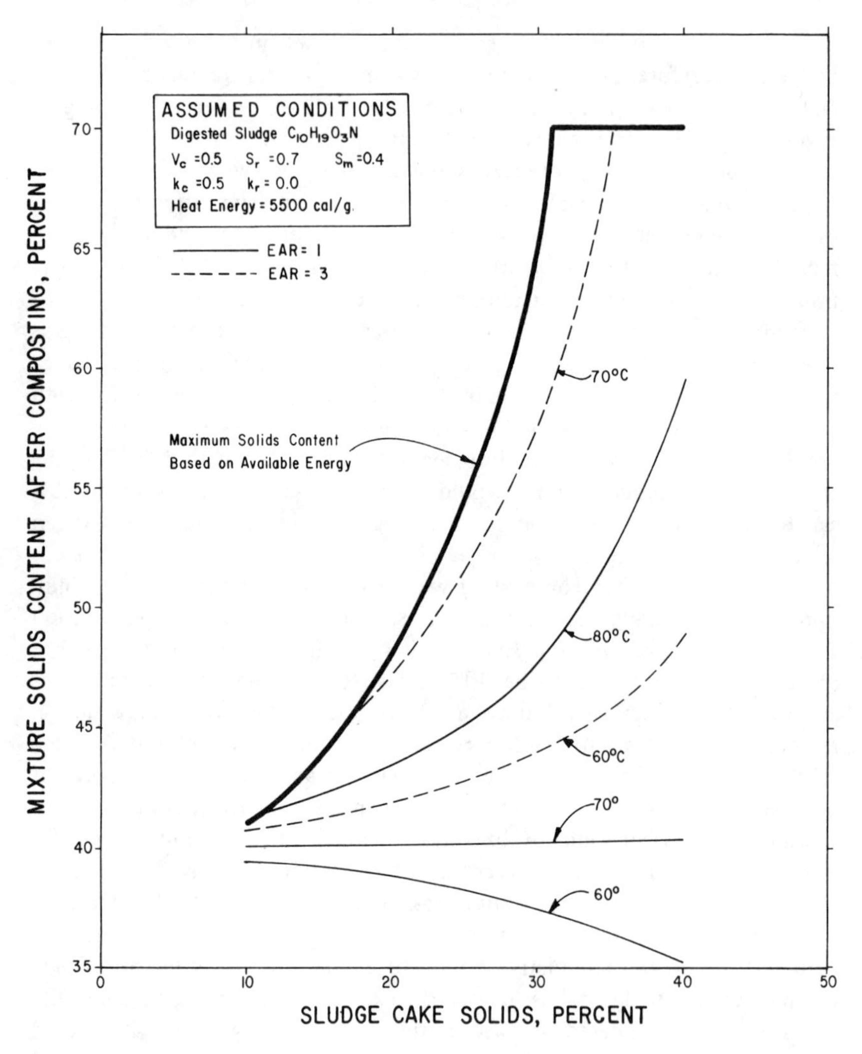

Figure 9-7. Mixture solids content after composting of digested sludge as a function of cake solids content under various conditions of exit air temperature and EAR. Recycle of compost or bulking agent adjusted to achieve an initial mixture of 40% solids (S_m = 0.40) [99].

Example 9-6

A 25% solids cake is to be composted using recycled product for moisture control. Other variables are as shown in Figure 9-7. Determine the maximum solids content achievable assuming no energy is lost to the surroundings.

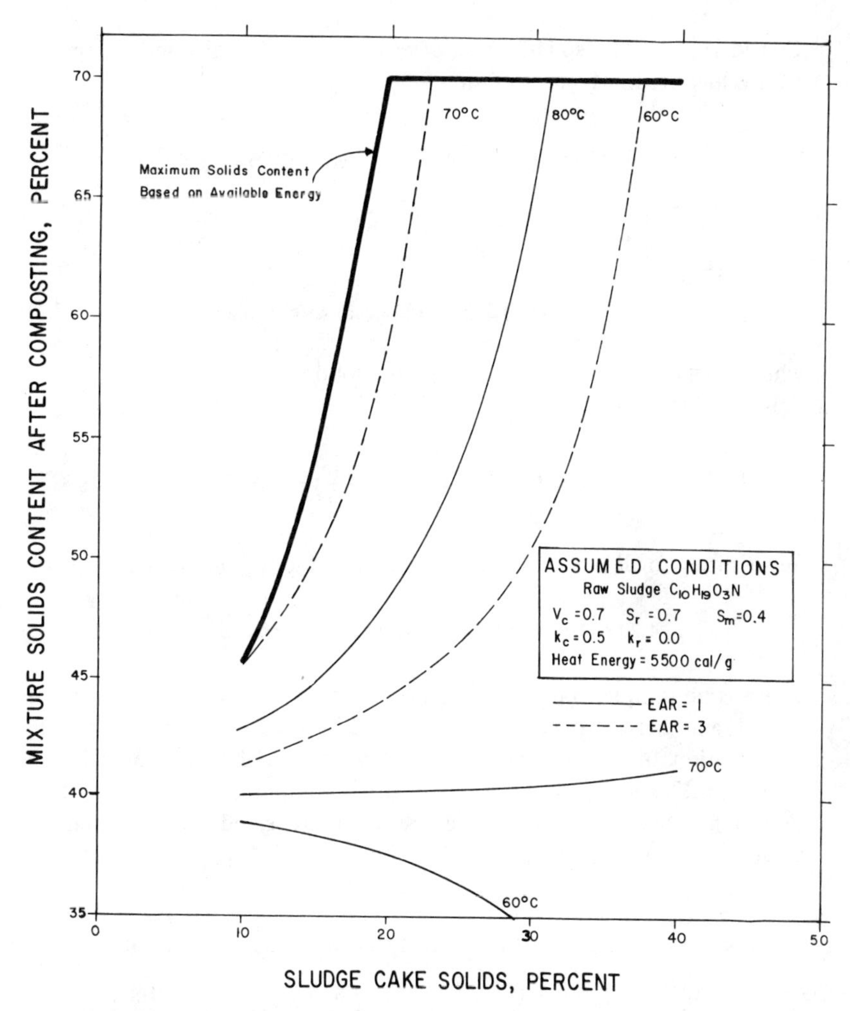

Figure 9-8. Mixture solids content after composting of raw sludge as a function of cake solids content under various conditions of exit air temperature and EAR. Recycle of compost or bulking agent adjusted to achieve an initial mixture of 40% solids ($S_m = 0.40$) [99].

Solution

1. A basic heat balance would be as follows:

$$\text{heat required} = \text{heat produced}$$

$$(q_s + q_w)_{cake} + q_v + (q_s + q_w)_{recycle} + q_a = (k_c V_c S_c X_c)\Delta H_R^0 \qquad (1)$$

Considering 1 g dry solids, a heat of reaction of 5500 cal/g and the specific heat values used in Figures 9-2 and 9-3:

$$\left[1.0\ (0.25)\Delta T + \left(\frac{1}{S_c} - 1\right)\Delta T\right]_{cake} + 540(\Delta H_2O)$$

$$+ \left[R_d(0.25)\Delta T + \left(\frac{R_d}{S_r} - R_d\right)\Delta T\right]_{recycle} + [G_a\ (0.25)\Delta T]_{air}$$

$$= 0.50\ (0.50)(1.0)\ 5500 = 1375 \text{ cal/g infeed solids} \quad (2)$$

where G_a = mass flow of air = g air/g cake solids.

2. Solve Equation 6-7 for R_d:

$$R_d = \left(\frac{0.40}{0.25} - 1\right) \Big/ \left(1 - \frac{0.40}{0.70}\right) = 1.40$$

3. Substituting for R_d, S_c and S_r into Equation 2 (above) and rearranging:

$$4.2\ (\Delta T) + 540\ (\Delta H_2O) + G_a\ (0.25)(\Delta T) = 1375 \quad (3)$$

There is no unique solution to Equation 3 because of the three unknowns ΔT, G_a and ΔH_2O. However, the value of ΔH_2O can be determined for any combination of ΔT and G_a values, allowing a trial and error solution. Values of ΔT and G_a will be assumed, ΔH_2O computed, and the left side of Equation 3 calculated. This procedure will be repeated using different G_a values until the left side equals 1375 cal/g.

G_a (g air/g cake solids)	ΔT (°C)	ΔH_2O (g H_2O/g cake solids)	Left Side Equation 3
10	40	1.37	1008
12	40	1.65	1179
14	40	1.92	1345
14.2	40	1.95	1363
14.4	40	1.97	1376

Values of ΔH_2O in the above table were calculated assuming inlet and exit air to be saturated at 20 and 60°C, respectively. From Example 9-2, part 4, about 0.137 g H_2O/g air would be removed under these conditions.

Repeating the above procedure for other values of Δt and G_a yields essentially the same answer for ΔH_2O. This is not surprising because q_v is the major term in the heat balance. Verification of this is left to the reader.

4. Final mixture solids after composting can be estimated as:

$$S_m = \frac{\text{cake solids} + \text{recycle solids}}{(\text{solids} + \text{water})_{\text{cake}} + (\text{solids} + \text{water})_{\text{recycle}} + \text{produced water} - \Delta H_2O}$$

$$S_m = \frac{0.75 + 1.40}{\left[0.75 + \left(\frac{1}{0.25} - 1\right)\right] + \frac{1.40}{0.70} + 0.18 - 1.97}$$

$$S_m = 0.543 \text{ or } 54.3\% \text{ solids}$$

Note that this value agrees with that shown in Figure 9-7 for $S_c = 0.25$. If losses to the surroundings are significant, airflow would have to be decreased to reduce the evaporative burden, and final cake solids would be less than that calculated above.

IMPLICATIONS FOR DESIGN

Based on the previous work, several observations can be made relative to design of full-scale composting facilities. Two general composting regions can be described based on the energy content of the sludge cake; (1) in which sufficient energy is available for both composting and drying, and (2) in which energy is sufficient for composting with only limited drying. Sludge cake solids is the critical parameter determining in which region a particular sludge lies. At cake solids above about 30% for digested sludge and 20% for raw sludge, energy should be available for both composting and drying to final compost solids of 60–70%. Higher cake solids may be necessary in cold climates or in cases where large heat losses to surroundings are expected. Within the higher energy region the windrow, aerated static pile and reactor systems should be thermodynamically capable of both composting and drying the sludge cake. However, with controlled aeration systems, such as the aerated static pile and many reactor systems, the actual drying obtained will depend on the EAR maintained in the system.

At lower cake solids content, moisture evaporation must be limited to conserve available energy resources for composting. This is most readily accomplished in controlled aeration systems where air supply can be regulated by the stoichiometric oxygen requirement. The windrow technique, because it allows little control over the air supply, may not be the system of choice within the lower-energy region. Of course, if a suitable organic amendment were added, W could be lowered below 10, removing this limitation. Alternatively, the sludge cake could be air dried to increase the solids content before composting. Site-specific factors will determine whether these alternatives are practical in a given situation.

With decreasing cake solids, the quantity of required conditioning agent, either compost recycle or other amendments, is greatly increased to maintain a porous and friable mixture. Because of the increased quantity of material to be handled, it is probably impractical to compost sludge cakes below about 10–15%. Fortunately, cake solids above this range can generally be obtained with the sludges and dewatering equipment common to wastewater treatment. If cake solids are <10%, the possibility of composting in the liquid state by processes similar to aerobic digestion should be investigated [118].

In any composting system where drying is limited, facilities must be provided for drying of the material after composting. Several approaches to drying were discussed in Chapter 6. Obviously, air drying is one feasible alternative in dry climates. In wet climates greater emphasis on obtaining higher dewatered cake solids is warranted so that energy resources are sufficient for both composting and drying to reduce the need for further air or heat drying.

From this discussion, it is obvious that sludge dewatering and composting are integrally related. The two unit processes cannot be viewed separately as has often been done in the past. Success in composting depends heavily on the ability to achieve dry cake solids. Optimization of the two processes to achieve the lowest net cost is beyond the scope of this chapter. However, considering the sensitivity of the compost operation to initial cake solids, designers are advised to achieve cake solids as dry as possible during dewatering.

SUMMARY

Energy released by organic decomposition is the driving force for organic stabilization, temperature elevation and moisture evaporation, all of which are desirable aspects of composting. Therefore, application of thermodynamic principles is a fundamental method for analyzing composting systems.

Sludge composting differs from composting of many other materials in that dewatered sludge cake is still mostly water. With such large quantities of water, recognition of thermodynamic constraints is essential. Sludge composting can be divided into two distinct thermodynamic regions; one in which sufficient energy is available for both composting and evaporative drying, and another in which energy is sufficient only for composting with limited drying.

For the case of composting and drying, the evaporative burden can represent over 75% of the total energy requirement. This energy must be supplied through biological oxidation of organics in the feed substrate. The weight of water per weight of degradable organic in the mixed feed, defined as W, has been shown to be a useful ratio to judge the overall thermodynamic balance. If W is <10, sufficient energy should be available for both composting and moisture evaporation. At $W > 10$, lower composting temperatures or less drying can be expected. With digested sludge, the W ratio should be <10 if sludge cake solids are $>30\%$. With raw sludge, cake solids $>20\%$ should be sufficient because of the higher volatile solids content and degradability of the feed substrate. If W is <10, the windrow, aerated static pile and reactor composting systems should be equally suitable from a thermodynamic standpoint for composting either digested or raw sludge.

Sludges with $W > 10$ can be composted successfully if drying is limited to reduce the evaporative burden. This is possible because two distinct air requirements can be described in composting: that required for moisture removal and drying, and that required for biological oxidation of organic materials (see Chapter 8). The air requirements become equivalent only at cake solids of 30–40% and exit air temperatures above about 70°C. At lower cake solids, the air required for moisture removal is considerably greater than that for biological oxidation. Drying can be limited by reducing the air supply to that required for biological oxidation. In this way, digested sludges with cake solids as low as 10–15% will contain sufficient energy for composting. Drying will be limited, however, and further drying by other methods will be necessary if a low-moisture product is desired.

Composting with limited drying can be most readily accomplished in controlled aeration systems, such as the aerated static pile and certain reactor composters. The conventional windrow technique allows little control over air supply and may not be the system of choice for sludges with insufficient energy for both composting and drying.

The solids content produced during dewatering is probably the single most important variable in determining the successful composting of sludge. Moisture and volatile solids control, and the energy budget for the system are largely influenced by this parameter. Implementation of any composting system should be coordinated with design of the sludge dewatering process to obtain cake solids with sufficient energy to drive the composting system selected. This conclusion is valid regardless of the composting system, whether windrow, aerated static pile or reactor type.

CHAPTER 10

KINETIC PRINCIPLES

INTRODUCTION

Kinetics deals with rates or velocities of reactions. Kinetics must be distinguished clearly from the related subject of thermodynamics. Thermodynamics is concerned with energy changes accompanying chemical reactions, but does not reveal how fast these reactions will occur. For example, oxidation of glucose is exothermic, but in the form of cellulose its biological oxidation proceeds slowly at best. If one should ignite the cellulose, such as by setting fire to this paper, oxidation proceeds rapidly indeed. The same energy is released whether the paper is burned or biologically oxidized to carbon dioxide and water. However, the kinetics of the two cases are remarkably different.

This chapter will investigate the kinetics of the composting system. This is a subject of vital interest to the design engineer, who must determine the type of reactor and detention times required to achieve a certain degree of organic stabilization. Observed oxygen consumption rates for composting materials will be investigated and the general types of biological systems will then be discussed. The chapter will conclude with a discussion of those factors probably responsible for limiting the rates of composting reactions. This should provide a better understanding of the composting system and the complexity of competing processes, any of which may limit the entire system if not properly controlled.

RATES OF OXYGEN CONSUMPTION

The subject of oxygen consumption during composting has been investigated extensively by numerous researchers using a variety of experimental procedures and feed materials. Both batch and continuous composters have been used. Most data have been developed using controlled, laboratory or bench-scale composters where various factors, such as temperature, pH, moisture and free airspace (FAS), can be held reasonably constant. Garbage and refuse materials have received the most intense study, probably because of the rather universal concern over their proper management. Unfortunately, there is a lack of similar data on organic sludges, which must be caused by the rather recent interest in sludge composting.

Rather than proceed with a thesis-type discussion of various researchers and their work, a few of the more notable studies will be examined in detail. This will give the reader a better feel for the experimental procedures involved and the reliability of the resultant data. Along the way a better understanding will be developed of the differences between batch and continuous reactors.

Among the more thorough studies of oxygen uptake rate were those conducted by Schultz [104,119-121] from about 1957 to 1962. Continuous composters were operated to achieve steady-state conditions. In addition, feed material consisted of various garbage/sludge mixtures; this was one of the few such studies to include sludge materials. Schultz used a rotating drum composter constructed from a 55-gal (208-liter) drum. The composter was normally filled to about two thirds of its volume. Feed material was added every 1-2 days and the composter was rotated for about 5 min before and after each feeding. Thus, conditions expected in a continuous-flow, well-mixed reactor were simulated. Data were collected after establishment of steady-state conditions within the reactor.

Schultz examined a number of feed mixtures composed of one or more of the following: (1) garbage consisting primarily of table scraps; (2) dewatered, digested sewage sludge containing 70-80% moisture which had been conditioned before dewatering with about 4% $FeCl_3$ and 12% lime on a dry-weight basis; (3) air-dried digested sludge; (4) air-dried compost; (5) wastepaper, mostly newsprint; and (6) vermiculite, an expanded mica consisting of particles 0.6-1.3 cm in size. The vermiculite served as a bulking agent in the same way that wood chips are used in the aerated static pile process. Data on the average composition of raw materials and mixtures are given in Table 10-1. Note that most mixtures contained a significant percentage of sludge cake. Garbage and wastepaper fractions were shredded before use.

During initial testing it was found that plain ground garbage, a mixture of garbage and dewatered sludge, or dewatered sludge cake by itself, would not successfully compost. Material in the drum became too dense and formed large balls which impeded oxygen transfer. Because the drum was rotated only intermittently, the tumbling action was apparently insufficient to supply oxygen to the wet material. In a separate study, Shell and Boyd [97] employed constant mixing and successfully composted raw dewatered sludge

Table 10-1. Average Analytical Data for Raw Materials and Mixtures Used by Schultz [104]

Item	Moisture (% Fresh Weight)	Ash (% Dry Weight)	pH	Wet Bulk Weight (g/1)	Specific Gravity (g/cm^3)	Dry Bulk Density (g/cm^3)	Porosity (%)	Free Airspace[a] (%)
Ground Garbage	63	10	5.9	740	1.064	0.237	77.7	27.4
Moist Sludge Cake	72	50	8.2	660	1.43	0.185	87.1	39.6
Dry Sludge Cake	6.0	50	8.4	390	1.43	0.367	74.3	72.0
Vermiculite	1.0	100	7.5	90	2.5	0.09	96.4	96.3
Dry Compost	10	60	8.0	290	1.563	0.261	83.3	80.4
Shredded Paper	8.0	8.0	5.0	25	1.0	0.023	97.7	97.5
Mixture A[b]	47.8	31.3	6.7	642	1.23	0.353	72.8	42.1
Mixture B[c]	56.5	49.4	5.9	410	1.42	0.178	87.4	64.3
Mixture C[d]	50.5	22.1[e]	6.2	410	1.154	0.203	82.4	61.7
Mixture D[f]	60.0	17.3[e]	6.0	410	1.152	0.164	85.8	61.2
Mixture D_1[g]	57.0	23.2	6.0	410	1.153	0.177	84.7	61.3

[a] Recalculated from original data according to Equation 7-10.
[b] Mixture A = 20 lb garbage, 10 lb air-dry sludge cake.
[c] Mixture B = 20 lb garbage, 10 lb moist sludge cake, 3 lb vermiculite.
[d] Mixture C = 20 lb garbage, 5 lb air-dry sludge cake, 4 lb paper.
[e] Computed from average data for components.
[f] Mixture D = 20 lb garbage, 10 lb moist sludge cake, 5 lb paper.
[g] Mixture D_1 = 20 lb garbage, 10 lb moist sludge cake, 5 lb paper, 2.5 lb air-dry compost.

in a mechanical system without addition of bulking agents. In any event, Schultz reduced the moisture content of the feed mixture to 50-60% using the dry materials previously mentioned. Note that the ground garbage and dewatered sludge cake had about 25 and 40% FAS, respectively, whereas the conditioning materials were characterized by values of 70-95%. For all mixtures listed, the FAS ranged from about 40 to 60%. It is felt that bulk weight measurements recorded by Schultz may have been low, particularly for dewatered sludge cake. Therefore, calculated values of FAS shown in Table 10-1 may be higher than those actually present in the reactor.

Schultz concluded that ground food waste and dewatered sludge cake are too high in bulk weight and in moisture to be composted as such and that "... those materials have to be mixed with a dry and bulky component such as refuse, waste paper, corncobs, wood shaving, rice hulls, etc. in order to obtain a suitable moisture content and bulkweight." This is a remarkable statement considering it was made in 1962, over a decade before the successful sludge composting operations by the Los Angeles County Sanitation Districts (since 1972) and the U.S. Department of Agriculture (USDA) at Beltsville, MD (since 1973).

Average operating data observed during the test periods are presented in Table 10-2 and Figures 10-1 to 10-3. Residence times varied between 7 and 18 days with volatile solids (VS) destructions of 37-45%. Detention times were estimated from the total weight of ash in the reactor divided by the daily input of ash. Moisture content of the outfeed approximated that of the infeed, indicating that moisture evaporation was compensated by the loss of volatile matter. Temperatures remained constant in the thermophilic range between 43 and 68°C.

Rates of oxygen consumption were determined throughout Run 2c with results shown in Figure 10-4. Oxygen consumption was strongly influenced by temperature and closely followed the relationship:

$$w_{O_2} = 0.11\,(1.066)^T \tag{10-1}$$

where w_{O_2} = rate of oxygen consumption, mg O_2/g volatile matter-hr
T = temperature, °C

Temperatures above 65-70°C were not investigated. However, experience has shown that the oxygen consumption rate would be expected to decrease at higher temperatures.

The pH of the outfeed was consistently >7, except in Run 2c (Figure 10-3). In the latter case about 25% of the volatile matter in the reactor was replaced with each daily feeding, the highest feed rate attempted. Apparently, the high loading rate shifted the process into the acid range, which indicated

Table 10-2. Average Operating Data Observed During Continuous Thermophilic Composting [104]

	Run No. 1	Run No. 2b	Run No. 2c	Run No. 3b
Feed Mixture	A	B	B	D,D_1
Feed Cycle, days	1	2	1	2
Test Duration, days		52	75	167
Data Collected, days[a]	11	34	23	35
Residence Time, days	8.9	12.7	7.0	18.3
Red. in Vol. Mat., %	36.8	43.1	42.4	45.2
Moisture, %				
In	47.8	56.5	58.6	57.0
Out	51.3	55.2	58.4	56.9
pH				
In	6.7	5.9	5.6	6.0
Out	7.6	7.8	6.6	8.1
Wet Weight, g/1				
In	642	410	404	412
Out	657	567	587	611
Free Airspace, %[b]				
In	42.1	64.0	64.8	61.1
Out	42.4	52.8	50.6	44.3
Air Supply, m^3/kg VS-day	0.28	0.55	0.73	0.31
Temperature Range, °C	43-64	53-68	59-68	62-68

[a]Indicates time over which data were collected. For Run 2b, for example, data collection began on day 18 and was completed on day 52.
[b]Recalculated from original data according to Equation 7-10.

that the material was not as close to a finished compost as material produced at lower feed rates.

Material removed from the drums was stored in open bins and for several days developed a temperature of about 38°C. After 2-3 weeks of storage the temperature decreased to ambient levels and the compost had a pleasant greenhouse odor. Odor from the sewage sludge was reported to be completely absent. Thus, a curing phase was required even after continuous thermophilic composting at residence times from 7 to 18 days.

Use of a continuous composter is necessary to determine temperature effects under steady-state conditions. However, many composting systems are operated on a batch basis and true steady-state conditions are never achieved. Numerous studies of oxygen consumption have been conducted using batch reactors. An example of the type of data derived from such studies is shown in Figure 10-5 based on the work of Snell [116]. Small, bench-scale digesters were used with temperature and moisture conditions held constant. Moisture control became difficult at higher temperatures and the samples became dried, which likely introduced a moisture rate-limitation.

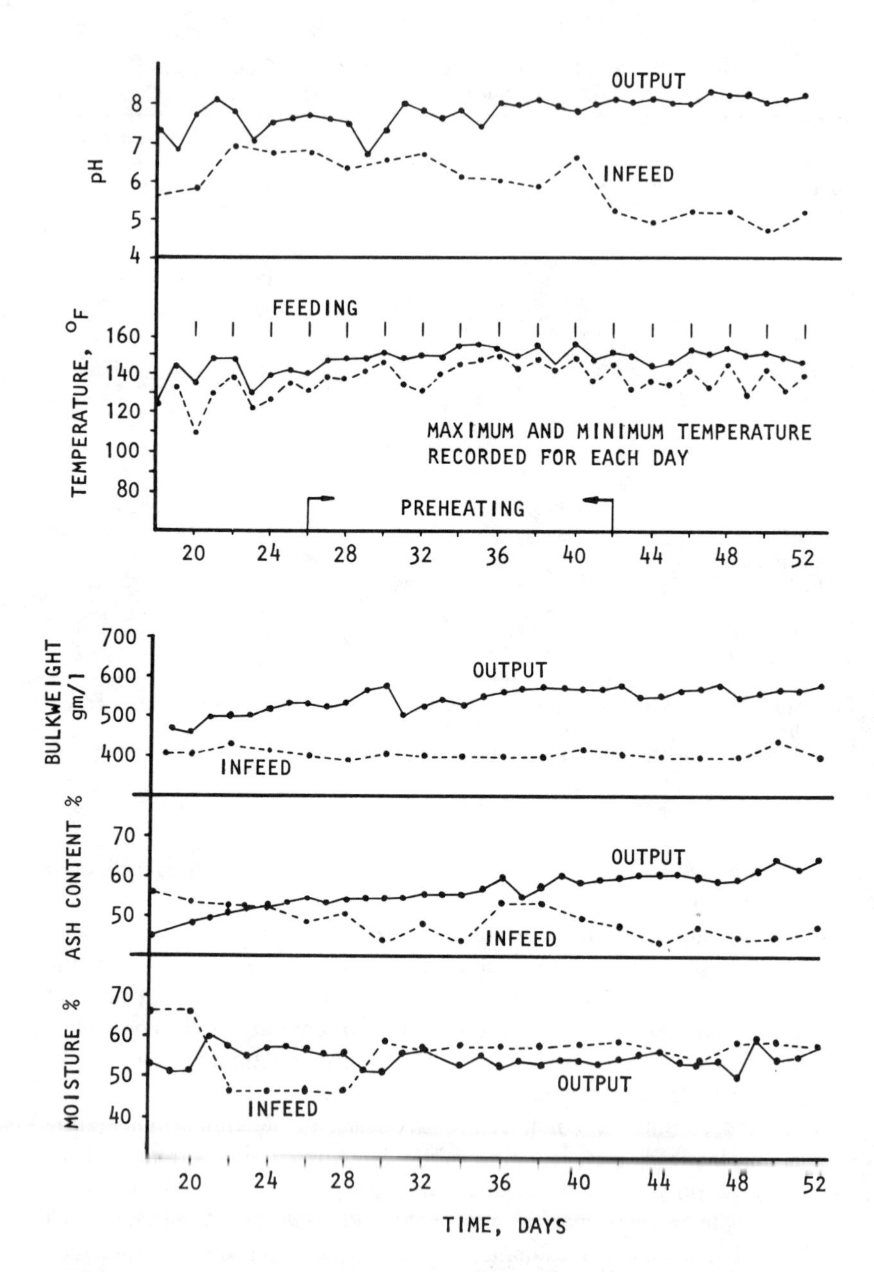

Figure 10-1. Average operating data recorded by Schultz [104] during continuous composting of a garbage/sludge mixture. Feeding was on a 2-day cycle, Run No. 2b, using Mixture B as identified in Table 10-1. Air supply was approximately 6.5 mg O_2/g VS-hr with a detention time based on an ash balance of 12.7 days. The data indicate that reasonably steady-state conditions were achieved.

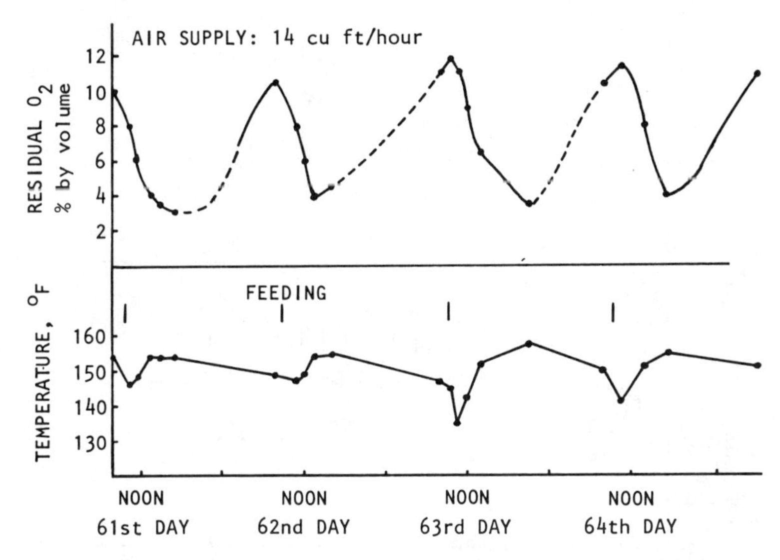

Figure 10-2. Relationship between feed cycle, temperature and residual oxygen during continuous composting as determined by Schultz [104]. Data correspond to Run No. 2c, Mixture B, as defined in Table 10-1.

No seeding or reseeding was practiced and the feed material consisted of ground garbage. Because of the batch nature of the process, a characteristic lag period was observed at the start of composting. As many as eight days were required to achieve maximum rates of oxygen uptake. The maximum rate would hold for several days and then begin to decrease as the more readily degradable feed components were exhausted.

A temperature curve typical of a batch windrow system is shown in Figure 10-6. The windrow consisted of a large mass of aerobically composting refuse. The difference between temperature curves for batch and continuous reactors becomes obvious upon comparison of Figures 10-1 and 10-6. Notice in Figure 10-6 that the temperature curve does not hesitate at the transition from mesophilic to thermophilic temperatures. In smaller masses of composting material there is sometimes a temporary temperature plateau as the mesophilic bacterial population declines and the thermophilic population develops. In large piles where heat loss from the interior is slow, the effect is dampened.

The reader should note the fundamental difference between continuous composters which achieve a steady-state condition as represented by Figures 10-1 to 10-3, and batch processes characterized by the data in Figures 10-5 and 10-6. If the feed schedule is continuous or semicontinuous, and if the reactor contents are well mixed, feed material will be quickly inoculated with

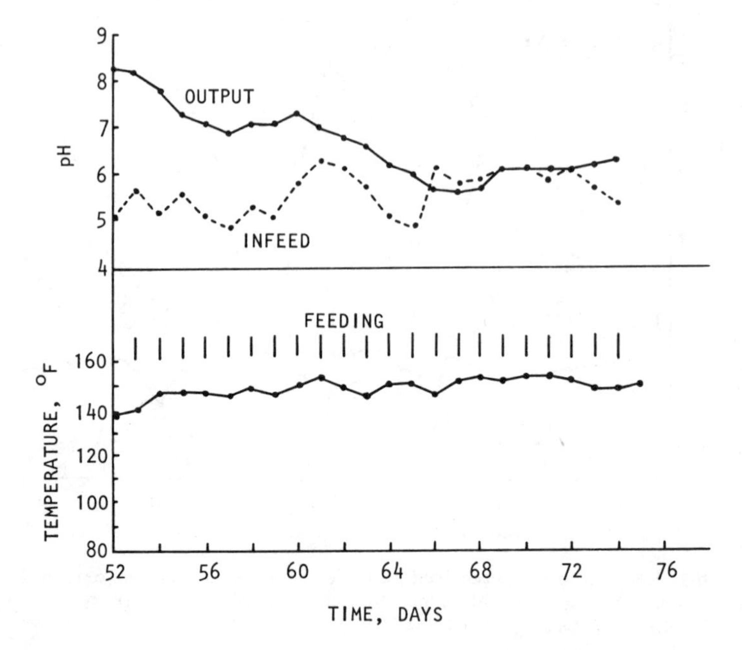

Figure 10-3. Effect of high organic loading on product pH (after Schultz [104]). Data correspond to Run No. 2c, Mixture B, as defined in Table 10-1. The daily feed schedule reduced the detention time to seven days. At high loading rates the rate of acid production was apparently higher than the rate of acid consumption and the pH shifted from the normal alkaline range to the acid range.

the mixture of microbes developed for the particular steady-state conditions. The material will also quickly be brought to conditions of temperature, pH, moisture and FAS established in the reactor. Thus, lag periods common to batch systems can be reduced or eliminated with continuous composters.

Most biological reactors used in wastewater treatment practice are operated on a continuous basis with varying degrees of mixing and, thus, are designed to achieve near steady-state conditions. Such reactors have the distinct advantage of operating with aqueous slurries or solutions for which continuous feeding and mixing are relatively straightforward. With solid materials, mixing in particular becomes rather difficult. It can also be expensive if detention times between 7 and 18 days are required as determined by Schultz [104]. Because composting deals with solid or semisolid materials, recourse is often made to the somewhat less efficient batch operation.

Numerous other rate studies have been conducted using batch and continuous composters on a variety of feed materials. A summary of some of

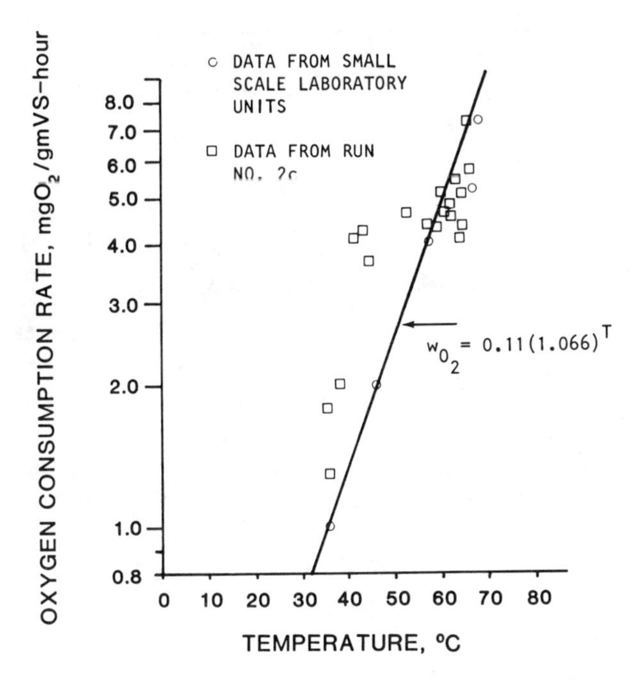

Figure 10-4. Relationship between temperature and oxygen consumption rate observed by Schultz [104] during continuous composting of garbage and digested dewatered sludge cake.

the available data is presented in Figure 10-7. Despite the variety of procedures and feed materials, the data are remarkably consistent. All studies show an increase in the rate of oxygen uptake with increasing temperature. Some of the studies show maximum rates at intermediate temperatures between 45 and 60°C with decreasing rates at higher temperatures. On the other hand, the studies by Schultz [104] and Wiley [122-125] show consistently increasing rates to about 70°C, the highest temperatures examined. Because Schultz used continuous composters, a degree of acclimation or development of a more optimum microbial flora may have occurred at the higher temperatures.

Jeris and Regan [103,126,127] conducted an interesting set of experiments using continuous composters fed with mixed refuse, composted mixed refuse and newsprint. As shown in Figure 10-7, compost and newsprint showed oxygen consumption rates about one order of magnitude less than mixed refuse. This illustrates that compost may be stable, but it is certainly not inert. It will continue to decompose, but at greatly reduced rates. The low rates observed for newsprint are caused by the structural resistance of mechanically pulped wood products (see Chapter 8). Mixed refuse used by

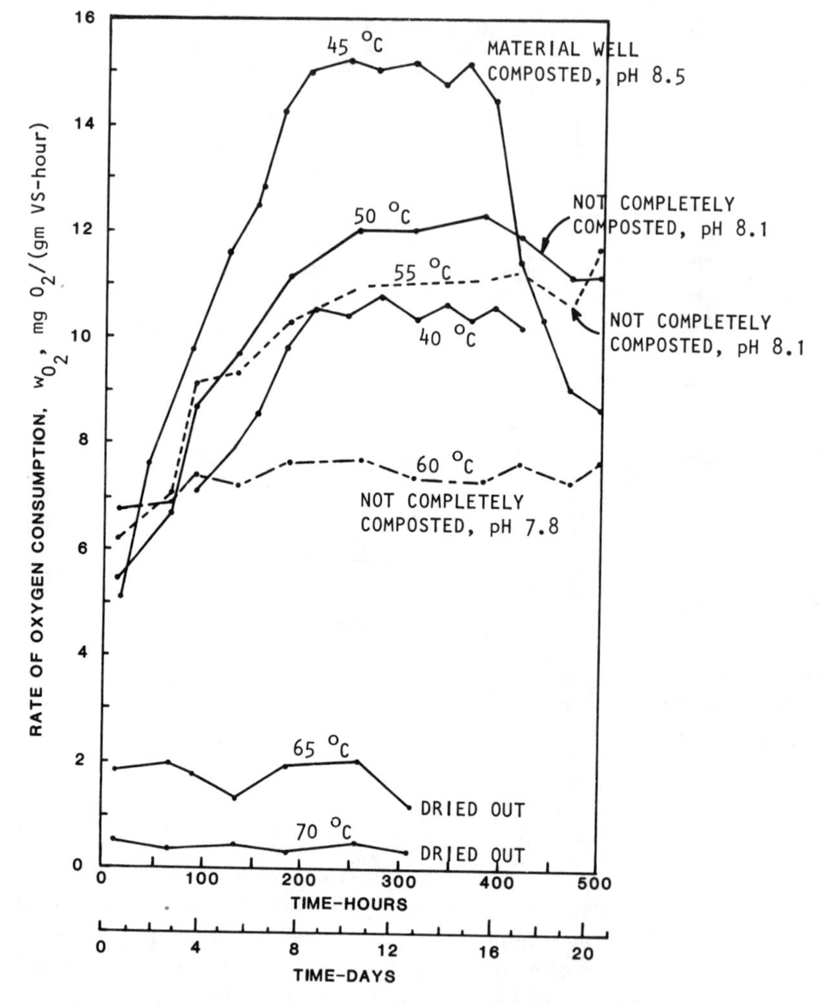

Figure 10-5. Rate of oxygen consumption for various temperatures using a batch composter with ground garbage. Note the somewhat linear increase in oxygen uptake rate to a maximum value for each temperature. Peak values are plotted in Figure 10-7 [116].

Jeris and Regan contained up to 50% crude fiber. Thus, the higher oxygen consumption was caused by decomposition of components other than newsprint.

In any composting system one must be prepared to meet the maximum oxygen consumption rate demanded by the microbial population. From

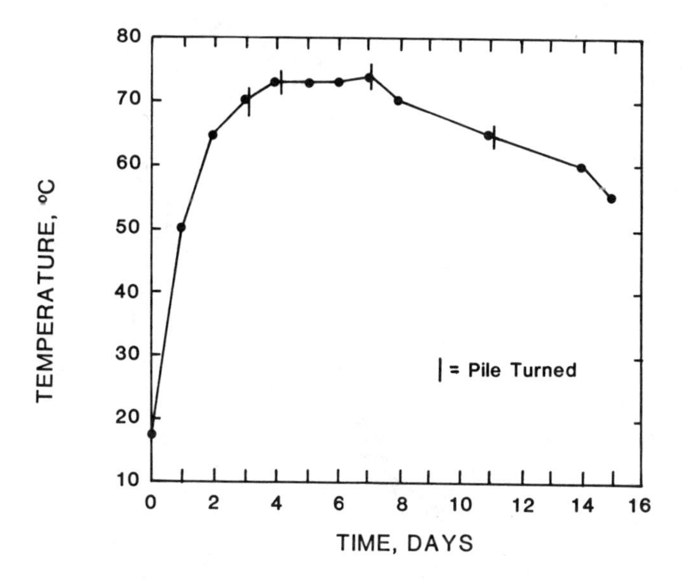

Figure 6. Temperature curve typical of a large mass of refuse material during aerobic composting in a batch operated windrow system [96].

Figure 10-7, a supply of 7-12 mg O_2/g VS-hr would appear to be sufficient in all but the most extreme cases. This equates to about 9-15 ft^3 air/lb VS-day (about 0.5-1.0 m^3/kg VS-day) and is consistent with the general range of maximum values reported in the literature. Again referring to Figure 10-7, the data by Schultz [104] are reasonably consistent with the general range of values observed. Therefore, Equation 10-1 can be used with reasonable success to predict peak oxygen consumption rates as a function of temperature.

The effect of aeration rate on the temperature profile was observed during aerated static pile composting of raw sludge at Beltsville [14]. An aeration rate of about 15-16 m^3/ton-h was recommended as usually adequate to meet the maximum oxygen consumption rate. Assuming the sludge cake to be about 30% ash, the recommended aeration rate equates to about 0.53 m^3/g VS-day which is reasonably consistent with previously discussed values.

The reader should note that three basic aeration requirements have been discussed. These are (1) the total quantity of oxygen required to satisfy the stoichiometric demand; (2) the air quantity required to remove moisture and thus dry the remaining sludge solids; and (3) the rate of air supply required to satisfy peak oxygen consumption rates. An additional air requirement to remove heat from the system and prevent excessively high temperatures will be discussed in Chapters 12 to 14.

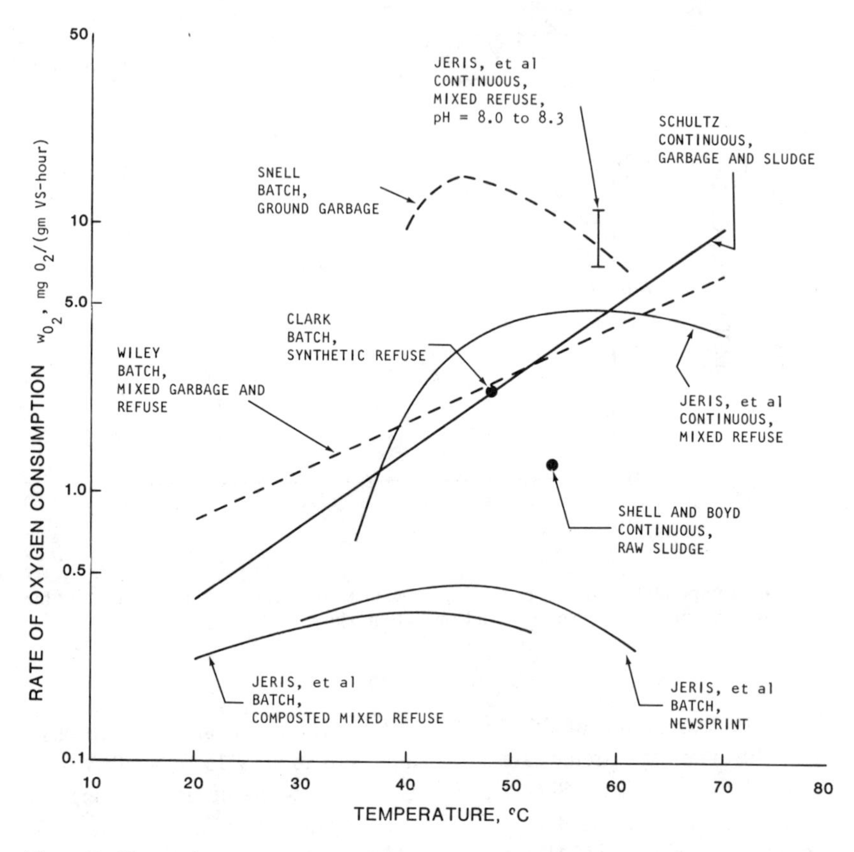

Figure 7. Observed oxygen consumption rates for various composting mixtures and reactor types as a function of temperature. Each curve represents the best fit of observed data.

Experimental data presented in this section are useful in indicating the general range of oxygen consumption rates expected in composting systems. However, the data do little to explain the mechanisms controlling rates of substrate use and oxygen consumption. Possible rate-controlling mechanisms in composting systems will be discussed in the remainder of this chapter.

CLASSIFICATION OF MICROBE-SUBSTRATE SYSTEMS

Microbe-substrate systems can generally be divided in two distinct types, homogeneous and heterogeneous. The systems are illustrated in Figure 10-8 and summarized in Table 10-3. In a homogeneous system, microbes are dispersed in an aqueous solution containing a soluble substrate. Thus, the mass of microorganisms is completely dispersed throughout the reactor volume. Substrate gradients between cells are minimized, and each cell sees virtually the same concentration if the system is well mixed. Such conditions are approximated in many industrial fermentations and waste treatment processes. The activated sludge process is an example of a nearly homogeneous system. It is not completely homogeneous, however, because flocculation causes some degree of separation between substrate and microbe. In addition, substrate concentration gradients may exist inside the floc particle so that each organism is not surrounded by the same substrate concentration. Nevertheless, each floc particle is randomly dispersed in the fluid phase and, from this standpoint, the activated sludge unit can be viewed as a nearly homogeneous system.

Homogeneous systems have traditionally been modeled using the Monod kinetics developed in Chapter 4 [128]. The assumption is usually made that

Table 10-3. Description of Biological Systems Normally Encountered in Biochemical and Sanitary Engineering Practice

I. HOMOGENEOUS SYSTEMS–Individual microbes are uniformly dispersed in a solution of soluble substrate. Model is also applied to cases of flocculated microbes in a solution of soluble and fine particulate solids. The activated sludge process treating industrial or municipal wastewater is typical of this system.

II. HETEROGENEOUS SYSTEMS

A. Attached Microbial Growth–Microbes are separated from the aqueous substrate usually by attachment to a solid surface.

1. Falling Film–Microbes are attached to a surface which is washed by a falling aqueous film containing the substrate. The trickling filter, oxidation tower, and rotating biological contactor are examples.

2. Submerged Film–Microbes are attached to a surface with the void spaces filled with fluid containing the substrate. The submerged filter, anaerobic filter and fluidized bed reactor are examples.

B. Particulate Substrates (Solid or Insoluble Substrates)

1. Aqueous Solution–Solid or insoluble substrate is immersed in an aqueous phase containing microbes, some of which attach to the substrate surface. Industrial growth of microbes on insoluble hydrocarbon substrates and anaerobic digestion of sludge solids could be classified in this category. However, the latter has also been analyzed using homogeneous kinetics.

2. Limited Moisture–Moisture required for microbial growth is limited to that associated with the solid organic substrate. Composting of organic residues and decomposition of organic solids in soils are examples.

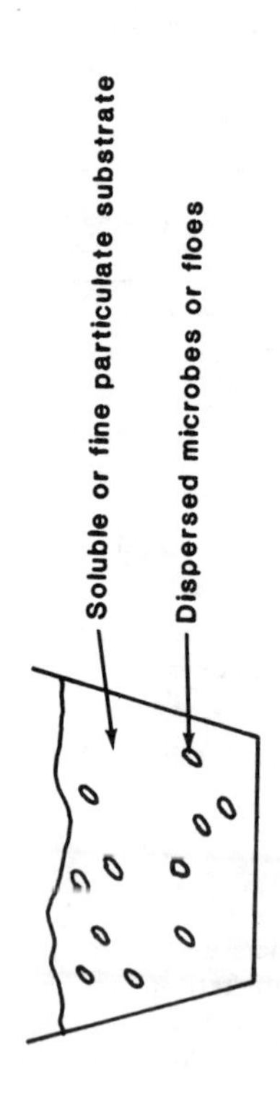

I Homogeneous System

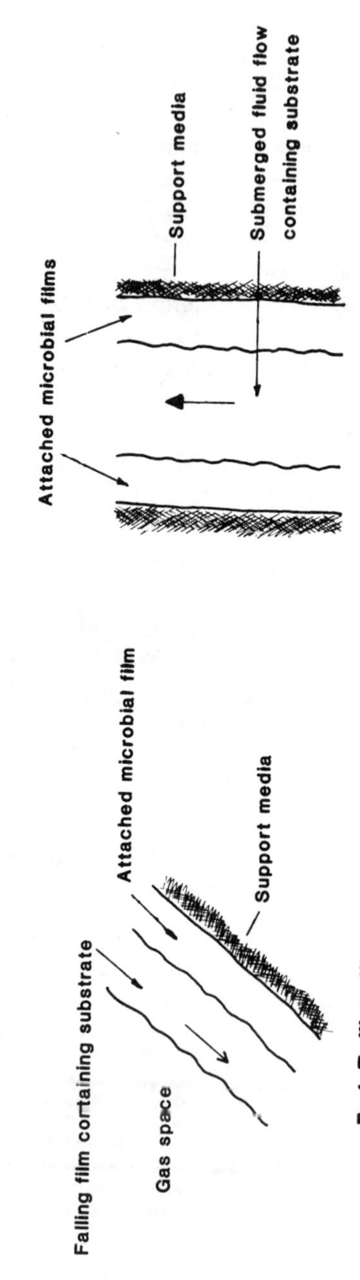

IIa1 Falling film

IIa2 Submerged flow

IIa Heterogeneous systems with attached microbial films

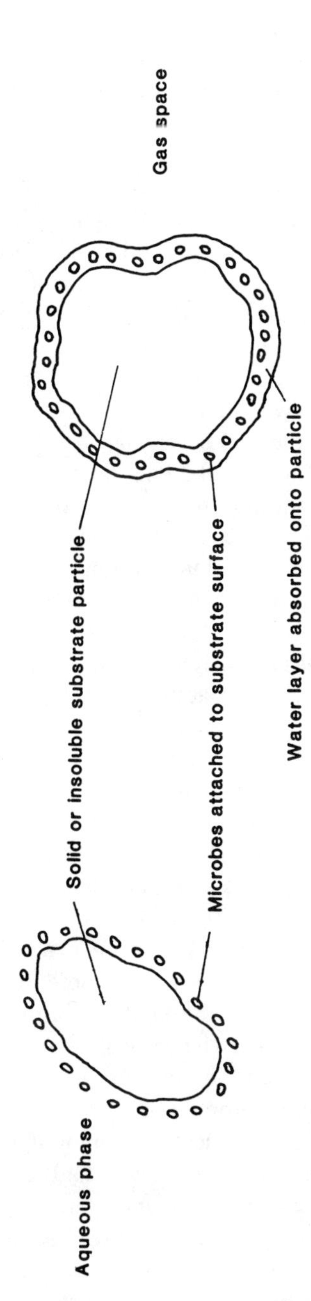

Figure 10-8. Types of biological systems normally encountered in biochemical and sanitary engineering practices.

mass transport of substrate to the cell is not limiting, so that Equation 4-1 can be applied directly. Other authors have examined the effects of substrate diffusion through suspended flocs of microbes [31,129]. The reader is referred to the cited references for a further discussion of this subject.

A heterogeneous system is one in which either the microbes or substrate are separated from the fluid phase containing the other component. Two distinct types of heterogeneous systems are possible. The type most common to waste treatment is a system in which microbes are separated from the fluid phase containing a soluble substrate. The trickling filter, oxidation tower, rotating biological contactor, submerged filter and anaerobic filter are examples of reactors which employ heterogeneous conditions. Each of these reactors contains an inert medium used to support the growth of microbes. Biological film constitutes one phase of the system while liquid containing the substrate constitutes the other. A definite interface exists between the microorganisms and the liquid phase. Substrate must move across this interface to be used in the biological film. A substrate gradient must exist between the microbial film and the bulk liquid to assure a mass flow of substrate into the film. In addition to the waste treatment reactors described above, several industrial processes employ enzymes isolated from living organisms and immobilized on or within solid supports such as polystyrene, porous glass or silica-alumina. Production of high-fructose syrups from cornstarch is an example [31].

The state of the art for mathematical description of heterogeneous systems in which microbes are attached to a fixed surface is reasonably well advanced. However, there are other types of heterogeneous systems for which the state of the art is not as well developed. If the reader has not already guessed, composting falls in this latter class. The system referred to is one in which the substrate is insoluble, existing in a particulate or solid form. Two subcategories of this system can be described: (1) the solid substrate is suspended in a bulk fluid phase, and (2) the aqueous phase is limited to bound water associated with the solid substrate. In either case, microbes must attach to the substrate surface. Hydrolysis of chemical components making up the solid substrate is then necessary before the cell can absorb the solubilized nutrient through its cellular membrane. In the limited-moisture case, available water is limited to that associated with the particulate organic substrate. Thus, an additional limitation may occur if moisture levels decrease to a point where moisture limitation becomes pronounced.

Because sludges consist of organic particulates removed during waste treatment, and because moisture is generally limited to that bound with the dewatered cake, sludge composting and the composting of other organic residues can be described as a heterogeneous system with solid substrate and limited moisture. Kinetics developed for such particulate-substrate systems should apply reasonably well to the case of composting.

Before beginning a discussion of heterogeneous systems with solid substrates, a heterogeneous system with an attached microbial film will first be considered. The latter is somewhat easier to visualize conceptually, and will serve to introduce important mass transfer limitations and methods for mathematical description.

HETEROGENEOUS SYSTEMS–ATTACHED GROWTH

In a heterogeneous reactor, mass transport limitations cannot be neglected. Numerous approaches have been used to account for mass transport limitations and to describe the kinetics of heterogeneous systems [130-133]. Perhaps the most fundamental approach is to describe the process as one of diffusion with simultaneous biochemical reaction. The kinetics of an individual cell are assumed to be governed by the environment immediately surrounding that cell and not by the external situation in the bulk liquid phase. Substrate molecules diffusing across the liquid-film interface are used by microbes in the film. This establishes a substrate gradient allowing for further diffusion of substrate into the film. At equilibrium, the rate of movement across the interface equals the rate of substrate use by the bacterial film. Kinetics of microbes in the film can be described with the Monod model (Equation 4-1) using the concentrations of substrate surrounding each cell. Knowing the substrate concentration as a function of depth into the biofilm, the gradient at the liquid-film interface can be determined. Fick's law of molecular diffusion can then be applied to determine the substrate flux across the interface.

Fick's law relates the rate of transfer of a mass of soluble substrate $\partial F/\partial t$ through a surface area A to the concentration gradient of that substance in the direction of transfer.

$$\frac{\partial F}{\partial t} = -AD_f \frac{\partial S}{\partial z} \tag{10-2}$$

where $\partial F/\partial t$ = mass rate of substrate transfer, mass/time
A = surface area, length^2
D_f = diffusion coefficient in the biofilm, length^2/time
$\partial S/\partial z$ = substrate gradient in a direction perpendicular to the surface layer, mass/volume/length

To develop an expression for the substrate concentration as a function of distance into a biological film, a material balance can be made of the substrate in an incremental volume of unit cross-sectional area and thickness dz. Such an element located in the biofilm is shown in Figure 10-9.

Mass balance terms which apply to this element are:

$$\text{inflow} = -AD_f \frac{\partial S_f}{\partial z} = -D_f \frac{\partial S_f}{\partial z}$$

$$\text{outflow} = -D_f \left[\frac{\partial S_f}{\partial z} + \frac{\partial}{\partial z} \frac{\partial S_f}{\partial z} dz \right]$$

$$\text{sinks} = \frac{kS_fX}{K_s + S_f} dz \qquad \text{(See Equation 4-1)}$$

From the mass balance about dz, an equation for steady-state is given by:

$$\frac{d^2S_f}{dz^2} = \frac{kS_fX}{D_f(K_s + S_f)} \tag{10-3}$$

No transfer of material occurs across the film-medium interface. Therefore, dS_f/dz at this interface must be zero. At the liquid-film interface the concentration of substrate must equal the concentration in the liquid phase, S_i. Therefore, boundary conditions (BC) for the problem become:

$$S_f = S_i \quad \text{at} \quad z = 0 \qquad \text{BC 1}$$

$$\frac{dS_f}{dz} = 0 = \frac{dS_0}{dz} \quad \text{at} \quad z = L \qquad \text{BC 2}$$

Equation 10-3 is a second-order, nonlinear, ordinary differential equation. As written it does not have an analytical, closed form solution, but is amenable to numerical solutions. Equation 10-3 defines the process of diffusion with simultaneous biochemical reaction. The mass concentration of microbes in the film, X, can be measured. k and K_s are the same coefficients as for the homogeneous case and can be determined by various experimental procedures [31] or from published literature values. Therefore, the terms appearing in Equation 10-3 can be determined reasonably well. The model has been applied to anaerobic filters [134], submerged filters [133], and was recently experimentally verified for films of nitrifying bacteria [135,136].

In many cases more than one substrate must be transported into the biofilm. For aerobic organoheterotrophic systems, for example, both organic substrate and oxygen must be supplied to the biofilm. The term S_f in Equation 10-3 refers to the concentration of the rate limiting substrate. If two or more substrates are being transported into the biofilm, however, which substrate will control the overall kinetics?

Consider a reaction involving two diffusing substrates such as:

$$v_1A_1 + v_2A_2 \rightarrow \text{products}$$

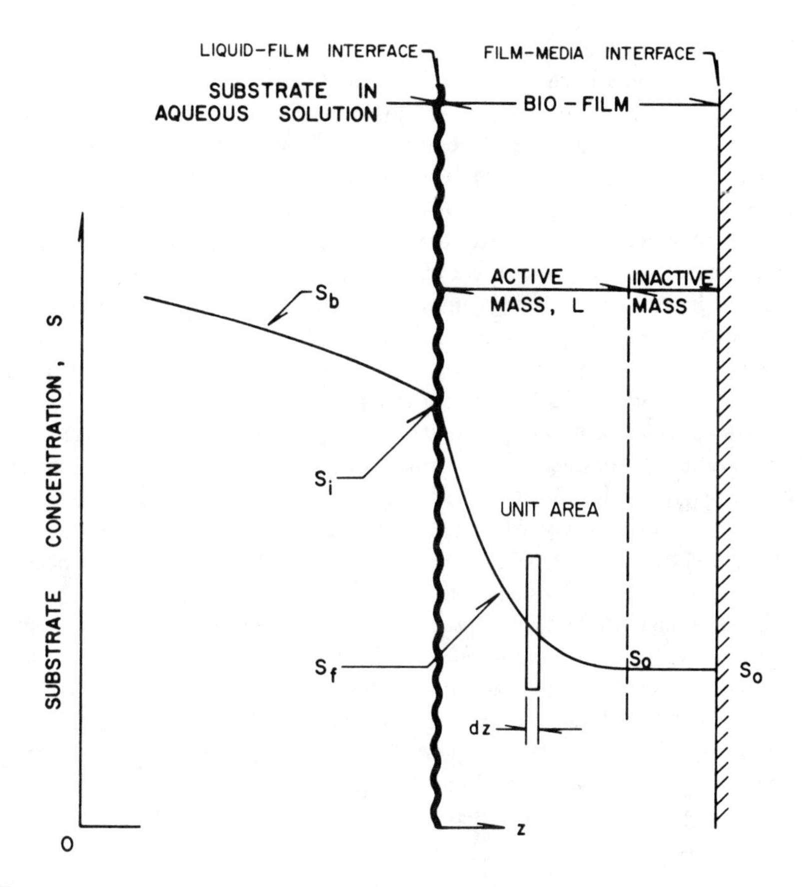

Figure 10-9. Potential substrate profiles in a heterogeneous system composted of substrate in aqueous solution and the microbial population in a biological film.

To assure that stoichiometric amounts of both substrates cross every boundary in the diffusion field the following relationship must hold:

$$\frac{D_1 C_1}{v_1} = \frac{D_2 C_2}{v_2} \tag{10-4}$$

where C_1 = molar concentration of substrate A_1
C_2 = molar concentration of substrate A_2
D_1 = diffusion coefficient of A_1
D_2 = diffusion coefficient of A_2

Consider a film of nitrifying bacteria for which both oxygen and ammonium must be supplied. The diffusion coefficient for oxygen is approximately

1.4 times the value of the diffusion coefficient for ammonium ion. Two moles of oxygen are required for oxidation of one mole of NH_3-N. Therefore, the rate of ammonium oxidation should not be oxygen-limited, provided about 3.2 mg of oxygen are present for each mg of NH_3-N in the solution adjacent to the liquid-film interface. About 4.57 mg oxygen/mg NH_3-N must be present to assure complete stoichiometric oxidation. If oxygen is present in the stoichiometric amount, therefore, it should not limit the rate of ammonium oxidation. In such a case ammonium can be considered as the rate-limiting substrate. If oxygen were present at much less than 70% stoichiometric, (i.e., 3.2/4.57) oxygen transport would likely be rate-controlling. This influence of oxygen to nitrogen ratio on substrate kinetics in nitrifying biofilms has been verified experimentally [133].

In many cases with multiple substrates it is difficult to predetermine which substrate is rate-limiting. Indeed, more than one substrate can influence the overall reaction rate. In such a case, diffusion of individual substrates can be followed mathematically with substrate kinetics defined by the relationship shown in Equation 4-9. While the mathematics are somewhat complicated, the system is still conceptually well defined.

Even though the attached growth heterogeneous system is not directly applicable to the composting process, the exercise has provided insight into the manner in which mass transport can be linked to chemical or biochemical reaction kinetics. It is necessary to first establish this link in a conceptual model amenable to mathematical development. This introduction to heterogeneous systems should be useful as the equally complex and less well defined process of composting is discussed.

HETEROGENEOUS SYSTEMS–SOLID SUBSTRATE

Before beginning a discussion of heterogeneous systems with particulate substrates, it would be helpful to discuss the sequence of events involved in metabolism of the substrate. Referring to Figure 10-8, the following processes can be conceptually described:

1. release of extracellular hydrolytic enzymes by the cell and transport of the enzymes to the surface of the substrate;
2. hydrolysis of substrate molecules into lower molecular weight, soluble fractions;
3. diffusion transport of solubilized substrate molecules to the cell;
4. diffusion transport of substrate into the microbial cell, floc or mycelia;
5. bulk transport of oxygen (usually in air) through the voids between particles;
6. transport of oxygen across the gas-liquid interface and the unmixed regions which lie on either side of such an interface;
7. diffusion transport of oxygen through the liquid region;
8. diffusion transport of oxygen into the microbial cell, floc or mycelia; and
9. aerobic oxidation of the substrate by biochemical reaction within the organism.

A rather complicated sequence of events is necessary before substrate can be composted successfully. The list could be expanded, but was summarized here to avoid unnecessary complication. (Can the reader think of other events in the sequence?) Any one of the events described above could limit overall process kinetics. Several of the more important processes will be discussed to gain better insight into rate limitations which can be imposed on composting systems.

Kinetics of Particulate Solubilization

Consider a hydrolytic enzyme which adsorbs to an active site on the substrate surface. The following equilibrium can be described:

$$E + A \underset{k_2}{\overset{k_1}{\rightleftharpoons}} EA \overset{k_3}{\rightarrow} P + A \tag{10-5}$$

where A is a vacant site on the substrate surface; E is a free hydrolytic enzyme in solution which can adsorb to the surface with a reaction rate constant k_1, and desorb with rate constant k_2; and EA is the enzyme substrate complex which can either desorb to the original constituents with a rate constant k_2 or react irreversibly to yield the original enzyme E and the desired product P with a rate constant k_3. If a_0 is the total number of adsorption sites per unit volume, then:

$$a_0 = a + (ea) \tag{10-6}$$

where a = number of free sites per unit volume
(ea) = number of sites with an adsorbed enzyme
e = number of free enzymes per unit volume of the reaction mixture

The change in concentration of the enzyme substrate complex can be described as:

$$\frac{d(ea)}{dt} = k_1(e)(a) - k_2(ea) - k_3(ea) \tag{10-7}$$

Under steady-state conditions d(ea)/dt will equal zero. Thus:

$$k_1(e)(a) = k_2(ea) + k_3(ea) \tag{10-8}$$

or

$$a = \frac{(ea)}{e}\left(\frac{k_2 + k_3}{k_1}\right) = \frac{(ea)}{e} K_a \quad (10\text{-}9)$$

Substituting Equation 10-9 into Equation 10-6 and rearranging:

$$(ea) = \frac{a_0(e)}{e + K_a} \quad (10\text{-}10)$$

The rate of product formation v was described in Equation 10-7, as $k_3(ea)$. Thus:

$$v = \text{product formation/unit time} = k_3(ea) = \frac{k_3(a_0)e}{K_a + e} \quad (10\text{-}11)$$

If e is the concentration of free enzyme, it can be related to the total concentration at the start of the experiment e_0 by:

$$e_0 = e + (ea) \quad (10\text{-}12)$$

In certain enzyme systems it is often the case that e_0 is much greater than a_0; therefore:

$$e_0 \simeq e \quad (10\text{-}13)$$

so that

$$v = \frac{k_3 a_0 e_0}{K_a + e_0} \quad (10\text{-}14)$$

Equations 10-11 and 10-14 are similar in form to Equation 4-1 developed for the case of homogeneous growth. In the present case, however, the rate of reaction should reach a maximum value at high enzyme concentrations. Physically, this corresponds to essentially complete adsorption of enzyme on the available substrate surface.

Equation 10-14 has been used to describe kinetics of solid substrate-enzyme systems with reasonable success. Data on enzyme-solid substrate reactions are presented in Figures 10-10 and 10-11. The latter is a Lineweaver-Burke double reciprocal plot ($1/v$ vs $1/e_0$), which from Equation 10-14 should be linear.

The form of Equation 10-11 can probably be adapted to the case of microbes growing on a solid substrate. The concentration of extracellular enzymes e_0 is likely a function of the mass concentration of microbes X in

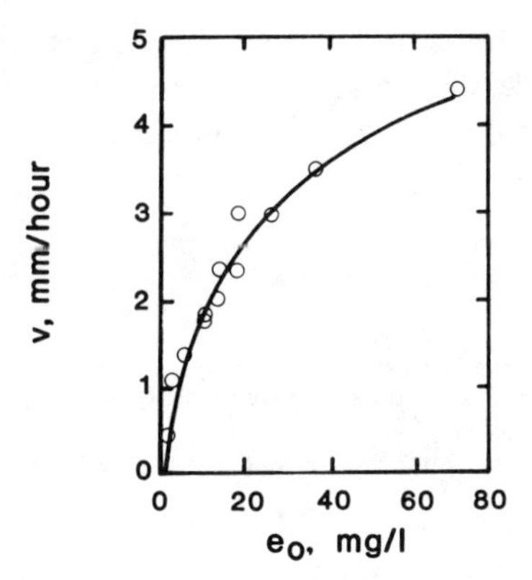

Figure 10-10. Dependence of the rate of disappearance of solid substrate (thiogel) on the concentration e_0 of enzyme in solution [205].

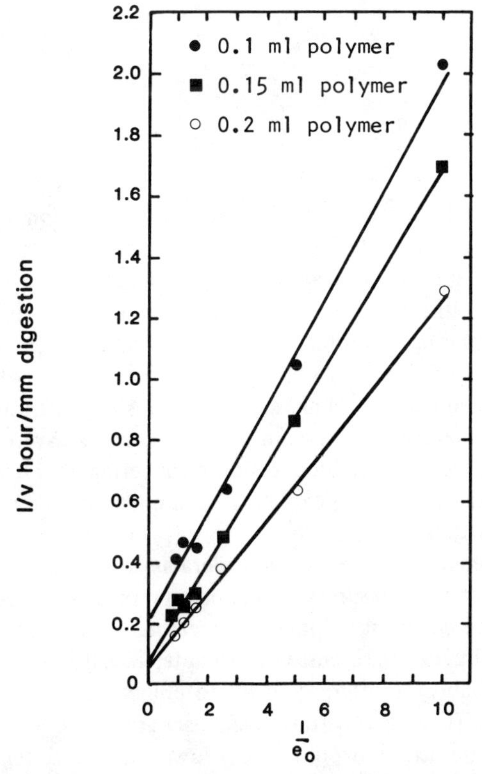

Figure 10-11. Double-reciprocal plot for digestion of an insoluble substrate (poly-β-hydroxybutyrate particles) by an enzyme (depolymerase of *P. lemoignei*) in solution [206].

the reaction mixture. Also, the total number of absorption sites a_0 is likely related to the available surface area per unit volume A_V. Substituting these terms into Equation 10-11, and dropping unnecessary subscripts:

$$v = -\frac{ds}{dt} = \frac{kA_V X}{K_X + X} \qquad (10\text{-}15)$$

where ds/dt = rate of hydrolysis of solid substrate
k = maximum rate of solid substrate hydrolysis which occurs at high microbial concentration
K_X = half velocity coefficient equal to the microbial concentration where ds/dt = k/2

The general form of Equation 10-15 is graphically illustrated in Figure 10-12. In the extreme cases where (1) X is very high ($X \gg K_X$), and (2) when X is very low ($X \ll K_X$), Equation 10-15 can be approximated by the following discontinuous functions:

$$\frac{ds}{dt} = -kA_V \qquad (X \gg K_X) \qquad (10\text{-}16)$$

$$\frac{ds}{dt} = -k'A_V X \qquad (X \ll K_X) \qquad (10\text{-}17)$$

where $k' = k/K_X$. Equation 10-16 is a zero-order reaction with respect to microbial concentration while Equation 10-17 is first-order.

There are no experimental data on actual composting systems which can be used to verify the form of Equation 10-15. For one thing, traditional methods of estimating microbial concentrations, such as use of volatile suspended solids (VSS) in the homogeneous system, are not applicable in the case of solid substrates. Thus, actual composting operations are not as "clean" as pure enzyme systems and reaction kinetics have not been developed to the same extent as in biochemical engineering. However, the form of Equation 10-15 can be used to explain several phenomena observed in actual composting operations.

Jeris and Regan [126] observed considerably lower oxygen consumption rates when composting newsprint compared to mixed refuse. Also, the effect of mechanical and chemical pulping in increasing the rate of degradation of wood and its products was discussed in Chapter 8. Differing rates of decomposition can probably be interpreted as differences in the value of kA_V for Equation 10-15. Thus, a substrate such as natural wood fiber, which is resistant to enzyme attack, would have a lower kA_V value relative to a substrate more amenable to solubilization by hydrolytic enzymes. This could be interpreted as a lower number of available enzyme binding sites or a lower number of successful enzyme reactions in the more resistant substrates.

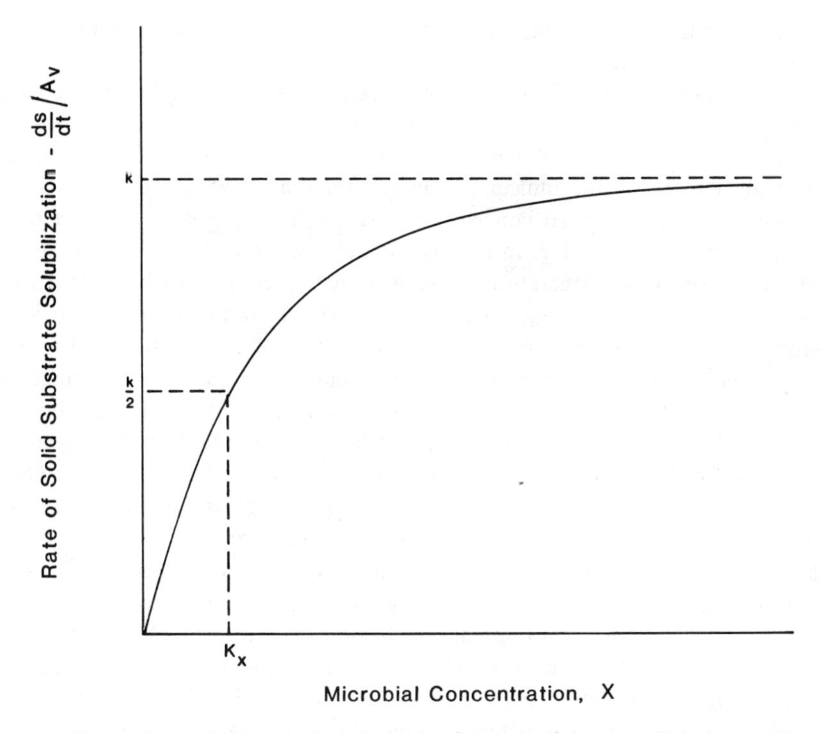

Figure 10-12. Rate of substrate hydrolysis as a function of the microbial concentration for a heterogeneous system with solid substrate.

Most refuse composting systems have been characterized by an initial period of high oxygen uptake followed by a longer period of low oxygen uptake. Composting refuse is commonly placed in curing or finishing piles during the latter period. Refuse is composed of a mixture of organics of varying kA_v values. During early stages, substrates with high values of kA_v are decomposing and the microbial population increasing. However, substrates with low kA_v values will continue to decompose at lower rates for a longer period of time. Also, the higher microbial concentration may not effect the rate of reaction of the more resistant substrates if $X \gg K_X$.

The Effect of Microbial Concentration

Increasing the mass concentration of microorganisms X in a composting mixture should increase the rate of solubilization as long as $X < K_X$. If the concentration should increase much beyond K_X, however, the rate will approach a maximum value. This may have practical implication with many composting substrates.

The concentration of microbes necessary to avoid rate limitations has been a subject of controversy for many years. There is no question that most organic wastes, including sludges and mixed municipal refuse, will decompose through activity of the indigenous microbial flora. However, this does not assure that the microbial concentration is not limiting, particularly in the early stages of composting. In fact, lag periods are often observed at the start of batch composting operations, although the lag could also be caused by other factors such as oxygen availability. Certainly if the waste material is sterile, reseeding should increase the kinetics according to Equation 10-15.

Golueke [44] distinguished between "minute" inoculation and "mass" inoculation with microbes. "Minute" inoculation referred to introduction of a relatively minute quantity of organisms into a large quantity of substrate, for example, 1 liter of a 10^6-bacteria/ml suspension into 1 ton of refuse. The comparative number of thermophilic actinomycetes isolated in various seeding materials is shown in Table 10-4. It would seem inconceivable that such a small inoculum could significantly increase the mass concentration of microbes in Equation 10-15, unless the starting substrate was sterile. Use of such small additions of inoculum to increase the rates of reaction has generally been discounted. Results of a comparative study of inoculated and uninoculated composting material is shown in Figure 10-13. The striking similarity between the temperature curves would indicate that the inoculum had essentially no effect. Golueke [44] concluded that if the addition of "minute" inoculum contributed anything to facilitate the compost process, it was so minute as to be undetectable. Even if the feed material were sterile, it would seem that rich soil or horse manure would be as effective as a commercial inoculum.

Table 10-4. A Comparison of Thermophilic Actinomycetes in Various Seeding Materials (Adapted from McGauhey and Gotaas [96])

Material	Organism Count (number of colonies per gram of material)
Commercial Inoculum	15.8×10^7
Rich Soil	13.4×10^7
Poor Soil	15.8×10^6
Horse Manure	15.0×10^7

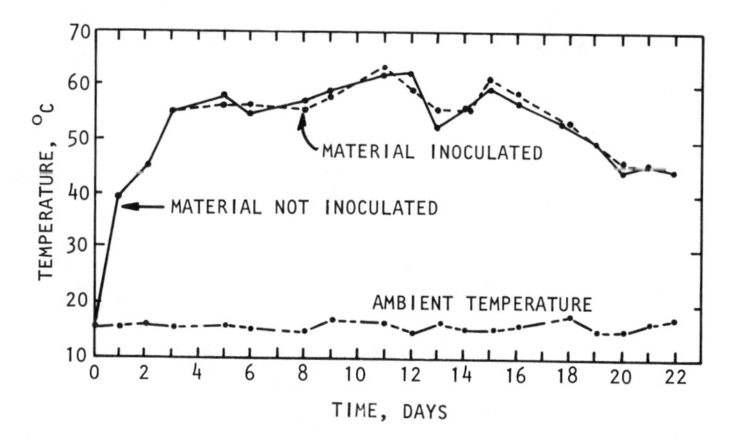

Figure 10-13. Temperature curves showing similarity between "minute" inoculated and uninoculated aerobically composting material. Feed material was composed of shredded vegetable trimming and paper. Inoculum was purported to be a culture high in thermophilic actinomycetes [96].

Mass inoculation refers to addition of large quantities of microbial culture. This can generally be accomplished by recycle of compost product or by use of a completely or partially mixed reactor. In the latter case the infeed is continually inoculated with the microbial population developed in the reactor. Using batch composters operated on refuse materials, Jeris and Regan [103] examined the effect of compost recycle ranging between 0 and 90% of the mixed infeed (i.e., a recycle ratio R_w of 0-9). In general, R_w values between 1.0 and 3.0 provided higher rates of oxygen consumption than lower ratios tested as shown in Figure 10-14. During these tests other environmental factors, such as moisture, temperature, FAS and aeration were held under conditions previously determined to be optimum. Therefore, the observed effect of recycle can probably be ascribed to the increased microbial concentration and not to other environmental factors influenced by compost recycle.

In deference to the above, the literature contains conflicting reports on the utility of mass inoculation by compost recycle. Golueke [44] reported no significant acceleration of windrow refuse composting through compost recycle. McGauhey and Gotass [96] examined the effect of recycle of compost product, addition of soil and addition of up to 30% horse manure to refuse. They concluded that none had any measurable effect on the rate of composting or the composition of the final product. On the other hand, compost recycle is definitely beneficial in windrow composting of dewatered sludge cake as described in Chapters 6, 7 and 8. However, the effect may be

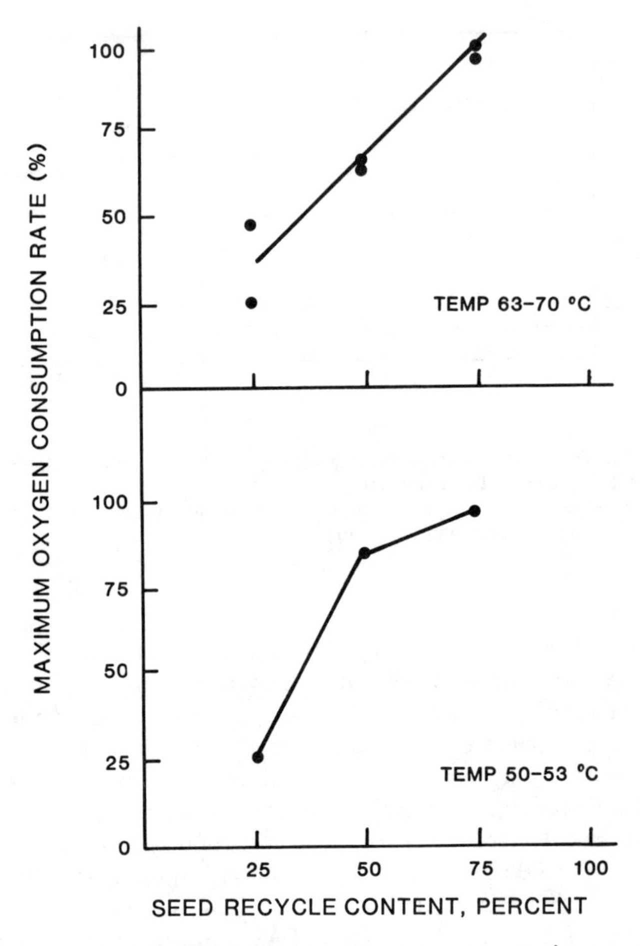

Figure 10-14. Effect of product recycle on the oxygen consumption rate measured during bench-scale composting of refuse [103].

related more to improved FAS and moisture control than to increased microbial concentrations. Senn [21] observed that recycle of at least 10% of finished product during composting of raw manure greatly facilitated production of a relatively odorless material that did not attract or produce fly larvae. Without product recycle, similar composting temperatures developed but the final product was odorous and led to development of fly larvae on rewetting. These conflicting literature results point to the difficulty of isolating effects of a single parameter during composting.

From the above discussion it is evident that the literature is unclear regarding the mass concentration of microbes required to avoid rate limitations. Indeed, values of K_X are likely a function of the type of substrate and should increase as the number of active sites per unit volume increase. Nevertheless, once the active sites on a given substrate have been saturated with enzymes ($e \gg K_a$ in Equation 10-11 or $X \gg K_X$ in Equation 10-15) the rate of solubilization should become constant. However, there are a number of possible ways to further increase the rate of reaction. The value of k (Equation 10-15) is undoubtedly a function of temperature and moisture content. The engineer can provide the most proper conditions for each and also avoid other rate limitations such as an inadequate oxygen supply. Beyond this, however, there is little else that can be done to improve the rate of solubilization beyond altering the substrate molecule itself. The effect of chemical and mechanical pulping on cellulose degradation has already been discussed. Both pulping techniques serve to increase the susceptibility of the wood structure to enzyme attack. Grinding the substrate to increase the number of sites per unit volume should also increase the rate of reaction. Pfeffer [108] noted that smaller particle sizes resulted in faster rates of reaction during anaerobic digestion of refuse slurries. Whether grinding would be effective with municipal sludges is doubtful, however, because of the small size of sludge particles.

Chemical and physical techniques are available to solubilize certain substrates and make the remaining solids more amenable to biological attack. Alkaline treatment to hydrolyze cellulosic material has been used to produce fermentable sugars. Gossett and McCarty [106] have noted that heat treatment of refuse can improve its degradability under anaerobic conditions. However, practical considerations will likely limit application of such techniques.

Kinetics in the Aqueous Phase

Once the particulate substrate has been solubilized, individual molecules can be transported by diffusion to the cell. Thereupon, the substrate is transported across the cell wall and is biochemically metabolized by the cell. Diffusion resistances across the cell wall, any internal diffusion resistances and the actual kinetics of metabolism are incorporated into the Monod kinetic model, as described in Chapter 4. Given the close proximity between microbes and substrate in the heterogeneous composting system, diffusion resistance through solution is likely low. Therefore, the microbial kinetics defined by Equation 4-1 probably govern:

$$\frac{dS}{dt} = -\frac{kSX}{K_s + S} \tag{4-1}$$

In the aqueous phase the rate of substrate use is a linear function of the microbial mass concentration, X, but nonlinear with regard to substrate concentration. As explained in Chapter 4, zero-order kinetics result when $S > K_s$, and occur as a result of saturation of the microbial metabolic system to the point where substrate is processed at the maximum possible rate.

This is exactly opposite to the kinetics of the heterogeneous system as represented by Equation 10-15. For the heterogeneous system, the kinetic rate is a linear function of the number of active sites, as represented by A_v in Equation 10-15, and nonlinear with the microbial concentration.

Kinetics of Oxygen Transport

Along with the organic substrate, oxygen must be available to complete aerobic metabolism. A conceptual view of the mass transport of major components consumed and produced during composting is illustrated in Figure 10-15. Air must be supplied to the FAS within the composting mixture. Oxygen must then be transported to the gas-liquid interface, diffuse across the interface and then diffuse through the liquid phase to the microbes. Consumption of oxygen by the microbial population produces a concentration gradient causing further diffusion from the FAS. Conversely, metabolic end products such as CO_2, H_2O and NH_3 will be at elevated concentrations in the liquid phase, will diffuse toward the airspace and ultimately be removed with the gas flow.

Ammonia produced from decomposition of proteinaceous material will also compete in at least two other reactions. Equilibrium will be established with the ionized ammonium form (NH_4^+). Furthermore, ammonia is a weak base and will tend to neutralize the weak acid, carbon dioxide, producing a buffer, ammonium bicarbonate. Ammonium can also be biologically oxidized to the intermediate nitrite, NO_2^-, and ultimately to nitrate, NO_3^-.

Gas transfer across an interface has been analyzed widely using the two-film model developed by Lewis and Whitman in 1924. An illustration of the idealized system is presented in Figure 10-16. Two laminar films are envisioned adjacent to the interface and provide resistance to mass transport of the gas molecules. Applying Fick's law:

$$\frac{\partial F}{\partial t} = -AD_l\left(\frac{\partial S}{\partial z}\right)_l = -AD_g\left(\frac{\partial S}{\partial z}\right)_g$$

$$\frac{\partial F}{\partial t} = -AD_l\left(\frac{C_i - C_l}{\delta_l}\right) = -AD_g\left(\frac{P_g - P_i}{\delta_g}\right)$$

where D_l and D_g are diffusion coefficients in the liquid and gaseous phases. Gas-phase diffusion coefficients are typically greater than those in the liquid

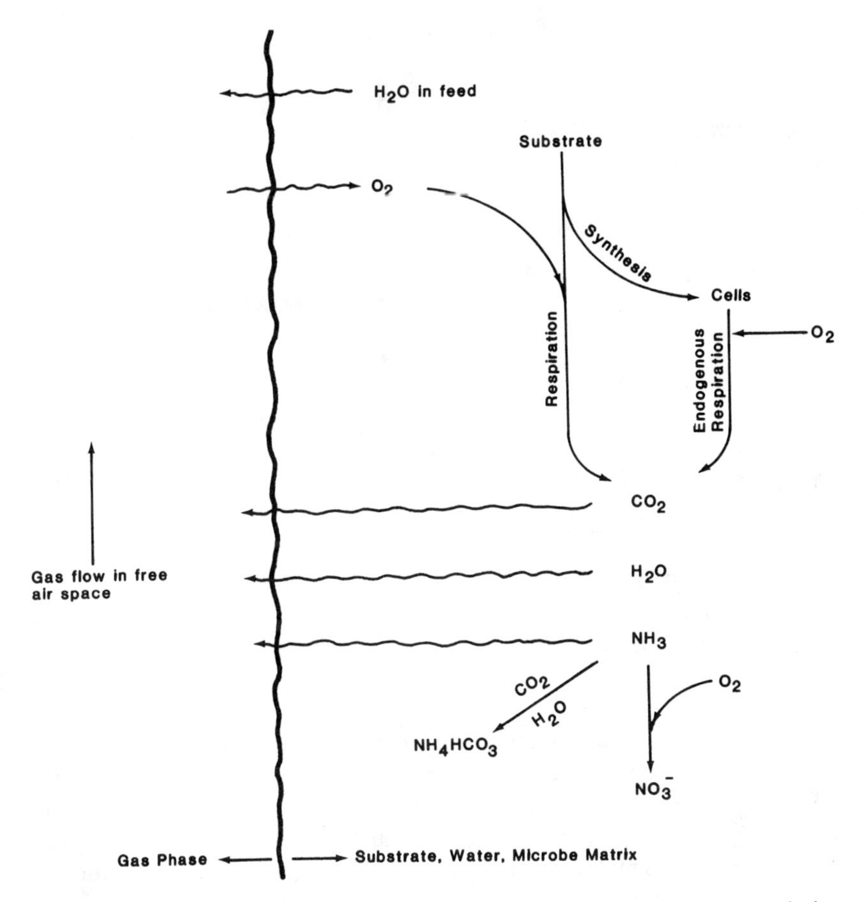

Figure 10-15. Conceptual illustration of mass transport of major components during composting.

phase by a factor of about 10^4. For sparingly soluble gases, therefore, it has been shown that essentially all resistance to mass transfer lies on the liquid film side [31]. Assuming this to be the case, one can neglect any resistance in the gas phase and concentrate on oxygen transport through the liquid phase which incorporates the solid substrate and its associated microbes.

As oxygen diffuses through the aqueous phase it will be consumed by microbes. This will produce a concentration gradient allowing further mass transport into the water/substrate/microbe matrix. Under steady-state conditions the concentration gradient should be similar in form to that shown in Figure 10-9. Because oxygen is in the aqueous phase, its rate of consumption should be governed by Monod kinetics previously discussed. Assuming

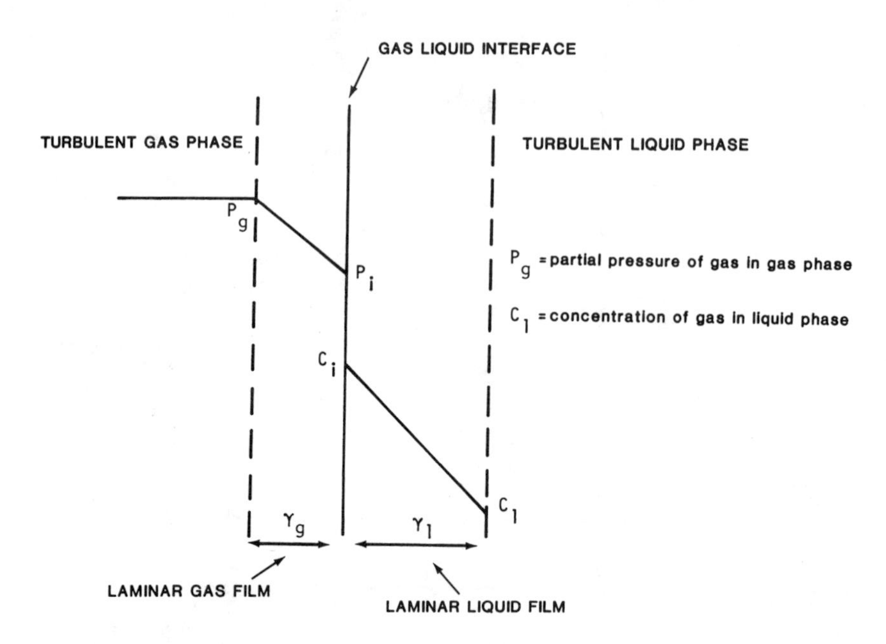

Figure 10-16. Idealized two-film model for gas transfer from gas to liquid phase.

this to be the case an equation similar in form to Equation 10-3 should define the combined process of diffusion and biochemical reaction. Unfortunately, without better information on active microbial concentrations in the composting matrix, application of such an equation would be difficult. Instead, a more simplified approach will be used to provide an "order of magnitude" estimate of the mass transfer rate for oxygen. These rates can then be compared with measured oxygen consumption rates to indicate conditions under which oxygen supply can be rate-limiting. Before developing this information, the value of the diffusion coefficient for oxygen will be examined.

The diffusion coefficient is an important parameter in mass transport. It determines the rate at which substrate materials diffuse into the matrix and the rate at which metabolized end products diffuse out of that matrix. Flow conditions within the water/substrate/microbe material are likely to be very quiescent. Therefore, mixing by turbulent or eddy diffusion would be negligible. In such cases the actual diffusion coefficient should approach the value of the molecular diffusion coefficient, the limiting value imposed solely by molecular motion of the diffusing materials.

Experimentally determined values for the molecular diffusion coefficient of nonelectrolytes such as oxygen can often be found. Reid and Sherwood

[139] and Perry and Chilton [25], presented values for numerous nonelectrolyte materials in water. Values for dilute solutions can also be estimated by using the Wilke-Chang equation given by Perry and Chilton [25]:

$$D = \frac{7.4 \times 10^{-8} T_K}{\eta} \frac{(FM)^{0.5}}{(V_0)^{0.6}} \tag{10-18}$$

where D = diffusion coefficient, cm^2/sec
η = solvent viscosity, cP
F = association factor for solvent (2.6 for water)
M = solvent molecular weight
V_0 = solute molal volume at normal boiling point (25.6 cm^3/g-mol for oxygen)
T_K = temperature, °K

Equation 10-18 is a useful correlation for estimating (usually to better than 10-15%) the diffusion coefficient of small molecules in low-molecular-weight solvents.

Actual diffusion coefficients in a composting particle and its water layer may be less than the molecular diffusion coefficient through water alone. This condition would arise from blocking of diffusing molecules by particulate matter or by changes in the viscosity of the fluid itself. If molecules were forced to diffuse around particulate matter, such as a bacterial cell, the diffusion coefficient would decrease because of the increased path length. If viscosity of the fluid increased, as might occur if the bacteria secreted a slime layer or extracellular polysaccharide, the resistance to passing molecules would increase, also resulting in a lowering of the diffusion coefficient.

Essentially no work has been done on determining actual diffusion coefficients into composting particles. However, a substantial amount of work has been conducted to determine diffusion coefficients through other biological materials. Table 10-5 summarizes some of this information along with experimentally determined values for oxygen in water. Most of the diffusion coefficients for various biological materials closely approximate the value found for water.

Atkinson et al. [130], working with glucose oxidation by a fixed film growth (falling film, heterogeneous system), assumed that the diffusion coefficient within the film was equal to the molecular diffusion coefficient for the glucose solution. The assumption worked well under their conditions of experimentation. Like Atkinson, most investigators have assumed a value equal to the corresponding value in water whenever a numerical value was necessary for their model. Mueller [129] experimentally determined the oxygen diffusivity through pure culture flocs of *Zoogloea ramigera*. Oxygen diffusion values ranged from 0.1 to 2 times that of the molecular diffusion coefficient, primarily because of difficulty in measuring the surface area of the floc particles. However, when the nominal diameter

Table 10-5. Oxygen Diffusion Coefficients Through Various Biological Materials (Adapted in Part from Mueller [129])

Substance	Temperature (°C)	Diffusion Coefficient (10^5 cm²/sec)	Original Reference
Muscle Tissue	20	0.86	185
Connective Tissue	20	0.71	185
Gelatin	20	1.73	185
Liver Tissue	37	0.83	186
Liver Tissue Homogenate	37	2.12	186
Rat Kidney Tissue	37	4.8	187
Grey Brain Matter	20	1.2	188
	37	1.7	188
Agar	25	0.96-1.21	189
Water	10	1.7	190
	18	1.98	191
	20	2.28	190
	30	2.82	190
Bacterial Slimes		0.04-1.5	31

of the floc particle was used in determining surface area, experimental values were reasonably close to that of the molecular diffusion coefficient. Based on oxygen profiles in an organoheterotrophic biofilm, Bungay et al. [140] estimated a value approximately 2% of the corresponding value in water, a result considerably lower than other data given above. In detailed experiments by Williamson and McCarty [135,136] diffusion coefficients were measured for NH_4^+, NO_2^-, NO_3^- and O_2 through nitrifying biofilms. Values ranged from 80 to 100% of corresponding values through water. It was concluded that a value of 80-90% of that in water would be an adequate estimate.

A composting particle consisting of water, particulate and dissolved substrates, and microbes is certainly as complex as the systems described above. Furthermore, the moisture content may be considerably less, particularly in the later stages of composting. Nevertheless, it would appear reasonable to assume the diffusion coefficient for oxygen through a composting matrix to equal the corresponding value in water. Remember that at this point we are interested in determining "order of magnitude" estimates.

The simplified model shown in Figure 10-17 was used to estimate oxygen diffusion rates through a water-saturated matrix of solid substrate and microbes. Diffusion was assumed to occur from both sides of the particle. The oxygen gradient decreased linearly from a saturation value at the outer

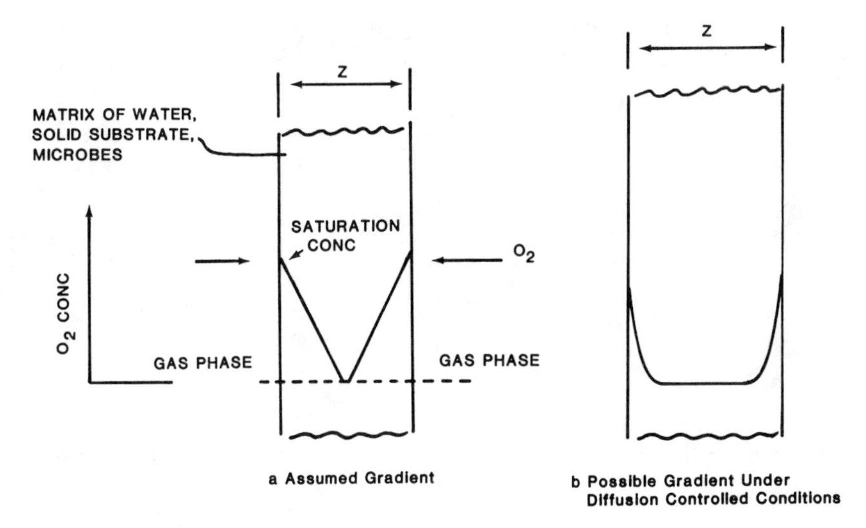

Figure 10-17. Simplified model of gas transfer used to estimate oxygen transport rates through a saturated matrix of solid substrate and microbes.

interface to zero at the particle midpoint. Obviously, such a model is a greatly simplified version of the actual composting matrix. However, it is unlikely that a more complex model would yield improved results because of the numerous ill-defined factors and unknown kinetic coefficients. Therefore, only this simple model will be considered to try to gain insight into the importance of oxygen diffusion as a rate-controlling mechanism. The method and assumptions used in calculating transfer rates are illustrated in the following example.

Example 10-1

Using the simplified model of Figure 10-17a, estimate the oxygen flux at 60°C for a particle thickness of 0.05 cm. Assume the saturation oxygen concentration at the gas-particle interface to be 6 mg/l. Estimate the time required to supply the stoichiometric quantity of oxygen assuming the matrix to be 50% solids ($S_m = 0.5$), with volatility of 50% ($V_m = 0.5$) and degradability of 50% ($k_m = 0.5$).

Solution

1. Estimate the liquid-phase diffusion coefficient D_l for oxygen using the Wilke-Chang correlation at 60°C. Assume the viscosity of water at 60°C to be 0.45 cP:

$$D_l = \frac{7.4 \times 10^{-8} (273 + 60)}{0.45} \frac{[2.6 (18)]^{0.5}}{(25.6)^{0.6}} = 5.35 \times 10^{-5} \text{ cm}^2/\text{sec}$$

$$D_l = 4.62 \text{ cm}^2/\text{day}$$

2. Estimate the specific gravity and bulk weight of the particle using Equations 7-1 and 7-3:

$$\frac{1}{G_s} = \frac{0.50}{1.0} + \frac{(1 - 0.5)}{2.5} \Rightarrow G_s = 1.43$$

$$\gamma_c = \frac{1.0}{\frac{0.5}{1.43} + 1 - 0.5} = 1.18 \text{ g/cm}^3$$

3. Determine the flux across the interface. The gradient of oxygen is given by:

$$\frac{\partial s}{\partial z} = -\frac{6 \text{ mg/l}}{0.025 \text{ cm}} = -240 \text{ mg/l-cm}$$

Using Fick's law and considering 1 cm^2 of area on each side of the particle:

$$\frac{\partial F}{\partial t} = -AD_l\left(\frac{\partial s}{\partial z}\right) = -(2.0 \text{ cm}^2)(4.62 \text{ cm}^2/\text{day})(-240 \text{ mg/l-cm})(\text{liter}/1000 \text{ cm}^3)$$

$$\frac{\partial F}{\partial t} = 2.22 \text{ mg } O_2/\text{day}$$

4. Each cm^2 of particle surface will contain $1.0 \times 0.05 = 0.05$ cm^3 of volume. The matrix mass in this volume is 1.18 (0.05) or 0.059 g:

$$\text{volatile solids} = 0.059 (S_m)(V_m)$$
$$= 0.059 (0.5)(0.5)$$
$$= 0.0148 \text{ g VS}$$

The oxygen flux can then be expressed as

$$\frac{\partial F}{\partial t} = \frac{2.22}{0.0148} = 150 \text{ mg } O_2/\text{day-g VS}$$

5. Time required to satisfy the stoichiometric demand can be estimated assuming 2 g O_2/g volatile matter oxidized (see Chapter 8):

$$\text{stoichiometric demand} = 0.0148\,(0.5)(2.0) = 0.0148 \text{ g } O_2$$

$$\text{time required} = \frac{0.0148\,(1000)}{2.22} = 6.6 \text{ days}$$

Note that the calculated flux and time requirement correspond to values in Figure 10-18.

Oxygen flux and time requirements were calculated for various particle thicknesses using the same approach and assumptions as in Example 10-1. Results are presented in Figure 10-18 along with the observed range of oxygen demands noted earlier in this chapter. Oxygen flux is shown to decrease as the particle thickness increases. This obviously leads to an increase

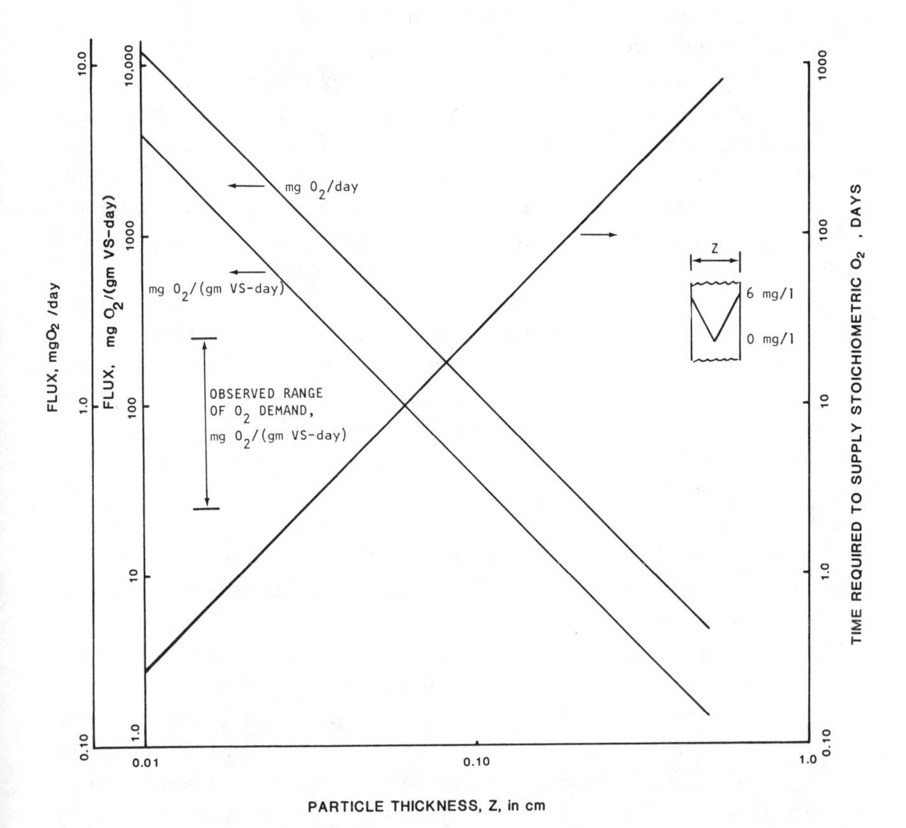

Figure 10-18. Effect of particle thickness on oxygen flux and time needed to satisfy the stoichiometric oxygen requirement. Values were calculated using the simplified model of Figure 10-17a and the procedure and assumed values of Example 10-1.

in the time necessary to supply the stoichiometric oxygen requirement. Conditions imposed by the model may be overly severe because the diffusion gradient decreases as particles size increases. Actually the gradient may remain constant leading to the concentration profile shown in Figure 10-17b. Such a profile would be expected when process kinetics are diffusion-controlled. Under such conditions the flux may no longer be a function of particle thickness. However, time required to satisfy the oxygen demand would still increase with increasing particle thickness but at a lower rate than shown in Figure 10-18.

Despite these uncertainties, it would appear that diffusion can match the oxygen consumption rate if the particle size is sufficiently small. Particle thickness on the order of 1.0 cm would appear to lead to large diffusion resistance which would dominate the process kinetics. Particle thickness of about 0.10 cm and lower would appear to be small enough that diffusion-supplied oxygen can begin to match the observed rates of oxygen demand. If particle thickness decreases below about 0.05 cm, oxygen diffusion would no longer appear to exert control on the overall composting rate. Golueke [44] noted that complete aeration of all particles "would involve reducing all particles to a size less than a millimeter or two because by its very dimensions a particle any larger could be anaerobic in its interior." Although no data were supplied to support this contention, the agreement with the simplified model presented here is both interesting and reassuring.

Several other observations can be drawn from the model results. The fact that oxygen is present in the pore spaces of composting material does not mean that oxygen diffusion is not rate-limiting. Oxygen is required *in* the matrix of water, solid substrate and microbes. The fact that it may be present in the pore space does not mean it is being supplied in adequate amounts within the matrix. It is common practice to use oxygen probes to measure the oxygen content of gases in the composting material. This is a useful practice but it only assures that oxygen is present in the pore space. It should not be misconstrued to mean that oxygen limitations have been removed. Indeed, process kinetics could still be controlled by diffusion transport within the compost particles.

The fact that diffusion transport is enhanced by smaller particle thickness leads to an unavoidable paradox. In the discussion of aeration mechanisms in Chapter 8, it was concluded that natural ventilation was a primary aeration mechanism in windrow composting systems. Natural ventilation is enhanced by larger particle sizes that increase the dimensions of the void spaces. Thus, we are presented with an interesting paradox. Ventilation is enhanced by larger particle size, but diffusion transport is not. Thus, a balance between these competing effects is necessary to assure that diffusion transport is not overly enhanced at the sake of oxygen supply, or vice-versa.

Finally, it should be noted that the model assumed a particle thickness within which the water/substrate/microbe matrix is continuous. In other words, there are no void spaces within the particle itself. In actual fact, this may not be the case. A schematic illustration of a composting particle with its associated bound water is shown in Figure 10-19. In one case the particle is saturated in the sense that water fills the voids and pores which may lie

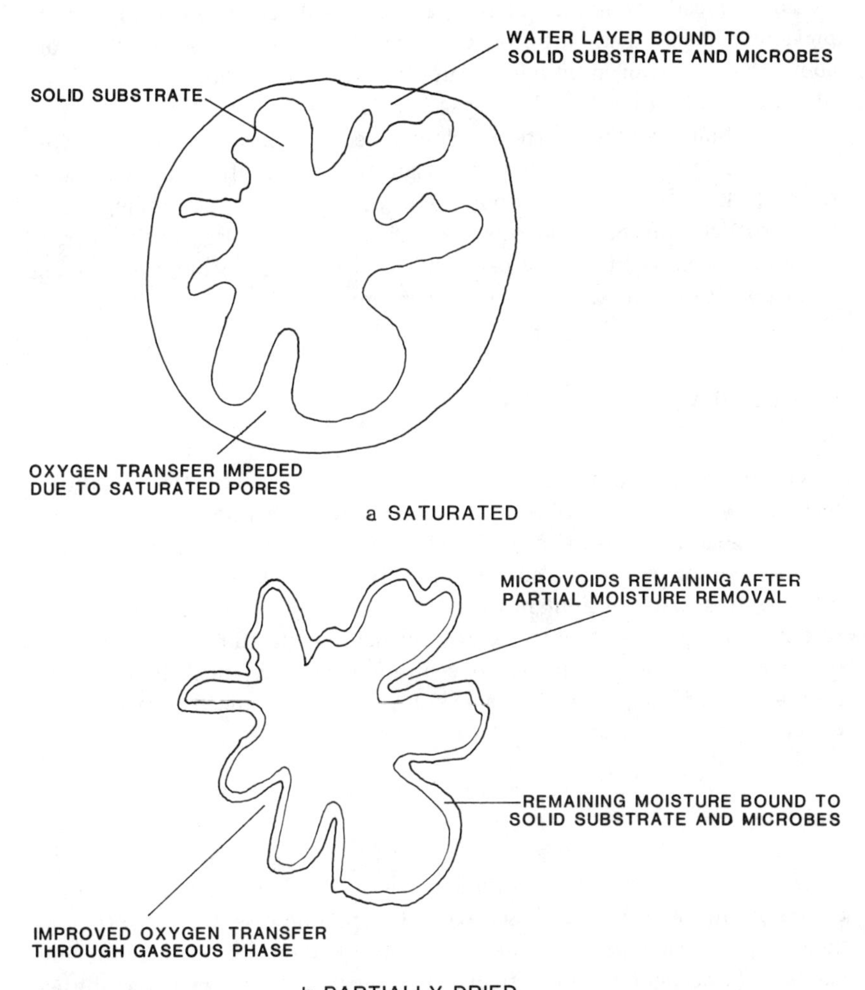

Figure 10-19. Schematic illustration of composting particles under saturated and partially dried conditions.

within the particle. Dewatered sludge cake would likely be described by such a condition. The model used to develop the data in Figure 10-18 would be applicable.

In the second case, water has been removed from micropores and voids within the particle. Evaporation or absorption of excess water by recycled product or into a bulking agent could be responsible for the moisture loss. Use of recycled product was discussed in Chapter 6 and moisture absorption into bulking agents in Chapter 7. Whatever the mechanism, oxygen diffusion would be greatly facilitated in the partially dried case of Figure 10-19. As previously mentioned, diffusion coefficients through a gaseous phase are about four orders of magnitude greater than through the liquid phase. Therefore, diffusion through small gas spaces remaining after moisture evaporation or absorption by other particles, should occur rapidly. Furthermore, the water/gas interfacial area would be considerably greater after partial moisture removal. Thus, diffusion transport of oxygen may be a controlling factor when particle moisture contents are high and assume less significance as moisture is removed from the particle. The significance of moisture content in the composting of wet materials is again emphasized.

OTHER RATE LIMITATIONS

The previous sections examined a number of possible rate limitations, including solubilization of solid substrate and mass transport of oxygen and solubilized substrate to the cell. The effect of moisture content on the mass transport of oxygen was also discussed. However, there are a number of other factors that can limit process kinetics. These include the presence of sufficient inorganic nutrients, excessively high temperature elevations, lack of moisture and hydrogen ion concentration. Some of these have already been mentioned in other chapters, but will be reexamined here for their effect on process kinetics.

Moisture

The effect of excessive moisture on available FAS and mass transport of oxygen has already been discussed. The opposite case, lack of sufficient moisture, can also impede the rates of reaction. A number of factors probably combine to account for this effect. First, most biological reactions during composting appear to be mediated by bacteria which require an aqueous environment. Second, mass transport limitations for soluble components may be encountered under low-moisture conditions. In general, moisture contents should be maintained as high as possible without violating requirements for FAS because of excessively high moisture. Because composting is a drying environment, moisture addition may be necessary at periodic intervals to avoid rate limitations.

The effect of moisture content on oxygen consumption rates for various composting materials is shown in Figure 10-20. The trend toward decreasing reaction rates at low moisture content is clearly evident. Below 20% moisture very little, if any, biological activity occurs. From that point rates of oxygen uptake increase in a more or less linear fashion to maximum values which begin at about 50-70% moisture. Rates begin to decrease again at high moisture contents, undoubtedly from loss of FAS. This points to the difficulty in isolating effects of moisture alone in such experiments because of the relationships between moisture, bulk weight and FAS. This probably accounts for much of the data scatter in Figure 10-20, as well as the slightly different trends observed for the various composting materials.

Nutrients

As described in Chapter 4, a number of inorganic nutrients are required for biological systems to maintain the cellular functions. The nutrient which has received the most attention in composting systems is the nitrogen content, and in particular the carbon/nitrogen (C/N) ratio. During active aerobic growth, living organisms use about 15-30 parts of carbon for each part of nitrogen. Hence, an initial C/N ratio of 30 or less would seem most favorable for rapid composting. Material with initial C/N ratios from 20 to 78 was composted in studies at the University of California [96]. An initial C/N ratio between 30 and 35 was recommended as optimum for rapid composting of municipal refuse. Below this range, rapid composting is likely to be accompanied by an increasing loss of excess nitrogen from ammonia volatilization. Above this range, composting time can be expected to increase with increasing C/N ratio. The C/N ratio for various organic materials is shown in Table 10-6.

The relationship between composting time and C/N ratio observed during studies at the University of California is as follows:

Initial C/N ratio = 20	Composting time about 12 days
Initial C/N ratio = 20-50	Composting time about 14 days
Initial C/N ratio = 78	Composting time 21 days

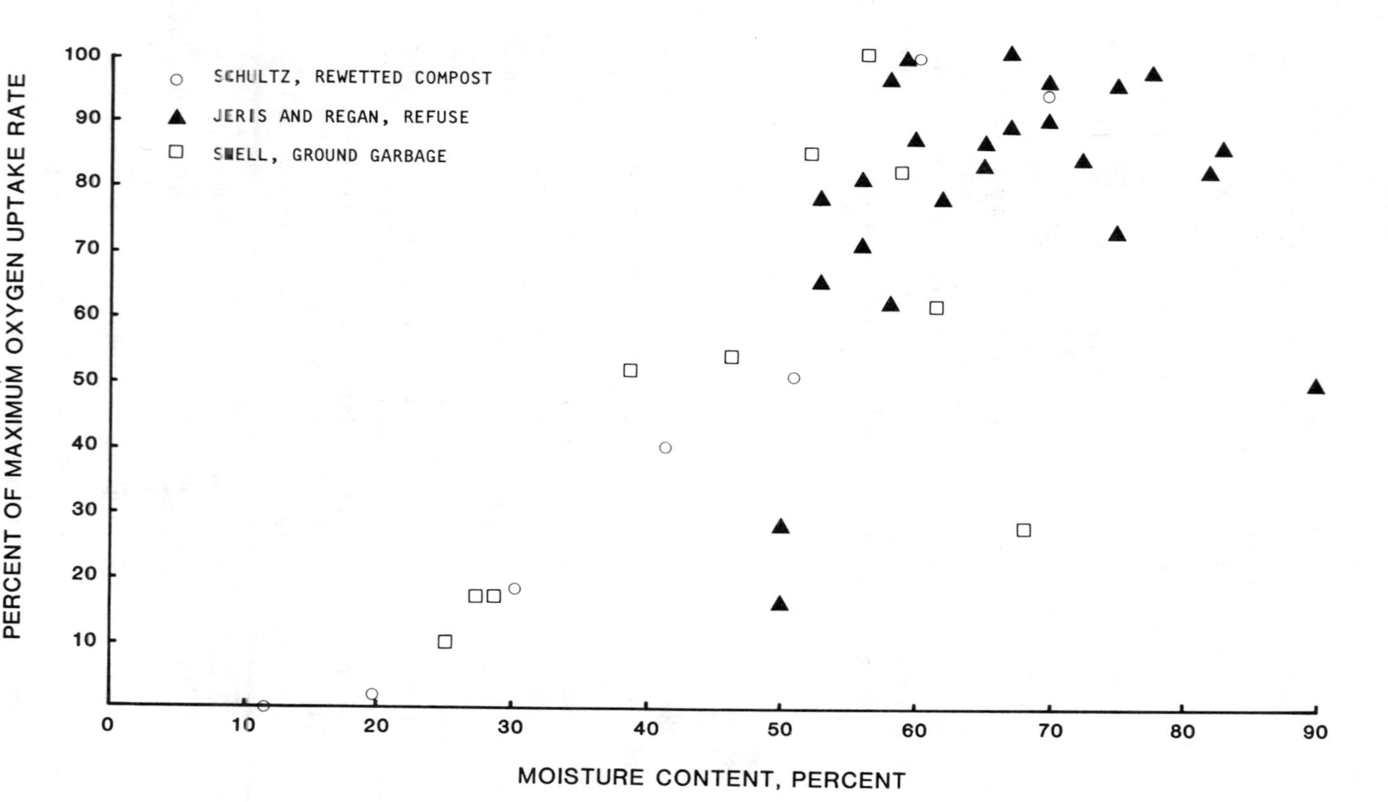

Figure 10-20. Effect of low and high moisture contents on oxygen consumption rates measured for various composting materials.

Table 10-6. C/N Ratio of Various Wastes [44]

Material	Nitrogen (% of dry weight)	C/N Ratio
Night Soil	5.5-6.5	6-10
Urine	15-18	0.8
Blood	10-14	3.0
Animal Tankage		4.1
Cow Manure	1.7	18
Poultry Manure	6.3	15
Sheep Manure	3.8	
Pig manure	3.8	
Horse Manure	2.3	25
Raw Sewage Sludge	4-7	11
Digested Sewage Sludge	2-4	
Activated Sludge	5	6
Grass Clippings	3-6	12-15
Nonlegume Vegetable Wastes	2.5-4	11-12
Mixed Grasses	2.4	19
Potato Tops	1.5	25
Straw, Wheat	0.3-0.5	128-150
Straw, Oats	1.1	48
Sawdust	0.1	200-500

Data were developed for mixed municipal refuse with a moisture content <70%. However, the trend toward increasing composting time with increasing C/N ratio would also apply to other feed materials.

The effect of low nitrogen concentration is caused by growth limitations imposed by lack of this particular nutrient. With growth limited, the overall kinetics of the process become limited in turn. The reader should refer back to Equation 4-1, the substrate utilization equation for aqueous solutions. Nitrogen can be the limiting substrate as well as oxygen or other solubilized organic substrate.

In many cases the value of the C/N ratio is not easily determined in the laboratory because of the difficulty in running the necessary test for carbon. New Zealand researchers [141] have suggested the following relationship:

$$\% \text{ carbon} = \frac{100 - \% \text{ ash}}{1.8} \qquad (10\text{-}19)$$

The relationship provided results accurate to within 2-10% during field studies, and should be sufficiently accurate for most practical purposes.

Example 10-2

Estimate the C/N ratio for digested sludge.

Solution

1. Assume a chemical composition of $C_{10}H_{19}O_3N$ for digested sludge:

$$\frac{C}{N} = \frac{10(12)}{14} = 8.6$$

2. Another approach to estimate the C/N ratio is to assume an average N concentration. From Table 10-6, digested sludge averages about 3% N based on total solids (TS). From Table 8-1 the carbon fraction for carbohydrates, proteins and fats is about 44, 55 and 75%, respectively. Using 50% as an average value and considering the solids to be 50% organic:

$$C/N = \frac{(1 \text{ g TS})(0.50)(0.50)}{(1 \text{ g TS})(0.03)} = 8.3$$

3. Using the New Zealand formula (Equation 10-19) and assuming an ash content of 50%, the percentage carbon is estimated as:

$$\% \text{ carbon} = \frac{100 - 50}{1.8} = 27.8\% \text{ of TS}$$

If nitrogen is 3% of the total solids:

$$C/N = \frac{27.8}{3} = 9.3$$

Based on Example 10-2, municipal sludge should have more than sufficient nitrogen to satisfy the growth requirement of the microbes. Furthermore, available information suggests that compost produced from digested sludge contains only about 1% nitrogen. Thus, a considerable fraction of the nitrogen, on the order of 2/3(100) = 67%, appears to be available under composting conditions. It seems that sufficient nitrogen is present and available to avoid any rate limitations. These comments can be extended to the case of raw sludge. About 50% of the organic N present in raw sludge is converted to ammonia during anaerobic digestion. Most of the soluble ammonia will be removed with the liquid sidestream during dewatering. Therefore, the percent of available nitrogen should be higher for raw compared to digested sludge.

High Temperature

All organisms have a temperature range over which biochemical functions can be maintained. Within certain limits the rate of these biochemical reactions will about double for each 10°C rise in temperature. However, an upper limit is placed by the thermal denaturation of proteins which catalyze the biochemical pathways of cellular metabolism. In a mixed population of microbes, such as will exist in a normal composting environment, a natural transition from mesophilic to thermophilic organisms can be expected. Thus, the exponential relationship between oxygen consumption rate and temperature as shown in Figure 10-4 is not an unreasonable expectation.

There is an upper limit to the exponential increase with temperature. At some point even the thermophilic microbes cannot overcome the effects of thermal denaturation of their enzymes. At this point, rates of reaction will decrease, assuming the kinetics to be limited by microbial activity and not some other factor such as oxygen supply. At what temperature the rates of reaction begin to decrease is not well defined. As shown in Figure 10-7, different investigators have found optimum values ranging from as low as 45 to above 70°C.

The point to remember here is that extremely high temperatures do not necessarily imply a high rate of reaction. This is a common mistake because one is inclined to measure composting effectiveness by high temperature elevation. Should the temperature develop to 75 or 80°C in a composting material, rates of reaction are quite likely depressed because of the high temperature. The kinetics of the process would probably be improved by reducing the temperature. To accomplish this, heat generated must be dissipated at a greater rate. Increasing the aeration rate or rate of turning will increase the evaporative heat loss and is probably the most effective approach.

It might be interesting at this point to mention that many industrial fermentations also face the problem of high heat generation rates. Because industrial fermentations are usually conducted in aqueous solution, cooling tubes can be placed in the fermentor to remove the excess heat. Unfortunately, this is not practical in a composting system.

Hydrogen Ion Concentration

Both hydrogen ion (H^+) and hydroxide ion (OH^-) are very toxic to microorganisms. For example, most bacteria will not survive in a pH 3 solution, but at that pH the H^+ concentration is only about 1 mg/l. The effect of high and low pH is to cause a change in the ionization state of various protein components such as amine and carboxyl groups. This in turn can cause a change in the physical structure of the protein and hence a loss of enzymatic activity.

Lime addition to achieve high pH conditions has been used to "stabilize" sludge by reducing the microbial activity. Obviously, the organics themselves are not stabilized and the effect is caused solely by the reduced microbial activity. The effect of lime treatment on typical pathogenic and indicator bacteria in liquid sludge and sludge cake was examined by Farrell et al. [68] and is shown in Tables 10-7 and 10-8. The total aerobic count was significantly reduced at pH ⩾11.5. The reduction was not nearly so extensive at pH 10.5.

Table 10-7. Reduction in Bacterial Count in Liquid Sludge as a Result of High pH Treatment [68]

Bacteria	Original Count (bacteria/100 ml)[a]	Fraction of the Original Count			
		pH 10.5 (0.5 hr)[b]	pH 11.5 (0.5 hr)[b]	pH 11.5 (24 hr)[b]	pH 12.5 (0.5 hr)[b]
Fecal Coliform	3.6×10^7	0.81	0.00028	0.00011	0.00011
Fecal Streptococci	2.2×10^6	10	0.17	0.059	0.009
Total Aerobic Count	4.9×10^9	0.11	0.00067	0.00025	0.00071
Salmonella sp.	>1,100	0.0033	<0.0027	<0.0027	<0.0027
Pseudomonas aeruginosa	1,100	0.0083	0.0033	<0.0027	<0.0027

[a]Approximately 5 g dry sludge solids.
[b]Contact time.

Table 10-8. Reduction in Bacterial Count in Sludge Cake as a Result of High pH Treatment [68]

Bacteria	Original Count (bacteria/g cake)[a]	Fraction of the Original Count	
		pH 11.5 (0.5 hr)[b]	pH 11.5 (24 hr)[b]
Fecal Coliform	3.8×10^6	0.005	0.022
Fecal Streptococci	0.28×10^6	0.72	0.77
Total Aerobic Count	6.1×10^8	0.0048	0.033
Salmonella sp.	>110	<0.0027	<0.0027
Pseudomonas aeruginosa	>46	<0.0027	0.014

[a]Approximately 0.15 g dry sludge solids.
[b]Contact time.

Most raw sludges have a pH in the range of 5-6.5, while digested sludges are well-buffered in the range of pH 7-8. pH values >9 would only be encountered where lime is used for coagulation in the wastewater treatment process or as a conditioning chemical during dewatering. Similarly, low pH could be expected where ferric chloride or alum are used for sludge conditioning. Under such conditions an initial lag in the composting rate could be expected. Eventually, however, CO_2 and NH_3 produced by the slow but persistent microbial activity will neutralize high or low pH, respectively, and overcome the rate limitation. Also, any lag period would likely be more significant in batch rather than continuous reactors. Using a well-mixed reactor, Shell and Boyd [97] composted a digested sludge conditioned with 4.5% ferric chloride ang 15.5% lime (based on sludge dry solids). The filter cake pH averaged 11.0 while the pH of the compost product averaged 6.5. Obviously, sufficient CO_2 was produced to effectively neutralize the added lime. It was noted in this study that sludge conditioned with lime and ferric chloride produced compost that mixed and handled better than sludge conditioned with organic polymers.

Example 10-3

Assume $C_{10}H_{19}O_3N$ as the chemical composition of the organic fraction of a raw sludge. The sludge is 30% ash by weight, and is conditioned by addition of 10% lime as $Ca(OH)_2$ based on dry weight. Estimate the fraction of organics which must degrade to produce sufficient CO_2 to neutralize the lime. Neglect any neutralization reactions which may occur during initial lime addition to the sludge.

Solution

1. Oxidation of the organics and neutralization by the produced CO_2 can be represented as:

$$\underset{201}{\overset{y}{C_{10}H_{19}O_3N}} + 12.5\ O_2 \rightarrow \underset{10(44)}{\overset{x}{10\ CO_2}} + NH_3 + 8\ H_2O$$

$$\underset{44}{\overset{x}{CO_2}} + \underset{74}{\overset{0.10}{Ca(OH)_2}} \rightarrow CaCO_3 + H_2O$$

Consider 0.10 g lime/g of dry solids. The CO_2 requirement is then given by:

$$x/44 = 0.10/74$$

$$x = 0.0595\ g\ CO_2$$

To supply this CO_2 the following organic decomposition is required:

$$y/201 = 0.0595/10(44)$$

$$y = 0.027 \text{ g}$$

2. The fraction of organic decomposition required can then be estimated as:

$$0.027/0.70 = 0.0386 \sim 3.9\%$$

Thus, only a very small percentage of organic decomposition is required to neutralize the high pH of lime conditioned sludge. Lime requirements as high as 30% have been reported [67], but even these large amounts would not severely tax the neutralizing capacity of the composting process. Rate limitations in the early stages of composting can still be possible until sufficient organic decomposition has occurred to lower the pH.

3. The reader should note that $Ca(OH)_2$ added to a sludge will react with carbon dioxide and bicarbonate alkalinity naturally present to form $CaCO_3$. Thus, the above calculations are somewhat conservative and tend to overestimate the required organic decomposition.

Example 10-4

Assume the sludge in Example 10-3 is conditioned by addition of 2% $FeCl_3$ based on dry weight. Estimate the fraction of organics which must degrade to produce sufficient ammonia to neutralize acids produced from the addition of ferric chloride. Neglect any neutralization reactions which may occur during initial ferric chloride addition to the sludge.

Solution

1. Production of acid from addition of ferric chloride can be described as:

$$\underset{162.5}{\overset{0.02}{FeCl_3}} + 3\,H_2O \rightarrow 3\,H^+ + 3\,Cl^- + Fe(OH)_3\downarrow$$

2. Oxidation of organics and neutralization by produced NH_3 can be represented as:

$$\underset{201}{\overset{y}{C_{10}H_{19}O_3N}} + 12.5\,O_2 \rightarrow 10\,CO_2 + \underset{17}{\overset{x}{NH_3}} + 8\,H_2O$$

$$\underset{3(17)}{\overset{x}{3\,NH_3}} + 3\,H^+ + 3\,Cl^- \rightarrow 3\,NH_4^+ + 3\,Cl^-$$

3. Consider 0.02 g $FeCl_3$/g dry solids. The NH_3 requirement is then given as:

$$\frac{x}{3(17)} = \frac{0.02}{162.5}$$

$$x = 0.0063 \text{ g } NH_3$$

To supply this NH_3, the following organic decomposition is required:

$$\frac{y}{201} = \frac{0.0063}{17}$$

$$y = 0.0745 \text{ g}$$

4. The fraction of organic decomposition can then be estimated as:

$$\frac{0.0745}{0.70} = 0.106 = 10.6\%$$

The organic decomposition required to neutralize low pH conditions will depend largely on the degradable protein content of the feed substrate, because this is the primary form of organic nitrogen. Substrates low in degradable protein will produce less ammonia and have a reduced neutralizing capacity against low pH conditions.

5. In comparing Examples 10-3 and 10-4, neutralizing reactions for both low and high pH conditions will occur during composting. It is common to observe a buffering of pH in the near-neutral range as composting proceeds towards a stable product.

For material not conditioned with lime, it is common in the early stages of batch composting to observe a decrease in pH from carbon dioxide and organic acid production. pH as low as 4.5-5.0 may at times be observed. However, the pH sag is rapidly overcome as organic acids are further decomposed and temperature rises. Protein decomposition will contribute NH_3 which will also tend to neutralize the acids. Increased rates of forced aeration or natural ventilation will tend to decrease CO_2 levels in the compost, which in turn will tend to increase pH. pH usually stabilizes in the near-neutral range of values.

McGauhey and Gotaas [96] examined the effect of calcium carbonate addition to reduce the normal pH sag observed during windrow composting of refuse. Although a slight reduction in the pH sag was observed, it was concluded that the practice was of dubious value to process kinetics and that pH control was not necessary in the composting of garbage and municipal refuse. The same conclusion can be extended to sewage sludge.

STABILIZATION–A MATTER OF DEGREE

To this point the term "stabilization" has been used to imply the oxidation of organic materials or their conversion to a more refractory (i.e., stable) form. The more stable compounds that remain at the end of composting are still degradable, but at a much reduced rate compared to the original feed substrate. The question which then arises is, how much stabilization is enough? There is no precise answer to this question. For example, digested sludges are generally considered as being stabilized, and in the liquid or cake form can be applied to land in controlled amounts without producing nuisance conditions. On the other hand, if digested sludge cake is allowed to sit in open piles, septic and odorous conditions can often develop.

As another example, raw sludges can be heat dried and stored for long periods without producing odorous conditions. The low moisture content limits the rate of biological activity resulting in a "stable" material. In the dried state, raw sludges are even bagged and sold by nurseries, perhaps the most demanding market in terms of product quality. When the material is rewetted, however, normal rates of biological activity will resume. By this time the material should be incorporated into the soil, thus reducing the nuisance potential.

We come back then to the original question, how much stabilization is enough? Certainly, there is no such thing as complete stabilization as long as any organic compounds remain, because they will all decompose at some rate. In a strict sense, complete stabilization would be the oxidation of all organic material to CO_2 and H_2O. Such a degree of stabilization would not be desirable because the value of compost as a soil amendment depends in part on its organic content. A working definition might be as follows: stabilization is sufficient when the rate of oxygen consumption is reduced to the point that anaerobic or odorous conditions are not produced to such an extent that they interfere with storage or end use of the product. Such a definition is not precise, but it is practical. Stabilization must be sufficient to reduce the nuisance potential but not so complete that organics are unnecessarily lost from the final product.

A number of approaches have been used to measure the degree of stabilization and to judge the condition of the compost process. Included among these are:

1. decline in temperature at the end of batch composting;
2. a low level of self-heating in the final product;
3. organic content of the compost as measured by the VS content, chemical oxygen demand (COD), percent carbon, ash content or carbon/nitrogen ratio (C/N);
4. oxygen uptake rate;
5. the presence of particular constituents such as nitrate, and the absence of others such as ammonia and starch;

6. lack of attraction of insects or development of insect larvae in the final product;
7. characteristic changes in odor during composting and odor producing potential of the final product upon rewetting;
8. rise in the redox potential;
9. experience of the operator.

Most of these approaches are qualitative in nature but can provide the operator a comparative tool by which to judge his own compost.

For a batch process, a declining temperature is a good indication that the compost is finished, provided the temperature drop is not caused by thermal kill, oxygen shortage, low moisture or lack of FAS. This approach is based on the fact that the rate of heat production is proportional to the rate of organic oxidation which decreases after the more degradable material is decomposed. A declining temperature is inevitable as the energy balance adjusts to the decreasing rate of heat input.

Reheating potential of the final product is another useful but comparative tool. In a method proposed by Niese [207], samples are adjusted to optimum conditions and placed in Dewar flasks. The flasks are placed in an incubator and the temperature regulated by a temperature-difference-sensing device. Niese observed temperatures >70°C with raw refuse, 40–60°C with partly stable refuse, and <30°C with fully stable material. One difficulty with this approach is that stability may result in part from other rate limitations such as low moisture content. These rate limitations are removed when the material is readjusted to optimum conditions.

The organic content of a compost will depend largely on organic characteristics of the feed substrate. Therefore, organic content can provide a measure of stability for a particular feed, but should not be used to compare composts produced from different feed substrates. A number of chemical tests have been used to measure indirectly the organic fraction including VS, ash content, COD, percent carbon and C/N ratio. The COD of compost samples taken during windrow composting of refuse is shown in Figure 10-21. The compost was considered stable and ready for use after eight weeks, which corresponds to a COD <700 mg/g compared to about 900 mg/g for fresh refuse. If the refuse is assumed to be primarily cellulose, the final COD would correspond to a VS content of about 70% (1 g glucose = 1.07 g COD). By contrast, compost produced from windrow composting of digested sludge usually has a VS content of only 30-40%. This emphasizes the point that organic content is a useful measure when applied to a particular feed substrate, but should not be used to compare different starting substrates.

The oxygen uptake rate of stabilized compost should be significantly reduced from that of the feed substrate. Chrometzka [143] indicated that under optimum conditions the uptake rate of fresh substrate was 30 times that of mature compost. Uptake rates were measured in a Warburg apparatus.

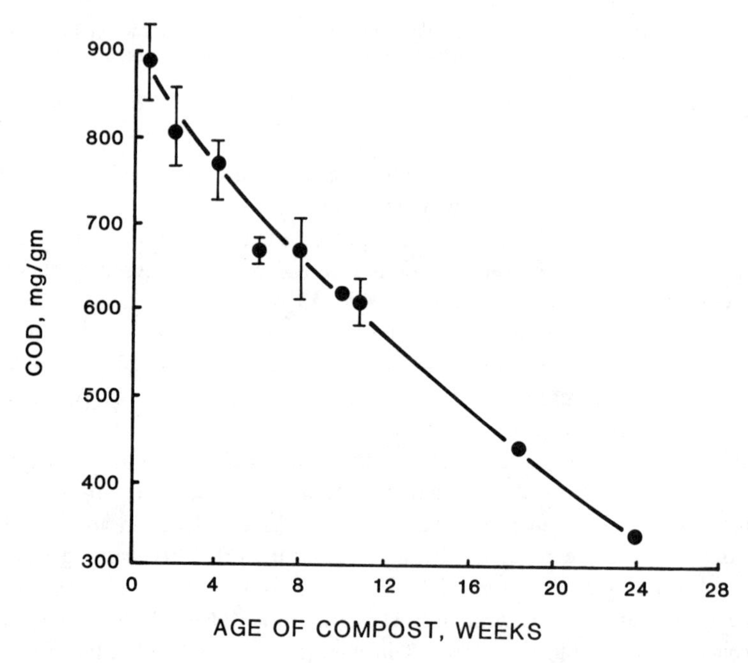

Figure 10-21. Effect of compost age on the COD of samples collected during windrow composting of refuse [142].

The data of Figure 10-7 would suggest a reduction of at least a factor of 10 between feed substrate and final compost. The oxygen consumption rate is a fundamental parameter by which to measure stability because it is proportional to the rate of organic decomposition. Unfortunately, there are few data by which to judge whether a particular oxygen consumption rate is low enough to avoid nuisance conditions. Thus, the parameter has not received widespread use. Oxygen consumption rates for raw sludge and compost produced at the Minamitama pilot composting facility are shown in Figure 10-22. See Chapter 2 for a description of the system.

Particular constituents in compost are often characteristic of the level of stability. Ammonia is usually present in the early stages of composting as organic nitrogen is decomposed. The ammonia concentration is eventually reduced through volatilization or oxidation to the nitrate form. Thus, presence of nitrate and absence of ammonia is indicative of a stabilized condition. Lossin [144] reported the use of starch content as a measure of stability. The rationale rests on the assumption that most feed substrates contain a measurable quantity of starch. Because starch is easily broken

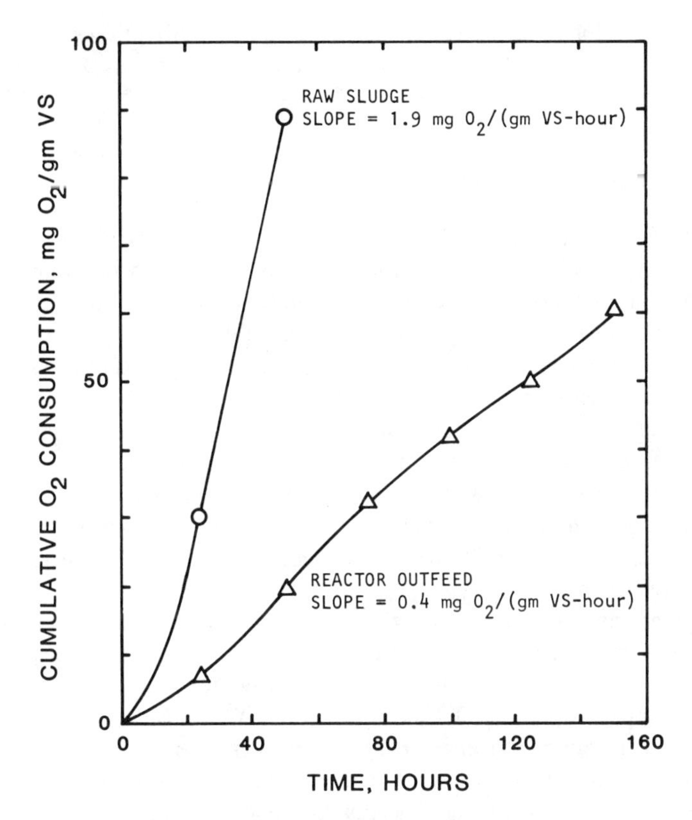

Figure 10-22. Cumulative oxygen consumption for raw sludge and reactor outfeed from the Minamitama bin composter pilot facility. Oxygen consumption was measured using 2- to 3-g samples in 300 ml of water. Temperature was not indicated but was probably about 25°C. Because samples were immersed in water, low moisture rate limitations were removed. Thus, oxygen consumption rates of relatively dry compost product may be less than that indicated [209].

down and metabolized, it should be degraded before the compost is considered acceptable. The test procedure involves formation of a colored starch-iodine complex in an acidic extract from the compost material. Lossin reported that well-stabilized compost never gave a positive starch reaction with this test.

The reader is referred to the cited references for further discussion of other stability tests. Bear in mind that most of these tests are qualitative in nature and are intended to provide a tool by which to measure stability. However, there is no substitute for the experience of the operator. An experienced operator can usually tell the condition of a compost by its appearance, color, odor, texture, particle size and other physical characteristics. It may not be the most scientific approach, but it is workable and usually accurate.

OVERVIEW AND CONCLUSIONS

The picture which emerges from the previous discussion is indeed complex. Would the casual observer have ever imagined the complexity of action within the common compost pile? Nevertheless, the sequence of events which must occur are reasonably defined, from the first solubilization of particulate substrate to the final aerobic metabolism of substrate. Major resistances to this sequence of events include solubilization of the solid substrate, diffusion transport of solubilized substrate along with oxygen to the microbe and the final kinetics of substrate utilization. Each resistance can become dominant under certain conditions. Many of the coefficients and factors necessary to describe each step in the sequence of events are unknown at this time. Nevertheless, various conditions can be described under which each factor may be controlling.

Consider first the composting of digested sludges. During anaerobic digestion of municipal sludges only about 50% of the solid organic substrate that enters the reactor is solubilized and fermented. These solubilization reactions apparently proceed fast enough to prevent this step from limiting the overall reaction sequence [31]. Indeed, most models of anaerobic digestion assume the subsequent fermentation to methane and carbon dioxide to be rate-controlling. The point is that outfeed from the digester which is subsequently captured during dewatering has not been solubilized and has resisted hydrolytic attack in the anaerobic reactor. Therefore, it must be concluded that digested, dewatered sludge consists of organic solids resistant to hydrolytic enzyme attack. Such resistance is likely maintained in the aerobic composting environment, although some differences could be expected. Therefore, the rate of hydrolysis is likely a serious rate-controlling step with digested sludge.

Raw sludge, on the other hand, contains the resistant fraction characteristic of digested sludge as well as the fraction more conducive to hydrolysis. Therefore, rates of reaction may not be limited by solubilization in the early stages of composting but may become so in later stages, after the readily hydrolyzed fraction has been degraded.

The above discussion may help to explain the lag period occasionally observed in many batch composting systems. With resistant materials, such as digested sludge and the paper fraction of refuse, early stages may be limited by a slow rate of hydrolysis. More readily hydrolyzed substrates, such as fresh manure, raw sludge and the garbage fraction of refuse, may shift the kinetics to control by solution kinetics or oxygen transfer. Once the readily hydrolyzed fraction is used, however, the process may again be limited by hydrolysis reactions.

During aerated pile composting of fresh manure, for example, Senn [21] noted temperatures as high as 70°C within six hours of startup. A similar batch system using digested sludge required several days to a week to develop full operating temperatures [17]. However, the literature is somewhat confusing in this area and more rapid temperature increases with digested sludge compared to raw sludge have also been noted [13]. With the number of possible rate limitations it is difficult to conduct experiments where the effect of a single variable can be isolated.

Hydrolysis or solubilization of the solid substrate is a function of the mass concentration of microbes if $X < K_X$. Should X become much greater than K_X, however, the rate becomes zero-order with respect to X (i.e., not a function of X). This result may help to explain the function of the curing stage which is often required to fully stabilize the compost. Organic feed to the compost process will usually contain a certain amount of solids resistant to hydrolytic enzymes, such as cellulose fiber and certain proteins. Furthermore, microbes synthesized on the feed organic will contain resistant fractions, such as the cell wall structure. If these organic structures become saturated with microbes ($X \gg K_X$) the rate of hydrolysis will become constant at $ds/dt = -kA_v$ (Equation 10-16). If the product of kA_v is small for the particular substrate, solubilization will require a considerable period of time. There is little the engineer can do to increase the rate other than to assure all environmental conditions are optimum, to keep k as large as possible. No particular reactor design or special inoculum of microbes is likely to reduce the time required as long as X is greater than K_X. Therefore, the function of the curing pile is to allow time for the more resistant reactions to occur. There would appear to be little that can be done to shorten the time required for these resistant reactions.

Numerous claims have been made that composting can be accomplished in as little as 24 hours. Such claims have usually been made by manufacturers of mechanical composting equipment and appear to be unfounded. Maintenance of high microbial populations and avoidance of oxygen transport limitations can improve overall kinetics for the readily solubilized components of the feed material. It would appear that any composting system, windrow, aerated pile or reactor, can accomplish these objectives to varying degrees. However, those components resistant to hydrolysis will require longer times for stabilization as Equation 10-15 suggests. Therefore, the curing stage would seem to be unavoidable if large amounts of resistant material is present in the original feed. Of course, uncured compost may be sufficiently well stabilized for certain applications.

SUMMARY

Batch reactors are characterized by a gradual buildup of temperature to a rather constant plateau level, followed by a gradual decline in temperature as the more degradable materials are decomposed and rates of reaction decrease. Thus, conditions within a batch composting system are constantly changing and steady-state conditions are never established. A continuous reactor operates with considerable mixing of materials within the reactor. Thus, conditions are more uniform, and steady-state conditions can be approximated.

Rates of oxygen consumption have been measured in both batch- and continuous-composting reactors on a variety of feed materials. In general, rates of oxygen utilization increase exponentially with temperature up to a peak value. Further temperature increases result in decreased rates of reaction from thermal denaturation of cellular enzymes. Temperature optima from about 45 to 70°C have been reported. Maximum rates of oxygen consumption are likely to vary from 7 to 12 mg O_2/g VS-hr, which equates to about 0.5-1.0 m^3 air/kg VS-day (9-15 ft^3/lb VS-day).

The composting system can be described as a heterogeneous system with solid substrate and limited moisture. Microbes either surround or are immersed in a solid substrate. Moisture is limited to that associated with the solid organic substrate. In such a case, a rather complex sequence of events is necessary for metabolism of the substrate. Solubilization of substrate, mass transport of oxygen to the cell and utilization of the solubilized substrate and oxygen by the cell are likely the predominant rate-controlling steps during aerobic composting. Kinetic models can be developed for each of these steps.

A model of substrate solubilization suggest that the process is a function of the mass concentration of microbes, X, at concentrations of X less than a constant K_X. The rate becomes independent of X at higher concentrations where X is greater than K_X. If $X < K_X$, such as in a sterile feed material, increasing the microbe concentration should increase the rate of reaction in batch reactors. In continuous reactors there is a continual inoculation of the feed with the microbial population developed within the reactor. When $X \gg K_X$, the rate of solubilization becomes constant. With materials resistant to solubilization, longer time periods will be required to allow solubilization and decomposition. This appears to be the function of the curing stage commonly used in composting systems.

Particle thickness on the order of 1 cm results in large oxygen mass transport limitations which would dominate the process kinetics. Particle thickness of about $\leqslant$0.10 cm is small enough that diffusion supplied oxygen can match the peak rates of oxygen demand. The fact that oxygen is present in the pore spaces of a composting material does not by itself assure that oxygen diffusion is not rate-limiting, because oxygen must still be transported within the composting particle. The moisture content of individual composting particles is significant because diffusion coefficients through a gaseous phase are about four orders of magnitude greater than through the liquid phase. Thus, diffusion transport of oxygen may be a controlling factor when particle moisture contents are high and assume less significance as moisture is removed from the particle.

Other factors such as C/N ratio, excessively high temperatures, lack of moisture and hydrogen ion concentration are potentially rate-controlling but their effects can be reduced through proper design and operation of the compost facility. With sewage sludge, the C/N ratio is sufficiently low to provide adequate nitrogen for cellular growth and avoid rate limitations. pH is not usually of concern with sludge composting. Even with lime-conditioned sludges, sufficient carbon dioxide is produced during composting to reduce the pH to neutral levels, although microbial growth may be limited until this occurs. Similarly, ammonia produced from protein decomposition can neutralize low-pH sludges that result from ferric chloride or alum conditioning.

There is no precise definition as to when sufficient organic stabilization has been achieved. Complete stabilization of all organic material to CO_2 and H_2O would not be desirable because the value of compost as a soil amendment depends in part on its organic content. A working definition is as follows: stabilization is sufficient when the rate of oxygen consumption is reduced to the point that anaerobic or odorous conditions are not produced to such an extent that they interfere with storage or end use of the product. A number of test procedures have been developed to help measure the degree of stabilization.

CHAPTER 11

PROCESS DYNAMICS I: DEVELOPMENT OF SIMULATION MODELS

INTRODUCTION

Physical, chemical and biological aspects of composting have been discussed in some depth in the preceding chapters. Concepts of mass and energy balances, aeration, moisture control, free airspace (FAS), thermodynamics and kinetics have all been discussed. Although relationships between these factors have been stressed, it is often difficult to mentally synthesize such a large volume of material. Understanding of the subject would certainly be improved by tying together the previous material into a more unified and understandable package.

Thermodynamic aspects of composting were introduced in Chapter 9. However, certain components of the total energy balance were necessarily excluded from the analysis. Energy inputs or losses that are a function of time could not be considered without a more detailed analysis. This would include energy inputs from agitation and energy losses to the surroundings, both of which are a function of the detention time in the composting process.

As soon as the concept of time is introduced, the kinetics of the process must be considered. Referring back to Figures 9-2 and 9-3, a fixed amount of organic decomposition was assumed in the thermodynamic analysis. This was necessary to avoid complicating the thermodynamic analysis by kinetic considerations. Although the assumed decomposition is reasonably consistent with field data, it would be interesting to incorporate process kinetics into the analysis. If kinetics and thermodynamics can be integrated into a single analysis, the thermodynamic analysis can be expanded to include time-dependent energy terms.

For a variety of reasons, therefore, it would be fruitful to integrate the previous subject matter into a more unified model of composting. Such a model by necessity will be mathematical in nature. Mathematical modeling in the design and analysis of systems is a powerful tool known to most engineers. Understanding the physics, chemistry and biology that describe the dynamic interactions of variables in the system is the key to successful modeling. This chapter will describe the approach, assumptions and equations used to model the dynamics of the compost process. Model applications to reactor, aerated static pile and windrow processes are discussed in Chapters 12, 13 and 14, respectively.

APPROACH TO THE PROBLEM

Two models of the composting process will be developed. The first simulates a system where material is fed on a continuous basis and the contents are completely mixed (continuous feed, complete-mix (CFCM) model). Certain reactor systems approach these conditions and the CFCM model should be reasonably applicable to them. The second model assumes a batch feeding of material at time 0 and plug flow of material during the remainder of the compost cycle (batch feed, plug flow (BATCH) model). The model should simulate the windrow and static pile processes, as well as certain reactor systems.

CFCM Model

Schematic diagrams showing mass and energy balances for a generalized compost process are shown in Figures 11-1 and 11-2. Because bulking particles have been used primarily in static pile processes, they have not been included in the analysis of the CFCM model. Bulking particles will be considered in the BATCH model, however. Terminology used to define the various energy terms is presented in Appendix II. The nomenclature is, for the most part, self-explanatory.

It is obvious from Figure 11-2 that the number of energy terms is considerably greater than previously allowed. However, there are other factors which complicate the analysis more so than the number of energy terms. The most important of these is the fact that as reactor temperature changes, so does the rate of decomposition. If the rate constant increases, for example, volatile solids (VS) in the outfeed will decrease, which alters the mass balance. Obvious, too, is the fact that an increase in the rate constant means more biological heat production in the reactor. This means that a trial-and-error solution to the problem is required. The number of calculations required is somewhat staggering and computer solutions are required. Nomenclature

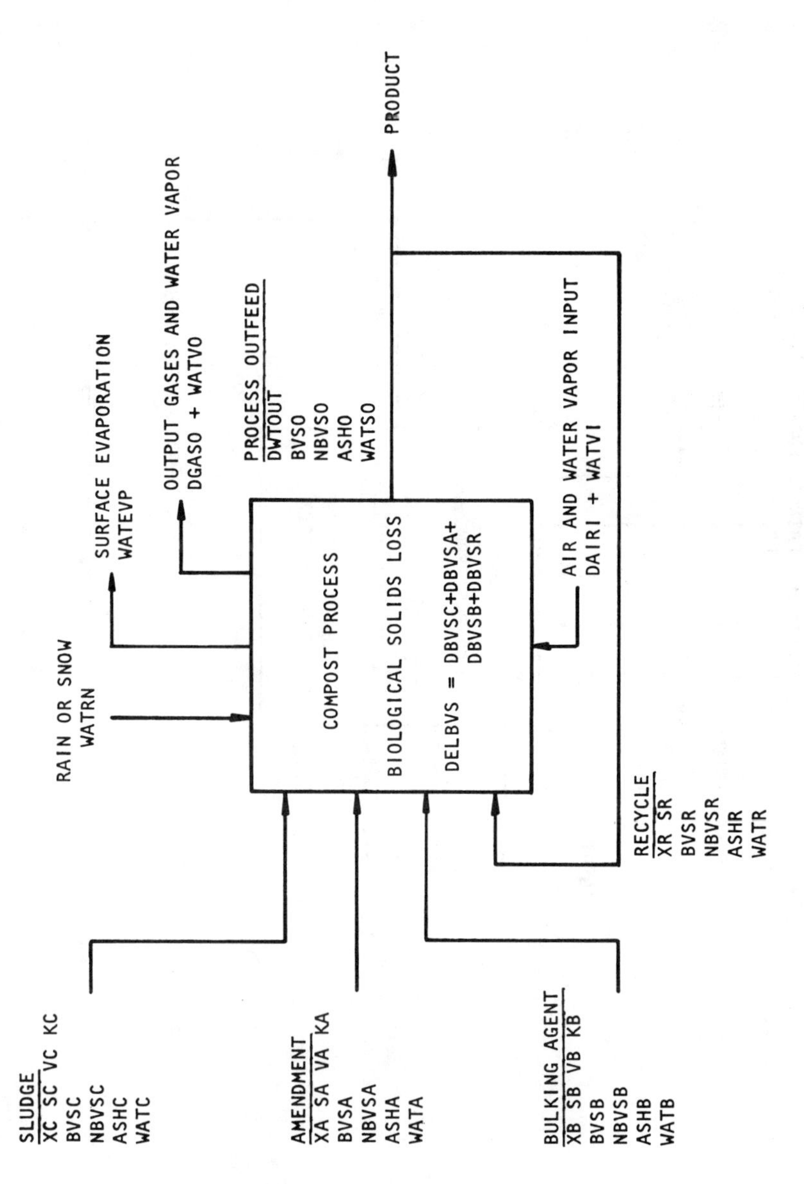

Figure 11-1. Mass and related terms for a generalized compost system.

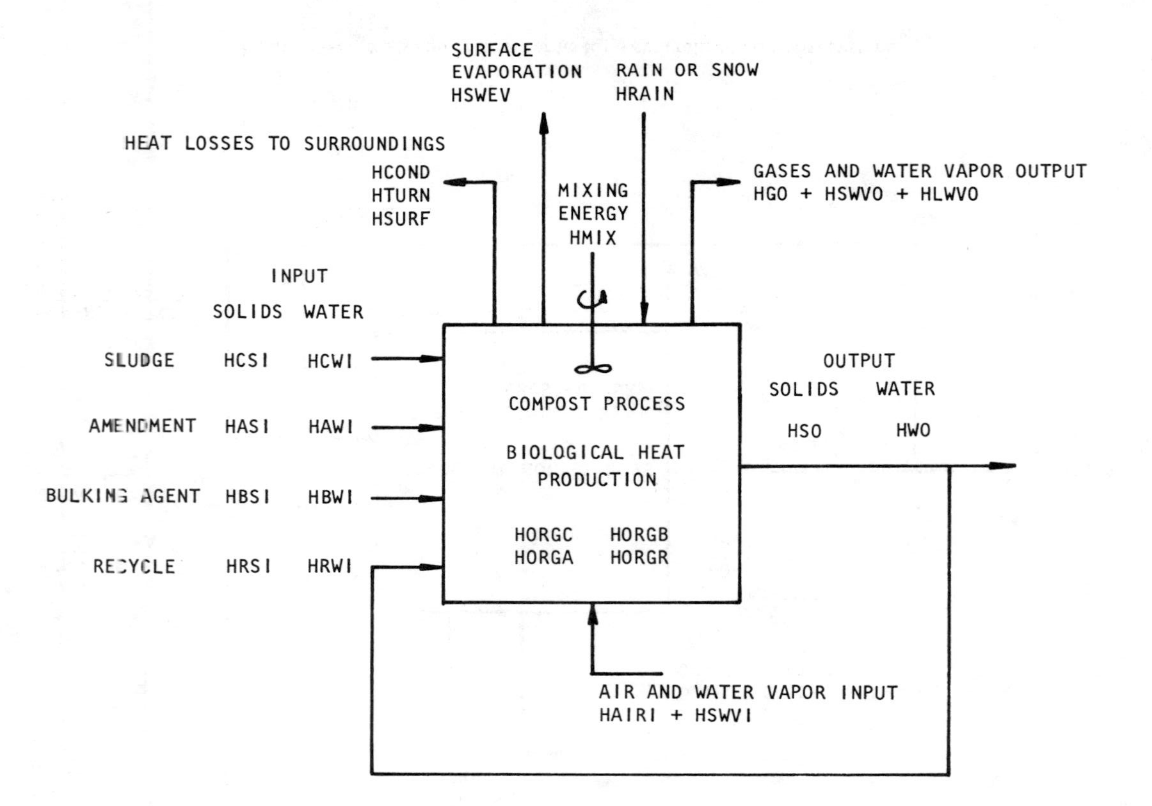

Figure 11-2. Energy terms for a generalized compost system.

shown in Figures 11-1 and 11-2 is the same as that used in the computer model.

The solution to our problem should include steady-state mass and energy balances as well as the temperature maintained in the reactor. The approach to obtaining such a solution is illustrated in Figure 11-3. The sum of heat input terms (HTOTI) will characteristically begin with an upward curve because of the exponential increase in biological activity with temperature. Increasing temperature will eventually result in thermal inactivation, a corresponding decrease in the rate constant and a reduction in total energy input. Output energy curves will generally curve upward as shown because of the exponential increase in water vapor pressure as a function of temperature. Solution of the problem occurs when total energy input balances total energy output.

Three different output energy curves are shown in Figure 11-3. Two of the curves, A and B, have solution points, while curve C does not. A number of factors could account for the higher output energies of curve B as compared to curve A. Included among these, for example, would be a higher heat loss rate to the surroundings or a higher aeration rate which increases latent heat losses. One might also imagine circumstances in which output energy is greater than input energy even at the beginning of the compost cycle (curve C). Extremely high aeration rates with cold ambient air might cause such a condition. In this case, compost material would tend to cool from its original input temperature. Obviously, this would not lead to composting as defined in this text. No solution point would exist except perhaps at a temperature below the inlet temperature of the feed materials.

The procedure for obtaining the solution point is as follows: (1) assume a reactor temperature; (2) determine the rate constant for organic decomposition and calculate the mass balance about the reactor; (3) sum all HTOTI and outputs (HTOTO) for the assumed temperature; (4) compare HTOTI with HTOTO and estimate a new temperature based on the difference between the two; (5) using the new temperature return to step 1 and repeat the procedure until the difference between HTOTI and HTOTO is sufficiently small. The procedure is well suited to computer analysis. The computational algorithm and pattern of information flow used in the computer program is shown in Figure 11-4.

Batch Model

The batch composting process was modeled as a large number of complete-mix cells arranged in series as shown in Figure 11-5. The compost cycle is divided into a large number of time steps, each of duration DELTAT. Composting material is considered to be homogeneous within each time step, and each time step is modeled as a complete-mix cell. Thus, the mass and

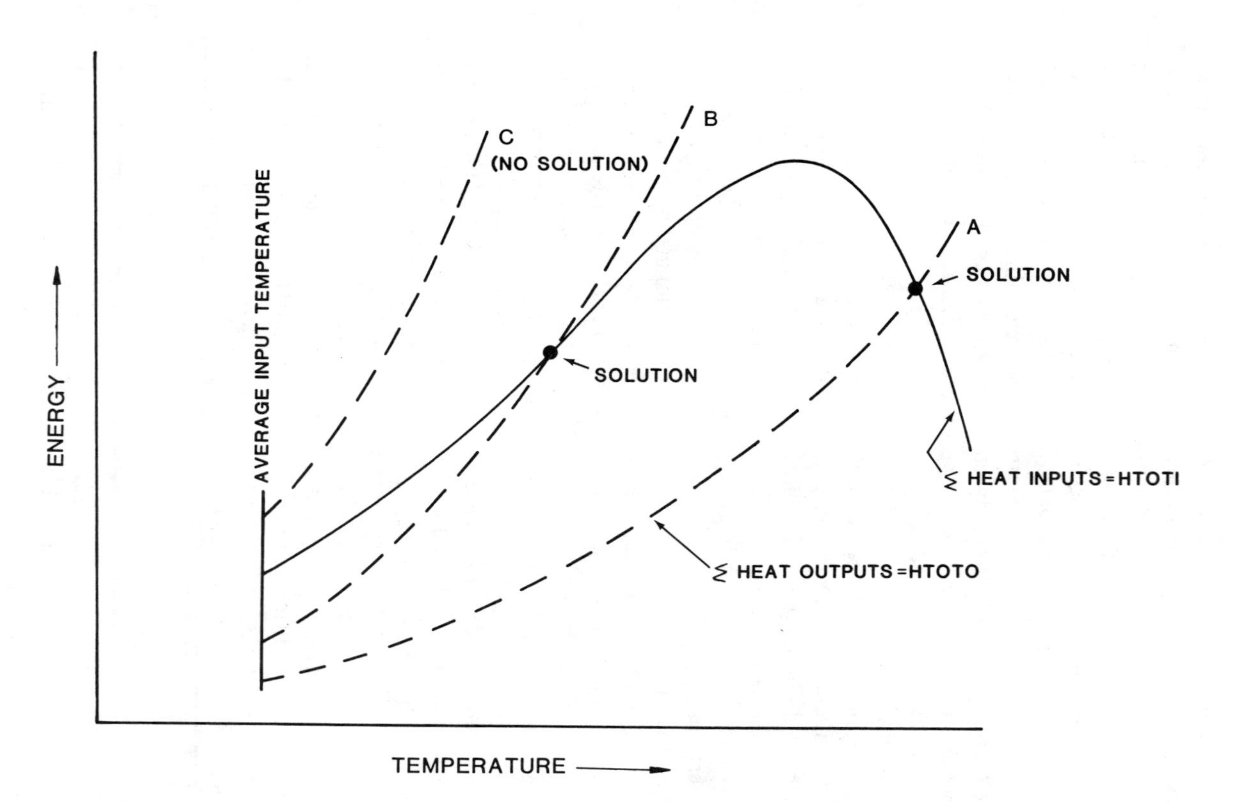

Figure 11-3. Schematic illustration of the characteristics of energy input and output curves and solution points.

energy balance and computational procedures illustrated in Figures 11-1 through 11-4 are applicable to each cell of the BATCH model. The one exception is compost product recycle, which occurs into the first cell only. Outfeed from one time step becomes the infeed to the next. Thus a complete solution is required in each time step, including mass and energy balances and steady-state temperature.

The reader may be somewhat reluctant to accept the fact that a batch process can be modeled as a series of complete-mix steps. However, this modeling technique is commonly applied to fluid systems. In such a case, it can be readily shown that both a plug flow reactor with continuous feed or batch process can be modeled as an infinite number of complete-mix cells arranged in series. In a plug flow reactor with continuous feed the reactor itself is divided into a series of cells. For a batch process the entire reactor contents are considered and the process divided into a series of time steps as shown in Figure 11-5. The proof of this is relatively straightforward, and the interested reader is referred to the large number of textbooks on reactor analysis in the chemical and sanitary engineering literature.

KINETIC ANALYSIS

From the discussion in Chapter 10 it is obvious that the subject of composting kinetics is complex. Reaction rates can be limited by solubilization of substrate, mass transport of oxygen, utilization of the solubilized substrate and oxygen and a variety of other factors such as moisture content, temperature and nutrient concentrations. We can discuss the variables involved in each of these processes, but a complete mathematical description of composting kinetics based entirely on first principles is probably not possible at this time. Therefore, reliance will be made on empirical expressions developed from experimental data, such as shown in Figure 10-7. Although one might wish for more complete information, particularly with sludge substrates, the data at hand must suffice.

First-Order Assumption

It is assumed that oxidation of biodegradable volatile solids (BVS) is first-order with respect to the quantity of BVS as follows:

$$\frac{d(BVS)}{dt} = -k_d(BVS) \tag{11-1}$$

where BVS = biodegradable volatile solids, kg
t = time, days
k_d = rate constant, day^{-1}

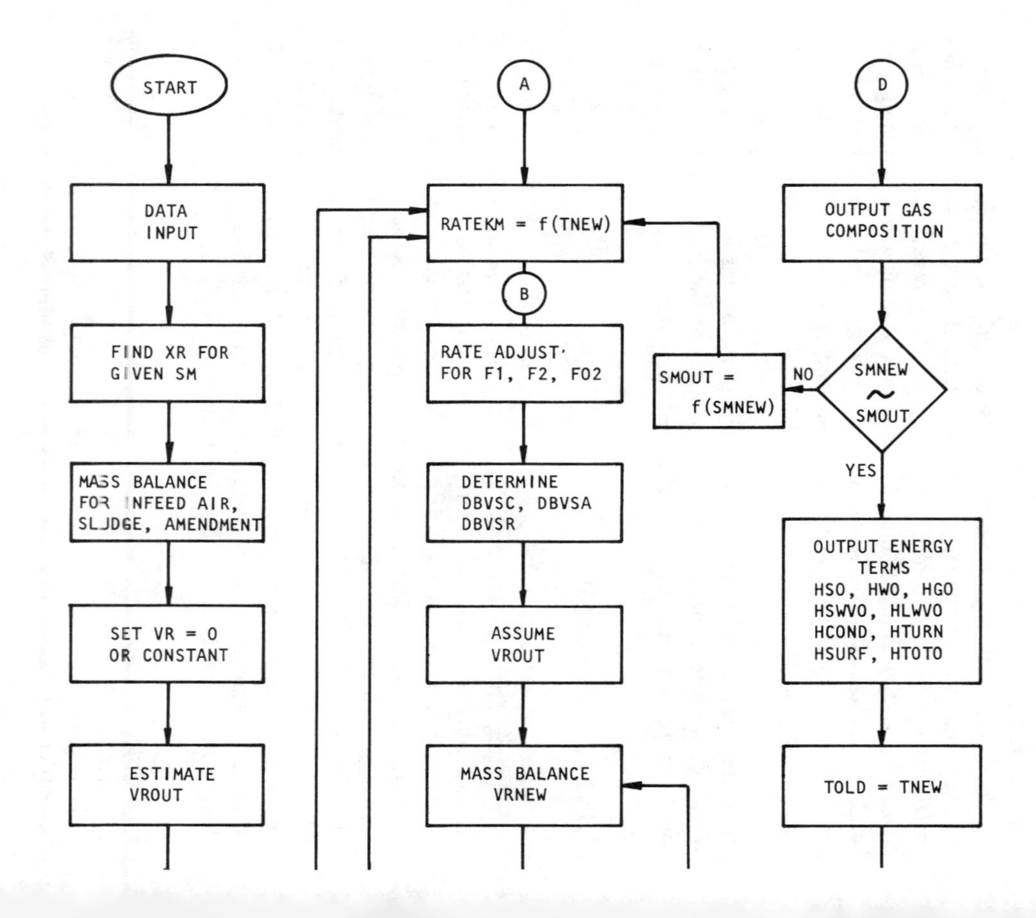

START
DATA INPUT
FIND XR FOR GIVEN SM
MASS BALANCE FOR INFEED AIR, SLUDGE, AMENDMENT
SET VR = 0 OR CONSTANT
ESTIMATE VROUT
A
RATEKM = f(TNEW)
B
RATE ADJUST. FOR F1, F2, FO2
DETERMINE DBVSC, DBVSA DBVSR
ASSUME VROUT
MASS BALANCE VRNEW
D
OUTPUT GAS COMPOSITION
SMNEW ~ SMOUT
NO
YES
SMOUT = f(SMNEW)
OUTPUT ENERGY TERMS
HSO, HWO, HGO
HSWVO, HLWVO
HCOND, HTURN
HSURF, HTOTO
TOLD = TNEW

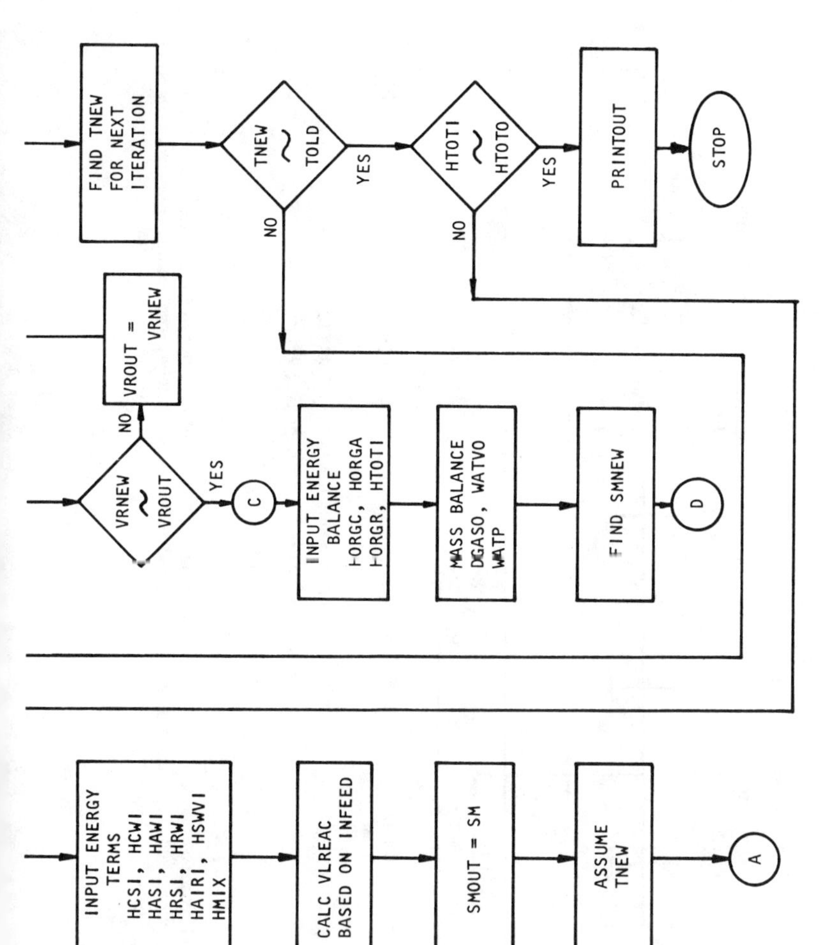

Figure 11-4. Computational algorithm and pattern of information flow for CFCM model.

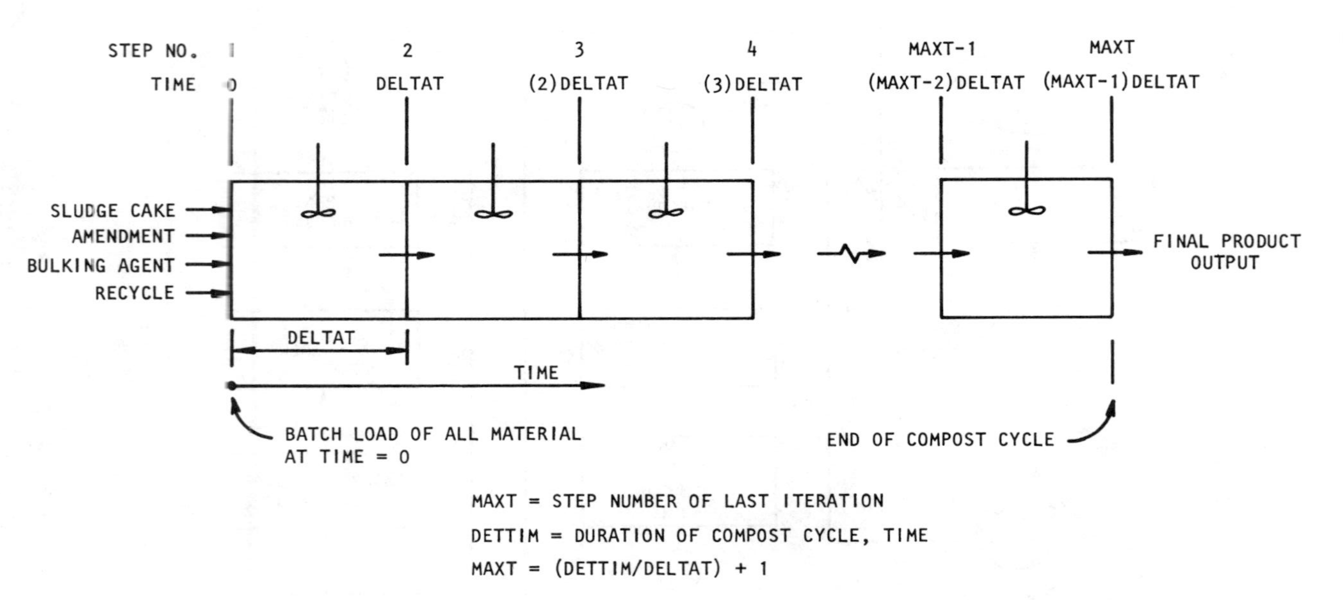

Figure 11-5. Model of batch composting process. The compost cycle is broken into a series of small time steps, DELTAT. Composting material is assumed to be homogeneous within each time step and is modeled as a completely mixed cell. Output from one time step becomes input to the next.

There is only limited scientific rationale for the assumption of first-order kinetics. Nevertheless, the approach has worked well in describing numerous processes involving biological oxidation. Included among these is the familar first order expression for organic oxidation in biological oxygen demand (BOD) bottles and streams. The approach has also worked well in describing more complex processes such as activated sludge, trickling filters (refer to the Velz equation in sanitary engineering textbooks), submerged filters for nitrification, anaerobic filters and aerobic digesters [145-147]. Without belaboring the point, the assumption of first-order kinetics with respect to BVS should provide a fairly reasonable description, but one for which further substantiating data would be highly desirable.

The general form of the first-order expression is shown in Figure 11-6. Total VS is divided between BVS and nonbiodegradable volatile solids (NBVS). This same distinction has often been assumed in other biological waste treatment systems. It is not so much a distinction between degradable and absolutely nondegradable fractions, but rather a distinction between

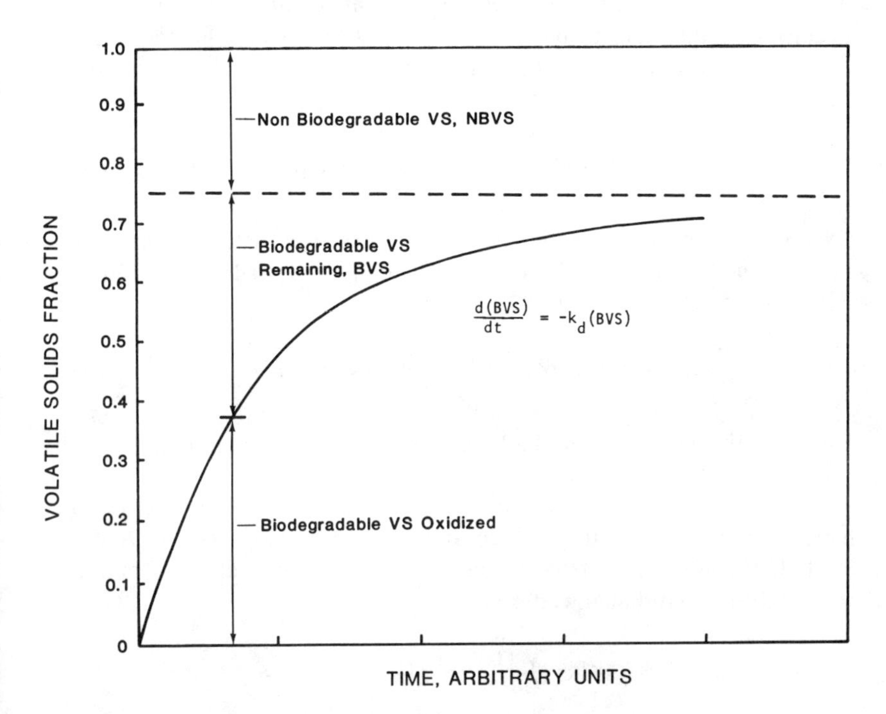

Figure 11-6. General form of the first-order expression for oxidation of BVS. Note the distinction between degradable and nondegradable fractions of the total volatile solids.

readily degradable materials and those which are slower to degrade. The latter are often referred to as refractory compounds. There is often no sharp dividing line between the two, particularly in composting systems where long detention times are sometimes used.

The Effect of Temperature

The rate constant k_d is a function of temperature. Experimental data shown in Figures 10-4 and 10-7, and previous theoretical discussions clearly suggest an exponential relationship. Referring to Equation 10-1, based on the data of Schultz [104], oxygen consumption followed the relationship:

$$w_{O_2} = 0.11\,(1.066)^{T} \quad (\text{mgO}_2/(\text{g VS-hr}) \qquad (10\text{-}1)$$

The feed material used in deriving Equation 10-1 consisted of a mixture of ground garbage and dewatered, digested sludge cake. Oxygen equivalent of the organics decomposed was not presented but was estimated from Schultz's original data to be about 1.5 mg O_2/mg volatile matter oxidized. Using this factor Equation 10-1 can be converted to a rate constant with the following form

$$k_d = 0.00632\,(1.066)^{T-20} \quad (\text{g BVS oxidized/g TVS-day}) \qquad (11\text{-}2)$$

Examination of Schultz's original data indicates a reduction in volatile matter ranging from about 37 to 45%. Assuming the degradable fraction to be 50% of the total, Equation 11-2 becomes:

$$k_d = 0.0126\,(1.066)^{T-20} \quad (\text{g BVS/g BVS-day}) \qquad (11\text{-}3)$$

Units on the rate constant in Equation 11-3 are now appropriate to the present analysis.

As temperature continues to increase, heat inactivation will become more pronounced and the rate constant will begin to decrease. Andrews et al. [147] used the following expression to describe similar effects during aerobic digestion of liquid wastes:

$$k_d = k_{dR1}\left[C_1^{(T-T_{R1})} - C_2^{(T-T_{R2})}\right] \qquad (11\text{-}4)$$

where k_{dR1} = rate constant at temperature T_{R1}, day^{-1}
C_1, C_2 = temperature coefficients
T_{R1}, T_{R2} = reference temperatures, °C

Substituting Equation 11-3 into 11-4 and assuming reasonable values for the other variables, the effect of temperature on k_d takes the final form:

$$k_{dm} = \text{RATEKM} = 0.0126\left[1.066^{(T-20)} - 1.21^{(T-60)}\right] \tag{11-5}$$

The subscript m is added to k_d to indicate that this is the maximum rate constant uncorrected for effects of moisture, FAS or oxygen content. Equation 11-5 is plotted in Figure 11-7. The exponential dependence of k_{dm} on temperature continues to about 65°C, beyond which effects of heat inactivation become pronounced. The maximum occurs at about 70°C. Referring back to Figure 10-7, optimum temperatures as low as 45°C have been reported. The data of Schultz [104] developed for continuous composters under steady-state conditions was weighted heavily in formulating

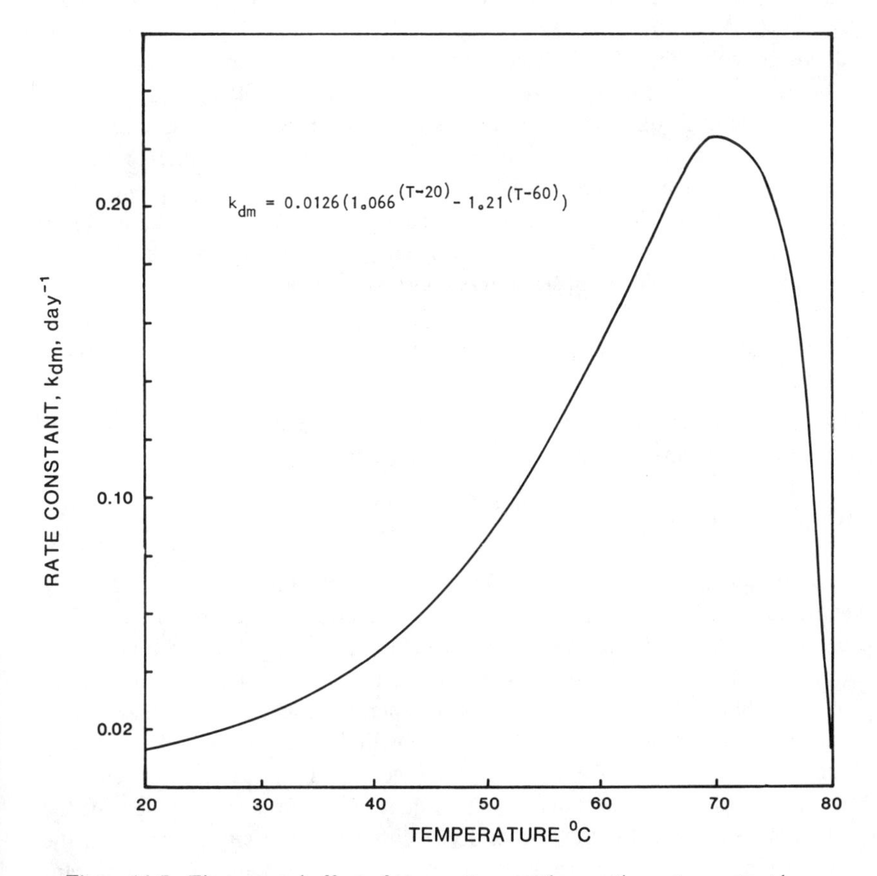

Figure 11-7. The assumed effect of temperature on the reaction rate constant k_{dm}.

Equation 11-5. The latter data indicate an exponential relationship to temperatures as high as 70°C. Figure 11-7 is also consistent with the observation that composting temperatures rarely exceed 80°C, a point at which the rate of heat inactivation must be overwhelming even for thermophiles.

Equation 11-5 was applied to sludge cake, amendment and recycle components. Recall that Equation 11-5 was developed from data on composting of garbage and sludge mixtures. Therefore, its application should not be extended to amendments quite different from this composition. Again, referring to Figure 10-7, Jeris and Regan [126] found similar oxygen consumption rates for mixed refuse which contained 60-70% paper. This is significant because many amendments, such as sawdust, rice hulls and tree trimmings, are largely cellulosic. Although the limitations of Equation 11-5 are recognized, experimental data summarized in Figure 10-7 suggest that it is reasonably applicable to a wide variety of solid substrates.

For bulking agents such as woodchips it is unlikely that reaction rates would approach those calculated by Equation 11-5. The problem of breakdown of complicated cellulosic structures such as wood was discussed in Chapter 8. Referring again to Figure 10-7, Jeris and Regan [126] found greatly reduced oxygen consumption rates for newsprint and composted refuse. Because newsprint is only mechanically pulped, its rate of decomposition should be more indicative of the rate of wood chip decomposition than other materials shown in Figure 10-7. Comparing the Jeris and Regan curve for newsprint with the Schultz curve for garbage and sludge, the rate of wood chip decomposition was approximated as:

$$k_{dm}^{b} = \text{RATEKB} = (0.2)\, k_{dm} \tag{11-6}$$

Obviously, more experimental data on wood chip decomposition are needed and the limitations of Equation 11-6 are recognized.

The Effect of Moisture

Experimental data on the effect of moisture content on oxygen consumption rates (and hence rate of BVS oxidation) were presented in Figure 10-20. Based on the trend of experimental data, the curves in Figure 11-8 were constructed and used to predict the effect of mixture moisture content on the rate constant. Because there is considerable scatter in the experimental data a great number of possible curves might be constructed. An S-shaped curve was selected since its shape is theoretically pleasing and generally consistent with the experimental data. Equations representing portions of the curve are shown in Figure 11-8 and were used to represent the curve in the simulation model. The procedure used was to first calculate the rate constant

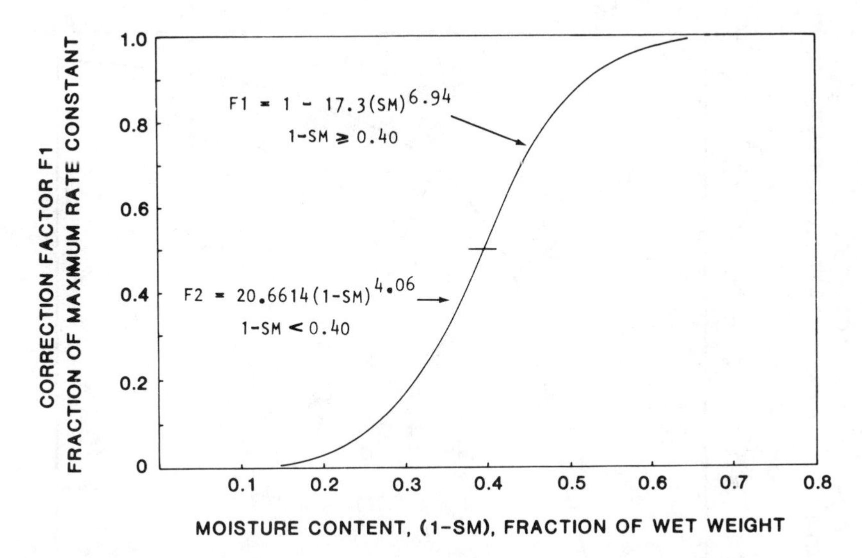

Figure 11-8. Assumed effect of mixture moisture content on the rate of organic oxidation. Large number of significant digits in lower equation is used to reduce the discontinuity at the intersection of the two curves.

from Equation 11-5 for the assumed conditions of temperature, and then correct the rate constant by the factor F1.

The Effect of Free Airspace

As discussed in Chapter 7, excessive moisture can reduce the FAS to the point where oxygen storage and transport through the void spaces is reduced. Reaction rates can then become oxygen-limited. This effect can be observed in the data of Figure 10-20, where oxygen consumption rates were reduced at high moisture levels. Considering the general trend of data in Figure 10-20 and recalling that optimum FAS is generally about 30%, the curve in Figure 11-9 was constructed and used to represent the effect of FAS on the reaction rate constant. Although data are limited, the selected curve is consistent with available data, and the general shape of the curve is theoretically reasonable.

The effect of FAS was considered only for those systems which did not use bulking agents. It was assumed that if bulking particles were used the material proportions would be adjusted to assure the presence of some FAS. Furthermore, the large pore sizes which result when bulking particles are used

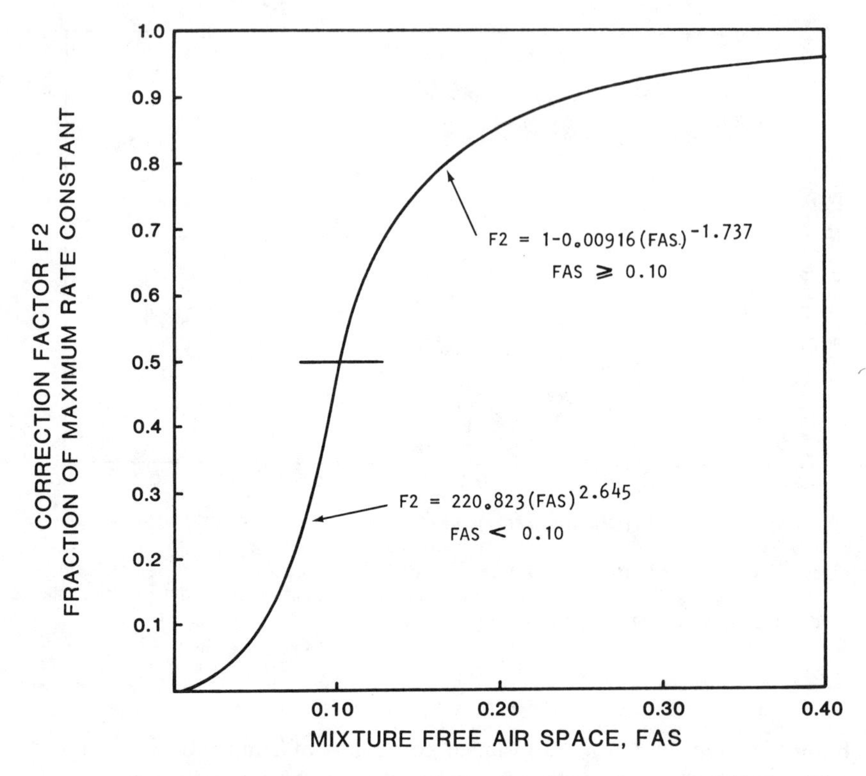

Figure 11-9. Assumed effect of mixture FAS on the rate of organic oxidation. The large number of significant digits in the equations is used to reduce the discontinuity at the intersection of the two curves.

should reduce resistance to oxygen transport through the pile. Therefore, it was felt that as long as some FAS was present, it would have a negligible effect on reaction kinetics.

Application of the correction factor F2 requires that the bulk weight be known to calculate the FAS. The following equations were used to estimate the bulk weight of mixtures of sludge, amendment and compost recycle:

$$\gamma_m = \text{GAMMAM} = \frac{1}{\frac{S_m}{G_m} + 1 - S_m} \qquad (S_m \leqslant 0.35) \qquad (7\text{-}3)$$

$$\gamma_m = 0.40/S_m \qquad (S_m > 0.35) \qquad (11\text{-}7)$$

where γ_m = GAMMAM = bulk weight of mixture, g/cm³
G_m = specific gravity of the mixed materials, calculated from Equation 7-1

Together, Equations 7-3 and 11-7 are consistent with the trend of data presented in Figures 7-2 and 7-3. Even with limited addition of very light weight amendments, such as dry rice hulls or refuse light fraction, the error introduced by use of Equations 7-3 and 11-7 should be small. However, the equations would not be applicable if amendment addition was increased to the point where it became the main component of the mixture.

Knowing the bulk weight and mixture solids content, FAS was calculated as:

$$FAS = f = 1 - \frac{\gamma_m S_m}{G_m \gamma_w} - \frac{\gamma_m (1 - S_m)}{\gamma_w} \qquad (7\text{-}10)$$

The correction factor F2 was then calculated using the relationships shown in Figure 11-9.

The Effect of Oxygen Content

Even though the influence of FAS has been accounted for, it does not by itself ensure that oxygen is not a rate-limiting substrate. The kinetics of oxygen transport were discussed in some detail in Chapter 10. The key points of that discussion are as follows: (1) transport is probably limited by diffusion through the particle matrix of solids and water; (2) even though oxygen partial pressure in the FAS is high, it does not necessarily mean that oxygen transport into the particle is not rate-limiting; (3) whether the mass flux of oxygen exceeds the oxygen demand depends on the size of the particle. On the latter point, particle thickness on the order of 1.0 cm would appear to present large diffusion resistances that would tend to dominate the process kinetics.

This leads to a rather complex picture of oxygen supply, because diffusion transport and particle size should both be considered. Such a sophisticated representation would seem to be beyond the present state of experimental data. Consider again the compilation of measured oxygen uptake rates shown in Figure 10-7. Different substrates with different particle sizes were used in these experiments. Can it be stated that oxygen was not rate-limiting in some cases? Actually, the answer to this question can be obtained indirectly by noting the rather significant effect of temperature on oxygen uptake. If oxygen diffusion were controlling, the influence of temperature would be much less significant (refer to Equation 10-18 for the effect of temperature on the diffusion coefficient). Therefore, the rate of oxygen diffusion probably did not control process kinetics for most of the data in Figure 10-7.

Because the simulation model has relied heavily on the data developed by Schultz [104], it seems advisable to reexamine his experimental conditions.

Schultz used continuous composters that were fed intermittently. Airflow through the composters was adjusted so that residual oxygen concentration in the exhaust gas remained between 5 and 10%. Because the composters were well mixed, oxygen levels in the FAS between composting particles should have been in the same range. Thus, there is some assurance that oxygen levels of 5% in the FAS do not impose severe oxygen limitations, provided, of course, that particle sizes are similar to those used by Schultz.

Because the effect of particle size is difficult to account for in a generalized model, a more simplified approach was adopted. The influence of oxygen concentration in the free air space was assumed to follow a Monod-type expression as shown in Figure 11-10. The half-velocity coefficient was assumed to be 1.0% oxygen by volume in the FAS. This assured that oxygen effects would be minimized at concentrations above about 5%. Furthermore, the rate of organic decomposition reduces to zero if the oxygen concentration is zero, a boundary condition known to be true for aerobic decomposition. An inherent assumption in this approach is that particle sizes are sufficiently small to avoid oxygen transport limitations.

Net Rate Coefficient

Once the maximum rate coefficient is determined from Equation 11-5, the factors F1, F2 and FO2 are applied to determine the actual rate constant under conditions of moisture content, FAS and oxygen content in the composting matrix. The actual rate constant for sludge, amendment and recycled organics is determined as:

$$k_{da} = \text{RATEK} = (\text{F1})(\text{F2})(\text{FO2})k_{dm} \tag{11-8a}$$

and for bulking agents as:

$$k_{da}^{b} = \text{RATEKB} = 0.2(\text{F1})(\text{F2})(\text{FO2})k_{dm} \tag{11-8b}$$

where for bulking agents F2 = 1.0 by definition.

MASS BALANCES

In any systems analysis, mass balances are required on all components entering and leaving the system. For the composting process, these components include solids (both organic and inorganic), water and gases. Each of these components will be discussed separately.

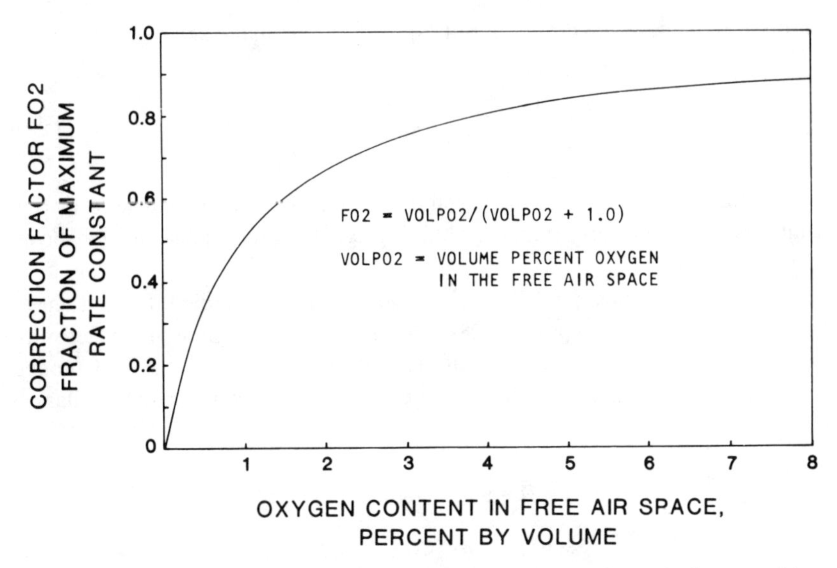

Figure 11-10. Assumed effect of oxygen content on the rate of organic decomposition.

Solids Balance

For purposes of analysis, components of the composting mixture are divided into the following fractions: BVS, NBVS, ash (ASH) and water (WAT). The position of these fractions in the mass balance is shown in Figure 11-1. For each component of the mixture (excluding any recycle) the wet weight (X), solids fraction (S), VS content (V) and degradability coefficient (k) must be known (i.e., input to the model). Using sludge cake as an example, the various fractions can be calculated as:

$$BVSC = k_c V_c S_c X_c \tag{11-9}$$

$$NBVSC = (1 - k_c) V_c S_c X_c \tag{11-10}$$

$$ASHC = (1 - V_c) S_c X_c \tag{11-11}$$

$$WATC = X_c - S_c X_c \tag{11-12}$$

Similar equations apply for amendment and bulking agent additions. Except where otherwise noted, mass terms have units of kg/day in the CFCM model and kg in the BATCH model.

For the recycle component only the solids content must be input. In this case, it has been assumed that additional drying may occur after the composting process which is typical of many systems. The wet weight of recycle (X_r)

required to achieve the desired input S_m is then calculated as:

$$X_r = \frac{X_c(S_m - S_c) + X_a(S_m - S_a)}{S_r - S_m} \qquad (11\text{-}13)$$

It is assumed that compost recycle and bulking particles are not used in the same system. Therefore, terms for bulking agent do not appear in the above expression.

Organic decomposition during composting is determined from a mass balance about the compost process. Recall that a complete-mix reactor is assumed in the CFCM model while each time step of the BATCH model is assumed well mixed. Considering the sludge component, a mass balance on BVS gives:

$$\text{rate of storage} = \text{input} - \text{output} + \text{sources} - \text{sinks}$$

$$V\frac{d(BVSC)}{dt} = BVSC - BVSCO + 0 - RATEK(BVSCO)\theta$$

where V = volume of reactor, (m^3) and θ = detention time, days. θ corresponds to the actual reactor detention time DETTIM in the CFCM model and to the time step DELTAT in the BATCH model. Assuming steady-state conditions where $d(BVS)/dt = 0$, the above expression can be rearranged to:

$$\text{sludge cake} \qquad BVSCO = \frac{BVSC}{1 + RATEK(\theta)} \qquad (11\text{-}14)$$

Similar expressions for amendment, bulking agent and recycle components are:

$$\text{amendment} \qquad BVSAO = \frac{BVSA}{1 + RATEK(\theta)} \qquad (11\text{-}15)$$

$$\text{bulking agent} \qquad BVSBO = \frac{BVSB}{1 + RATEKB(\theta)} \qquad (11\text{-}16)$$

$$\text{recycle} \qquad BVSRO = \frac{BVSR}{1 + RATEK(\theta)} \qquad (11\text{-}17)$$

When compost recycle is used, a small problem arises because the volatile fraction of the recycle depends on the extent of decomposition within the compost reactor. This requires a trial-and-error solution at each assumed temperature to complete the mass balance. The algorithm used to calculate the mass balance is shown in Figure 11-11. Alternatively, constant values

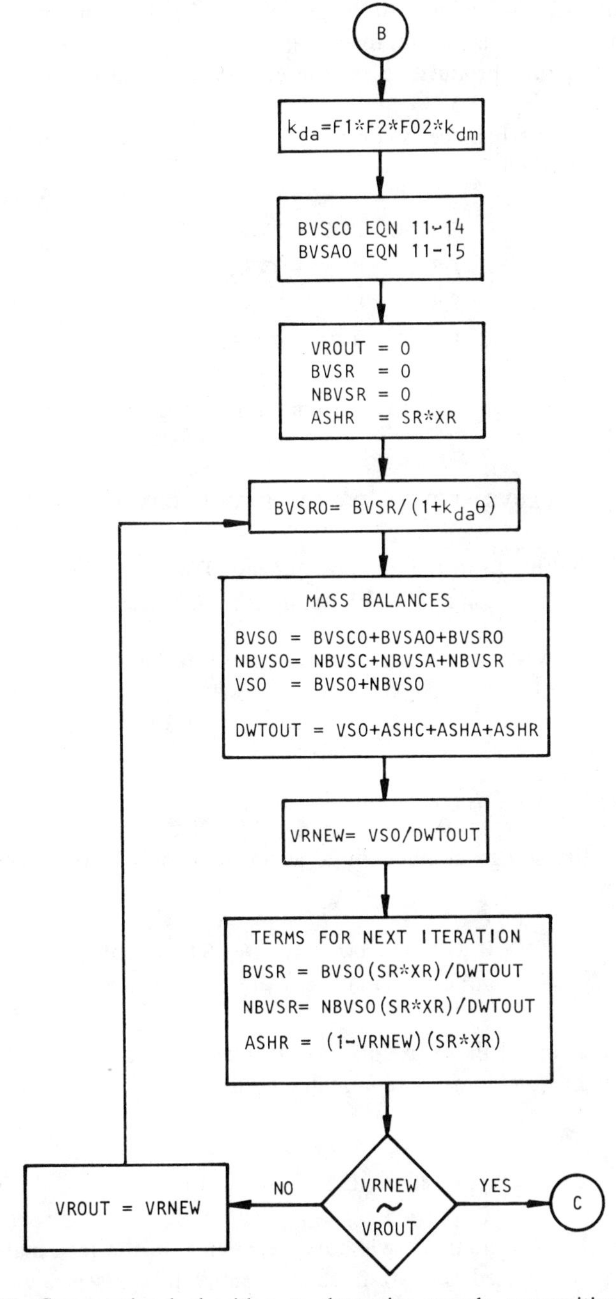

Figure 11-11. Computational algorithm to determine recycle composition. See Figure 11-4 for location of this algorithm in the CFCM computational model.

for KR and VR can be input to the model, and the algorithm of Figure 11-11 is not applied. The latter approach can be used to simulate the case where subsequent curing produces a recycle component with constant characteristics.

Once the mass balance is completed, the loss of BVS during composting is determined as:

$$\text{DBVSC} = \text{BVSC} - \text{BVSCO} \tag{11-18}$$

$$\text{DBVSA} = \text{BVSA} - \text{BVSAO} \tag{11-19}$$

$$\text{DBVSB} = \text{BVSB} - \text{BVSBO} \tag{11-20}$$

$$\text{DBVSR} = \text{BVSR} - \text{BVSRO} \tag{11-21}$$

Summing the change in individual component masses, the total change in BVS is:

$$\text{DELBVS} = \text{DBVSC} + \text{DBVSA} + \text{DBVSB} + \text{DBVSR} \tag{11-22}$$

Dry weight output from the compost process, DWTOUT, can then be determined, assuming the ash and NBVS fractions to be conservative:

$$\text{DWTOUT} = \Sigma\text{ASH} + \Sigma\text{NBVS} + \Sigma\text{BVS} - \text{DELBVS} \tag{11-23}$$

Gas Balance

Stoichiometry of Solids Oxidation

BVS in the sludge component is assumed to be oxidized according to Equation 8-1:

$$\begin{array}{ccccc} C_{10}H_{19}O_3N + 12.5\ O_2 & \rightarrow & 10\ CO_2 + & 8\ H_2O + & NH_3 \\ 201 \qquad 400 & & 440 & 144 & 17 \end{array}$$

Amendments and bulking agents are assumed to be primarily cellulosic in nature, with stoichiometry represented by the oxidation of glucose:

$$\begin{array}{cccc} C_6H_{12}O_6 + & 6\ O_2 \rightarrow & 6\ CO_2 + & 6\ H_2O \\ 180 & 192 & 264 & 108 \end{array}$$

Obviously, the stoichiometry will vary depending on the feed material that is used. Equations used here are reasonable and will represent a wide variety of sludge and amendment materials. Nevertheless, the assumed stoichiometry should be reexamined for each case.

If compost product is used in the mixed material, the recycle will consist of a mixture of undegraded sludge organics, undegraded organics remaining from any other added amendments, and ASH and NBUS components. Obviously, composition of final product cannot be known until the solution procedure is complete. A trial-and-error solution is required at each assumed temperature to complete the mass balance. This poses no difficulty because the recycle composition is determined during each iteration in the CFCM model. The algorithm used to calculate the mass balance is shown in Figure 11-11.

For the BATCH model, however, all time steps of the compost cycle (step 1 to MAXT in Figure 11-5) must be solved before the final product composition is known. One solution would be to use the calculated product composition and repeat the entire solution procedure until the product composition is constant. The drawback of this approach is the large number of calculations and computer time required. Alternatively, a solution can be obtained using the algorithm of Figure 11-11 which assumes a complete-mix reactor. An average temperature of 50°C is assumed and the detention time corresponds to that assumed in the BATCH model. Product composition from the complete mix system is then used to represent the recycle composition in the BATCH model. This approach worked well and any errors introduced were found to be acceptably small. Alternatively, constant values for KR and VR can be input to the model.

Based on the assumed stoichiometry and knowing the organic loss for each component, the quantity of oxygen consumed and carbon dioxide and water produced are calculated as shown in Table 11-1. If compost product is used in the mixture, the recycle will contain undegraded organics from the original sludge feed as well as any other added amendments. Stoichiometry of solids loss for product recycle is estimated using a weighted average based on the quantity of sludge cake and amendment in the final product.

Input Dry Air and Water Vapor

The aeration rate is input to the model as the flowrate of air (m^3/min) adjusted to atmospheric pressure (760 mm Hg) and ambient temperature (TAIR). The mass of input dry air and water vapor is then calculated from the equation of state for an ideal gas:

$$PV = mRT_a \tag{11-24}$$

where P = absolute pressure, 1.01×10^5 N/m^2 corresponding to 760 mm Hg
V = volume, m^3
m = mass, kg
R = universal gas constant, 287 J/kg-°K for air, 461.6 J/kg-°K for water vapor.
T_a = absolute temperature, °K

Table 11-L. Expressions Used to Calculate Oxygen Consumption and Carbon Dioxide and Water Production

Component	Oxygen Consumed	Carbon Dioxide Produced	Ammonia Produced	Water Produced
Sludge Cake	DBVSC(1.990)	DBVSC(2.189)	DBVSC(0.085)	DBVSC(0.716)
Amendment	DBVSA(1.067)	DBVSA(1.467)		DBVSA(0.600)
Bulking Agent	DBVSB(1.067)	DBVSB(1.467)		DBVSB(0.600)
Recycle	DBVSR(RCYCF3)	DBVSR(RCYCF4)		DBVSR(RCYCF2)

$$RCYCF2 = \frac{BVSCO\ (0.716) + BVSAO\ (0.600)}{BVSCO + BVSAO}$$

$$RCYCF3 = \frac{BVSCO\ (1.990) + BVSAO\ (1.067)}{BVSCO + BVSAO}$$

$$RCYCF4 = \frac{BVSCO\ (2.189) + BVSAO\ (1.467)}{BVSCO + BVSAO}$$

Rearranging Equation 11-24 and converting volume to flowrate gives:

$$\frac{m}{t} = \frac{P(QAIR)1440}{R(273 + TAIR)} \tag{11-25}$$

where m/t = weight flowrate, kg/day
QAIR = flowrate of ambient air including water vapor, m^3/min
TAIR = ambient air temperature, °C

The weight flowrate of water vapor in both inlet air and outlet gases is required to complete the mass and energy balances. This is a particularly important term because heat of vaporization is usually a major component of the energy balance. Water vapor pressure is a function of temperature and can be approximated by an equation of the form:

$$\log_{10} PVS = \frac{a}{T_a} + b \tag{11-26}$$

where PVS = saturation water vapor pressure, mm Hg
a = constant (−2238 for water)
b = constant (8.896 for water)

Knowing the relative humidity, the actual water vapor pressure can then be estimated as:

$$PV = (RH)\,PVS \tag{11-27}$$

where PV = actual water vapor pressure (mm Hg) and RH = relative humidity, fraction of saturation vapor pressure.

The specific humidity, shown in Figure 8-2 as a function of temperature, is defined as the ratio of the mass of water vapor to the mass of dry air in a given volume of mixture:

$$w = \frac{m_v}{m_a} \tag{11-28}$$

where w = specific humidity
m_v = mass of water vapor
m_a = mass of dry gas

Applying the equation of state for both vapor and gas in a unit volume:

$$w = \frac{P_v/R_vT_a}{P_a/R_aT_a} = \frac{R_a}{R_v}\frac{P_v}{P_a} = 0.622\,\frac{P_v}{P_a} \tag{11-29}$$

Equation 11-25 can now be applied to determine the weight flow of dry air and water vapor.

$$\text{DAIRI} = \frac{(1.01 \times 10^5)\left(\frac{760 - \text{PV}}{760}\right)\text{QAIR}(1440)}{(287.0)(273 + \text{TAIR})} \tag{11-30}$$

$$\text{WATVI} = \frac{(1.01 \times 10^5)\left(\frac{\text{PV}}{760}\right)\text{QAIR}(1440)}{(461.6)(273 + \text{TAIR})} \tag{11-31}$$

$$\text{WATVI} = \text{DAIRI}(0.622)\left(\frac{\text{PV}}{760 - \text{PV}}\right) \tag{11-32}$$

where DAIRI = input dry air (kg/day) and WATVI = input water vapor (kg/day). Equations 11-30 to 11-32 are used as shown in the CFCM model. For use in the BATCH model each equation is multiplied by the time step, DELTAT, to give the input weight during each time interval.

Output Gases and Water Vapor

Output dry gas quantity is determined from the input gas quantity, the amount of organic decomposition and the reaction stoichiometry. The change in dry gas quantity will equal the production of CO_2 and NH_3 minus the consumption of oxygen. Adding appropriate terms from Table 11-1, the following equation can be developed:

$$\begin{aligned}\text{DGASO} = \text{DAIRI} &+ \text{DBVSC}(0.284) + \text{DBVSA}(0.400)\\ &+ \text{DBVSB}(0.400) + \text{DBVSR}(\text{RCYCF1})\end{aligned} \tag{11-33}$$

where DGASO = output dry gas

$$\text{RCYCF1} = \frac{\text{BVSCO}(0.284) + \text{BVSAO}(0.400)}{\text{BVSCO} + \text{BVSAO}} \tag{11-34}$$

Output water vapor, WATVO, is determined by first using Equation 11-26 to predict the saturation water vapor pressure at the exit gas temperature. Exit gas temperature is assumed to be the same as that in the composting reactor or pile. It is reasonable to assume that output gases are saturated if the compost is reasonably moist, say ⩾50% moisture. At some point, however, the gases will no longer exit under a saturated condition. Suppose the composting material were only 20-30% moisture. It is hard to imagine the exit gases to be saturated under such dry conditions. Transfer of moisture to the gas phase during composting is a difficult mass transfer problem to describe because of the complexity of the physical system. To avoid further complications to the simulation model, the partial pressure of water vapor in the exit gas was assumed to be a function of the solids content and is calculated as follows:

$$PVO = PV + (PVSO - PV)\,F1 \qquad (11\text{-}35)$$

where PVO = actual water vapor pressure in the exit gas, mm Hg
PV = water vapor pressure in inlet gas, mm Hg
PVSO = saturation vapor pressure in exit gas, corresponding to the exit gas temperature, mm Hg
F1 = adjustment factor as shown in Figure 11-8.

Note that the adjustment factor F1 is also that used to adjust the reaction rate constant for the effect of moisture. This was simply a matter of convenience because the curve shown in Figure 11-8 seemed reasonable for the present purposes. It does not imply any fundamental relationship between the two. Equation 11-35 should provide a reasonable approximation of the exit water vapor pressure. However, the heat of vaporization is such a significant term in the total energy balance that additional work is certainly warranted to better define the exit gas condition.

Having estimated the exit water vapor pressure, the output weight flow of water vapor can be calculated using the form of Equation 11-32 as follows:

$$WATVO = (DGASO)(0.622)\left(\frac{PVO}{760 - PVO}\right) \qquad (11\text{-}36)$$

The composition of output gases is computed in a rather straightforward manner. Referring again to Table 11-1, the weight flowrates of oxygen, carbon dioxide and nitrogen in the output gas become:

$$\begin{aligned}
\text{wt } O_2 &= (0.23)DAIRI - DBVSC(1.990) - DBVSA(1.067) - DBVSB(1.067) \\
&\quad - DBVSR(RCYCF3) \\
\text{wt } CO_2 &= (0.0005)DAIRI + DBVSC(2.189) + DBVSA(1.467) \\
&\quad + DBVSB(1.467) + DBVSR(RCYCF4) \\
\text{wt } N_2 &= (0.77)DAIRI
\end{aligned}$$

In developing these equations it was assumed that air is 23% oxygen, 77% nitrogen and 0.05% carbon dioxide by weight. It is common practice to express gas composition in terms of the percent by volume of each component. For example:

$$O_2\ (\%\text{ by volume}) = VOLPO2 = \frac{(\text{wt } O_2/32)(100)}{(\text{wt } O_2/32) + (\text{wt } N_2/28) + (\text{wt } CO_2/44)} \qquad (11\text{-}37)$$

Development of similar expressions for N_2 and CO_2 is left to the reader.

Water Balance

Having completed balances for solids and gases, the only remaining balance is that for water. Water vapor in inlet and outlet gases has already been accounted. Here the subjects of produced water, rainfall, surface evaporation and water contained in the output solids will be considered.

Produced Water

Water produced during organic decomposition is determined from the assumed stoichiometric equations for sludge, amendments and bulking agents. Referring to Table 11-1, the following expression for produced water can be developed:

$$\text{WATP} = \text{DBVSC}(0.716) + \text{DBVSA}(0.600) + \text{DBVSB}(0.600) + \text{DBVSR}(\text{RCYCF2}) \quad (11\text{-}38)$$

where WATP = produced water, kg/day in CFCM model and kg in the BATCH model.

Rainfall

In the case of windrow or static pile composting the piles may be exposed to rainfall or snowfall. The BATCH model allows for rainfall of any duration or intensity. Rainfall intensity, cm/day, can be input at any time step of the compost cycle. Water addition from the rain or snow is calculated as:

$$\text{WATRN} = \text{RAIN(I)} \times \text{DELTAT} \times \text{AREARN} \times 10.0 \quad (11\text{-}39)$$

where WATRN = water supplied by rain in time step I, kg
RAIN(I) = rainfall intensity in time step I, cm/day
AREARN = area of pile subject to rainfall, m^2 (see Figure 11-15)

Rainfall is not considered in the CFCM model.

Surface Evaporation

If composting material is exposed to the ambient environment, evaporation of water from the surface can occur provided the relative humidity of ambient air is <1 (i.e., <100%). Surface evaporation does not depend on biological heat production and can have a significant effect on moisture removal in open air windrow systems, particularly in arid climates. Therefore, surface evaporation is considered in the BATCH model, but not the CFCM model. The latter was assumed to be partially or completely enclosed and surface evaporation of minor significance.

The surface evaporation rate, cm H_2O/day, can be input at any time step of the BATCH compost cycle. This allows the evaporation rate to be reduced to zero during periods of rainfall or varied to reflect changing conditions during the compost cycle. Water removal is calculated as:

$$\text{WATEVP} = \text{EVAP(I)} \times \text{AREAEV} \times \text{DELTAT} \times 10.0 \times \text{F5} \tag{11-40}$$

where WATEVP = water removed by surface evaporation in time step I, kg
EVAP(I) = surface evaporation rate in time step I, cm/day
AREAEV = area of pile subject to surface evaporation, m^2 (see Figure 11-15).
F5 = factor to correct measured pan evaporation rates to evaporation rates from compost surface, $0 \leqslant F5 \leqslant 1.00$

In assessing the significance of surface evaporation, the most reliable information is probably pan evaporation data collected in the area of question. Such data are usually available from local agencies or the National Climatic Center. The term EVAP(I) can be assumed to equal measured pan evaporation rates. The question which then arises is whether evaporation rates from a water surface (pan rates) are equal to removal rates from a compost surface. To answer this we must discuss some physical factors involved in drying.

In studying rates of drying it is useful to consider the drying rate, in mass of water removed per unit area and unit time, as a function of the moisture content. It is assumed that the material is exposed to constant conditions of air temperature, humidity, and velocity and direction of airflow. Note that only conditions in the airstream are constant because moisture content and other factors in the solid are changing. Such a drying rate curve for paper-pulp material is shown in Figure 11-12 and is typical of the general shape of curves obtained for a variety of porous, hydroscopic materials. Each curve has a horizontal segment AB in which the rate of water removal is constant. During this period a continuous film of water is present at the surface of the solid, and water removal occurs as if the solid were not present.

As moisture content decreases, a point B is reached where further drying occurs at a reduced rate. This is termed the critical point and in porous solids occurs when the rate of water flow to the surface of the solid by capillary forces drops below the rate of evaporation from a free water surface. Beyond this point a continuous film of water can no longer be maintained at the surface of the solid and the drying rate falls (falling rate period). Critical moisture content varies with the type of material, its thickness and the rate of drying. It is best determined experimentally and can vary over a wide range for different types of solids. Furthermore, the shape of the curve in the falling rate period is also variable as indicated in Figure 11-12.

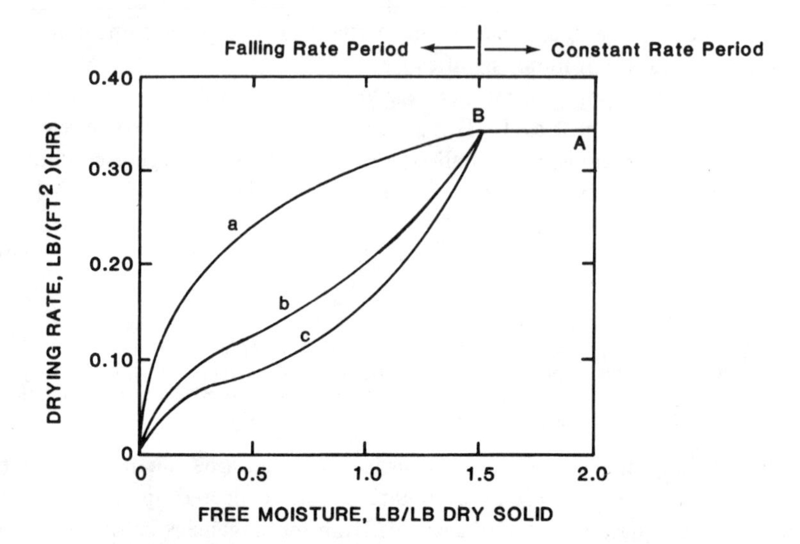

Figure 11-12. Drying-rate curves for paper-pulp slabs: (a) thickness = 0.108 cm, (b) thickness = 0.648 cm, (c) thickness = 2.37 cm [148].

Note in Figure 11-12 that moisture content is expressed in terms of weight of water per weight of *dry* solids. This is the usual convention adopted for analysis of drying processes. In this book moisture content has always been understood to mean the weight of water per weight of *wet* solids unless otherwise noted as in this case.

Livingston et al. [112] measured the effect of solids concentration on the surface evaporation rate during windrow composting of digested sludge blended with recycled product. Special evaporation pans, 0.6 m square and 7.6 cm deep, were filled with windrow material, weighed and placed flush to the windrow surface. The pans were reweighed the next day and the moisture loss determined from the weight difference. Unfortunately, corresponding pan evaporation rates from a water surface were not measured. It can only be stated that the measured rates from the compost surface were comparable to water evaporation rates recorded at nearby weather stations at the same time of year. Livingston's data were normalized against rates recorded at solids contents of about 45%, the minimum solids content that was examined. The data are shown in Figure 11-13 and clearly indicate a trend of decreasing surface evaporation rate with increasing solids content.

In a somewhat similar system, the rate of moisture evaporation from sand drying beds is known to decrease with increasing cake dryness. Furthermore, the rate is nearly equivalent to the pan evaporation rate when the cake

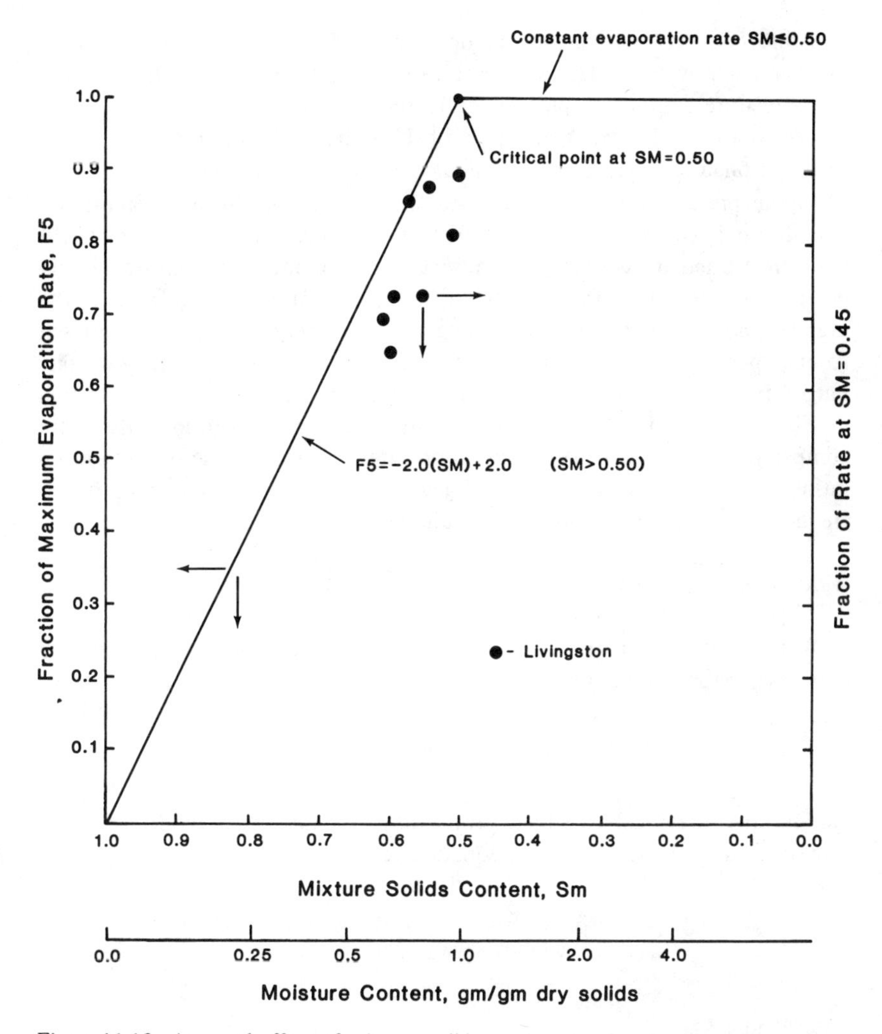

Figure 11-13. Assumed effect of mixture solids on the rate of surface evaporation from a compost material. Experimental data were obtained from a windrow system using digested sludge blended with recycled compost.

is wet (i.e., a continuous film of water at the surface). Another useful datum is the fact that solids contents as high as 90% have been achieved during ambient air drying of compost. Apparently evaporation continues even at very low moisture levels although probably at greatly reduced rates.

Faced with limited experimental data, it was assumed that drying during the constant rate period would occur at the pan evaporation rate. The critical moisture content was assumed to occur at 50% solids (i.e., 100% moisture based on dry weight). The drying rate decreased linearly from this point to zero moisture content. A plot of the factor F5 as used in Equation 11-40 is shown in Figure 11-13. Again, it is obvious that although the assumptions seem reasonable, better experimental data are certainly needed.

In the present model it is assumed that the rate of surface evaporation is constant between pile turnings. In fact, the rate may not be constant if piles are turned infrequently, because the rate of moisture removal should decrease as the outer surface dries. Livingston [112] noted a decrease in the average evaporation rate if pile turning was once every three days as opposed to once per day. Average evaporation rates with once every three-day turning ranged from about 55 to 80% of average rates with daily turning. Data shown in Figure 11-13 simulated a daily pile turning and evaporation rates were comparable to recorded pan evaporation rates. Therefore, use of pan evaporation data is probably justified if daily turning is used. If turning is less frequent, a lower evaporation rate should be assumed.

Water with Output Solids

Water contained in the output solids can be determined by applying the basic mass balance equation:

$$\text{rate of storage} = \text{input} - \text{output} + \text{sources} - \text{sinks}$$

Adding the previously defined terms under steady state conditions (i.e., rate of storage = 0) the following equation results:

$$0 = (\text{WATC} + \text{WATA} + \text{WATB} + \text{WATR} + \text{WATVI} + \text{WATRN}) - (\text{WATSO} + \text{WATVO} + \text{WATEVP}) + \text{WATP}$$

Rearranging:

$$\text{WATSO} = \text{WATC} + \text{WATA} + \text{WATB} + \text{WATR} + \text{WATVI} + \text{WATRN} + \text{WATP} - \text{WATVO} - \text{WATEVP} \quad (11\text{-}41)$$

where WATSO = weight of water in the output solids, kg/day in the CFCM model and kg in the BATCH model. Recall again that in the CFCM model WATSO represents moisture in the outfeed material from the reactor. In the BATCH model, however, WATSO is the water output from the particular time step under consideration. WATSO then becomes input to the next time step. After the first time step, water inputs can no longer be attributed to individual components in the mixture. In this case Equation 11-41 is no longer applicable and the water balance becomes:

$$\mathrm{WATSO = WATSI + WATVI + WATRN + WATP - WATVO - WATEVP} \qquad (11\text{-}42)$$

where WATSI = water contained in solids input to time step I (kg), equal to output water from time step I-1.

Volume Relationships

In the case of amendments and/or compost recycle addition, the volume of mixed material is determined from the mixture wet weight and the unit bulk weight as determined by Equations 7-3 or 11-7. Thus for the BATCH model:

$$\mathrm{VOLUME = (XC + XA + XR)/(GAMMAM \times 1000.0)} \qquad (11\text{-}43)$$

where XC, XA, XR = wet weight of sludge cake, amendment and compost recycle components, respectively (kg in BATCH model), and VOLUME = mixture volume (m^3). The calculation is performed at the beginning and end of the compost cycle to determine the reduction in cross-sectional area during composting.

For the CFCM model, reactor volume is based on output solids characteristics. Recall that in a complete-mix reactor the outfeed characteristics are the same as for material within the reactor. Therefore, reactor volume is based on the outfeed characteristics, and is calculated as:

$$\mathrm{VLREAC = (DETTIM \times DWTOUT/SMOUT)/(GAMMAM \times 1000.0)} \qquad (11\text{-}44)$$

where VLREAC = required reactor volume, m^3
DETTIM = detention time in the CFCM reactor, days
DWTOUT = output dry weight, kg/day
SMOUT = output solids content
GAMMAM = unit bulk weight of output solids, calculated from Equation 7-3 or 11-7

If bulking agents such as wood chips are used in the BATCH model, it is assumed that procedures described in Chapter 7 (see Equations 7-15 and 7-18) are used to assure sufficient FAS in the mixture. It was further assumed that sludge would not add to the volume of bulking particles. Unit bulk weight of a wood chip bulking agent was assumed to follow the expression:

$$\mathrm{GAMMAB = 0.2886/SB} \qquad (7\text{-}20)$$

where SB = solids content of the bulking agent and GAMMAB = unit bulk weight of bulking agent (g/cm^3). The volume of mixture is then calculated as the volume of bulking agent component only:

$$\text{VOLUME} = \text{XB}/(\text{GAMMAB} \times 1000.0) \quad (11\text{-}45)$$

This volume is assumed to remain constant throughout the compost cycle.

ENERGY BALANCES

Once the mass balance is completed, the corresponding energy balance can then be determined. Because many components, including solids and gases, enter the composting system, there are likely to be differences in feed temperatures. Therefore, a reference temperature, TREF, was established against which all energy inputs and outputs are measured. This in no way affects the final energy balance, but merely provides a convenient datum against which all sensible heat values are measured. In most calculations the reference point was established at 0°C.

Energy Inputs

Solids, Water and Gases

Sensible heat of solids, water and gas components is determined from the basic heat equation:

$$q_p = mc_p \, \Delta T = \Delta H \quad (3\text{-}12)$$

or

$$q_p = mc_p \, (T - \text{TREF}) \quad (11\text{-}46)$$

Recall that the heat flow in isobaric (constant pressure) processes is equal to the enthalpy change. The value of specific heat at constant pressure c_p varies depending on the component. The following values were assumed for analysis; water, 1.00 cal/g-°C; solids, 0.25; dry gases, 0.24; and water vapor, 0.444. Values for water, dry gas (air) and water vapor are available in handbooks and the values chosen are applicable over the temperature range common to composting. Specific heat for all dry gases was assumed to be equal to that for air. This is reasonable given the large N_2 component which acts as a conservative substance throughout the process. Specific heat for solids is based on the work of Mears et al. [117], but was increased slightly to be conservative (see Chapter 9).

Using Equation 11-46, the following equations for the sludge cake component can be developed:

$$HCSI = (S_cX_c)\ CPSOL\ (TC - TREF) \tag{11-47}$$

$$HCWI = (X_c - S_cX_c)\ CPWAT\ (TC - TREF) \tag{11-48}$$

where HCSI = sensible heat content of infeed sludge cake solids, kcal/day in CFCM model and kcal in BATCH model
HCWI = sensible heat content of infeed sludge cake water, kcal/day in CFCM model and kcal in BATCH model
CPSOL = specific heat of solids, cal/g-°C
CPWAT = specific heat of water, cal/g-°C
TC = input sludge temperature, °C

Development of similar expressions for amendment, recycle and bulking agent components is left to the reader. Except where otherwise noted, energy terms have units of kcal/day in the CFCM model and kcal in the BATCH model.

In the BATCH model, individual water and solids components are not considered after the first time step. In this case the sensible heat of solids and water is determined as:

$$HTSI = DWTIN \times CPSOL \times (TIN - TREF) \tag{11-49}$$

$$HTWI = WATSI \times CPWAT \times (TIN - TREF) \tag{11-50}$$

where DWTIN = input dry weight to time step I, kg, equal to output dry weight from time step I-1
HTSI = total sensible heat of input solids, kcal
HTWI = total sensible heat of input water, kcal
TIN = temperature of infeed components to time step I, °C, equal to temperature of outfeed components from time step I-1

Sensible heat inputs for dry air and water vapor are determined as:

$$HAIRI = DAIRI \times CPGAS \times (TAIR - TREF) \tag{11-51}$$

$$HSWVI = WATVI \times CPWATV \times (TAIR - TREF) \tag{11-52}$$

where HAIRI = sensible heat content of infeed dry air
HSWVI = sensible heat content of input water vapor
CPGAS = specific heat of dry gases, cal/g-°C
CPWATV = specific heat of water vapor, cal/g-°C
TAIR = temperature of input air, °C

For the BATCH model, if rainfall occurs during the time step being considered the associated heat input is calculated as:

$$HRAIN = (WATRN)CPWAT(RAINTP(I) - TREF) \tag{11-53}$$

where HRAIN = sensible heat input from rainfall or snowfall
CPWAT = specific heat of water, cal/g-°C
RAINTP = average rainfall temperature in time step I, °C

Organic Decomposition

Biological heat production is estimated from the change in BVS and heat of combustion for each component as follows,

$$\text{sludge} \quad HORGC = DBVSC(HC) \tag{11-54}$$

$$\text{amendment} \quad HORGA = DBVSA(HA) \tag{11-55}$$

$$\text{bulking agent} \quad HORGB = DBVSB(HB) \tag{11-56}$$

$$\text{recycle} \quad HORGR = DBVSR(HR) \tag{11-57}$$

HC, HA, HB and HR are heats of combustion (kcal/kg) for sludge cake, amendment, bulking agent and recycle, respectively. HC, HA and HB are input values to the model and are estimated assuming 3.4 kcal/g chemical oxygen demand (COD) (see Table 3-2). Assuming the stoichiometry of Equation 8-1, for example, the heat of combustion for sludge organics would be:

$$\frac{3.4 \text{ kcal}}{\text{g COD}} \times \frac{12.5(32)}{201} \times 1000 = 6766 \text{ kcal/kg}$$

Because recycle material is composed of a mixture of components, HR is estimated as the average of HC and HA weighted on the basis of outfeed quantities of cake and amendment biodegradable organics. Recall that bulking agent and recycle are not used in the same process.

Oxidation of nonorganic substrates, such as nitrification of ammonium to nitrate, was not considered in the energy balance.

Mixing Energy

Mechanical agitation of the composting material and/or compression of infeed gases results in the input of energy to the reactor or pile. This heat input is estimated from the power of the agitation and/or aeration equipment, the kcal equivalent of the power (859.2 kcal/kWh), and the daily hours of operation as follows:

$$\text{HMIX} = \text{MIXPW}\,(859.2)\,\text{HRSDAY}\ (\times\ \text{DELTAT in BATCH model}) \qquad (11\text{-}58)$$

where HMIX = agitation and/or aeration energy input
MIXPW = agitator and/or aerator power, kW
HRSDAY = hours per day in which the agitator and/or aeration equipment is operated

Note that if induced draft fans are used for aeration (i.e., suction), energy input will be into the exit gas stream and therefore should not be considered in the energy balance.

Total Input Energy

Total energy input is the summation of all input terms, or:

$$\begin{aligned}\text{HTOTI} = {} & (\text{HCSI} + \text{HASI} + \text{HBSI} + \text{HRSI}) + (\text{HCWI} + \text{HAWI} + \text{HBWI} \\ & + \text{HRWI} + \text{HRAIN}) + (\text{HAIRI} + \text{HSWVI}) + (\text{HORGC} + \text{HORGA} \\ & + \text{HORGB} + \text{HORGR}) + \text{HMIX} \qquad (\text{CFCM model}) \qquad (11\text{-}59)\end{aligned}$$

where HTOTI = total input energy. In the BATCH model, total energy input to the time step is determined in a similar manner:

$$\begin{aligned}\text{HTOTI} = {} & \text{HTSI} + \text{HTWI} + \text{HRAIN} + \text{HAIRI} + \text{HSWVI} + \text{HORGC} + \text{HORGA} \\ & + \text{HORGB} + \text{HORGR} + \text{HMIX} \qquad (\text{BATCH model}) \qquad (11\text{-}60)\end{aligned}$$

Energy Outputs

Solids, Water and Gases

Output energy terms for solids and water are computed from Equation 11-46 using the DWTOUT and WATSO components as determined from the mass balance:

$$\text{HSO} = \text{DWTOUT} \times \text{CPSOL} \times (\text{TNEW} - \text{TREF}) \qquad (11\text{-}61)$$

$$\text{HWO} = \text{WATSO} \times \text{CPWAT} \times (\text{TNEW} - \text{TREF}) \qquad (11\text{-}62)$$

where HSO = sensible heat content of the output solids
HWO = sensible heat content of the output water
TNEW = temperature of outfeed solids, water and gases, °C

The value of TNEW corresponds to the output temperature as determined using the algorithm of Figure 11-4. Sensible heat in the output gases is determined as:

$$\text{HGO} = \text{DGASO} \times \text{CPGAS} \times (\text{TNEW} - \text{TREF}) \tag{11-63}$$

$$\text{HSWVO} = \text{WATVI} \times \text{CPWATV} \times (\text{TNEW} - \text{TREF}) + (\text{WATVO} - \text{WATVI}) \times \text{CPWAT} \times (\text{TNEW} - \text{TREF}) \tag{11-64}$$

where HGO = sensible heat content of output gases and HSWVO = sensible heat in the output water vapor. All components, including solids, water, gases and vapor, are assumed to exit the process at the same temperature.

The latent heat of water vaporization is determined from the water vapor content of the inlet and outlet gases as follows:

$$\text{HLWVO} = (\text{WATVO} - \text{WATVI})(\Delta H_{fg}) \tag{11-65}$$

where HLWVO = latent heat of output water vapor minus latent heat of the input water vapor and ΔH_{fg} = enthalpy change from liquid to vapor at temperature TNEW, kcal/kg. ΔH_{fg} is a function of temperature and over the temperature range from 0 to 100°C varies by about 10%. The heat of vaporization is such a large energy output that a 10% variation is larger than many of the other energy terms. Therefore, the effect of temperature on ΔH_{fg} must be considered. Based on data available in steam tables the following equation can be developed:

$$\Delta H_{fg} = 597.0 - 0.581\,(\text{TNEW}) = \text{DELHFG} \tag{11-66}$$

The rationale behind Equation 11-64 may not be readily apparent to the reader. Recall that enthalpy is a property of a system and is not a function of the path followed in moving between two equilibrium states, i.e., inlet to outlet conditions. In these calculations it is assumed that entering water vapor is heated to the exit temperature TNEW which accounts for the first half of Equation 11-64 and the use of CPWATV. It is further assumed that entering water is first heated to TNEW and then a portion is vaporized. Thus, CPWAT is used in the second half of Equation 11-64 instead of CPWATV even though the water exits as a vapor. Given these previous assumptions it is then necessary that the heat of vaporization in Equation 11-65 be evaluated at temperature TNEW, the assumed temperature of vaporization. Other paths could be assumed, but the change in enthalpy would remain the same regardless of path (refer to Example 3-1). This is fortunate because it is unlikely that all entering water follows the same thermodynamic path in moving from inlet to outlet conditions. The path chosen here is simply convenient and easily visualized.

Water removed by surface evaporation in the BATCH model is assumed to occur independently of processes in the pile interior. Latent heats are assumed to be supplied by the ambient air. Thus, water is removed without placing additional demands on energy resources in the system. Certainly after the pile is turned, surface evaporation will reduce the temperature of the surface material. However, convective and radiative mechanisms will also result in heat losses and the effects cannot be distinguished. Therefore, heat losses from surface material are accounted for as heat lost to the surroundings (see next section). In the case of surface evaporation only the sensible heat of the departing water is considered as follows:

$$\text{HSWEV} = \text{WATEVP} \times \text{CPWAT} \times (\text{TAIR(I)} - \text{TREF}) \tag{11-67}$$

where HSWEV = sensible heat in the surface evaporated water, kcal (BATCH model only). This approach is consistent since energy supplied by ambient air in evaporating the water is also not considered.

Losses to Surroundings

The magnitude of heat loss to the surroundings is difficult to estimate accurately because of the lack of information on actual systems. Recourse was made to engineering estimates which should provide at least "order of magnitude" answers. The problem is lessened considerably by the fact that conductive heat losses are already largely accounted for in the heat content of output dry gas and water vapor. Here only additional heat losses that are not already covered by previously defined energy terms are being considered. This includes conductive heat loss through the walls of a reactor, conductive and convective heat loss from the surface of a static pile or windrow, and heat losses during turning of a windrow.

For the case of heat loss from a surface not exposed to ambient air, it is assumed that heat transfer is limited by conduction (see Example 9-1). This applies to heat loss through concrete walls of a reactor or from the bottom of a windrow or pile in contact with earth, concrete or other material. Any convective transport is assumed to be fully accounted for in the output gases. Conductive heat losses are estimated by applying the conventional heat loss equation as follows:

$$\text{HCOND} = \text{U} \times \text{AREACL} \times (\text{TNEW} - \text{TAIR}) \times 24.0 \tag{11-68}$$

where HCOND = heat loss by conduction from a surface not exposed to ambient air
U = overall heat transfer coefficient, kcal/m^2-hr-°C
AREACL = surface area over which conductive heat transfer occurs, m^2

For the CFCM model, Equation 11-68 is used as shown, whereas in the BATCH model it is multiplied by DELTAT to give the heat loss in the incremental time step DELTAT. For heat loss from a reactor surface or from the bottom surface of a static pile or windrow the value of U is taken as 1.5 kcal/m^2-hr-°C, consistent with the general range of values discussed in Chapter 9.

A cylindrical reactor is assumed in the CFCM reactor as shown in Figure 11-14. AREACL is calculated as follows:

$$\text{AREACL} = \text{bottom area} + \text{sidewall area}$$

$$\text{AREACL} = (\text{VLREAC}/\text{HEIGHT}) + 2 \times (\pi \times \text{HEIGHT} \times \text{VLREAC})^{0.50} \quad (11\text{-}69)$$

where HEIGHT = height of compost material in the reactor, m. HEIGHT is the only required input value as all other dimensional characteristics are calculated from this value.

In the BATCH model AREACL is calculated as:

$$\text{AREACL} = \text{DIMCL} \times \text{LENGTH} \quad (11\text{-}70)$$

where DIMCL = characteristic dimension for conductive heat transfer, and LENGTH = length of windrow or pile, m.

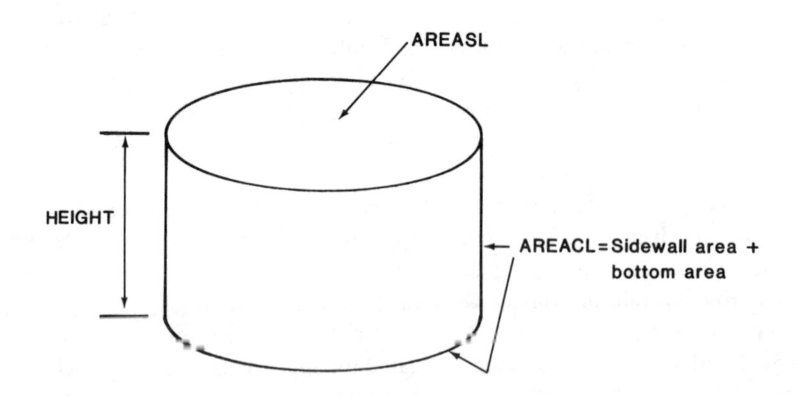

AREASL = VLREAC/HEIGHT

AREACL = VLREAC/HEIGHT + 2(π · HEIGHT · VLREAC)$^{1/2}$

Figure 11-14. CFCM model. Characteristic dimensions assumed for enclosed reactor composting.

Characteristic dimensions assumed in the BATCH model are described in Figure 11-15. The pile WIDTH, DIMCL, DIMSL and AREAXS are input to the model. The volume of the pile is then calculated as previously described and remaining dimensions determined as per Figure 11-15.

For cases where the compost is turned and exposed to ambient air, additional heat transfer from the compost mixture to the ambient air can be expected. Normally, air is exchanged in the void volume on the order of

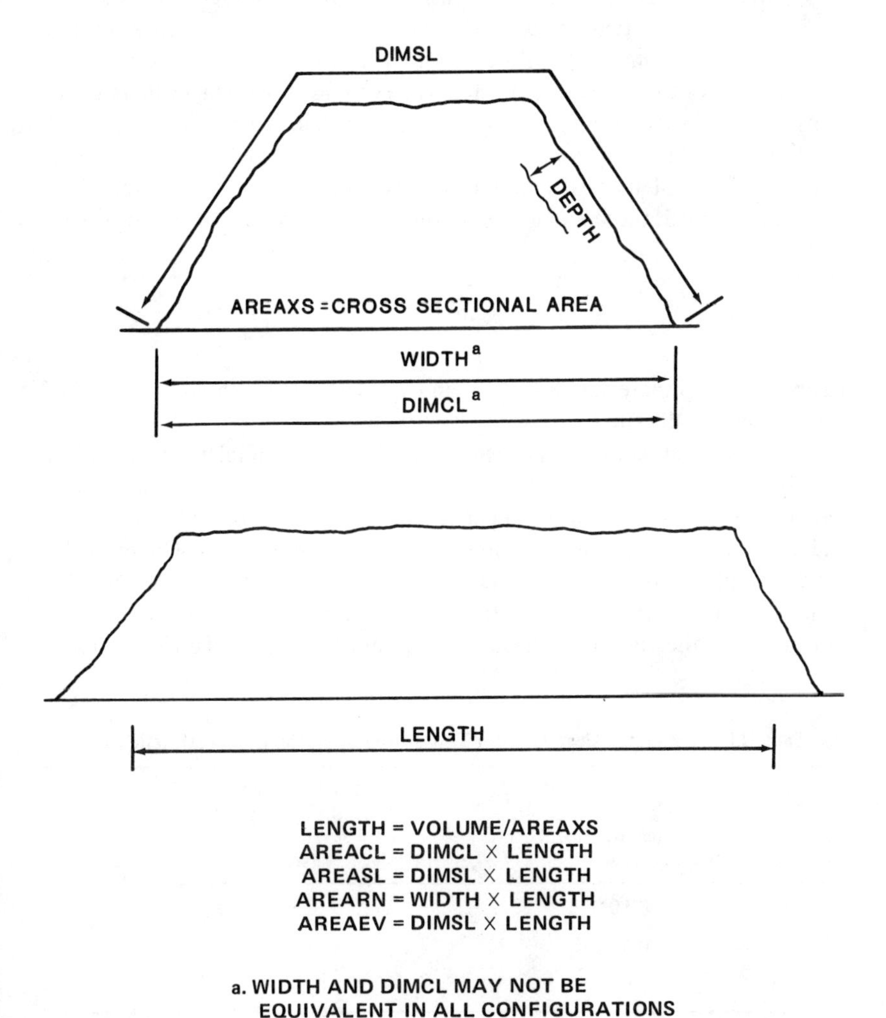

Figure 11-15. BATCH model. Characteristic dimensions of a compost windrow or pile.

10^2-10^3 times, so it is unlikely that the additional air exchange upon turning would be a significant loss. However, the solids can cool during the turning and this should be examined further.

When windrows are turned with modern equipment, considerable exposure to ambient air occurs for a brief period of time. The question arises as to whether the exposure time is sufficient to allow significant heat loss from the compost particles. This question can be approached by referring back to Figure 5-15, which shows dimensionless temperature profiles within a sphere of a radius R as a function of dimensionless time. Consider a particle with an initial uniform temperature of 60°C and an ambient temperature of 20°C. Let us estimate the time required for dimensionless temperature $T - T_0/T_1 - T_0$ to decrease to 0.5 at a value of r/R of 0.8. Under these conditions half the particle volume would be greater than temperature T and half would be less. In other words, the time is being estimated for the particle to lose about half of its heat content, relative to the ambient temperature T_0. From Figure 5-15, the value of dimensionless time under these conditions is:

$$\frac{kt}{\rho c_p R^2} = 0.03$$

Assuming appropriate values for k, ρ and c_p, the data presented in Table 11-2 can readily be calculated.

The time that compost particles are exposed to ambient conditions is usually on the order of a few seconds at most. Nevertheless, considerable temperature loss can occur if the particles are less than about 0.5 cm in radius and are uniformly exposed to lower ambient temperatures. On the other hand, complete exposure to ambient conditions is unlikely during any rapid turning of a windrow or pile. Most particles remain in close proximity to others during the agitation period and are somewhat insulated from ambient

Table 11-2. Estimated Heat Transfer Times from Spherical Compost Particles[a,b]

Particle Radius, R (cm)	Time to Reach $(T - T_0)/(T_1 - T_0) = 0.5$ @ r/R = 0.8 (sec)
0.10	0.135
0.50	3.375
1.00	13.5

[a] $\rho = 0.8$ g/cm³; $c_p = 0.625$ cal/g-°C; and k = 4 cal/hr-cm²-°C/cm.
[b] Calculations assume conductive heat transfer to be rate-controlling.

conditions. Furthermore, the above calculations assume conductive transport within the particle to be the rate-limiting step. In actual fact heat transport across laminar boundary layers in the gas phase surrounding the particle may be rate controlling. This would have the effect of increasing the cooling times shown in Table 11-2.

The potential for temperature loss as a result of turning is illustrated by Figure 11-16. In these studies a mixture of shredded vegetable trimmings and paper was composted in bins approximately 1 m square by 1.5 m high. Turning was accomplished by removing material from the bin and then replacing it, care being taken to work the original outer layer into the interior. Rather significant temperature decreases at the bin center are evident immediately after turning from, in part, heat losses to surroundings during turning. Some of the temperature drop results from cooled surface material being turned into the pile interior. A rapid temperature recovery is also evident following each turning. Temperature losses shown in Figure 11-16 may be somewhat greater than those expected using more conventional turning equipment which should provide less exposure to ambient conditions.

In the BATCH model, heat losses during turning are estimated as:

$$\text{HTURN} = \text{F3} \times \text{TURNUM} \times \text{DELTAT} \times [\text{DWTOUT} \times \text{CPSOL} \times (\text{TNEW} - \text{TAIR}) + \text{WATSO} \times \text{CPWAT} \times (\text{TNEW} - \text{TAIR})] \quad (11\text{-}71)$$

and in the CFCM model as:

$$\text{WEIGHT} = \text{VLREAC} \times \text{GAMMAM} \times 1000.0$$

$$\text{HTURN} = \text{F3} \times \text{TURNUM} \times [\text{WEIGHT} \times \text{SMOUT} \times \text{CPSOL} \times (\text{TNEW} - \text{TAIR}) + \text{WEIGHT} \times (1 - \text{SMOUT}) \times \text{CPWAT} \times (\text{TNEW} - \text{TAIR})] \quad (11\text{-}72)$$

where WEIGHT = wet weight of composting material, kg
F3 = factor to account for relative exposure of compost to ambient conditions during turning; usually taken as 0.0 for enclosed reactor or static pile composting
TURNUM = number of windrow or pile turnings per day

Immediately after turning of a windrow or pile, material near the surface will lose heat rapidly by both radiation and convection. Even though convective losses are largely accounted for by controlled, forced aeration, additional convective transport near the surface is probable, in response to ambient wind currents. In time these losses will diminish and further heat loss will be limited by conductive heat transport as described by Equation 11-68 (see Example 9-1). To estimate this heat loss it was assumed that a linear temperature gradient would exist from the surface to a depth, DEPTH, of about 0.20 m below the surface. The value of 0.20 m is reasonably

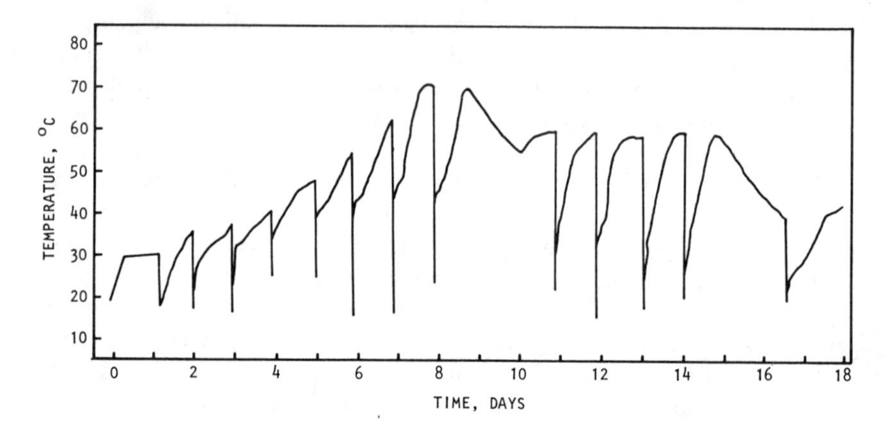

Figure 11-16. Temperature loss and recovery during turning of material in a bin-type composter. Temperatures were recorded at the bin center [149].

consistent with field observations. Surface losses in the BATCH model are then estimated as:

$$\text{HSURF} = \text{F4}\left(\frac{\text{AREASL} \times \text{DEPTH}}{\text{VOLUME}}\right)(\text{TURNUM})(0.5)\,[\text{DWTOUT} \times \text{CPSOL} \times (\text{TNEW} - \text{TAIR}) + \text{WATSO} \times \text{CPWAT}\,(\text{TNEW} - \text{TAIR})]\,(\text{DELTAT}) \tag{11-73}$$

where AREASL = area of windrow or pile subject to surface losses, m^2
DEPTH = depth from pile surface influenced by ambient convective and radiative heat losses, assumed to be about 0.20 m
F4 = factor to account for relative exposure of surface to ambient conditions after turning, assumed to be 1.0 for windrow system and 0.0 for enclosed reactor and static pile systems
VOLUME = volume of windrow or pile, m^3

In the CFCM model, surface loss is taken as:

$$\text{SURFWT} = \text{WEIGHT} \times \text{DEPTH/HEIGHT} \tag{11-74}$$

$$\text{HSURF} = \text{F4} \times \text{TURNUM} \times 0.50 \times \text{SURFWT} \times (\text{TNEW} - \text{TAIR}) \times (\text{SMOUT} \times \text{CPSOL} + (1 - \text{SMOUT}) \times \text{CPWAT}) \tag{11-75}$$

where SURFWT = wet weight of material in reactor subject to surface heat losses, kg.

Total Energy Output

Total energy output HTOTO is the sum of individual output terms as follows:

$$\text{HTOTO} = \text{HSO} + \text{HWO} + \text{HGO} + \text{HSWVO} + \text{HLWVO} + \text{HCOND} + \text{HTURN} + \text{HSURF} \tag{11-76}$$

As illustrated in Figure 11-3, energy output must balance energy input at the solution point. In other words:

$$\left|\frac{\text{HTOTI} - \text{HTOTO}}{\text{HTOTI}}\right| < \text{ASSUMED ERROR} \tag{11-77}$$

In most cases an error of $<0.1\%$ is adequate for computational purposes. If the error is greater than this, a new output temperature TNEW is assumed and the mass and energy balances recalculated. This procedure is repeated until Equation 11-77 is satisfied. At this point the problem is solved in the case of the CFCM model. In the BATCH model, the program advances to the next time step and the mass and energy balances recalculated until the last time step is reached.

SUMMARY

Mathematical models have been developed to study the dynamics of composting processes. These models allow integration of the thermodynamic and kinetic principles discussed previously in this text. The present chapter describes the approach, assumptions and equations used in the modeling.

Two models of the compost process are developed. One simulates a system where material is fed on a continuous basis and the contents are completely mixed (CFCM model). The second assumes a batch feeding of material at time = 0 and plug flow during the remainder of the cycle (BATCH model). The batch process is mathematically described as a large number of complete mix cells arranged in a time series.

In the case of the CFCM model the solution procedure is as follows:

1. assume an outfeed temperature;
2. determine the rate constant for organic decomposition and calculate the mass balance about the reactor;
3. sum all energy inputs and outputs for the assumed temperature;
4. compare energy inputs and outputs and estimate a new temperature based on the difference between the two; and
5. using the new temperature return to step 1 and repeat the procedure until the energy balance is closed.

For the BATCH model the above procedure is repeated for each time step in the compost cycle. Computer solutions are required because of the large number of calculations.

It is assumed that oxidation of BVS is first-order with respect to the quantity of remaining BVS. Effects of temperature, moisture content, FAS and oxygen content on the first-order rate constant are considered in the model. The computational procedure allows for more sophisticated mass and energy balances compared to those presented in previous chapters.

CHAPTER 12

PROCESS DYNAMICS II: SIMULATION OF THE CONTINUOUS FEED, COMPLETE-MIX REACTOR

INTRODUCTION

The previous chapter described an approach to mathematical analysis of the composting process. Mass and energy balances were related to known kinetics to produce a dynamic model of the process. Two separate models were developed. The first was termed a continuous feed, complete-mix model (CFCM) and can be applied to the case of a reactor composter where materials are well mixed and fed on a continuous or intermittent but frequent basis. The second is termed a BATCH model and is applicable to batch processing of materials in systems such as the windrow or aerated static pile. The purpose of Chapters 12 through 14 is to discuss the application of these models, their verification through comparison with field data, results of the analysis and implications for design and operation of actual compost systems.

It should be restated at this point that dynamic modeling of the composting process is a new area of investigation. Thus, the reader will be exploring new ideas, new tools for analysis and new results. It would be unfair to expect complete verification of model results or a level of development comparable to other aqueous biological systems, such as activated sludge or fixed film reactors. Such will come if sufficient time and effort are applied to the problem. These statements are not meant to detract from results of the analysis. Indeed, the reader will likely be surprised at the close correlation of model results with field data. Perhaps more valuable will be the fuller realization of the effects of design and operating variables on the process.

BASIC ASSUMPTIONS

The CFCM model simulates a system where material is fed on a continuous basis and the contents are completely mixed. Both conditions are seldom achieved in actual practice, particularly with solid substrates. Nearly continuous feeding or regular feeding at intermittent intervals is readily achievable in some reactor systems, however, and model results should be reasonably applicable to these conditions. Complete mixing is a difficult condition to attain even in aqueous systems let alone one with a solid substrate. However, a condition loosely described as "well mixed" should be achievable, provided feed materials are distributed evenly throughout the reactor, final product is withdrawn uniformly from the reactor volume and high levels of agitator mixing are maintained. Results of the CFCM model should be reasonably applicable to these conditions.

In all cases the cylindrical reactor shape described in Figure 11-14 was assumed with a compost height, HEIGHT, of 2.0 m. Dry weight of sludge cake was held constant at 10,000 kg/day (10 ton/day). Solids content of the mixed feed was adjusted by recycle of compost product. The latter was assumed to have a solids content SR of 0.70. Other conditions and values assumed in the CFCM model are listed in Table 12-1. These conditions should be assumed unless other values are noted on individual graphs and figures.

Typical output from the CFCM model is presented in Appendix IV. It was on the basis of numerous model studies of this type that the following results were obtained.

REACTOR DYNAMICS (VR = VROUT)

In the first analysis, volatile solids (VS) content of the product recycle VR was assumed to be equal to that of the reactor outfeed, VROUT. Such would be the case if reactor output is used in a reasonably short time or dried sufficiently to restrict further biological activity. If additional aging or curing is used following the process, the assumption that VR = VROUT would not be valid, a case to be examined in later sections.

Digested Sludge

Predicted reactor temperature, reaction rate constant, output solids content, output VS fraction, output oxygen consumption rate and other factors are presented as a function of air supply flowrate in Figures 12-1 and 12-2. Feed sludge was assumed to be digested with a cake solids SC of 0.30. Cake, recycle and air temperatures were assumed to be 20°C.

Table 12-1. Conditions and Variables Assumed in CFCM Model[a]

Feed Rates	
Sludge Feed Dry Weight, SC(XC)	10,000 kg/day
Product Recycle Dry Weight, SR(XR)	Adjusted as necessary to achieve desired mixture solids content.
Amendment Weight, XA	Amendment not used.
Bulking Agent Weight, XB	Bulking agent not used.
Feed Characteristics	
Product Recycle Solids Content, SR	0.70
Solids Content of Infeed Mixture, SM	0.45
Digested Sludge Volatile Solids Fraction, VC	0.50
Raw Sludge Volatile Solids Fraction, VC	0.70
Digested Sludge Degradability Coefficient, KC	0.50
Raw Sludge Degradability Coefficient, KC	0.786
Thermodynamic Constants	
Specific Heat of Solids, CPSOL	0.25 cal/g-°C
Specific Heat of Water, CPWAT	1.00 cal/g-°C
Specific Heat of Dry Gases, CPGAS	0.24 cal/g-°C
Specific Heat of Water Vapor, CPWATV	0.44 cal/g-°C
Specific Gravity of Volatiles, GV	1.00
Specific Gravity of Ash, GF	2.50
Heat of Combustion of Sludge Organics, HC	6766 cal/g
Heat of Combustion of Amendment Organics, HA	3739 cal/g
Heat of Combustion of Bulking Agent Organics, HB	3739 cal/g
Operational Characteristics	
Detention Time, DETTIM	7.0 days
Height of Compost in Reactor, HEIGHT	2.0 m
Heat Loss Turning Factor, F3	0.0
Heat Loss Surface Factor, F4	0.0
Pile Turnings per Day, TURNUM	10.0
Heat Transfer Coefficient, U	1.5 $kcal/m^2$-hr-°C
Airflow Relative Humidity, RHAIR	0.50
Reference Temperature, TREF	0.0°C
Agitator Power, MIXPW	75.0 kW
Agitator Operation, HRSDAY	6.0 hr/day

[a]Values in this table were held constant for most of the analysis presented in this chapter. Exceptions are noted on individual figures and graphs where applicable.

Reactor temperature is strongly influenced by the air supply rate, decreasing with increasing air supply. However, the reaction rate constant actually increases with specific air supply to a maximum value near 5-6 m^3/min-dry ton/day of feed sludge, and then begins to decrease. This means that reaction rates are temperature limited at lower air supply rates. Referring back to Figure 11-7, temperatures beyond about 70°C are assumed to result in decreased reaction rates. Obviously, the reactor operates at very high

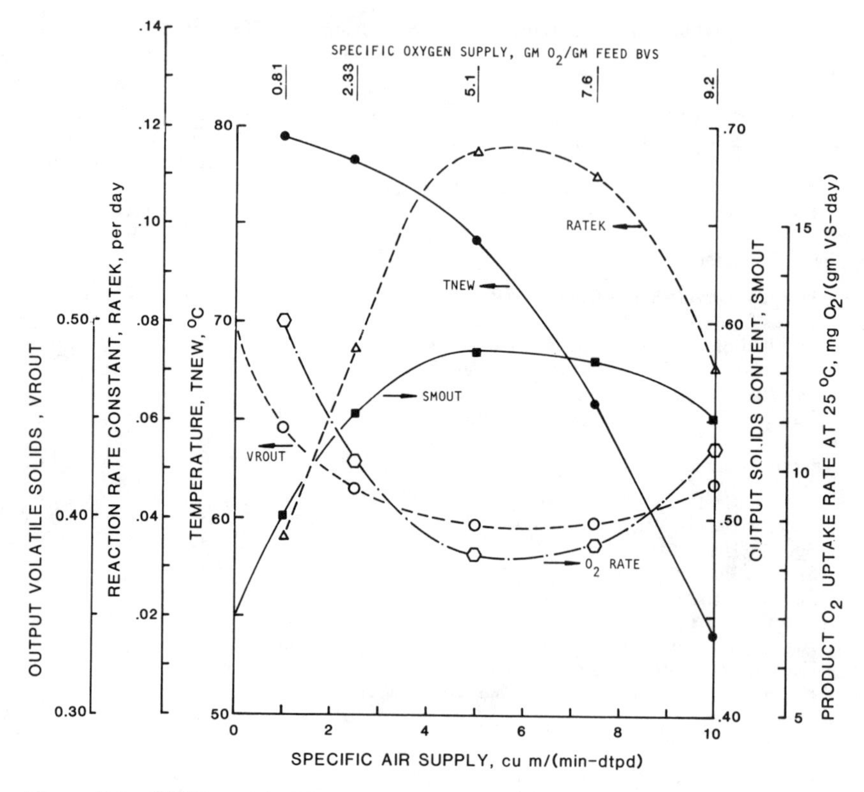

Figure 12-1. CFCM model. Effect of air supply flowrate on reactor temperature, reaction rate constant, output solids content, output volatile solids fraction and output O_2 consumption rate. Digested sludge, SC = 0.30, TC = TR = TAIR = 20°C, VR = VROUT.

temperatures to the point where significant thermal limitations are imposed at air supply rates less than about 3 m^3/min-dry ton/day of feed sludge. This is reflected in the fact that VS content and O_2 consumption rate of the outfeed actually decrease with increasing air supply up to about 5-7 m^3/min-dry ton/day of feed sludge. The lesson here is that a compost process actually can operate with too high a temperature. Reaction rates decrease and stability of the output product suffers as a result.

Output solids content SMOUT is predicted to increase in a near linear manner up to an air supply of 3-4 m^3/min-dry ton/day of feed sludge. A plateau is eventually reached with maximum output solids about 0.58. Output solids begin to decrease at high air supply in response to the decreased reactor temperature. Considering that infeed solids content was assumed to

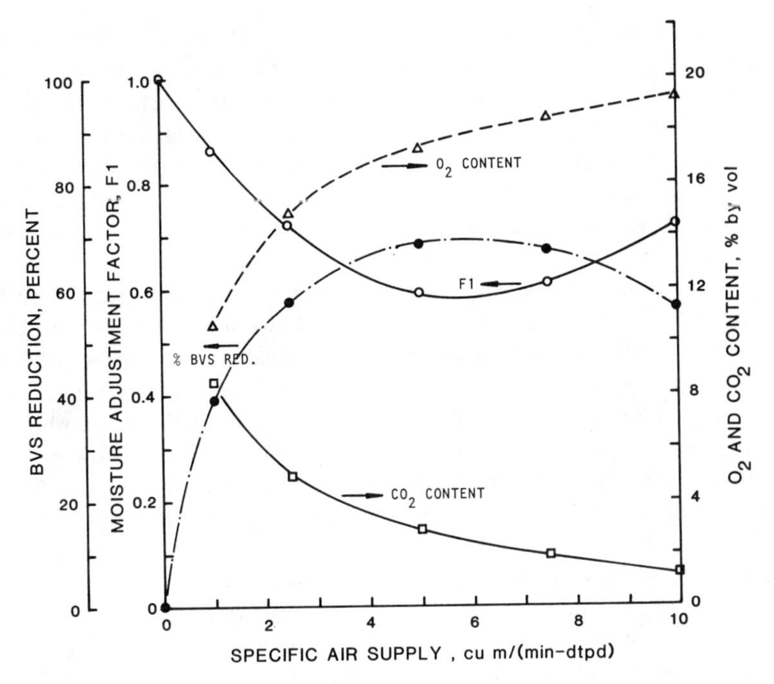

Figure 12-2. CFCM model. Effect of air supply flowrate on BVS reduction, moisture adjustment factor, and oxygen and carbon dioxide content. Digested sludge, SC = 0.30, TC = TR = TAIR = 20°C, VR = VROUT.

be 0.45, considerable drying can be achieved in the reactor, which is in keeping with earlier conclusions drawn in Chapter 9. The pattern of outfeed solids content answers one frequently asked question–should the air supply be increased to produce a drier product? Stated another way, if air supply is increased will the decrease in reactor temperature and hence saturation vapor pressure be offset by the increased mass flowrate of air? Apparently at low air rates, up to about 4 m^3/min-dry ton/day of infeed sludge the advantage is with increased air supply. Beyond about 8 m^3/min-dry ton/day the effects of decreased reactor temperature become dominant and outfeed solids content decreases.

Volatile solids content of the outfeed mixture VROUT decreases to a minimum of about 0.40, corresponding to the peak values for the rate constant. Biodegradable volatile solids (BVS) reduction also increases to a maximum at this point. Because infeed cake was assumed to be 50% VS, with a degradability coefficient of 0.50, the minimum possible VROUT would be 0.33. With the predicted minimum of about 0.40 as shown in Figure 12-1, the reduction in biodegradable volatile solids would be about 68%. Greater reductions in BVS can be achieved by increasing the detention time (a case to be examined shortly) or increasing the reaction rate constant. The latter could be effected through moisture addition to the reactor because the moisture adjustment factor F1 decreases to about 0.6 at the peak reaction rates. The advantage of adding moisture to increase the rate constant would have to be carefully considered in light of increased requirements for subsequent drying.

A mass balance for solids, water and gaseous components is shown in Figure 12-3. Conditions correspond to those shown in Figures 12-1 and 12-2 for an air supply of 5 m^3/min-dry ton/day. Similar balances were produced for each air flowrate examined in Figure 12-1, but the present example will adequately serve for purposes of illustration. The relatively large contribution of BVS in the product recycle should be noted. Recall that the present analysis assumes no further BVS loss after material leaves the reactor. Thus, the reactor outfeed, final product and recycle have the same VS percentage. BVS reduction across the entire system is about 68 percent (i.e., (2500-792)/2500), corresponding to the value shown in Figure 12-2.

The solids content of the product recycle was assumed to be 70%, and some additional drying would be required beyond that achieved in the compost reactor. Nevertheless, about 71% of the moisture removal is predicted to occur during the compost process, leaving about 29% for other means such as air or heat drying. It should be further noted in Figure 12-3 that the ASH and NBVS fractions act as conservative materials throughout the process, exiting in the same quantity as originally present in the feed sludge.

A bar graph showing energy contributions for two different air flowrates is shown in Figure 12-4. Energy inputs are dominated by organic decomposition of both the sludge cake HORGC and recycle components HORGR. All other energy inputs are relatively minor by comparison. The relatively high energy contribution from product recycle may be somewhat surprising in light of previous analysis in Chapter 9. This arises largely from the assumption that volatility of product recycle is equal to volatility of reactor outfeed. Even though significant BVS destruction occurs in the reactor, a high rate of recycle can return a considerable quantity of BVS to the system. For Case A in Figure 12-4 the recycle component accounts for about 49% of BVS input to the reactor. This decreases to about 35% for Case B because of increased BVS destruction in the reactor at the higher airflow.

As expected, energy output is dominated by the latent heat of water vaporization, which accounts for about 50 and 63% of total output energy for Cases A and B, respectively. Other output energy terms are of equivalent significance among themselves and change their contribution to the overall

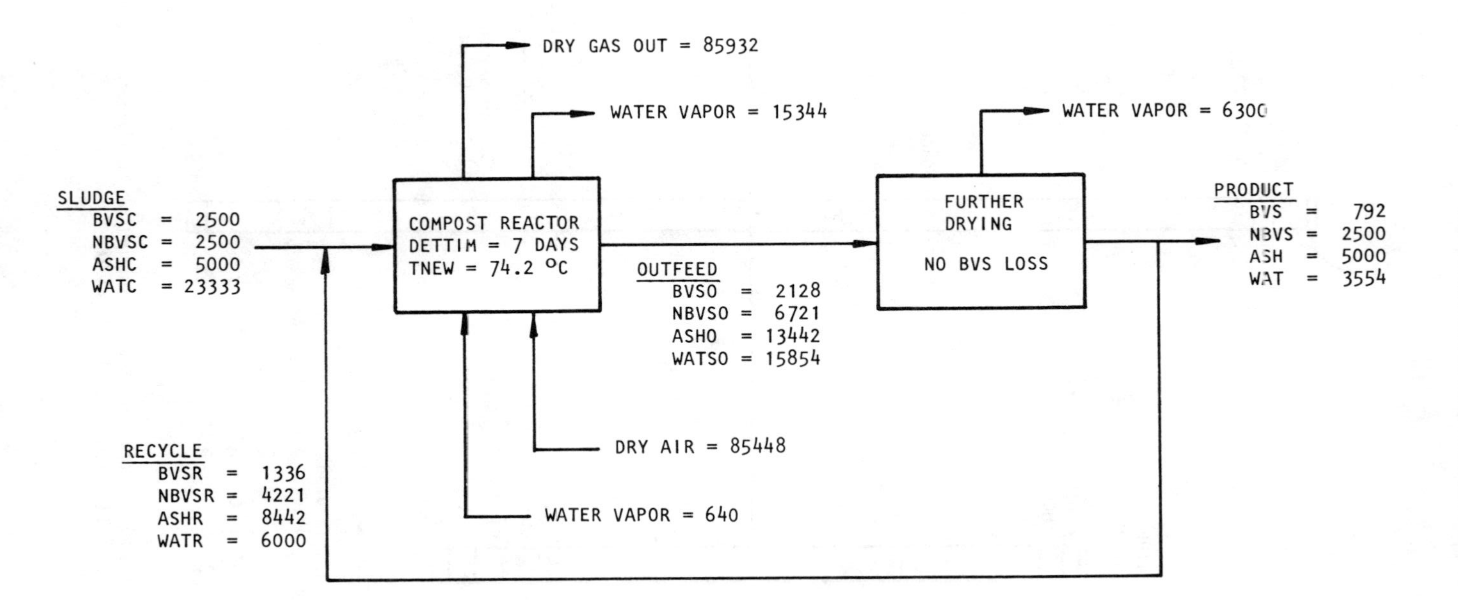

Figure 12-3. CFCM model. Mass balance for solids, water and gas components for a specific air supply of 5.0 m^3/min-dry ton/day and specific oxygen supply of 5.12 g O_2/g mixture BVS. Digested sludge, SC = 0.30, TC = TR = TAIR = 20°C, VR = VROUT. Units on mass terms are kg/day.

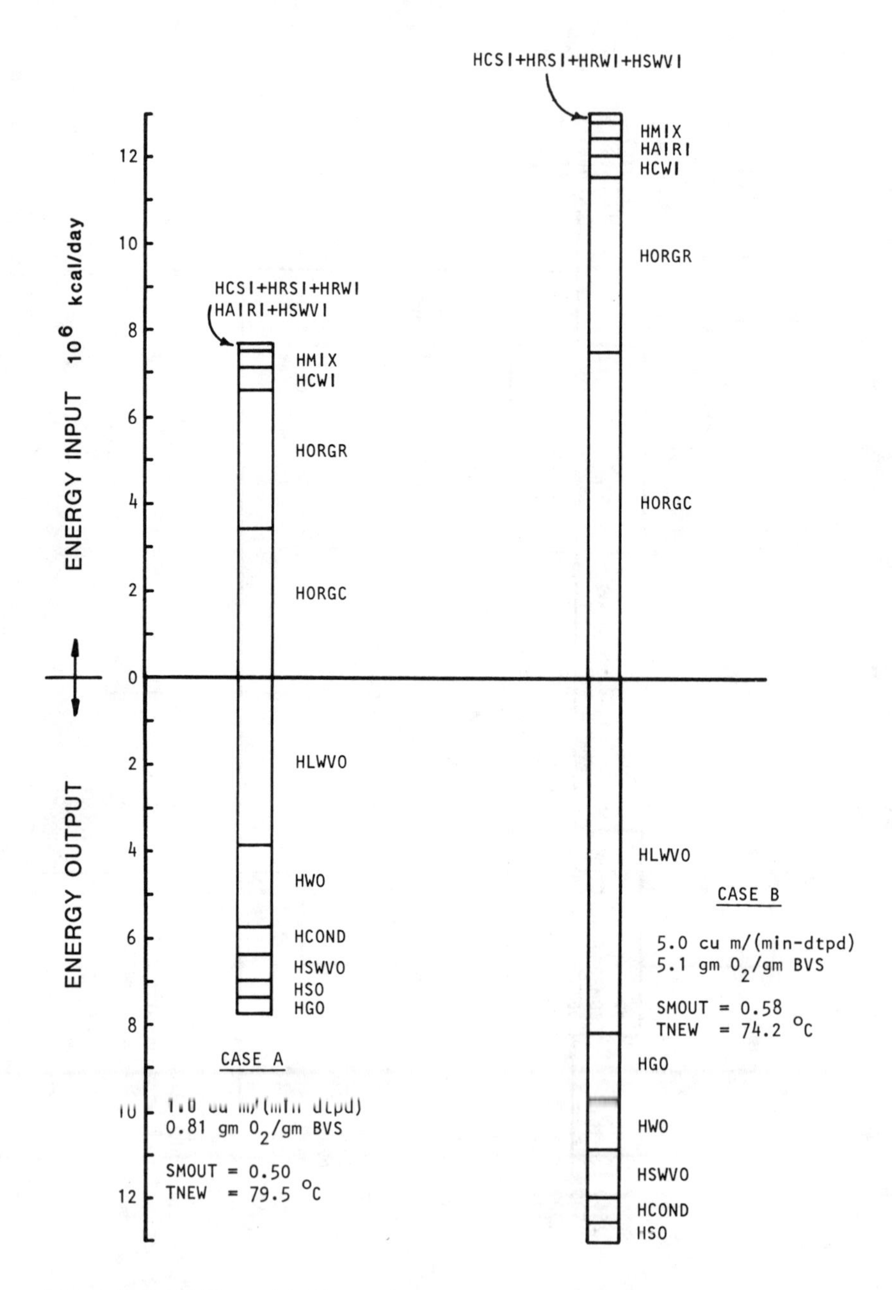

Figure 12-4. CFCM model. Energy balances for selected air supply rates. Digested sludge, SC = 0.30, TC = TR = TAIR = 20°C, VR = VROUT.

energy balance depending on operating conditions. For example, sensible heat in output gases HGO is the smallest output energy for Case A but increases to the second largest for Case B where air flowrate is greater. Thus, the latent heat of vaporization can be expected to remain a major output energy with others varying in importance depending on the particular operating conditions within the system.

The effect of specific oxygen supply, infeed cake solids and feed temperature on reactor temperature are shown in Figure 12-5. Specific oxygen supply η is defined as the g O_2 supplied/g BVS in the mixed feed. Suppose that maintenance of 60°C is defined as indicative of successful

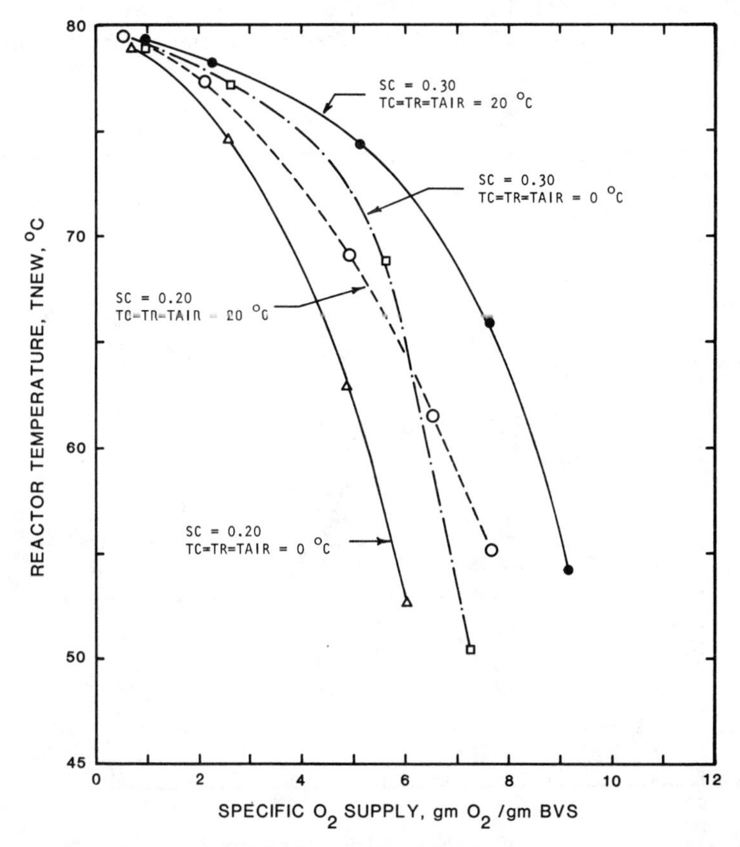

Figure 12-5. CFCM model. Effect of specific oxygen supply on reactor temperature for various conditions of infeed cake solids and infeed temperatures. Digested sludge, VR = VROUT.

composting. With this definition, successful composting is achieved with cake solids of 20-30%, with input temperatures from 0 to 20°C, i.e., TC = TR = TAIR = 0-20°C. Successful operation outside this range might also be possible but was not examined here. The reactor temperature decreases as air supply is increased for all conditions tested. Process failure would occur over a range of specific oxygen supply from about 5 to 8.5 g O_2/g feed BVS. For a system resembling the CFCM model, successful composting can be achieved over a wide range of input variables. However, system failure seems to occur rapidly if oxygen supply exceeds the values stated above.

Predicted output solids content from the reactor is shown in Figure 12-6 for the same operating conditions shown in Figure 12-5. Maximum output solids occur with infeed cake solids of 30%, decreasing markedly as cake solids are reduced to 20%. In either case, infeed temperature has only a

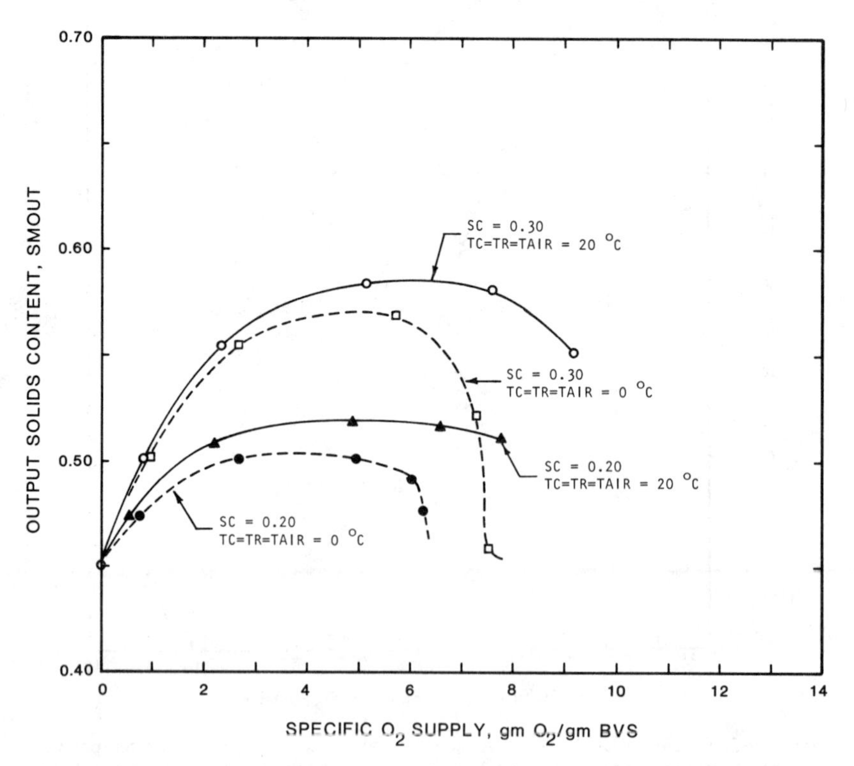

Figure 12-6. CFCM model. Effect of specific oxygen supply on output solids content for various conditions of infeed cake solids and infeed temperatures. Digested sludge, VR = VROUT.

nominal effect up to a specific oxygen supply of about 5 g O_2/g BVS. Maximum output solids SMOUT range from about 0.57 to 0.58 for 30% cake solids, decreasing to about 0.50-0.52 for 20% solids, Beyond about 6 g O_2/g BVS, process failure occurs rapidly when input temperature is 0°C, regardless of cake solids. With an infeed temperature of 20°C, process failure does not appear to be as rapid. Subsequent analysis will reinforce the observation that compost systems become more sensitive to high oxygen (air) supplies as infeed temperatures are reduced.

Raw Sludge

Results of a similar analysis for the case of raw sludge are presented in Figures 12-7 to 12-12. For this analysis, raw sludge was assumed to have a VS content of 70% (VC = 0.70) and a degradability coefficient of 0.786. The three decimal places afforded the degradability coefficient do not imply any high level of knowledge regarding the coefficient. Rather, it was reasoned that the nondegradable fraction should be reasonably similar regardless of whether raw or digested sludge is input to the reactor. For digested sludge a volatile fraction of 0.50 and degradability coefficient of 0.50 were assumed. Under these conditions the minimum volatile content, assuming complete decomposition of all degradable organics, would be 33%. To achieve this same minimum for raw sludge, the degradability coefficient must be adjusted to 0.786 for an infeed volatility of 0.70. Thus, the number of decimal places reflects a desire for internal consistency and implies nothing as to the state of actual knowledge.

Numerous operational parameters are shown in Figure 12-7 and 12-8 for the case of raw sludge feed at 30% cake solids and input temperature of 20°C, Reactor temperature is again influenced by the air supply rate. However, adequate temperatures are maintained with considerably higher air flowrates than was the case with digested sludge. This is, of course, caused by the greater input of degradable organics with raw sludge. Although higher air flowrates can be maintained with raw sludge, the specific O_2 supply at which temperatures drop below 60°C is remarkably similar for both cases. For digested sludge this occurred at about 8 g O_2/g BVS, whereas for raw sludge the corresponding value is between 9 and 10 g O_2/g BVS. This suggests that specific oxygen supply η can be a useful parameter for controlling composter operation regardless of the feed substrate.

Output solids content is predicted to increase with air flowrate to a maximum value near 67% solids. Significant moisture limitations occur at such high solids contents. The moisture adjustment factor F1 decreases to a minimum of about 0.25, corresponding to the peak output solids content.

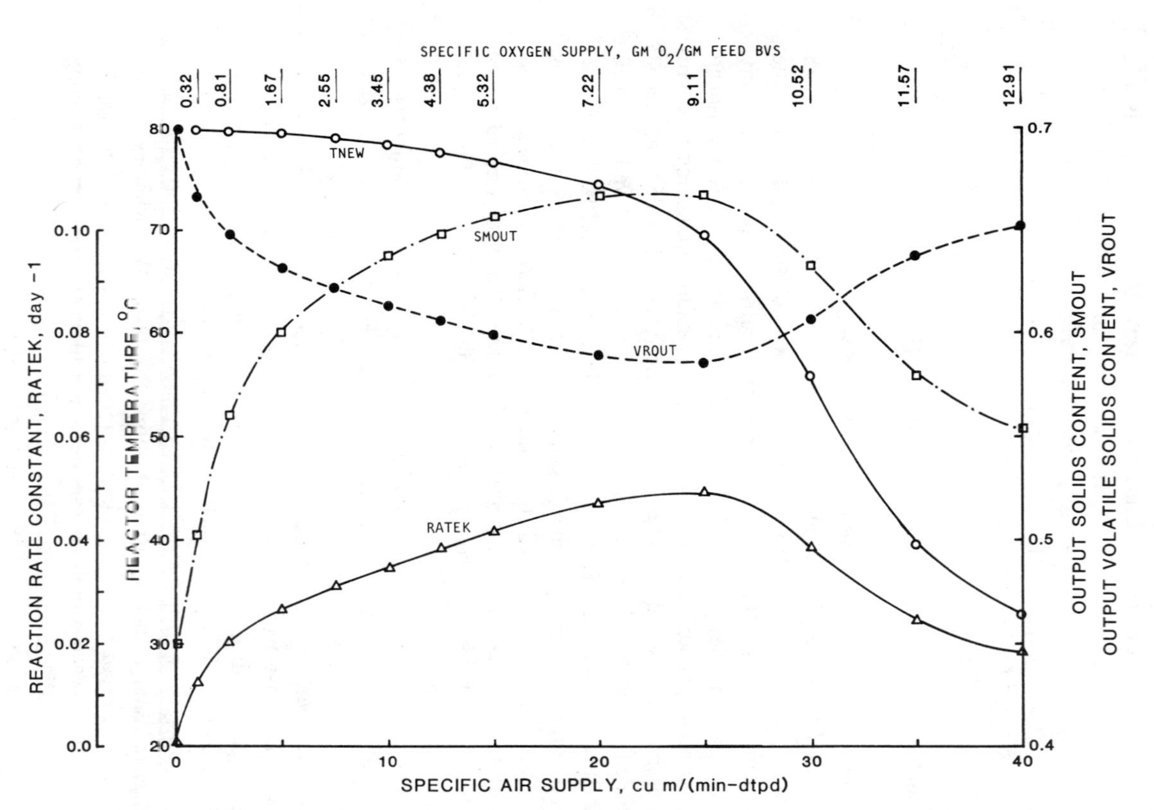

Figure 12-7. CFCM model. Effect of specific air supply on reactor temperature, reaction rate constant, output solids content and output VS fraction. Raw sludge, SC = 0.30, TC = TR = TAIR = 20°C, VR = VROUT.

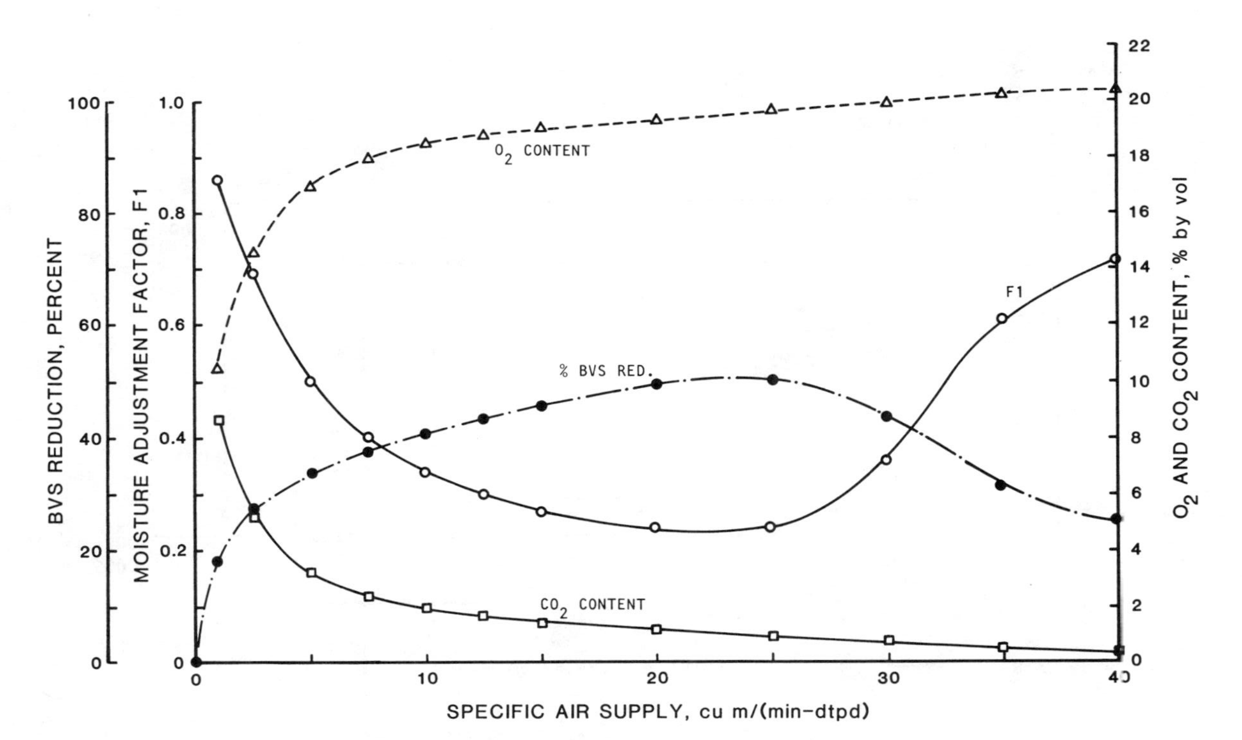

Figure 12-8. CFCM model. Effect of air supply on BVS reduction, moisture adjustment factor, and oxygen and carbon dioxide contents. Raw sludge, SC = 0.30, TC = TR = TAIR = 20°C, VR = VROUT.

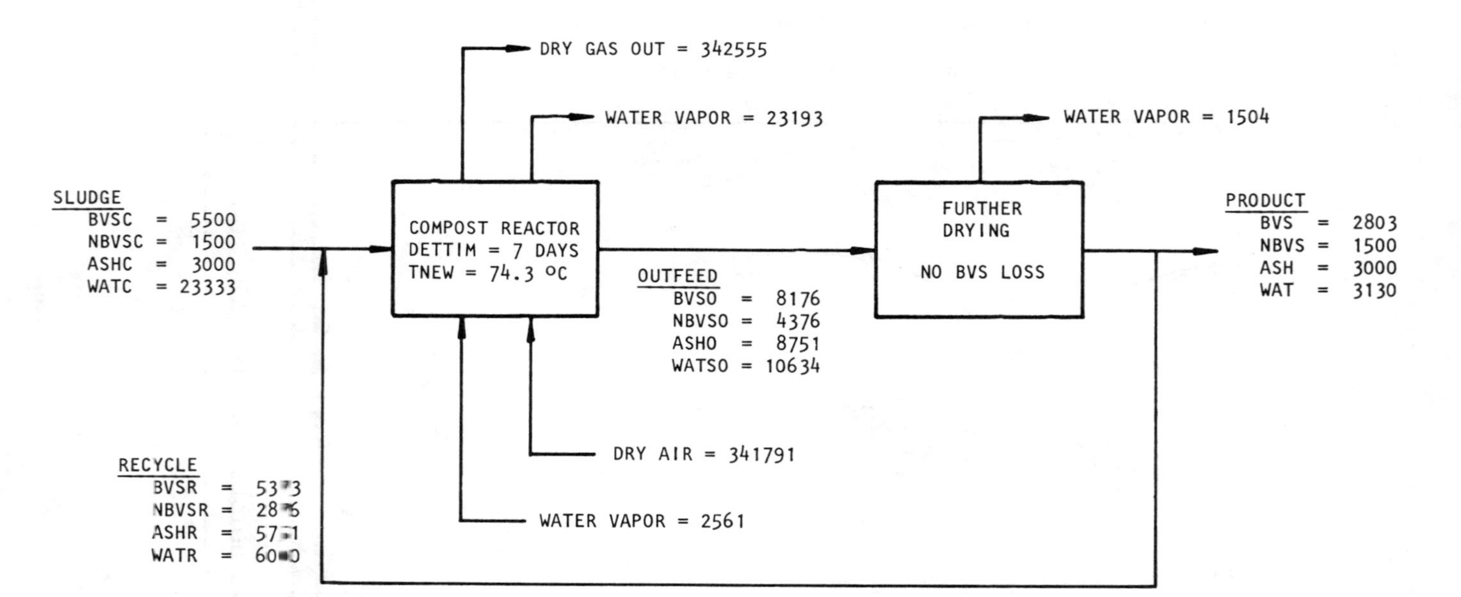

Figure 12-9. CFCM model. Mass balance for solids, water and gas components for a specific air supply of 20 m^3/min-dry ton/day and specific oxygen supply of 7.[illegible]2 g O_2/g mixture BVS. Raw sludge, SC = 0.30, TC = TR = TAIR = 20°C, VR = VROUT. Units on mass terms are kg/day.

Moisture limitations result in a peak reaction rate constant which is actually less than for the case of digested sludge (compare Figure 12-1 and 12-7). Because of the rate limitations, outfeed VS content remains relatively high, never decreasing below about 59%. BVS reduction remains below about 50% for all air flowrates examined.

The picture that emerges from information in Figures 12-7 and 12-8 is one of an oversupply of energy from the raw sludge organics. Energy availability is so large that high reactor temperatures can be maintained at high air flowrates. This leads to excessive drying and the low reaction rates and reduced BVS stabilization which follow. In effect, energy supplies are so great that the system is driven to severe moisture limitations even at specific O_2 supplies only slightly greater than stoichiometric. Further reducing the air supply limits the drying and alleviates the moisture limitation. However, reactor temperatures then reach levels where temperature limitations predominate and BVS reduction is not improved. In short, the potential energy supply is so large that the system is difficult to maintain under conditions to achieve optimum BVS reduction.

Actions can be taken to increase reaction rate constants and increase the stabilization of organics. Perhaps the most obvious approach is to add water to the reactor to increase moisture content and relieve some of the rate limitations. Although effective, input of additional water seems somewhat self-defeating considering that high moisture content is one of the major difficulties in composting wet organic substrates. A less obvious and more subtle approach is to decrease infeed mixture solids to the reactor. This has the benefit of reducing the requirements for product recycle. A more complete discussion of this approach is presented in subsequent sections of this chapter.

The mass balance for a specific air supply of 20 m^3/min-dry ton/day is shown in Figure 12-9. Because of previously discussed rate limitations, BVS reduction is less than for the case of digested sludge shown in Figure 12-3. As a result, BVS contributed by product recycle is nearly equivalent to that contributed by raw sludge feed. Total BVS reduction across the system is about 49% and final product VS content is about 59%. Another noteworthy difference is the greater moisture removal potential with raw sludge. In this case almost all required moisture removal occurs in the reactor, with very little dependence on subsequent drying. Of course, this occurs at the expense of reaction kinetics, which are greatly reduced at the low moisture content.

The mass balance shown in Figure 12-9 is comparable to that observed during raw sludge composting in a Fairfield-Hardy composter (see Figure 2-19). In these tests a heat dryer was used to dry the reactor outfeed. Thus, VR should have equalled VROUT, which is a condition of the present analysis. In both cases a large measure of the required moisture removal is

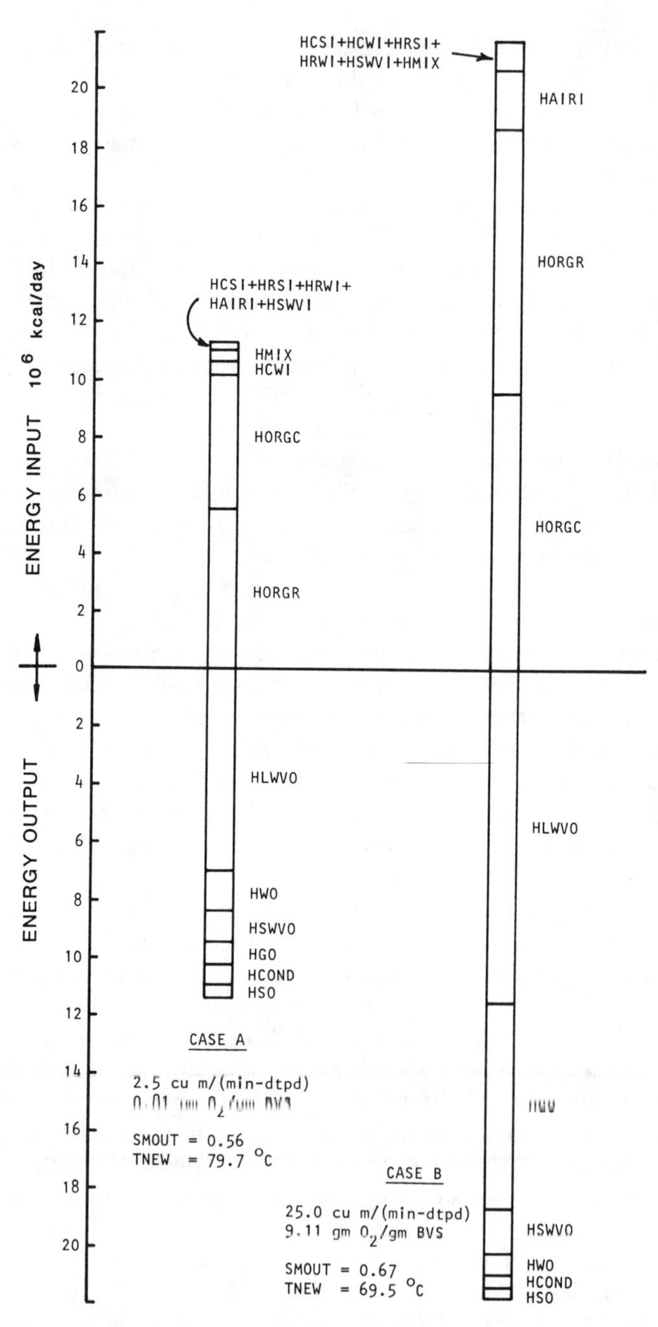

Figure 12-10. CFCM model. Energy balances for selected air supply rates. Raw sludge, SC = 0.30, TC – TR – TAIR = 20°C, VR = VROUT.

achieved in the composter. This relieves the burden on downstream drying processes and greatly reduces energy requirements if heat drying is used.

A bar graph showing energy contributions for two different air flowrates is shown in Figure 12-10. The general pattern of energy contributions is similar to that previously discussed for digested sludge. However, there are a few differences. For Case A energy input from recycled organics is actually greater than from the input sludge cake. This is caused by the higher input of degradable organics with raw sludge, limited BVS reduction achieved in the reactor and the high product recycle rate required to maintain a mixed infeed solids of 45%. The situation is reversed for Case B where air

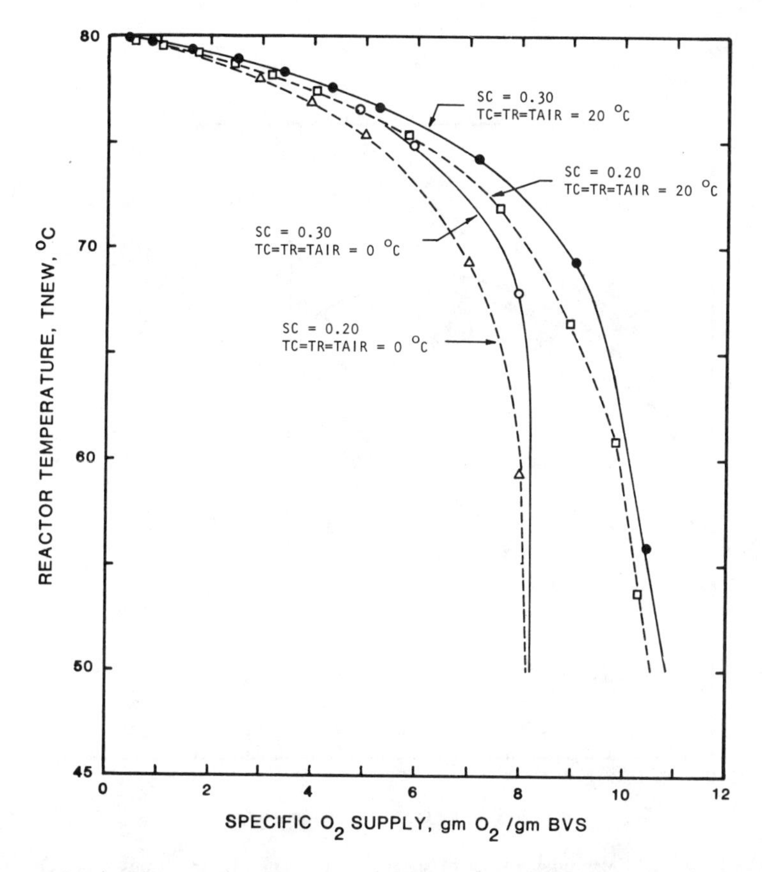

Figure 12-11. CFCM model. Effect of specific oxygen supply on reactor temperature for various conditions of infeed cake solids and infeed temperatures. Raw sludge, VR = VROUT.

flowrate is increased along with the reaction rate constant. Energy contribution from recycled organics remains significant, however. Energy output is again dominated by the latent heat of water evaporation HLWVO. However, for Case B with its high air flowrate, sensible heat removed by the dry gas HGO becomes a very significant factor.

The effect of other input conditions on reactor performance is shown in Figure 12-11. Reactor temperature >60°C can be maintained for all conditions examined with specific oxygen supply less than about 8 g O_2/g BVS. Beyond this, the system is predicted to fail with an infeed temperature of 0°C. It is interesting to note that the point of failure is not significantly affected by infeed cake solids. The explanation is that greater moisture input associated with lower cake solids is balanced by reduced moisture limitations and an increased reaction rate constant. With an infeed temperature of 20°C,

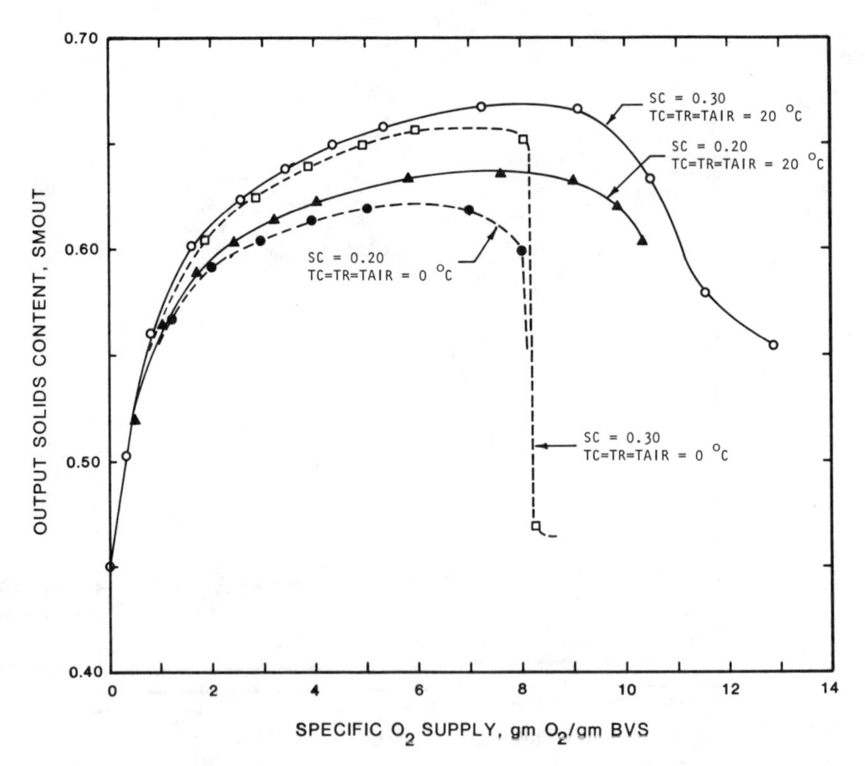

Figure 12-12. CFCM model. Effect of specific oxygen supply on output solids content for various conditions of infeed cake solids and infeed temperatures. Raw sludge, VR = VROUT.

system failure does not occur until the specific oxygen supply is about 10 g/g BVS. Again, infeed cake solids do not significantly alter this point.

Predicted output solids content from the reactor is shown in Figure 12-12 for the same conditions shown in Figure 12-11. Infeed temperatures do not significantly affect output solids content until the point of process failure. As was observed for the case of digested sludge, process failure occurs rapidly with low input temperatures of 0°C. Failure is much less dramatic at higher infeed temperatures.

REACTOR DYNAMICS (VR < VROUT)

In many cases, the VS content of recycled product will not equal that of the reactor output. Compost produced by many processes is frequently stockpiled or cured for long periods, allowing continued decomposition of more resistant organics. Product recycled after such curing should be low in degradable organics, resulting in a reduced energy input compared to the previous analysis. The purpose of this section is to examine reactor performance when recycled product has been cured or aged and contains a minimum of degradable organics. For purposes of analysis, recycled product was assumed to have a degradability coefficient of 0.05 and VS content of 0.35. The same values were used regardless of whether feed sludge was raw or digested. Again, these values are consistent with previous estimates of the minimum achievable volatility if all degradable organics are decomposed.

Digested Sludge

Predicted results for a digested sludge feed with the VS fraction of product recycle set at 0.35 (VR < VROUT) are shown in Figures 12-13 to 12-18. These figures correspond to similar results developed for the case where VR = VROUT and shown in Figures 12-1 to 12-6. Reference should be made to the two sets of figures to understand better the comparative differences.

Figures 12-13 and 12-14 (compare to Figures 12-1 and 12-2) show that the reduced contribution of BVS in product recycle decreases the range of air flowrate over which composting can be maintained. Using a reactor temperature >60°C as the criterion, maximum air supply is reduced from about 9 to 5 m^3/min-dry ton/day. Furthermore, reactor temperatures are reduced for the same air flowrate. Nevertheless, there is sufficient energy to maintain good composting, but in a somewhat more narrow range of operation.

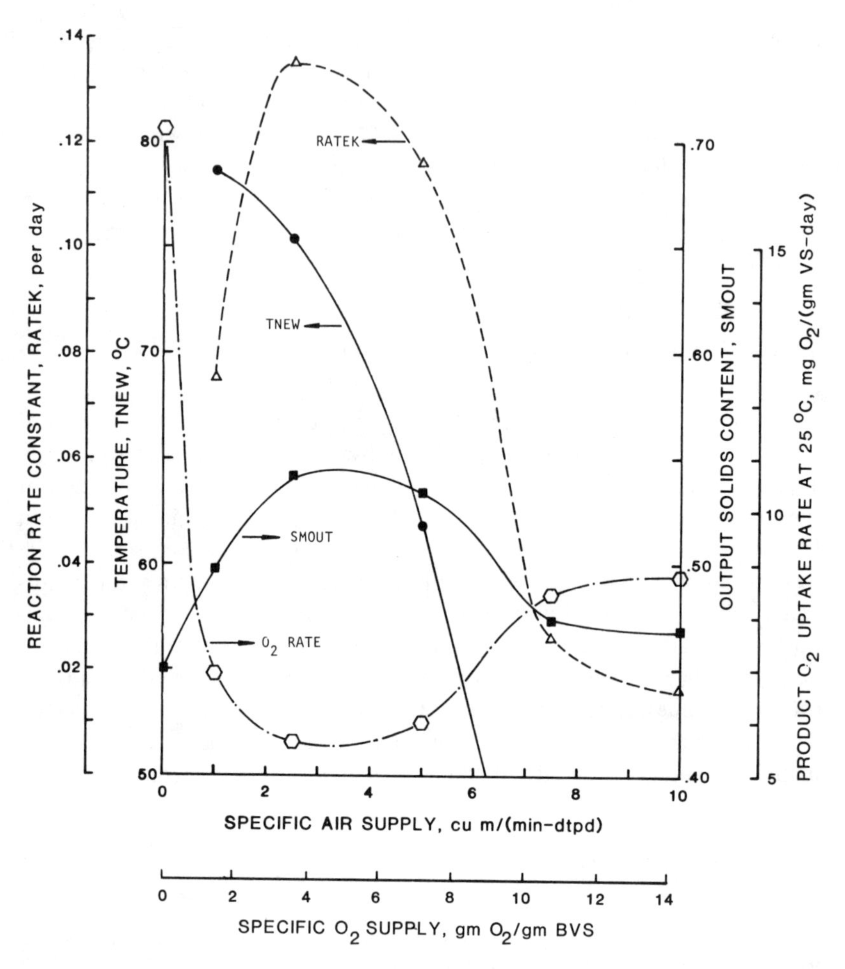

Figure 12-13. CFCM model. Effect of specific air supply on reactor temperature, reaction rate constant, output solids content and output VS fraction. Digested sludge, SC = 0.30, TC = TR = TAIR = 20°C, VR constant at 0.35.

Maximum output solids fraction SMOUT is predicted to be about 0.54, again reduced from the previous analysis where the maximum was 0.58. This has an interesting effect in that the maximum rate constant is actually increased because of reduced moisture limitations. The moisture adjustment factor F1 never drops below about 0.75.

The mass balance for a specific air supply of 2.5 m^3/min-dry ton/day is shown in Figure 12-15. This particular case was selected for illustration because the reaction rate constant is near the maximum value observed. BVS

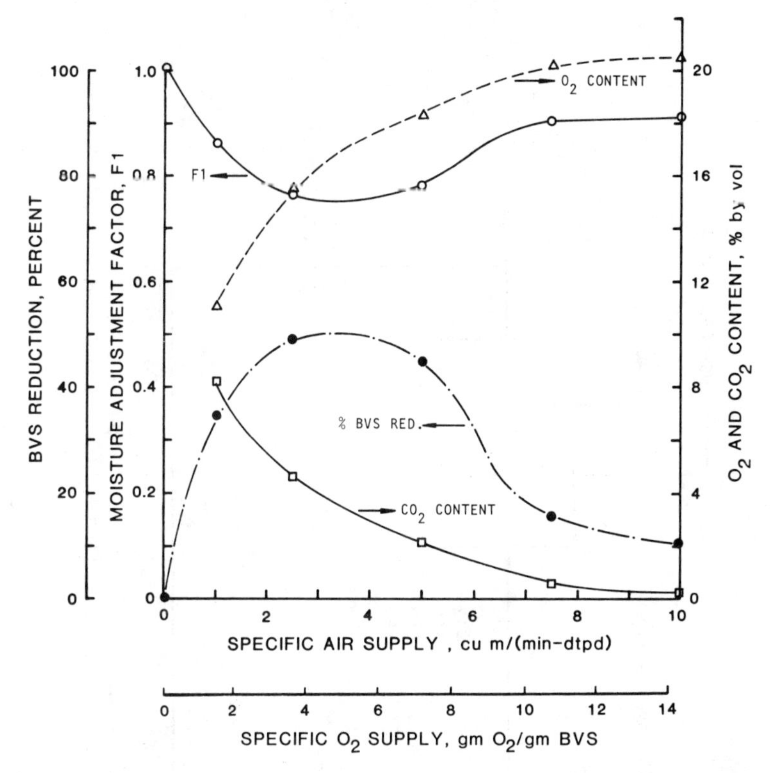

Figure 12-14. CFCM model. Effect of specific air supply on BVS reduction, moisture adjustment factor, and oxygen and carbon dioxide content. Digested sludge, SC = 0.30, TC = TR = TAIR = 20°C, VR constant at 0.35.

contributed by product recycle is significantly reduced compared to the previous analysis. However, BVS reduction in the reactor is not sufficient by itself to reduce the VS fraction of the final product to the assumed value of 0.35. Therefore, additional BVS loss during curing is required. As it turns out, BVS loss during curing is nearly equivalent to that observed in the compost reactor. This may be somewhat surprising, but it is not totally unrealistic. It is known, for example, that curing piles can maintain elevated temperatures for long durations, a fire which can only be fueled by significant BVS decomposition.

The curing stage shown in Figure 12-15 brings up a point which should be clarified. Referring to the previous analysis with VR = VROUT and in particular Figures 12-3 and 12-9, BVS reduction was calculated across the the entire system from feed sludge to final product. In the present case, final product volatility is set at 0.35 which means that BVS reduction would be

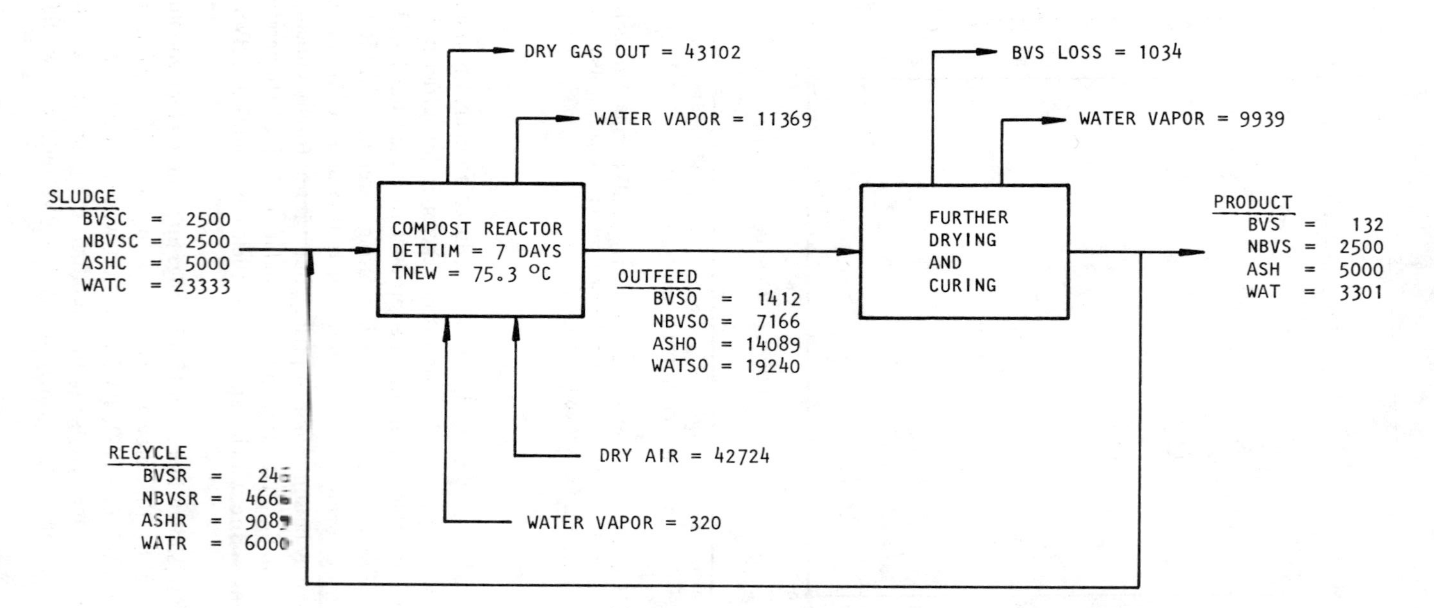

Figure 12-15. CFCM model. Mass balance for solids, water and gas components for a specific air supply of 2.5 m^3/min-dry ton/day and specific oxygen supply of 3.5 g O_2/g mixture BVS. Digested sludge, SC = 0.30, TC = TR = TAIR = 20°C, VR constant at 0.35. Units on mass terms are kg/day.

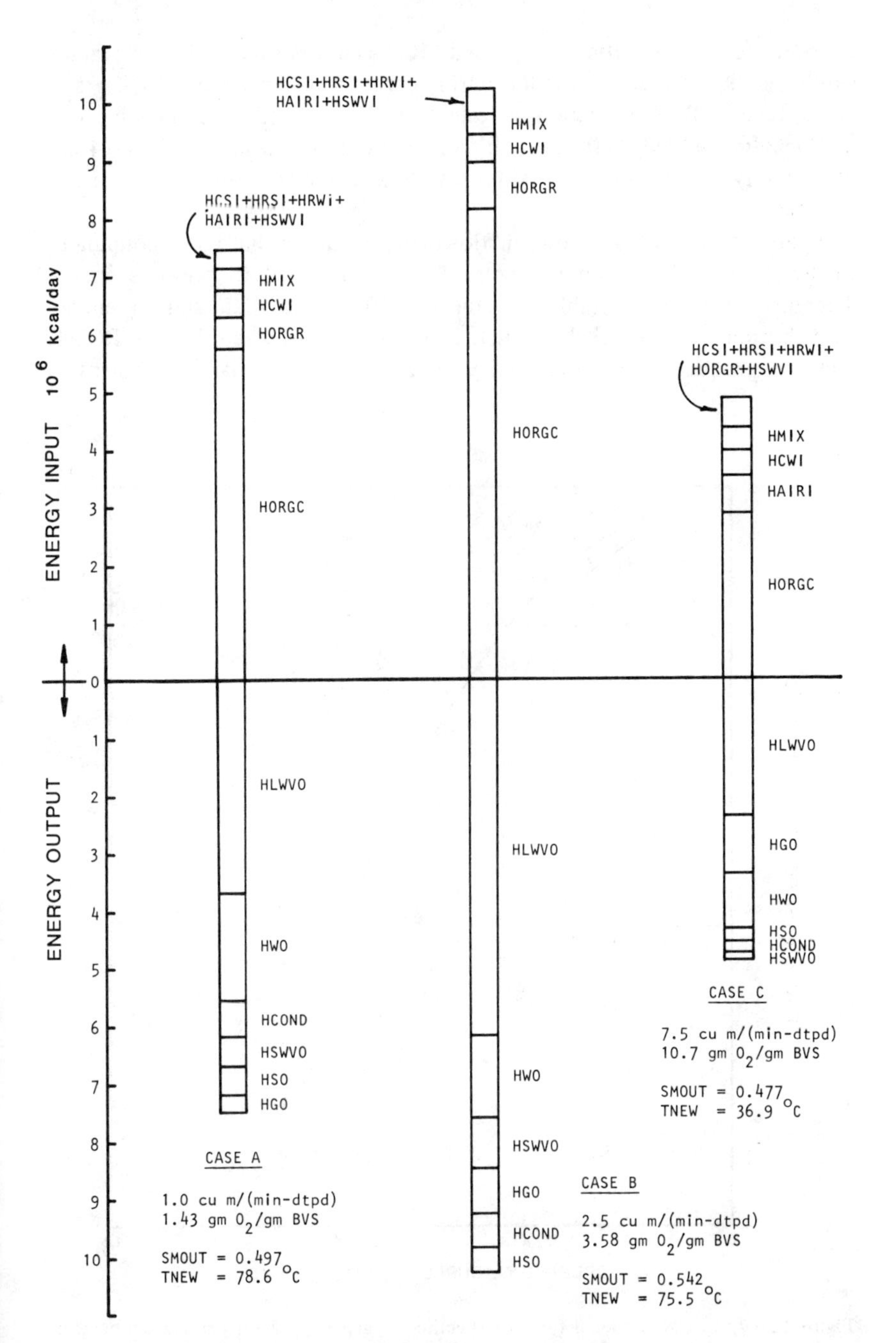

Figure 12-16. CFCM model. Energy balances for selected air supply rates. Digested sludge, SC = 0.30, TC = TR = TAIR = 20°C, VR constant at 0.35.

constant for all conditions examined. Reduced destruction in the reactor would require greater destruction during curing to balance the mass diagram. As a result, BVS reduction is calculated across the reactor only, i.e., (2500+246−1412)/(2500+246). BVS reductions shown in Figure 12-14 are not directly comparable to previous results where subsequent curing was not assumed.

Energy balances for various air flowrates, including that corresponding to Figure 12-15, are shown in Figure 12-16. Compared to previous energy diagrams the most notable difference is the large reduction in energy contribution from recycled product. In previous cases where VR = VROUT, recycled product contributed energy to the reactor essentially equivalent to

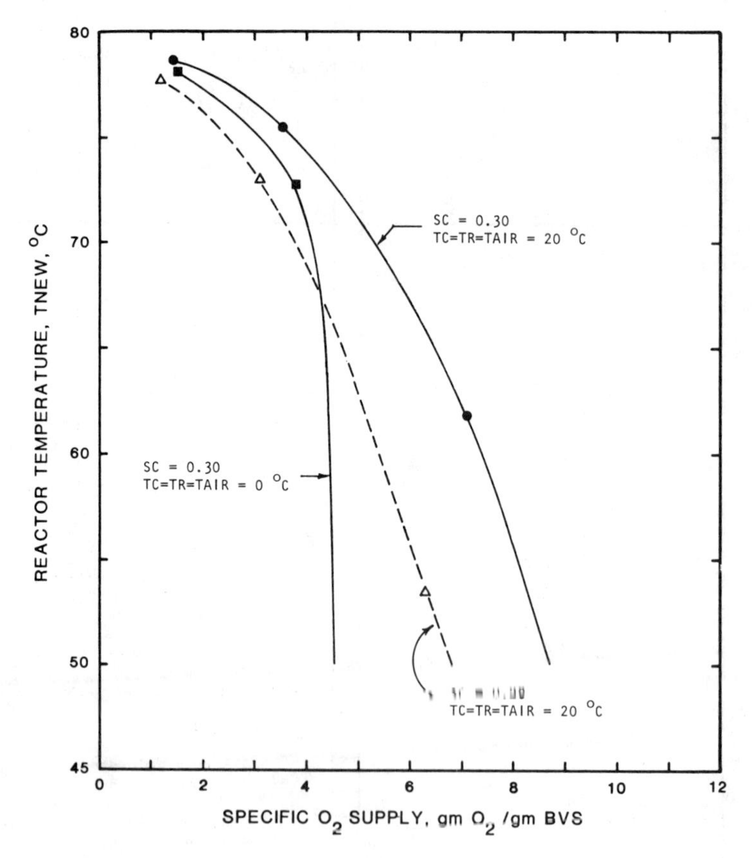

Figure 12-17. CFCM model. Effect of specific oxygen supply on reactor temperature for various conditions of infeed cake solids and infeed temperatures. Digested sludge, VR constant at 0.35.

that of feed sludge. In the present case, however, the contribution is quite small and of comparable magnitude to sensible heat inputs. For all practical purposes one could assume recycled product to be inert without introducing significant error in the energy balance.

Steady-state reactor temperatures for various feed solids and input temperatures is shown in Figure 12-17. For the case of 30% feed solids, temperatures in excess of 60°C are maintained with infeed temperatures of 0 and 20°C. Once again, very rapid failure is observed at the colder temperature, a small change in specific oxygen supply causing a sudden and precipitous drop in reactor temperature. This is not particularly alarming, but it does mean that operating air supply must be maintained below the critical value to assure stable performance. Critical O_2 supply is significantly reduced compared to the case where VR = VROUT as shown in Figure 12-5. In the

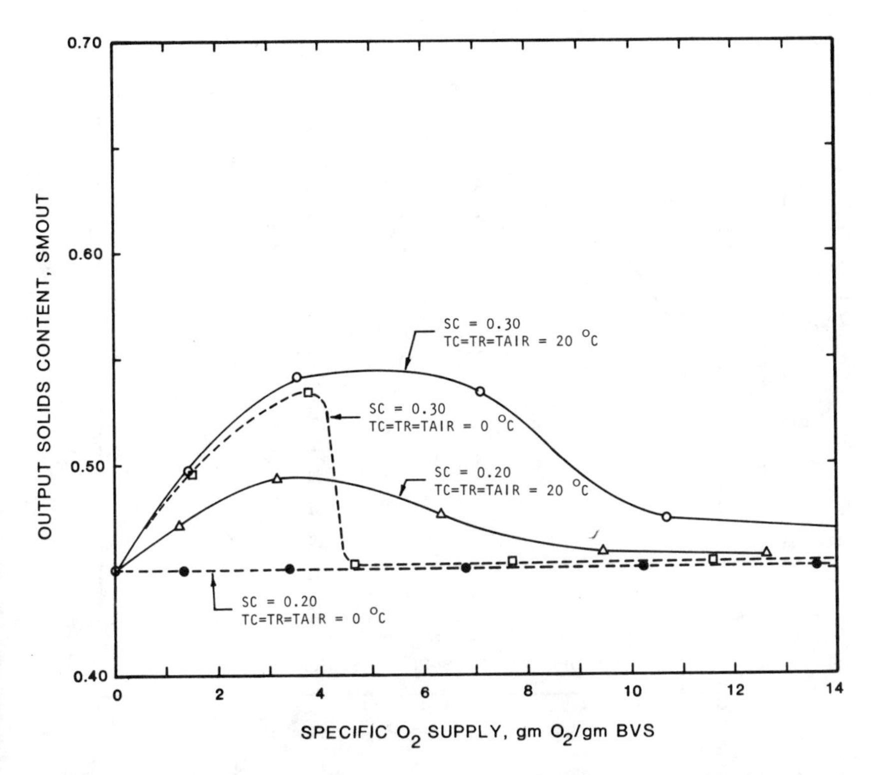

Figure 12-18. CFCM model. Effect of specific oxygen supply on output solids content for various conditions of infeed cake solids and infeed temperatures. Digested sludge, VR constant at 0.35.

present case, the critical value is about 4.5 g O_2/g BVS for 0°C infeed, compared to about 6.5 for similar conditions in Figure 12-5.

With infeed solids of 20% and infeed temperatures of 0°C, predicted reactor temperature never exceeded 20°C over the range of air flowrates examined. This underscores comments made in Chapter 9 concerning the importance of infeed solids to the composting process.

The picture which begins to emerge is one in which BVS of product recycle affects the operating range over which composting can occur. Successful composting can be maintained, but the range of operating variables is reduced as BVS of product recycle is reduced. On a practical basis this reduction in operating leeway becomes more significant as infeed temperatures become colder. Under such conditions one should consider addition of biodegradable amendment to extend the operating limits of the system. The

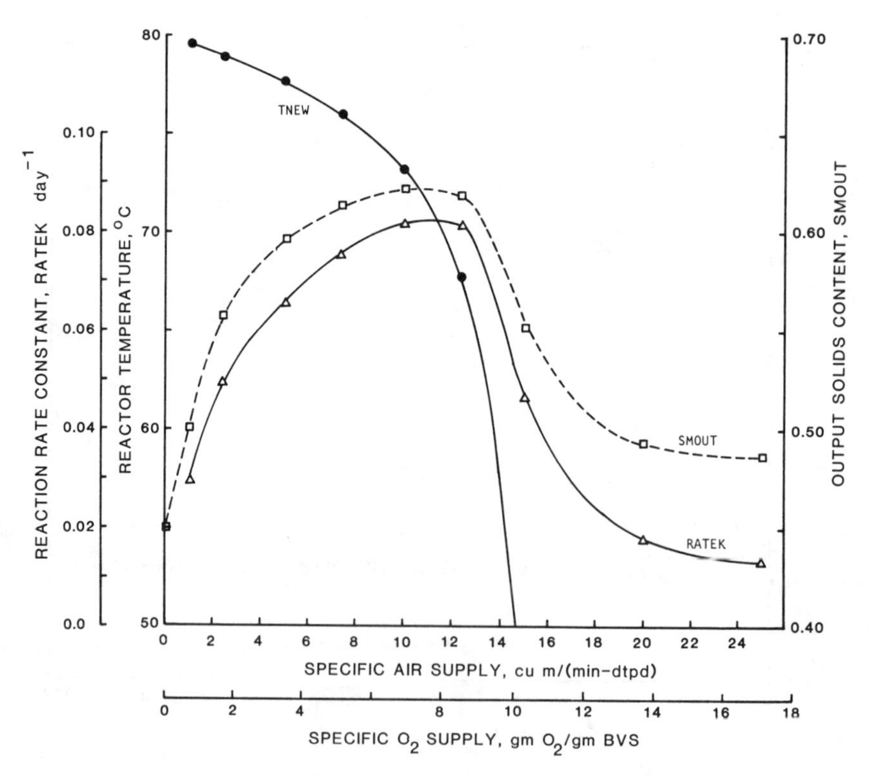

Figure 12-19. CFCM model. Effect of specific air supply on reactor temperature, reaction rate constant and output solids content. Raw sludge, SC = 0.30, TC = TR = TAIR = 20°C, VR constant at 0.35.

two cases examined, i.e., VR = VROUT and VR < VROUT, are extremes and should cover the likely range observed in real situations.

Raw Sludge

Reactor dynamics predicted for the case of raw sludge are illustrated in Figures 12-19 to 12-24. The pattern for results is consistent with those just discussed for digested sludge. Comparing to Figures 12-7 to 12-12 for the case of raw sludge with VR = VROUT, the range of operating variables is reduced in the present case. Again, using a reactor temperature >60°C as the criterion, maximum specific air supply is reduced from about 27.5 to about 14 m^3/min-dry ton/day in the present case. However, this range is still significantly greater than for the case of digested sludge discussed above.

The mass balance for an air supply of 10 m^3/min-dry ton/day is shown in Figure 12-21 and the corresponding energy balance in the right half of

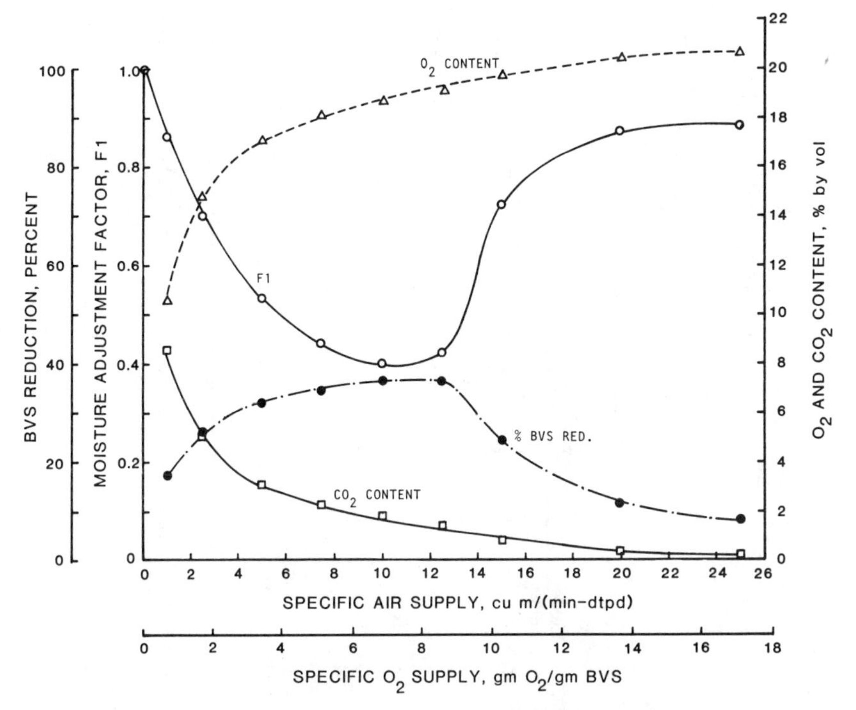

Figure 12-20. CFCM model. Effect of air supply on BVS reduction, moisture adjustment factor, and oxygen and carbon dioxide content. Raw sludge, SC = 0.30, TC = TR = TAIR = 20°C, VR constant at 0.35.

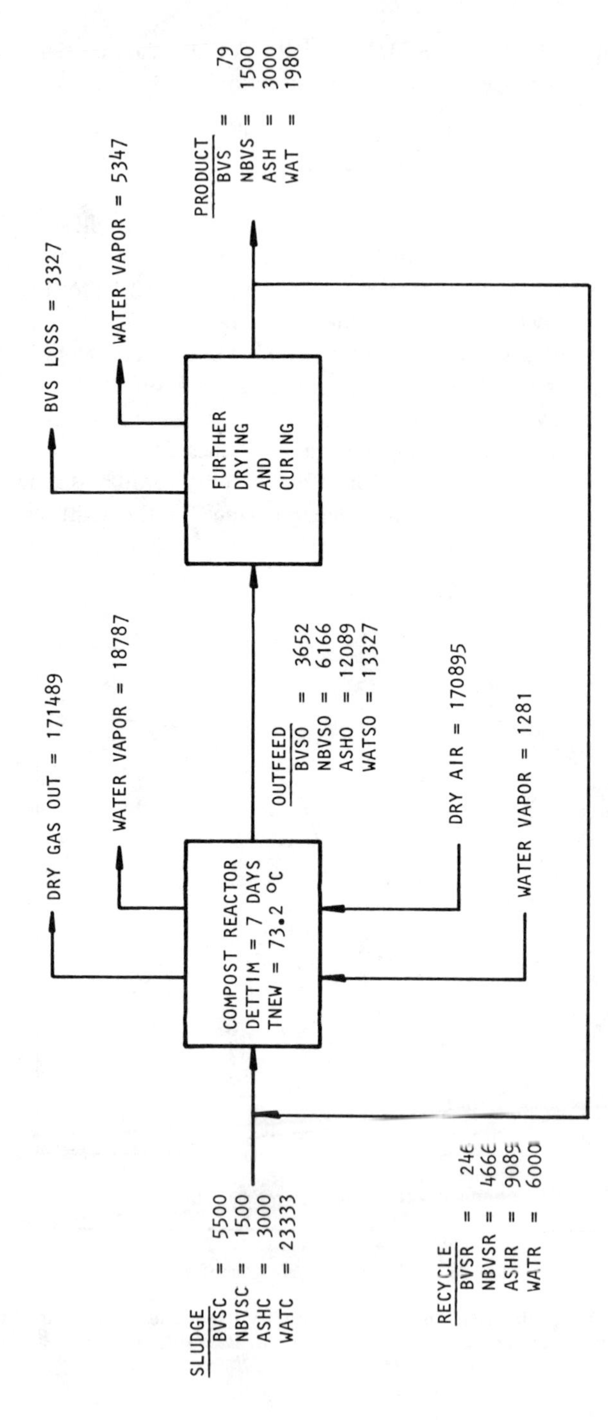

Figure 12-21. CFCM model 1. Mass balance for solids, water and gas components for a specific air supply of 10 m^3/min-dry ton/day and specific oxygen supply of 6.84 g O_2/g mixture BVS. Raw sludge, SC = 0.30, TC = TR = TAIR = 20°C, VR constant at 0.35. Units on mass terms are kg/day.

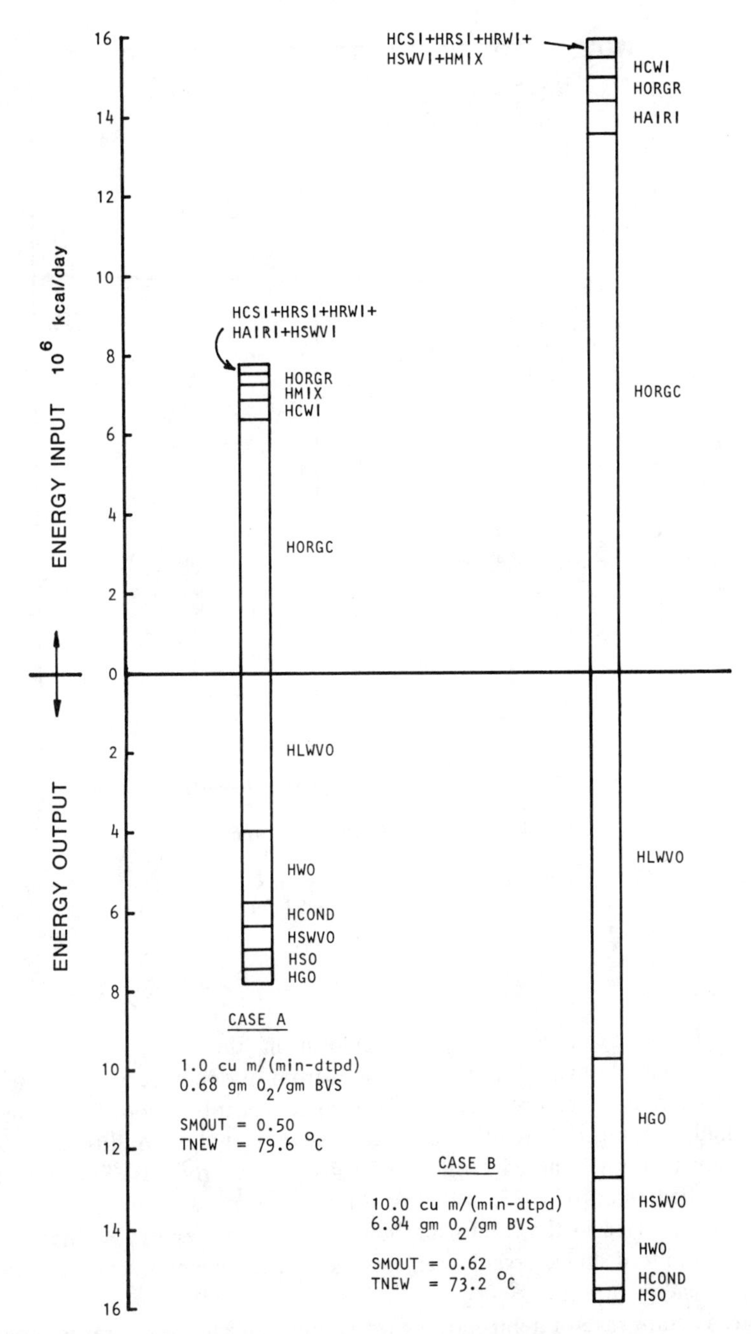

Figure 12-22. CFCM model. Energy balances for selected air supply rates. Raw sludge, SC = 0.30, TC = TR = TAIR = 20°C, VR constant at 0.35.

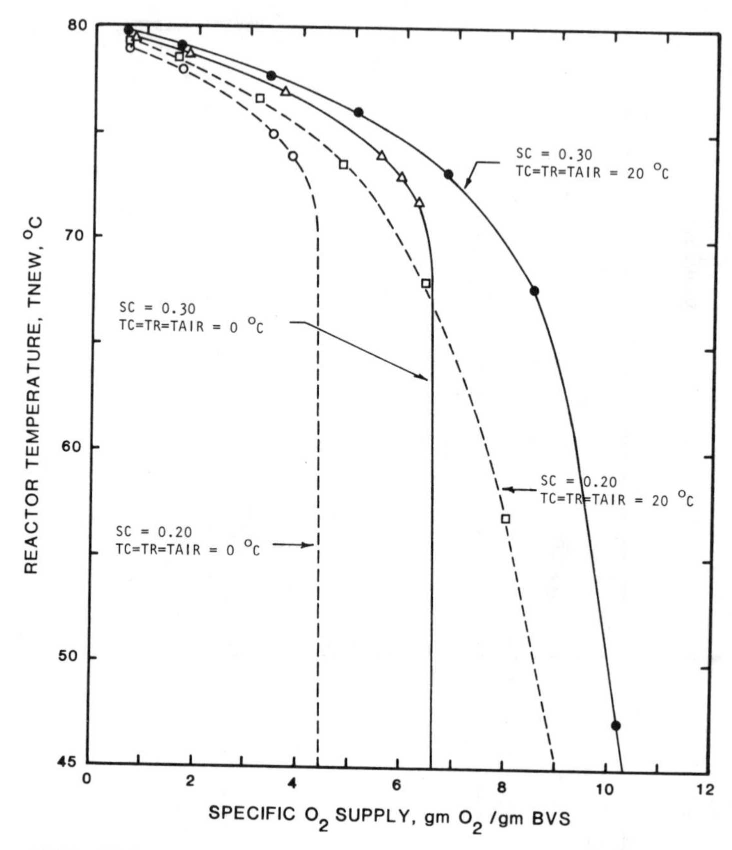

Figure 12-23. CFCM model. Effect of specific oxygen supply on reactor temperature for various conditions of infeed cake solids and infeed temperatures. Raw sludge, VR constant at 0.35.

Figure 12-22. Significant moisture removal in the compost reactor is possible even though BVS contribution from product recycle is much reduced. This is, of course, caused by the higher contribution of BVS in feed raw sludge compared to the case of digested sludge. Additional BVS reduction is again required during curing to reduce product VS content to the specified value of 0.35. BVS loss during curing is actually greater than that accomplished in the reactor itself. Referring to the energy balance, heat release from feed sludge organics remains the dominant energy term, dwarfing all other energy inputs.

Steady-state reactor temperatures for various feed temperatures is shown in Figure 12-23. Unlike the case with digested sludge, reactor temperatures

>60°C are maintained under all conditions examined, i.e., feed solids 20 to 30% and infeed temperatures from 0 to 20°C. However, the range of operating variables is reduced as cake solids and input temperatures are decreased. With a feed sludge of 20% solids and infeed temperature of 0°C, reactor temperature is >60°C only at air supplies less than about 4.0-4.5 g O_2/g BVS. Again, the rather dramatic reactor failure under low temperature conditions should be noted.

Output solids content is shown in Figure 12-24. For a given feed cake solids, infeed temperature does not significantly affect output solids content until the point of process failure. Output solids near 60% are maintained with infeed cake solids of 30%, decreasing to about 55% with 20% cake solids.

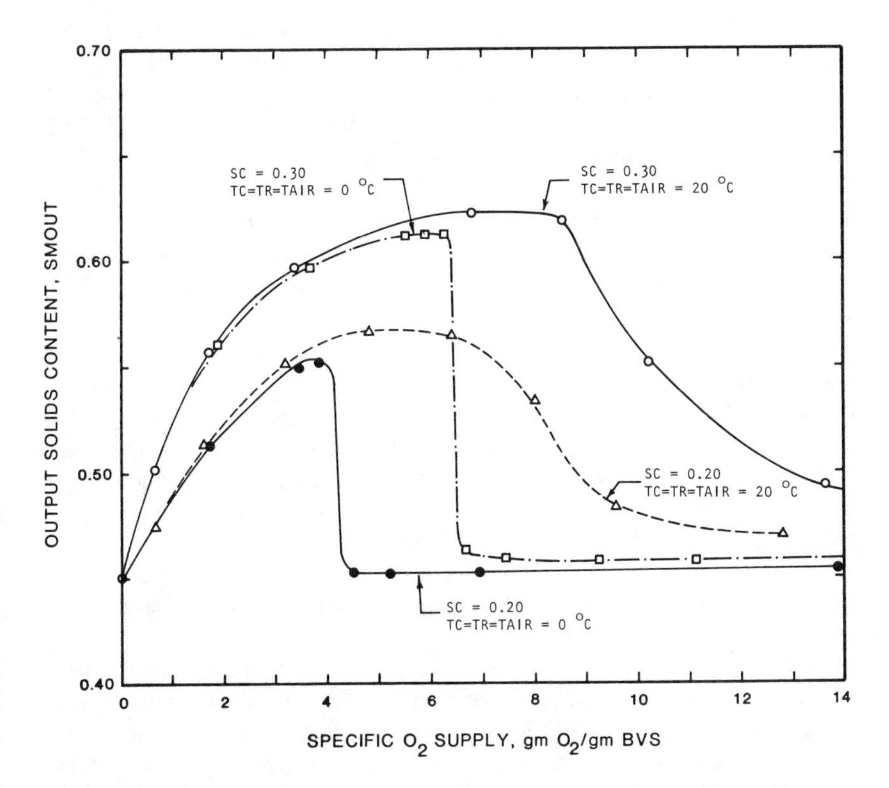

Figure 12-24. CFCM model. Effect of specific oxygen supply on output solids content for various conditions of infeed cake solids and infeed temperatures. Raw sludge, VR constant at 0.35.

EFFECT OF DETENTION TIME

Previous calculations are based on a reactor detention time of seven days. Recall that detention time is based on outfeed characteristics and reflects the average detention time of a particle in the reactor. A question then arises as to the effect of detention time on reactor performance and the range of operating variables. To reduce the quantity of results to be reported, the following analysis is limited to the case of digested sludge with an infeed solids content of 30%. This will be sufficient to illustrate the effect of detention time as an operating variable.

Case Where VR = VROUT

In keeping with the previous organization of this chapter, the case where VR = VROUT will be discussed first. In other words, no further BVS reduction occurs as a result of subsequent processing, and the volatile content of any product recycle is equivalent to that discharged from the compost reactor. Predicted reactor temperature, output solids content and BVS reduction as a function of air supply are presented in Figures 12-25 to 12-27 for input temperatures of 20°C. The effect of increasing reactor detention time is to increase the range of air supply over which adequate composting temperatures can be maintained, in other words, to increase process stability. Output solids content is increased for a given air supply, as is the stabilization of BVS.

It would seem that adequate composting temperatures can be maintained with detention times as low as two days. Shorter detention times might also be possible but were not examined here. The advantage of a shorter detention time is, of course, the reduced reactor volume required and the associated capital cost savings. However, this savings is offset by the smaller range of operating parameters, i.e., specific air supply, the reduction in maximum output solids and lower BVS stabilization. With a two-day detention time, maximum BVS reduction is about 45% compared to values above 70% at longer detention times. Clearly, the tradeoff is between capital cost savings and operational performance.

With a two-day detention time, reactor outfeed would probably require subsequent curing to produce a sufficiently stable material for most uses. Such may not be the case at longer detention times. Indeed, a requirement for subsequent curing has been observed in most compost systems operated at short detention times. This need is certainly borne out by results of the simulation model.

It is interesting to note that several reactor systems have been designed for detention times on the order of seven days. This would appear to be a

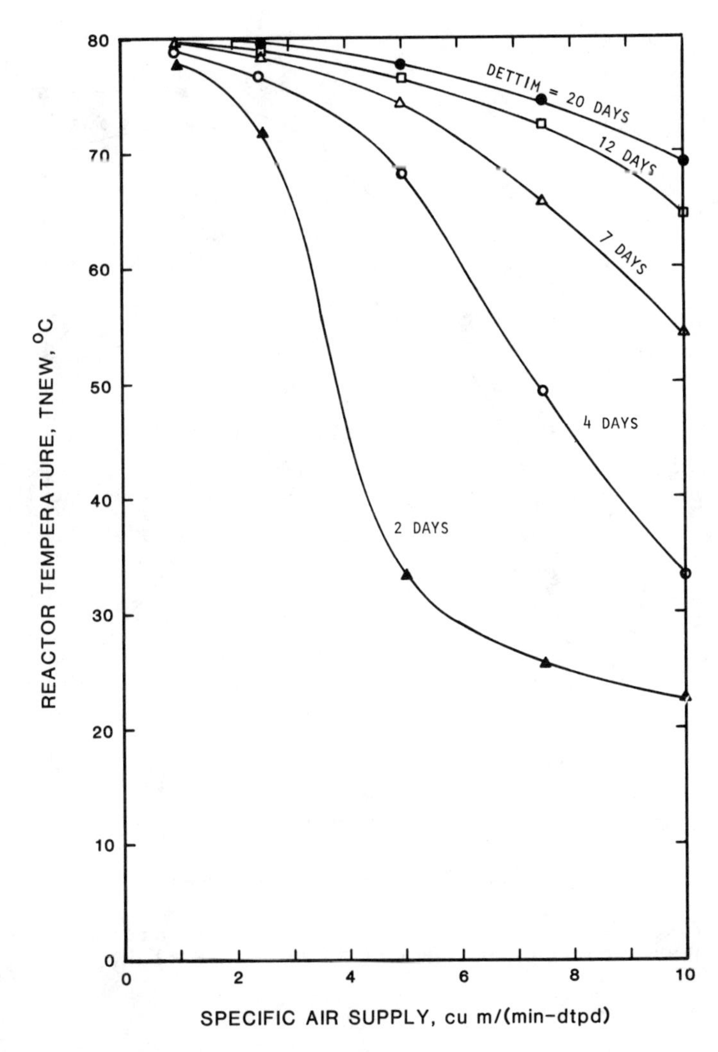

Figure 12-25. CFCM model. Effect of specific air supply and reactor detention time on steady-state reactor temperature. Digested sludge, SC = 0.30, TC = TR = TAIR = 20°C, VR = VROUT.

wise choice because the operating range is wide, and output solids and BVS reduction are only slightly reduced compared to longer detention times. For this reason a detention time of seven days was used for most of the analysis in this chapter. It is encouraging that the predicted range of detention times corresponds to that observed in actual systems. Recall that the simulation

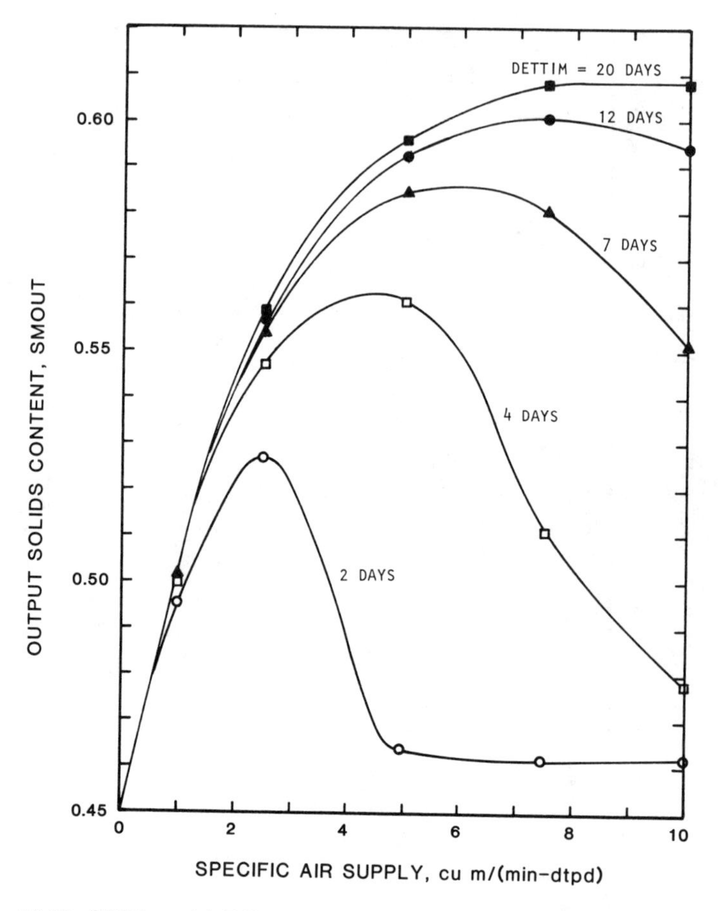

Figure 12-26. CFCM model. Effect of specific air supply and reactor detention time on output solids content. Digested sludge, SC = 0.30, TC = TR = TAIR = 20 °C, VR = VROUT.

models are based on first principles to the extent possible. Therefore, correlation with actual operating data is a strong indication that fundamental principles of composting are embodied in the model.

Similar analysis for the case of 0°C input temperatures is presented in Figures 12-28 to 12-30. The trend of results is similar to those discussed above. Longer detention times result in a greater range of operating air supply, higher output solids content and improved BVS stabilization. For a given detention time, however, the range of air supply for reactor temperatures >60°C is reduced compared to the case with an infeed temperature of 20°C. At a two-day detention time, for example, temperatures >60°C are

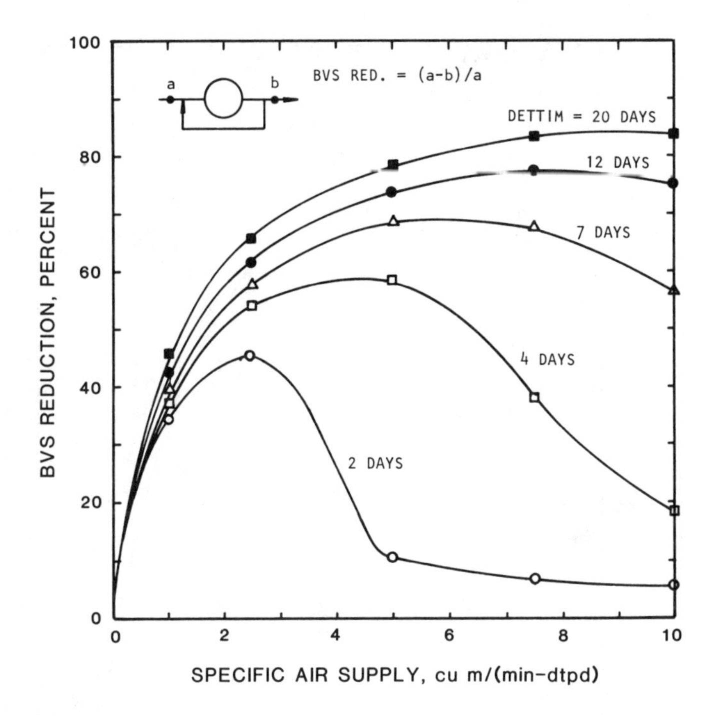

Figure 12-27. CFCM model. Effect of air supply and reactor detention time on BVS reduction. Digested sludge, SC = 0.30, TC = TR = TAIR = 20°C, VR = VROUT.

maintained at air supplies less than about 3.5 m^3/min-dry ton/day if the infeed temperature is 20°C. This is reduced to about 1.5 m^3/min-dry ton/day with an infeed temperature of 0°C. The effect appears to be more pronounced at shorter detention times, indicating an increased sensitivity to operating changes.

In the case of output solids content, peak values are consistently reduced compared to the 20°C case. Significant drying is still achieved, however, and the reductions are only slight.

BVS stabilization does not appear to be reduced to any significant extent by lowering input temperature to 0°C. In fact, for detention times above two days, peak BVS reduction is actually greater at the lower input temperature. Differences are slight, however, and are probably caused by lower reactor temperatures and a subsequent increase in reaction rate constant. It would appear that lower infeed temperatures may reduce the operating range and output solids from the system, but will not significantly affect production of a stable compost.

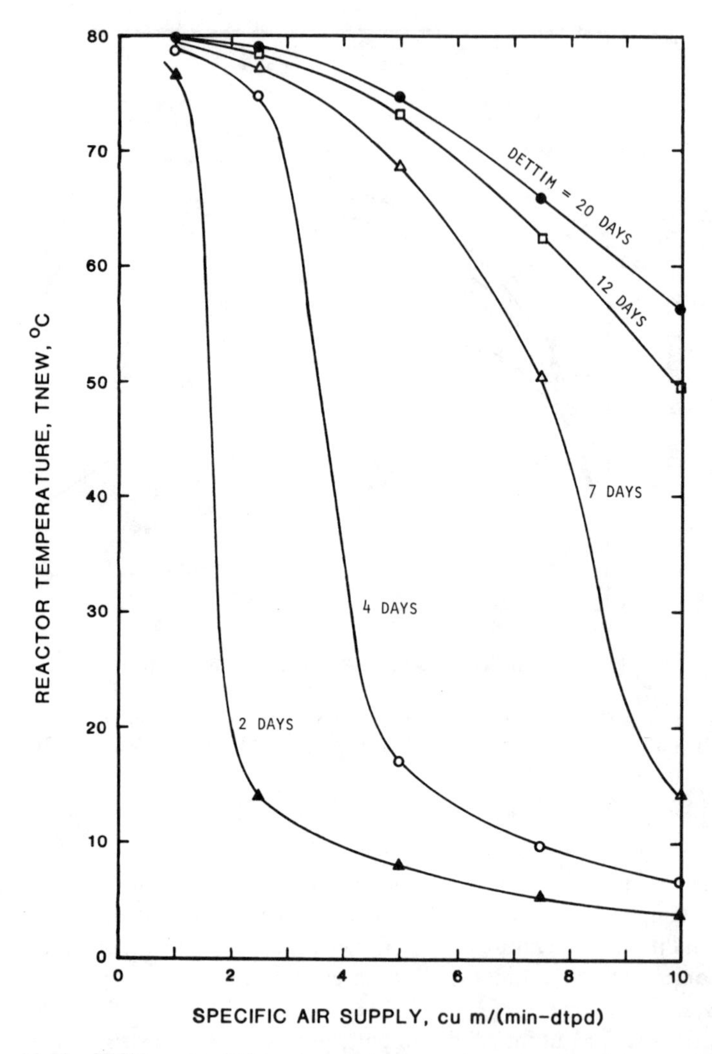

Figure 12-28. CFCM model. Effect of specific air supply and reactor detention time on steady-state reactor temperature. Digested sludge, SC = 0.30, TC = TR = TAIR = 0°C, VR = VROUT.

Case with VR < VROUT

The case was examined where the VS content of reactor output VROUT is not equal to the volatility of product recycle. Recall that this simulates a system where subsequent curing further reduces the BVS content. Recycled

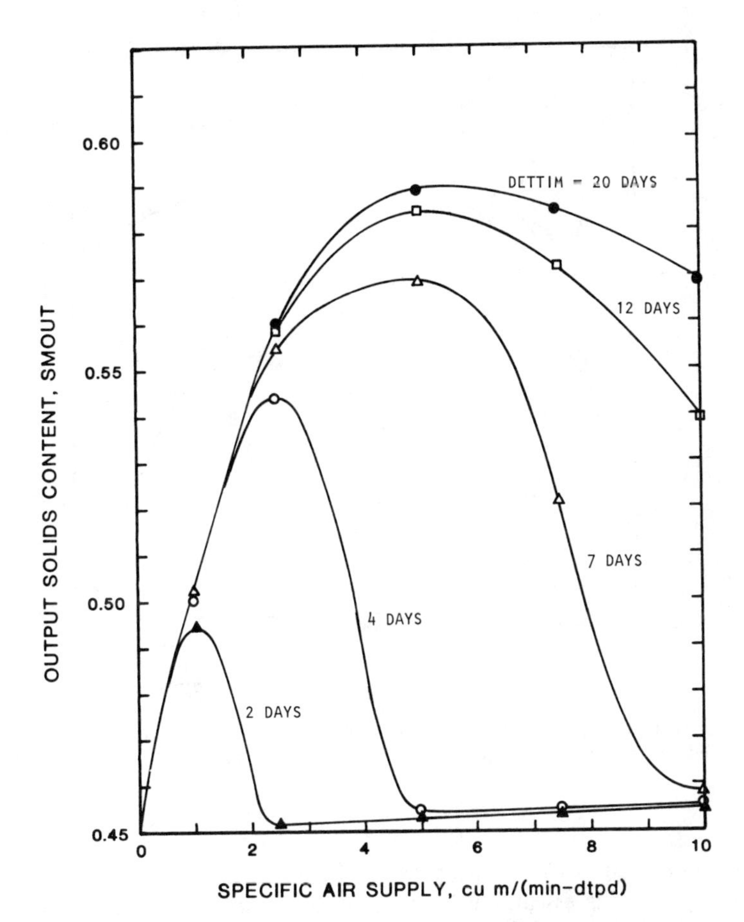

Figure 12-29. CFCM model. Effect of specific air supply and reactor detention time on output solids content. Digested sludge, SC = 0.30, TC = TR = TAIR = 0°C, VR = VROUT.

product was again assumed to have a degradability coefficient of 0.05 and a VS content of 0.35, consistent with assumptions made previously in this chapter. Results are presented in Figures 12-31 to 12-33 for the case of 20°C infeed temperature and Figures 12-34 to 12-36 for 0°C infeed.

The same general patterns emerge as in the previous analysis with VR = VROUT. However, the energy contribution from recycled product is significantly reduced. As one would expect, the range of operating air supply is reduced, as is the peak output solids content for a given detention time. As was observed before, production of a dry reactor product is beneficially affected by longer detention times. Since moisture is a major problem with

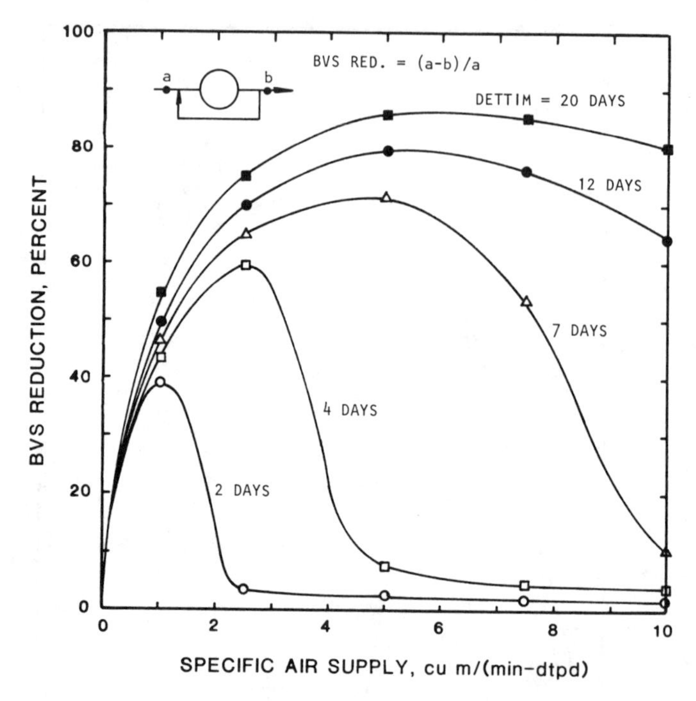

Figure 12-30. CFCM model. Effect of specific air supply and reactor detention time on BVS reduction. Digested sludge, SC = 0.30, TC = TR = TAIR = 0°C, VR = VROUT.

sludge composting, this would appear to be a rather significant benefit. If compost product is essentially inert, as in the present case, a drier product can be achieved by operating at a longer detention time.

If infeed temperatures are reduced to 0°C, a rather interesting result is predicted. Steady-state solutions for reactor temperature never exceed 30°C at detention times of two or four days. Although this may be somewhat surprising, consider that detention times cannot be reduced indefinitely. Eventually process failure must occur. Apparently, detention times of several days are not sufficient with low infeed temperatures. Even at a seven-day detention time, successful composting is limited to an air supply less than about 2.7 m^3/min-dry ton/day or a specific oxygen supply of about 4.5 g O_2/g BVS.

It is apparent that process stability is favored by long detention times, particularly when recycled product or other additives contribute little to the energy budget and when infeed temperatures are low. It is also apparent that climatological factors must be considered in reactor sizing and system

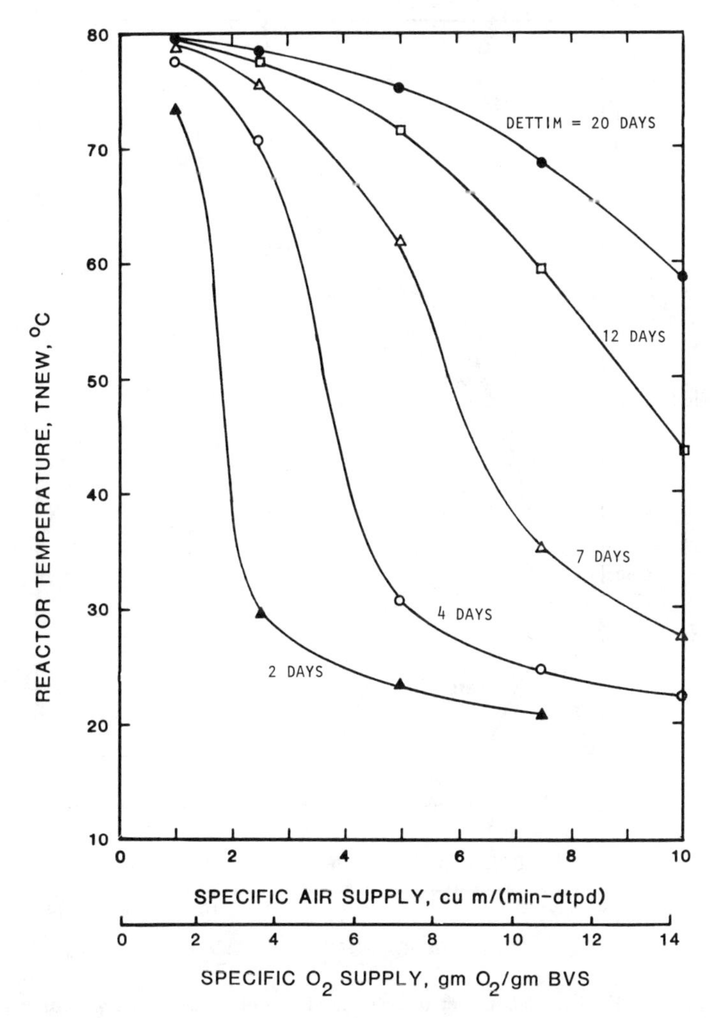

Figure 12-31. CFCM model. Effect of specific air supply, specific oxygen supply and reactor detention time on steady-state reactor temperature. Digested sludge, SC = 0.30, TC = TR = TAIR = 20°C, VR constant at 0.35.

design. Longer detention times and use of higher-energy substrates and amendments become more important in colder climates. Another approach which has been included in recent designs involves use of a recuperative heat exchange system to transfer heat from the exhaust gases to the incoming air stream. As far as the compost reactor is concerned, the effect of such heat

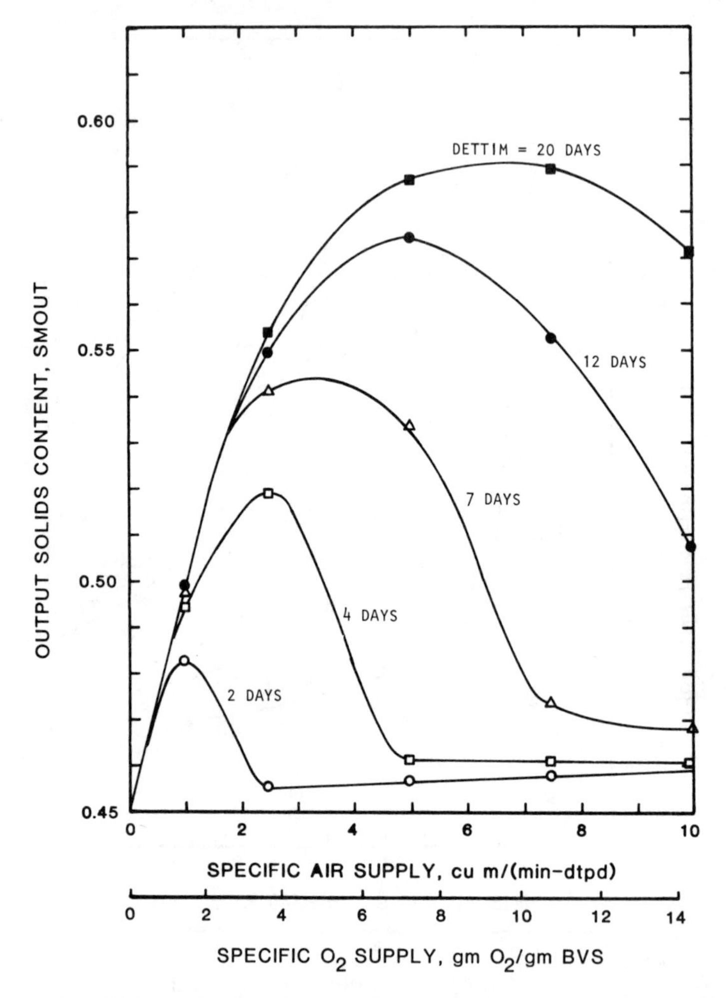

Figure 12-32. CFCM model. Effect of specific air supply, specific oxygen supply and reactor detention time on output solids content. Digested sludge, SC = 0.30, TC = TR = TAIR = 20°C, VR constant at 0.35.

exchange would be to increase the infeed air temperature. This will have a beneficial effect from a thermodynamic standpoint and should improve process stability. Referring back to the energy balance in Figure 12-16, however, the sensible heat of the infeed sludge water HCWI is greater than that of infeed air HAIRI, except at high air supply rates. Therefore, it is equally beneficial to conserve heat contained in the sludge feed, recycled product or other amendments used.

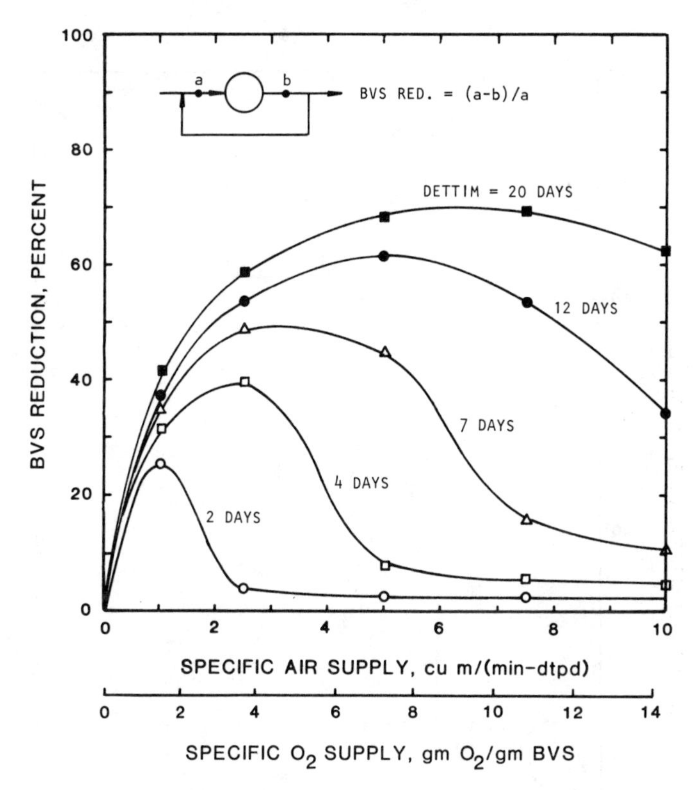

Figure 12-33. CFCM model. Effect of specific air supply, specific oxygen supply and reactor detention time on BVS reduction. Digested sludge, SC = 0.30, TC = TR = TAIR = 20°C, VR constant at 0.35.

REACTOR RESPONSE TO INFEED MIXTURE SOLIDS

All cases examined to this point have assumed an infeed mixture solids of 45%. This should be sufficient to provide some free airspace (FAS) to the infeed material. However, it should be obvious by now that reactor contents can be considerably drier than the feed mixture. This suggests the possibility that infeed mixture solids to a CFCM reactor can be reduced, because composting is governed by conditions in the reactor itself and not in the feed material.

Let us approach this from another standpoint. For a batch process, infeed conditions must be capable of supporting composting. Hence, with the aerated static pile and windrow systems (i.e., batch processes) bulking particles, recycled product and/or amendments are necessary to adjust the

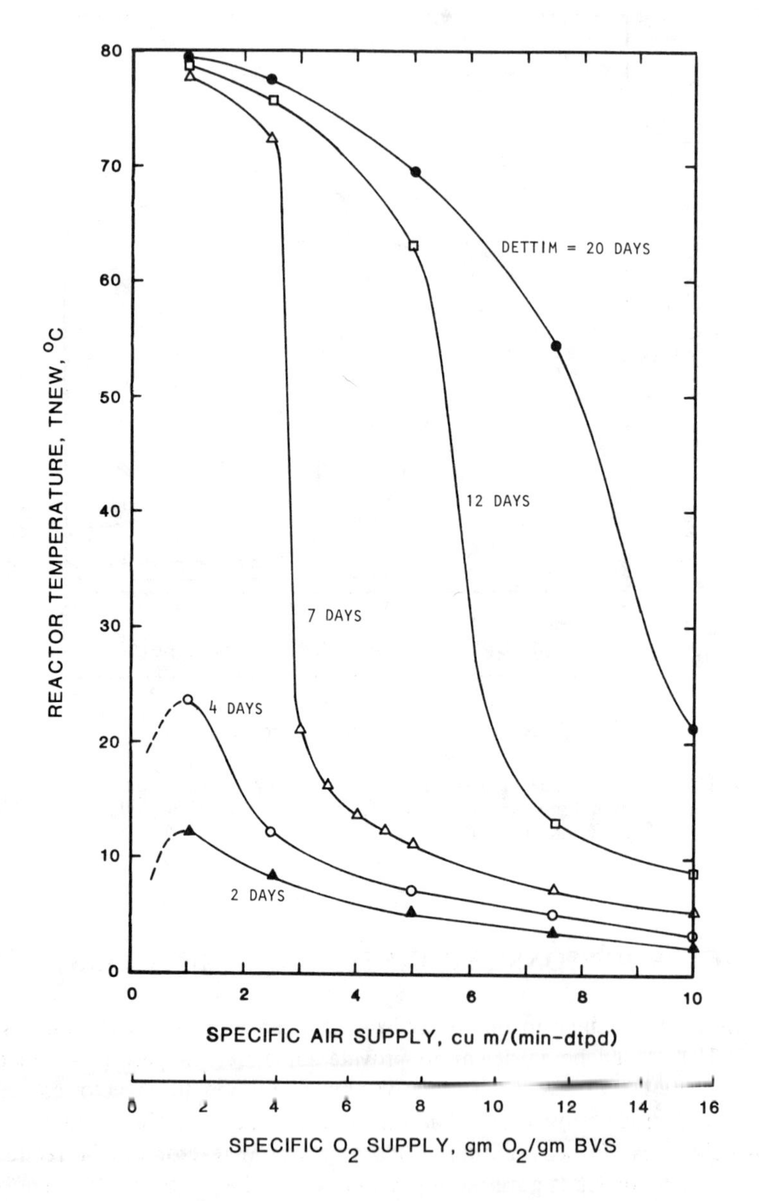

Figure 12-34. CFCM model. Effect of specific air supply, specific oxygen supply and reactor detention time on the steady-state reactor temperature. Digested sludge, SC = 0.30, TC = TR = TAIR = 0°C, VR constant at 0.35.

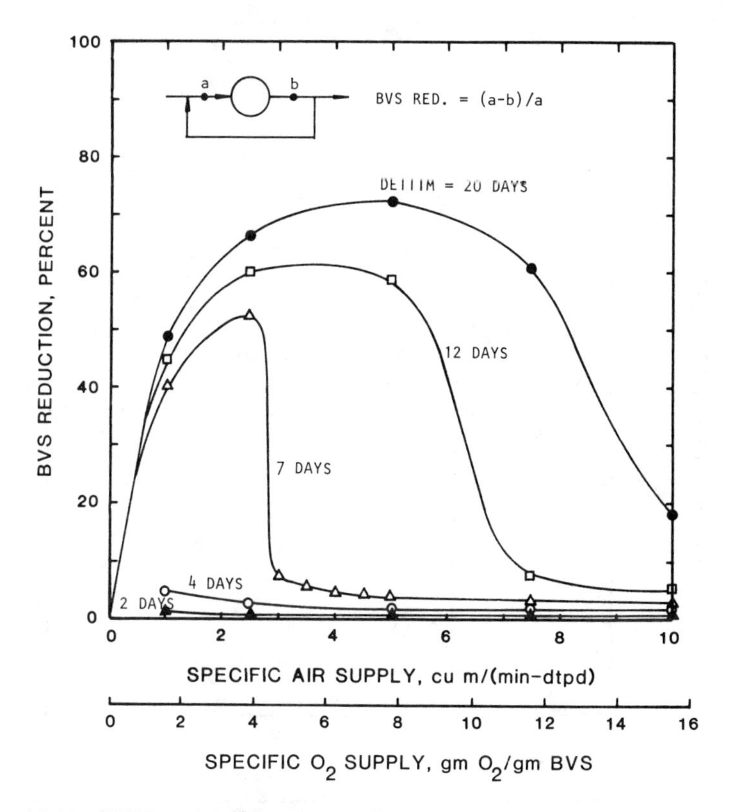

Figure 12-35. CFCM model. Effect of specific air supply, specific oxygen supply and reactor detention time on BVS reduction. Digested sludge, SC = 0.30, TC = TR = TAIR = 0°C, VR constant at 0.35.

starting material to a proper range of moisture and FAS. This is not necessarily the case for a CFCM reactor. For the latter it is only necessary that conditions within the reactor be capable of supporting composting. The moisture content of the feed may be higher than would normally be permitted in a batch process. This could significantly reduce the required quantity of product recycle or amendment and would be of significant advantage, particularly in wet weather when the latter materials may be difficult to obtain in a dry state.

To analyze reactor response to infeed mixture solids it was assumed that compost product at 70% solids is recycled to adjust the infeed mixture. Sludge cake was assumed to be 30% solids and infeed mixture solids of

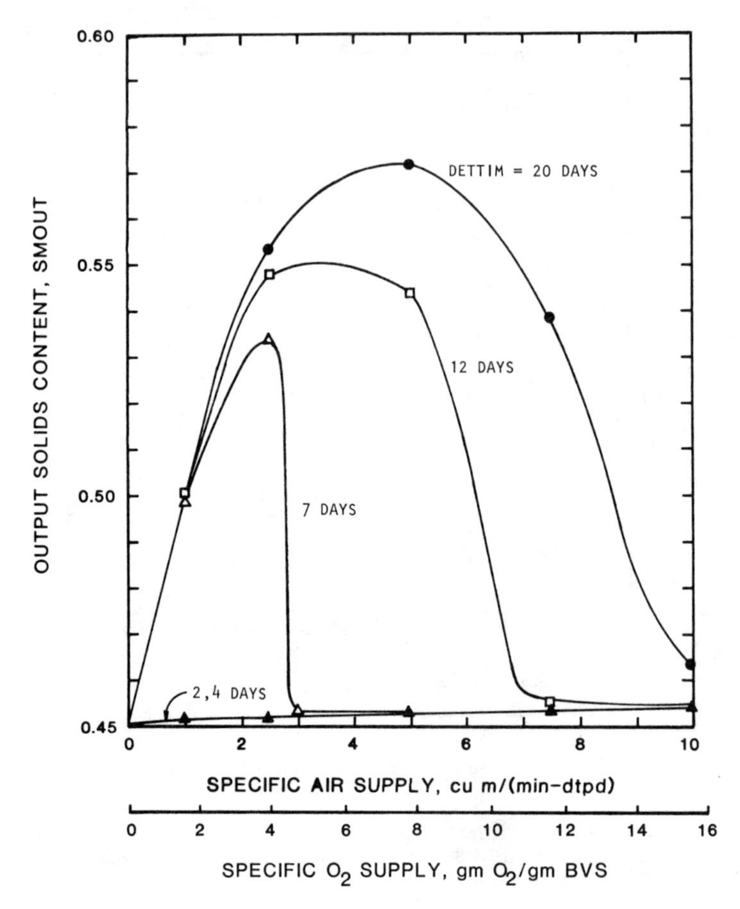

Figure 12-36. CFCM model. Effect of specific air supply, specific oxygen supply and reactor detention time on output solids content. Digested sludge, SC = 0.30, TC = TR = TAIR = 0°C, VR constant at 0.35.

30-50% were considered. In all cases the recycled product was assumed to have a VS content of 0.35 and degradability coefficient of 0.05. This is a conservative case because it results in little energy input from recycled product.

Raw Sludge

Results of the analysis for the case of raw sludge are presented graphically in Figures 12-37 to 12-40. Reactor temperature and the range of

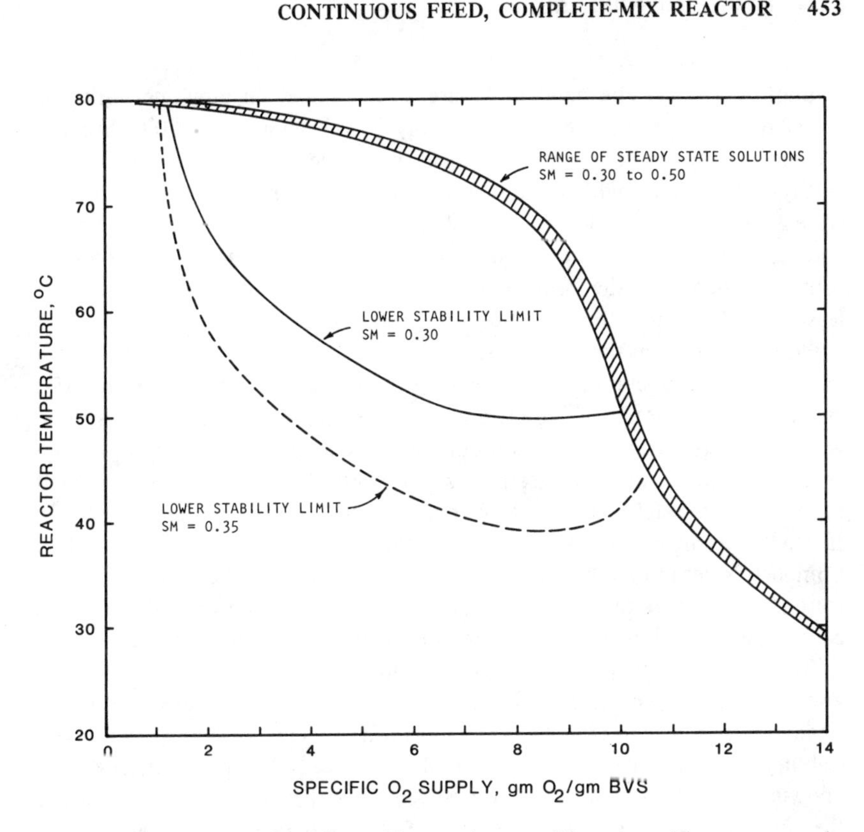

Figure 12-37. CFCM model. Effect of input mixture solids and specific oxygen supply on the steady-state reactor temperature and zone of stable operation. Raw sludge, SC = 0.30, TC = TR = TAIR = 20°C, VR constant at 0.35.

operational stability are shown in Figure 12-37 as a function of specific oxygen supply. A narrow band of steady-state reactor temperatures is predicted regardless of infeed mixture solids. For example, at a specific oxygen supply of 5 g O_2/g BVS, the steady-state reactor temperature is predicted to vary from about 76 to 77°C over the range of input mixture solids from 30 to 50%. It is obvious from this that mixture solids content can be reduced substantially without significantly affecting the steady-state reactor temperature. The rather minor impact on reactor temperature is explained in part by the small contribution of recycled product to energy inputs and outputs.

For the cases where SM = 0.40, 0.45 and 0.50, the initial reactor temperature was assumed to be 20°C. Reactor temperature increased iteratively until the steady-state solution was achieved. In other words, as long as the reactor temperature was 20°C or above, the reactor would seek the steady-state solution shown by the band in Figure 12-37. Based on previous work, the initial reactor temperature could probably have been as low as 0°C, but this was not examined in the present analysis. If the reactor was operating at steady-state and a sudden reduction in temperature occurred, the reactor would tend to reestablish the steady-state temperature. This assumes, of course, that the initial reduction in temperature is not caused by biological toxicity. Therefore, the zone of stable operation for SM = 0.40-0.50 is at least from 20°C up to the predicted steady-state value.

When infeed mixture solids are reduced to 30-35%, a reduction in the range of stable operation is predicted. Consider an infeed mixture of 35% and an oxygen supply of 7 g O_2/g BVS. Under these conditions, the system would seek the steady-state solution only as long as reactor temperature remained >40°C. Any decrease below this causes the system to seek a second solution where reactor temperature approximates that of the infeed materials. If reactor temperature did decrease below 40°C, the system would fail. Composting could then be reestablished only by artifically heating the reactor contents to greater than 40°C or adjusting the mixed feed to a higher solids content, one which would allow a broader range of operating conditions.

The lower stability limit referred to in Figure 12-37 is defined as the lowest starting temperature for which the system still seeks the upper steady-state solution. If reactor temperature decreases below the lower stability limit, process failure will result. The reader might be interested in knowing that the CFCM model was solved more than 150 times to construct the curves shown in Figure 12-37. It is safe to say that such results would not be achievable without the computer simulation model.

It is apparent from Figure 12-37 that the zone of process stability continues to decrease with decreasing infeed mixture solids. An infeed mixture as low as 30% solids can be maintained as long as stability limitations are recognized. Infeed solids below 30% might also be possible but were not examined. The tradeoff is increased sensitivity to changes in operating conditions.

The pattern that emerges is one in which composting can be maintained over a wide range of infeed moisture levels. Restrictions in the range of operation are slight at solids contents as low as 40%. With 35% infeed solids, the constraints become noticeable, increasing to rather severe limitations at 30% solids. With feed mixture solids <40%, a number of conditions must be satisifed to assure high reactor temperatures. First, sufficient water must be removed in the exhaust gases to balance the infeed water and allow a proper combination of FAS and moisture in the reactor contents. This assures a high reaction rate constant, which is essential for proper temperature conditions. Second, the reactor must be well mixed so that feed material is distributed uniformly throughout the reactor volume. Third, the reactor must be fed on a continuous or semicontinuous basis. The latter condition can be approximated by withdrawal and feeding on an intermittent but frequent basis, i.e., several times daily. Fourth, feed material must have sufficient energy to accommodate the required moisture removal. Fifth, the air supply must be sufficiently high to meet the requirements for moisture removal. Sixth, reactor temperature must be maintained above the lower stability limit. If

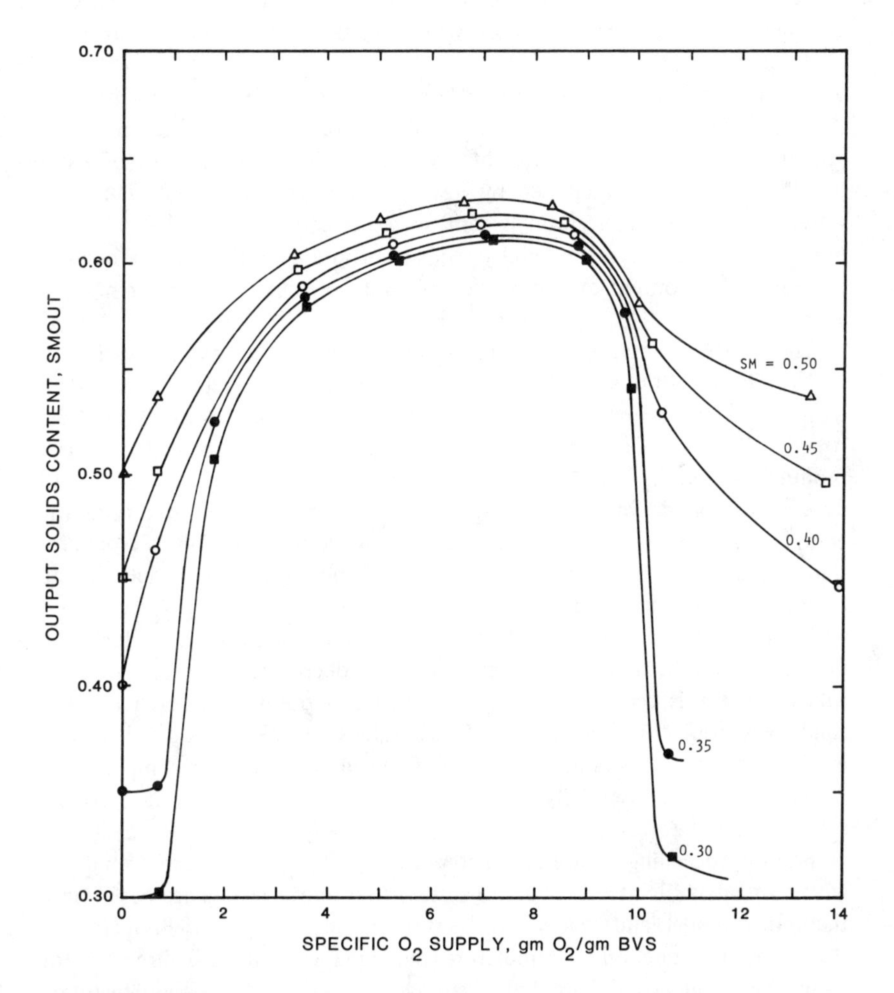

Figure 12-38. CFCM model. Effect of input mixture solids and specific oxygen supply on output solids content. Raw sludge, SC = 0.30, TC = TR = TAIR = 20°C, VR constant at 0.35.

all those conditions are met, the reactor can operate with a high moisture feed, but with a reduced zone of stability.

The reader should not interpret these remarks to mean that constrained areas of operation should be avoided. There are several reasons why operation with lower feed solids is desirable. One of the more obvious is the reduced requirement for dry product recycle or other dry amendment addition. Another is that comparatively dry output can be maintained even with a high moisture infeed. This is illustrated in Figure 12-38. Output solids content is remarkably similar regardless of feed solids over a wide specific air supply range. Reactor operation within the range of stability was assumed in all cases.

The phenomenon is easier to understand when one recalls that many of the steady-state solutions with raw sludge involved rather severe moisture limitations. For a given quantity of dry sludge solids, the mass rate of water to the reactor is not increased by lower mixture solids content. Thus, the water/degradable organic ratio W remains rather constant regardless of infeed mixture solids. The system thereby tends to maintain a rather narrow band of outfeed solids content except, of course, at the extremes of low or high air supply.

The above advantages are illustrated by the mass balance shown in Figure 12-39. The conditions are the same as those shown in Figure 12-21, except that infeed solids are 0.30 instead of the 0.45 previously assumed. Mass balances are remarkably similar between the two cases, except that no dry product recycle is required for the case shown in Figure 12-39. All dependence on downstream curing and drying phases is removed, and the compost reactor can operate independently of other amendments. Note also that the quantity of water vapor removed from the reactor is actually increased over that shown in Figure 12-21.

Another advantage of operating with a reduced solids input is illustrated in Figure 12-40. Required reactor volume decreases as infeed mixture solids content is reduced. A range of values is observed at each mixture solids level, reflecting the range of QAIR values assumed. The trend of data is clearly evident, however, and can be of considerable economic importance. Reactor volume is essentially halved in going from an infeed mixture of 45% solids to one of 35%.

Before proceeding with a discussion of results found for digested sludge, it would be advisable at this point to restate that above results are based on a theoretical model, and have yet to be verified fully by actual field operation. But isn't one function of simulation modeling to point the direction for future field endeavors? Potential advantages for wet feeds such as dewatered sludge, particularly in wet climates, are so obvious that one hopes for field verification in a timely manner.

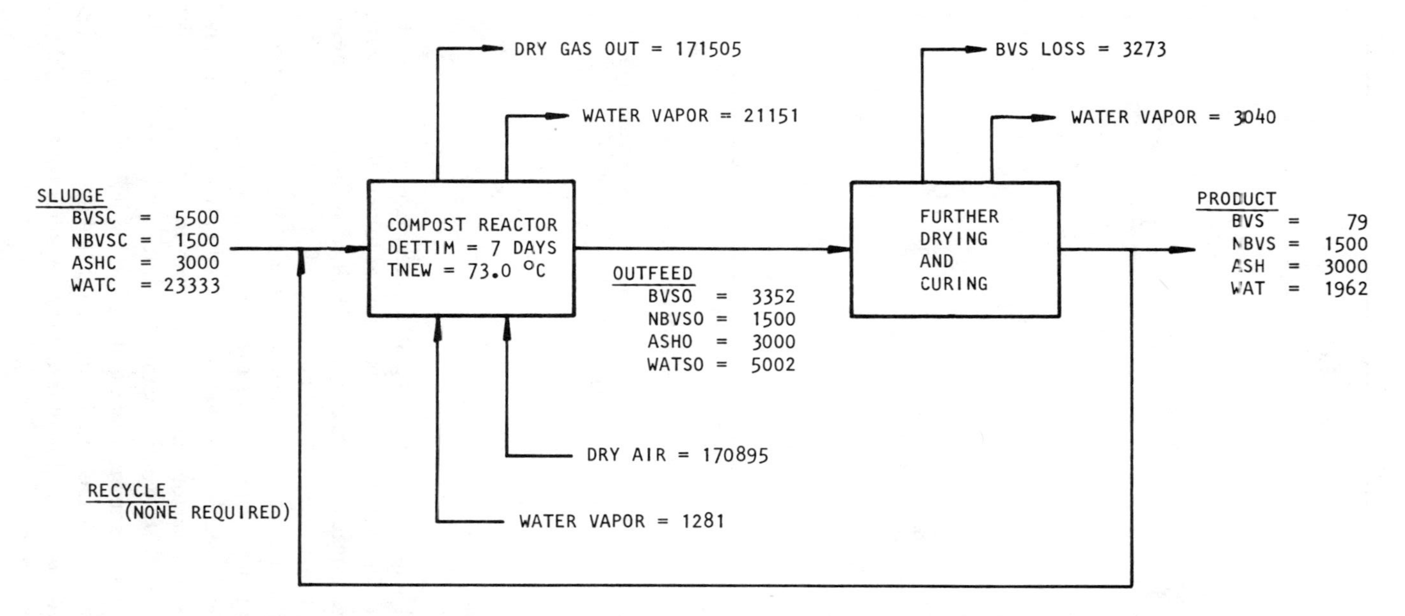

Figure 12-39. CFCM model. Mass balance for solids, water and gas components for an infeed mixture solids content, SM, of 0.30. Raw sludge, SC = 0.30, TC = TR = TAIR = 20°C, VR constant at 0.35. Specific air supply = 10.0 m^3/min-dry ton/day. Specific oxygen supply = 7.15 g O_2/g mixture BVS. Units on mass terms are kg/day.

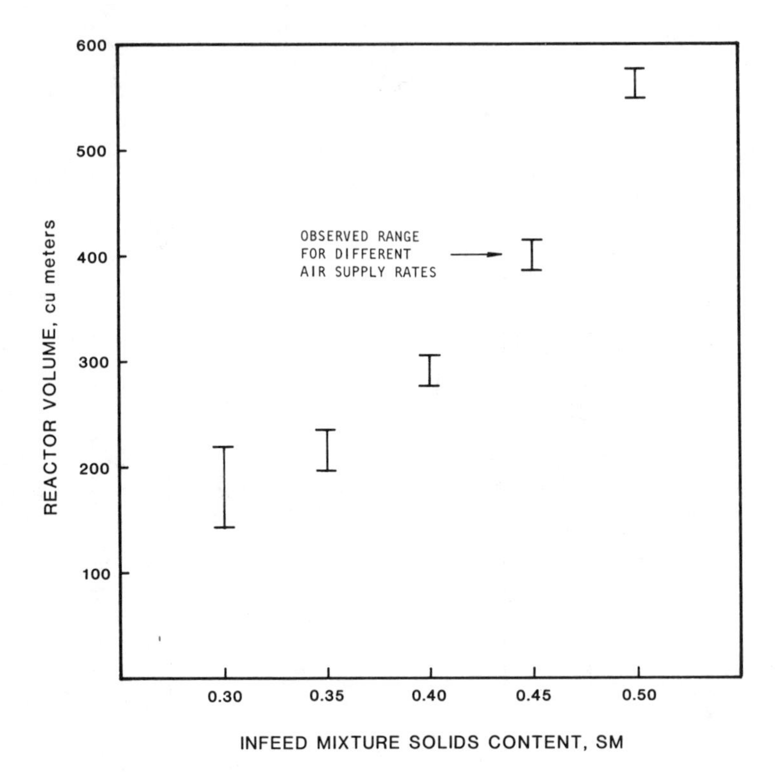

Figure 12-40. CFCM model. Effect of infeed mixture solids, SM, on the required reactor volume for a 7-day detention time. Raw sludge, SC = 0.30, TC = TR = TAIR = 20°C, VR constant at 0.35.

Digested Sludge

Model results for the case of digested sludge are presented in Figures 12-41 to 12-44. The pattern of results is similar to that already discussed for raw sludge. Stable reactor operation is maintained at infeed mixtures of 40, 45 and 50% solids. Below this the range of stable operation becomes constrained. Because of lower available energy in digested sludge, the constraints are more severe than for the case of raw sludge. With 30% infeed solids, for example, stable operation is predicted only in a very narrow band of air flowrates and reactor temperature. Such operation may not be possible on a practical basis. It would appear that infeed mixture solids between 35 and 40% represent a more practical minimum for digested sludges.

From review of Figures 12-41 through 12-44 it is apparent that the advantages of operation at low infeed mixture solids content apply also to

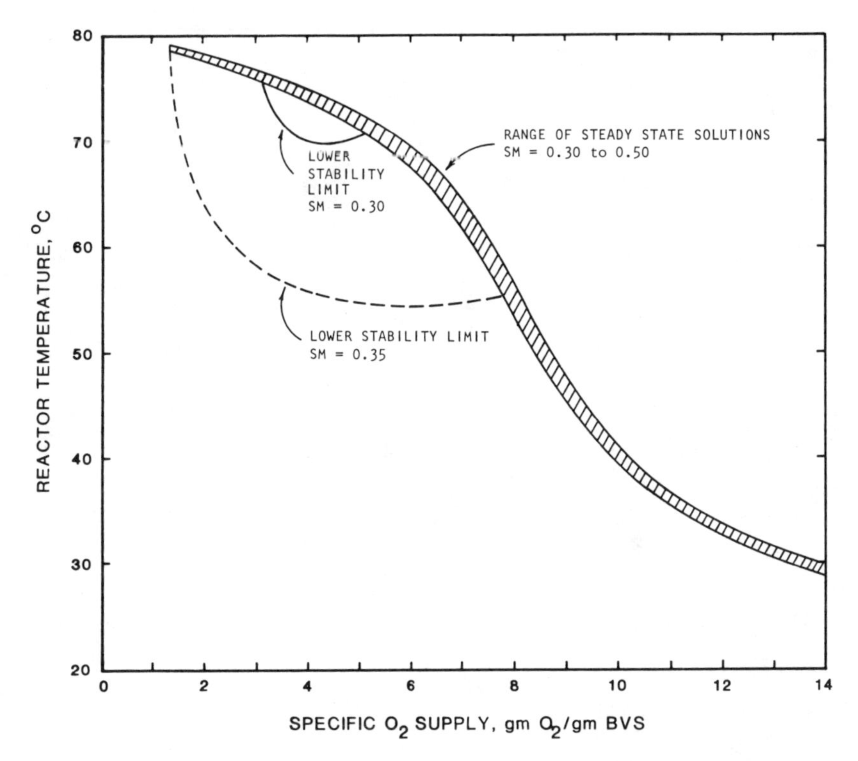

Figure 12-41. CFCM model. Effect of input mixture solids and specific oxygen supply on the steady-state reactor temperature and zone of stable operation. Digested sludge, SC = 0.30, TC = TR = TAIR = 20°C, VR constant at 0.35.

the case of digested sludge. However, full realization of these advantages is more difficult to attain because of the reduced energy supply in digested sludge.

SUMMARY

The CFCM reactor refers to a system operated with continuous feed and withdrawal and with reactor contents under complete-mix conditions. Such conditions probably cannot be attained on a practical basis with solid materials. The conditions are approximated, however, by semicontinuous or intermittent feeding and well-mixed conditions.

Both digested and raw sludges can be composted in such a system. For each sludge type two cases were examined: one in which VS content of

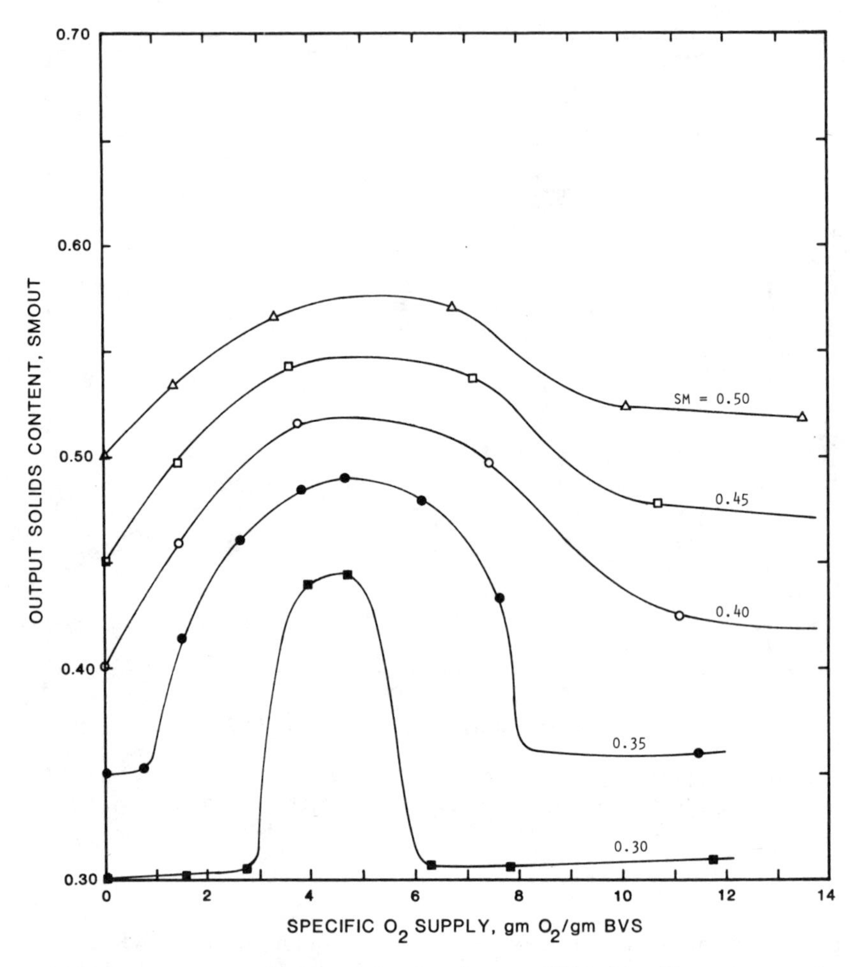

Figure 12-42. CFCM model. Effect of input mixture solids and specific oxygen supply on output solids content. Digested sludge, SC = 0.30, TC = TR = TAIR = 20°C, VR constant at 0.35.

product recycle equalled that of the reactor outfeed (VR = VROUT), and one in which it was significantly less (VR < VROUT). The first case simulates a system in which further curing of reactor outfeed is reduced by either rapid recycle or subsequent drying to reduce biological activity. The second case simulates a system where reactor outfeed is allowed to cure. BVS loss during curing reduces the volatile content of any recycled product. Of the two cases examined the latter is more conservative because energy

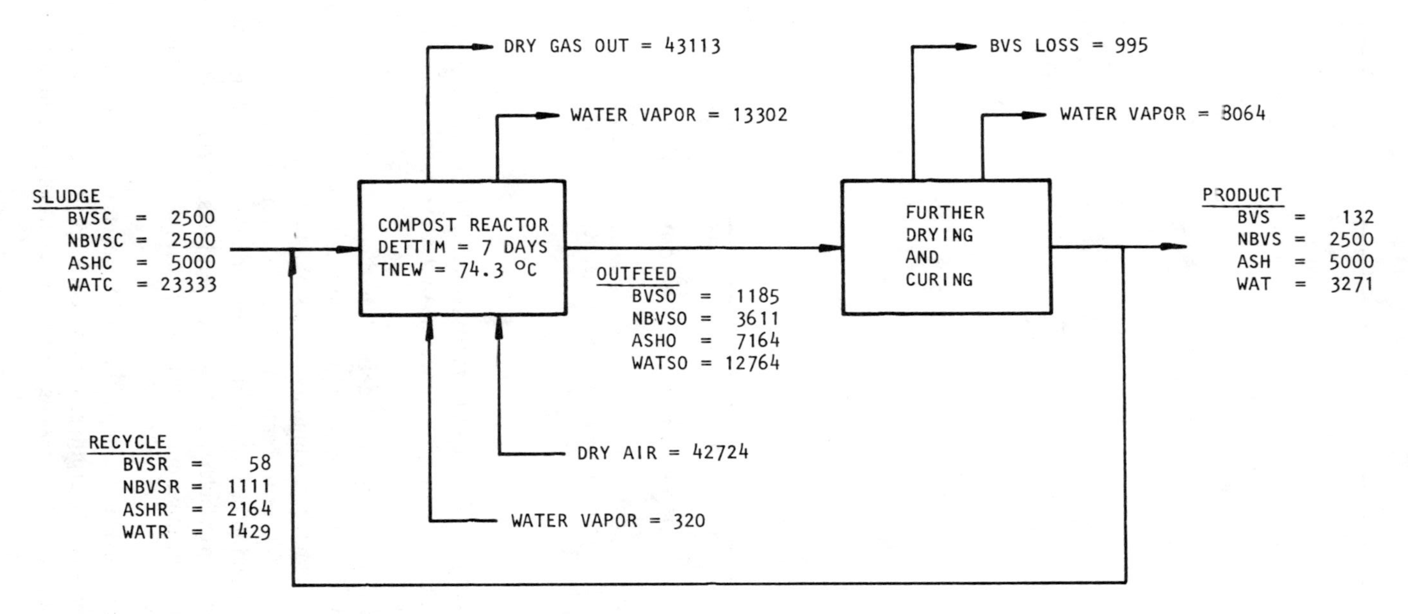

Figure 12-43. CFCM model. Mass balance for solids, water and gas components for an infeed mixture solids content, SM, of 0.35. Digested sludge, SC = 0.30, TC = TR = TAIR = 20°C, VR constant at 0.35, specific air supply = 2.5 m^3/min-dry ton/day, specific oxygen supply = 3.84 g O_2/g mixture BVS. Units on mass terms are kg/day.

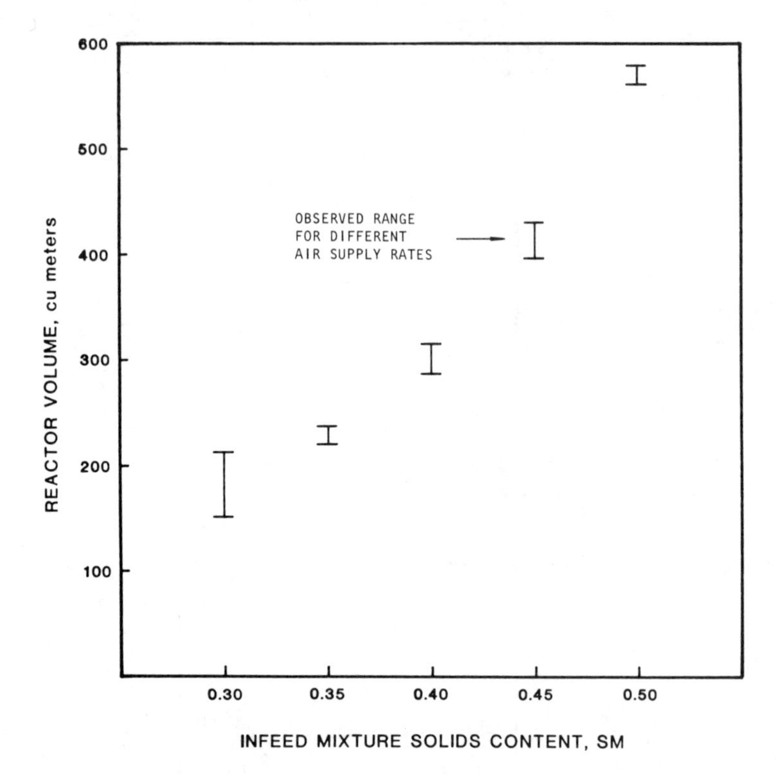

Figure 12-44. CFCM model. Effect of infeed mixture solids, SM, on the required reactor volume for a 7-day detention time. Digested sludge, SC = 0.30, TC = TR = TAIR = 20°C, VR constant at 0.35.

supplied by recycled solids is considerably reduced. Successful composting is predicted to occur for both cases, however, over a reasonably wide range of operating variables.

The specific oxygen supply η, defined as the g O_2 supplied /g BVS of mixed infeed, was found to be a useful parameter for defining the operating range of a CFCM reactor. A summary of maximum η values to maintain reactor temperature >60°C is presented in Table 12-2 for various conditions of sludge cake solids and infeed temperatures. Beyond these values, rapid process failure is predicted, particularly at lower temperatures. Despite the variety of conditions examined, maximum η values fall within a range of only about 5-10 g O_2/g BVS.

Significant drying can occur in the CFCM reactor as water vapor exits with the exhaust gases. Output solids content is predicted to increase in a nearly linear manner with increasing air supply, eventually reaching a plateau

Table 12-2. Predicted Maximum Specific Oxygen Supply for Various Operating Conditions (g O_2/g feed mixture BVS)[a]

	Recycle Product Conditions			
	VR = VROUT		VR < VROUT[b]	
	Infeed Temperatures[c]		Infeed Temperatures[c]	
Sludge Condition	0°C	20°C	0°C	20°C
Digested				
SC = 0.20	5.2	6.7	no solution >60°C	5.4
SC = 0.30	6.4	8.5	4.4	7.3
Raw				
SC = 0.20	8.0	9.8	4.4	7.7
SC = 0.30	8.2	10.2	6.6	9.4

[a]Assumed reactor operating conditions are DETTIM = 7 days, SM = 0.45.
[b]VR constant at 0.35.
[c]Infeed temperatures for water, solids and air.

and finally decreasing at high air flowrates approximately equal to those shown in Table 12-2. Maximum solids output was determined to be a function of the energy content and solids content of dewatered cake, energy content of product recycle, air supply rate and temperature of infeed materials to the reactor. A compilation of maximum output solids predicted for various operating conditions is presented in Table 12-3.

Reactor detention time is an important parameter that affects composting temperature, output solids content and the stabilization of BVS. At a given air supply rate, all of these factors are predicted to increase with increasing reactor detention time within detention times of 2 to 20 days. Composting temperatures >60°C were generally maintained with detention times as low as two days. However, the range of operating parameters (i.e., air supply), maximum output solids and BVS reduction were significantly less than with longer detention times. In fact, with digested sludge at SC = 0.30, infeed temperature of 0°C and VR < VROUT, composting temperatures >60°C were maintained only with detention times of seven days or greater. It is apparent that process stability is favored by long detention times, particularly when recycled product or other amendments contribute little to the energy budget and when infeed temperatures are low. Apparently, a design detention time of about seven days would be a wise choice because the system is stable over a wide range of air supply, and maximum output solids and BVS reduction are only slightly reduced compared to longer detention times.

For a CFCM reactor, feed moisture content may be higher than would

Table 12-3. Predicted Maximum Output Solids for Various Operating Conditions[a] (Output Solids Content is Expressed as a Fraction of Total Solids)

	Recycle Product Condition			
	VR = VROUT		VR < VROUT[b]	
	Infeed Temperature[c]		Infeed Temperature[c]	
Sludge Condition	**0°C**	**20°C**	**0°C**	**20°C**
Digested				
SC = 0.20	0.50	0.52	0.45	0.49
SC = 0.30	0.57	0.59	0.53	0.54
RAW				
SC = 0.20	0.62	0.64	0.55	0.57
SC = 0.30	0.66	0.67	0.61	0.62

[a]Assumed reactor operating conditions are DETTIM = 7 days, SM = 0.45.
[b]VR constant at 0.35.
[c]Infeed temperatures for water, solids and air.

normally be permitted in a batch process. For a batch process, infeed conditions of FAS and moisture must be capable of supporting composting. For the CFCM reactor, however, it is only necessary that conditions within the reactor be capable of supporting composting. In this analysis, raw sludge at 30% cake solids was able to support composting with a mixture solids content as low as 30%. For digested sludge at 30% cake solids, the practical minimum was about 35%. The ability to operate with reduced mixture solids is of significant advantage when composting wet organic sludges. The need for dry product recycle or other dry amendment is reduced, outfeed solids content can still be maintained at a high level and the required reactor volume is significantly reduced. The tradeoff is that the range of stable operation, in terms of air supply and reactor temperature, is reduced with infeed solids contents below about 40%. At 35% infeed solids the constraints become noticable, increasing to rather severe limitations at 30% solids. At both 30 and 35% solids the constraints are more severe with digested sludge compared to raw sludge.

CHAPTER 13

PROCESS DYNAMICS III: SIMULATION OF THE AERATED STATIC PILE PROCESS

BASIC ASSUMPTIONS

The BATCH model is used to simulate a compost process in which all compost materials are assembled at time = 0. Composting then proceeds on a batch basis with no further input of materials other than air and possibly water. The BATCH model was applied to the case of the aerated static pile process using dewatered sludge cake and wood chips for bulking particles. Use of other amendments or recycle product in the mixture was not assumed. The quantity of dry sludge solids was constant at 10,000 kg. Other conditions and variable, assumed in the analysis are presented in Table 13-1.

The quantity of required bulking agent was calculated in accordance with procedures described in Chapter 7. Mixture free air space f_m was adjusted to about 0.20 resulting in a volumetric mixing ratio M_{bc} of about 2.4. This calculates to about 56,000 kg of wood chips at an assumed moisture content of 60%. These values are consistent with the range observed in field studies conducted thus far, and were held constant throughout the simulation analysis.

Static pile cross-sectional dimensions used in the analysis are shown in Figure 13-1. A single triangular pile was assumed with dimensions similar to those commonly used in practice. The volume of mixed material was calculated in accordance with Equation 11-45, assuming that sludge does not add to the total mixture volume.

A 20-day detention time was assumed. For purposes of integration, the detention time was broken into consecutive 0.25-day time steps. Thus, 80 integration steps were used to span the entire compost cycle. By trial and error this was determined to be a sufficient number to assure accurate prediction of the temperature profile. In other words, essentially the same results were observed at shorter integration step sizes, making a larger number of integration steps unnecessary.

TEMPERATURE PROFILE RESPONSES–RAW SLUDGE

The Effect of Air Supply and Infeed Temperature

Predicted temperature profiles throughout the compost cycle are shown in Figures 13-2 through 13-4 for various input temperature conditions. In all cases a raw sludge solids content of 20% was assumed. For the case shown in

Table 13-1. Conditions and Variables Assumed in Batch Model Simulation of Aerated Static Pile Composting[a]

Feed Rates	
Sludge Feed Dry Weight, SC (XC)	10000 kg
Bulking Agent Wet Weight, XB	56000 kg
Amendment Weight, XA	Amendment not used.
Recycle Product, XR	Product recycle not used.
Feed Characteristics	
Bulking Agent Solids Content, SB	0.60
Digested Sludge Volatile Solids Fraction, VC	0.50
Raw Sludge Volatile Solids Fraction, VC	0.70
Bulking Agent Volatile Solids Fraction, VB	0.85
Digested Sludge Degradability Coefficient, KC	0.50
Raw Sludge Degradability Coefficient, KC	0.786
Bulking Agent Degradability Coefficient, KB	0.20
Pile Dimensions	
Cross-Sectional Area, AREAXS	7.50 m^2
Pile Width, WIDTH	5.0 m
Pile Dimension for Conductive Heat Loss, DIMCL	5.0 m
Pile Dimension for Surface Loss, DIMSL	7.8 m
Depth for Surface Heat Loss, DEPTH	0.2 m
Operational Characteristics	
Pile Turnings per Day, TURNUM	0.0
Heat Loss Turning Factor, F3	0.0
Heat Loss Turning Factor, F4	0.0
Detention Time, DETTIM	20 days
Reference Temperature, TREF	0.0°C
Agitator Power, MIXPW	0.0 kW
Agitator Operations, HRSDAY	0.0 hr/day
Heat Transfer Coefficient, U	1.5 kcal/m^2-hr-°C

[a]Values in this table were held constant for most of the analysis presented in this section. Exceptions are noted on individual figures and graphs where applicable. Values of specific heat, specific gravity and heat of combustion are as shown in Table 12-1.

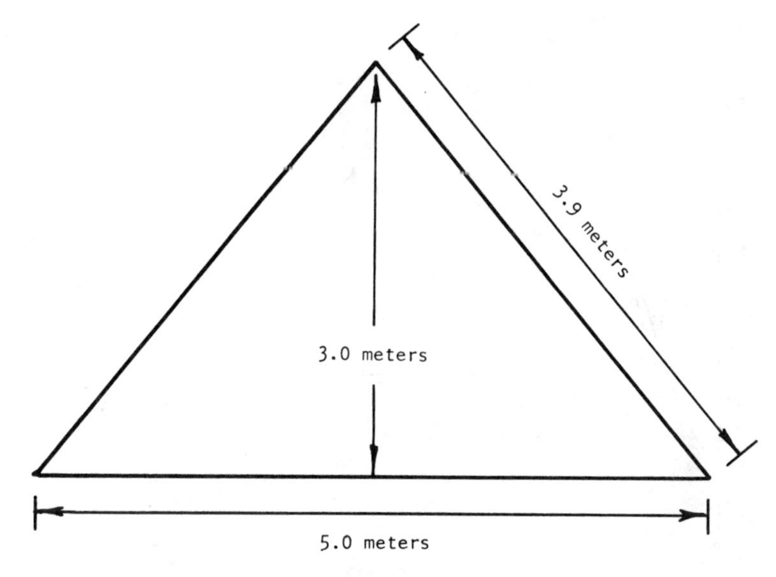

Figure 13-1. BATCH model. Static pile dimensions assumed in the simulation model analysis.

Figure 13-2 all input temperatures were assumed to be 20°C (i.e., TC = TB = TAIR = 20°C). Four different temperature profiles are shown for different air supply rates. For batch processing the specific air supply rate λ is defined as the cubic meters of air supplied per minute per dry metric ton of sludge solids (m^3/min-dry ton). Specific air rates ranging from 0.1 to 3.0 are shown in Figure 13-2.

To verify the accuracy of predicted temperature profiles, it is necessary to compare them with field values collected under conditions closely approximating those assumed in the simulation model. There is something less than an overabundance of published data in this area. However, Willson [14] did publish the curves shown in Figure 13-5 for piles somewhat smaller than those assumed here, but with similar sludge cake solids and input temperatures. The similarity between measured and predicted profiles is quite remarkable. Rapid temperature elevation with very little lag period is noted for the first 2-4 days, provided that λ is less than about 1.5 m^3/min-dry ton. Temperature then peaks at a value between 70 and 80°C. After this initial period, the shape of the temperature profile depends on the air supply rate to the pile. A rather constant, high temperature profile is predicted at 0.1 m^3/min-dry ton, corresponding approximately to curve 4 in Figure 13-5. A tailing-off of the profile after about 12 days becomes noticeable at 1.0 m^3/min-dry ton, which corresponds approximately to curve 2 in Figure 13-5.

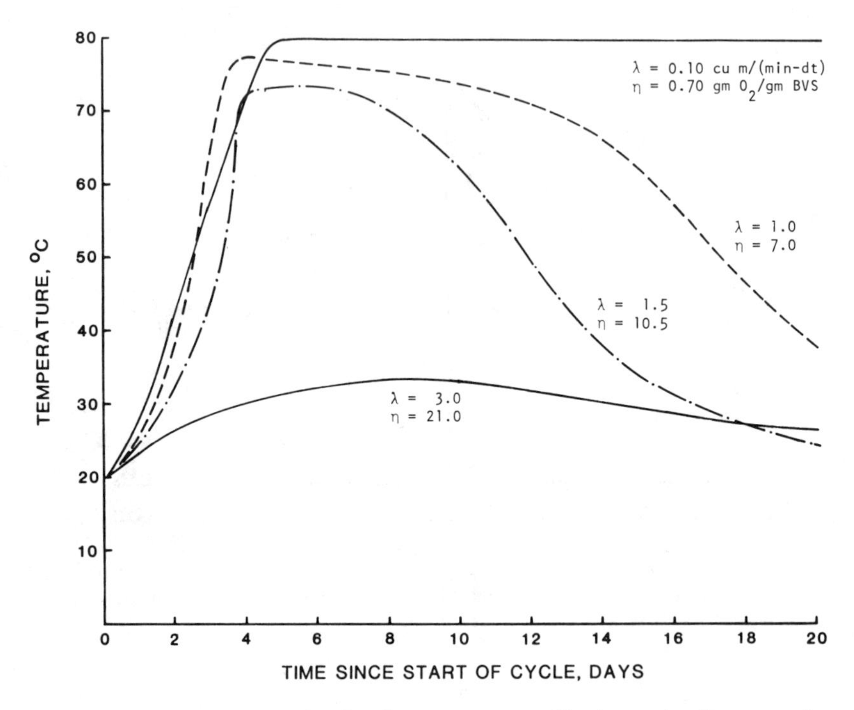

Figure 13-2. BATCH model. Predicted temperature profiles for static pile composting. Raw sludge, SC = 0.20, TC = TB = TAIR = 20°C.

The tailing-off occurs sooner and is more noticeable with an air supply of 1.5 m^3/min-dry ton. Again, this trend is substantiated by the data of Figure 13-5, corresponding approximately to curve 1. The similarity between predicted and observed results, in terms of shape and duration of the initial temperature elevation period, peak temperature achieved and subsequent shape of the temperature profile, is most encouraging.

Before leaving Figure 13-2, it should be noted that inadequate temperature elevations are predicted at an air supply of 3.0 m^3/min-dry ton. As was the case with the CPCM reactor, there is an upper level of air supply beyond which process failure results. For conditions assumed in Figure 13-2, this should occur at an air supply between 1.5 and 3.0 m^3/min-dry ton.

Mass and energy balances corresponding to conditions of Figure 13-2 are presented in Figures 13-6 and 13-7. In both cases the balances refer to cumulative input and output over the 20-day compost cycle. The mass balance corresponds to a specific air supply of 1.0 m^3/min-dry ton, which is near the peak for output solids content. This is reflected in the large quantity

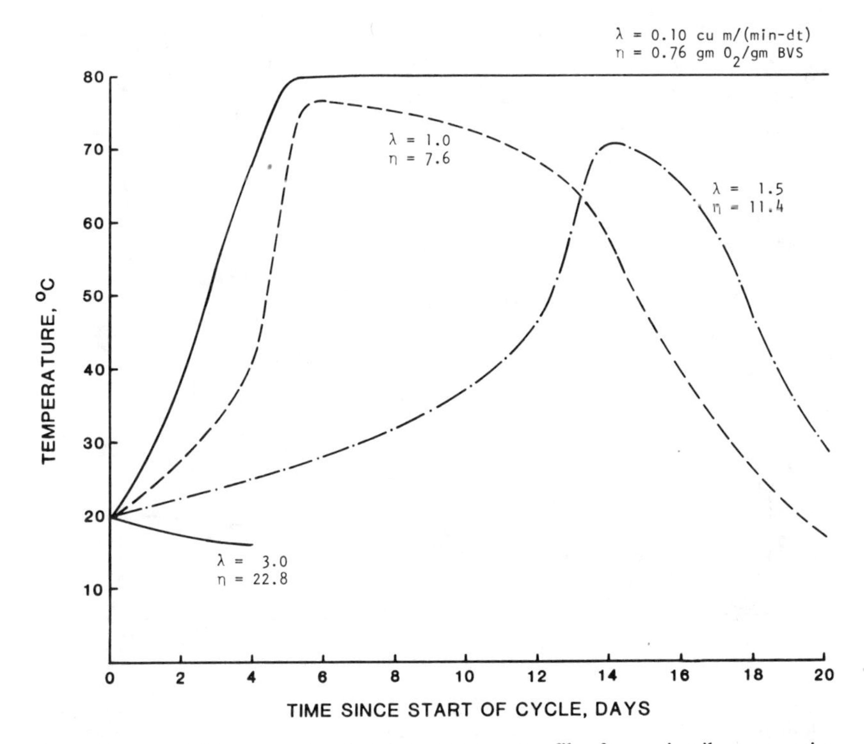

Figure 13-3. BATCH model. Predicted temperature profiles for static pile composting. Raw sludge, SC = 0.20, TC = TB = 20°C, TAIR = 0°C.

of water removed from the process in the form of water vapor. It is interesting to note that the largest mass terms are for dry air input and dry gas output. Both terms dominate the mass balance at the air flowrate shown. The next largest term is water vapor output, which emphasizes the drying potential inherent in the compost process. This is further illustrated by the outfeed solids content, which is about 66%. If this is evenly distributed between sludge and wood chips there would be no difficulty in meeting the 60% solids requirement for bulking agent assumed in the analysis.

Energy balances for specific air supply rates of 0.1 and 1.0 m^3/min-dry ton are shown in Figure 13-7. Energy input from biological oxidation is considerably less at the lower air supply rate, because the system is driven to very high and sustained temperatures with resultant rate constant limitations. Nevertheless, heat released from sludge organics dominates the energy input in both cases. The reader may be somewhat confused by the fact that BVS contribution from sludge and bulking agent are comparable in Figure 13-6. Why

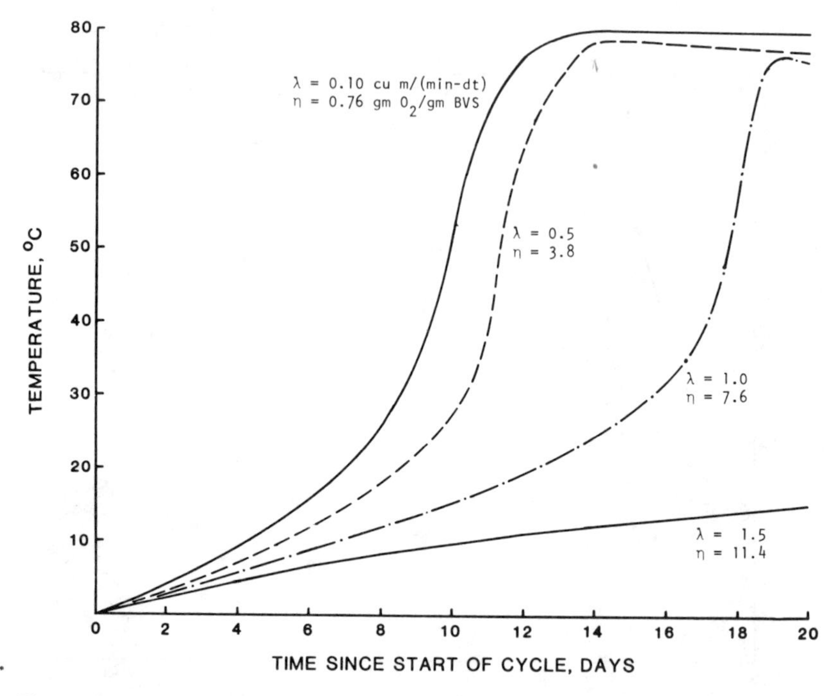

Figure 13-4. BATCH model. Predicted temperature profiles for static pile composting. Raw sludge, SC = 0.20, TC = TB = TAIR = 0°C.

then are the energy input terms different? Recall that the rate constant for BVS oxidation was reduced for wood chips because of their cellulosic nature and structural integrity (see Equation 11-6).

Latent heat of exhausted water vapor is the largest energy output term, even at the lower air flowrate where drying is much reduced. Heat lost by conduction from the bottom of the pile, although not insignificant, is certainly not a dominating factor. In cold climates, therefore, composting is more likely to be affected by the low temperature of infeed materials rather than by increased conductive heat loss.

The simulation model was used to investigate the effect of low input temperatures on system dynamics. Profiles presented in Figure 13-3 were developed with sludge cake and bulking agent input at 20°C and air supply at 0°C. Comparing results with those of Figure 13-2, very little difference in predicted temperature profiles is observed at low air supply rates. Even the rate of temperature development within the pile is not affected in any significant way. Where differences do appear is at higher air supply rates.

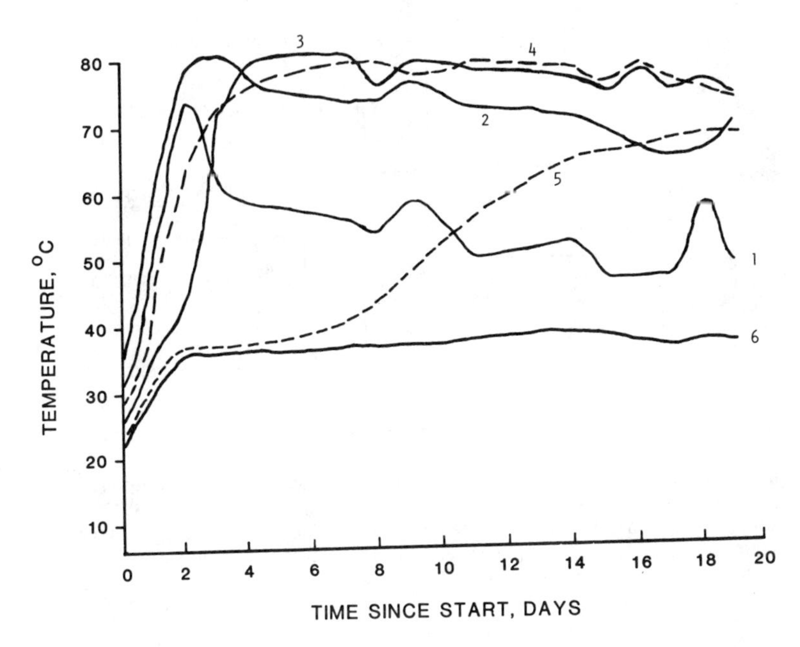

Figure 13-5. Observed effect of aeration rates on the average temperature of static piles composed of a mixture of wood chips and raw sludge at ~20% cake solids. Tests were conducted in July and August at Beltsville, MD, so input temperatures correspond most closely to those of Figure 13-2. Aeration rates were as follows: (1) 1.70 m^3/min-dry ton; (2) 0.63 m^3/min-dry ton; (3) 0.25 m^3/min-dry ton; (4) 0.12 m^3/min-dry ton; (5) 0.05 m^3/min-dry ton; (6) 0.0 m^3/min-dry ton [14].

At 1.5 m^3/min-dry ton, temperature development is greatly protracted because of the lower inlet air temperature. When elevated temperatures finally occur, a peak is achieved followed by a rather rapid decline in temperature. For all practical purposes, maximum air supply should be limited to about 1.0 m^3/min-dry ton under conditions of Figure 13-3.

A more severe case is presented in Figure 13-4, where all input temperatures were assumed to be 0°C. In this case, temperature profiles are markedly affected at all air supply rates. The most obvious impact is the long lag period observed before temperature elevation occurs. For the lowest air supply examined a period of 10-12 days is required to develop peak temperatures. Once developed, however, temperatures remain constant at the elevated levels. The duration of the lag period increases with the air supply rate. At 1.0 m^3/min-dry ton, the temperature profile develops just before termination of the compost cycle. For all practical purposes, the maximum air supply is limited to about 0.5 m^3/min-dry ton.

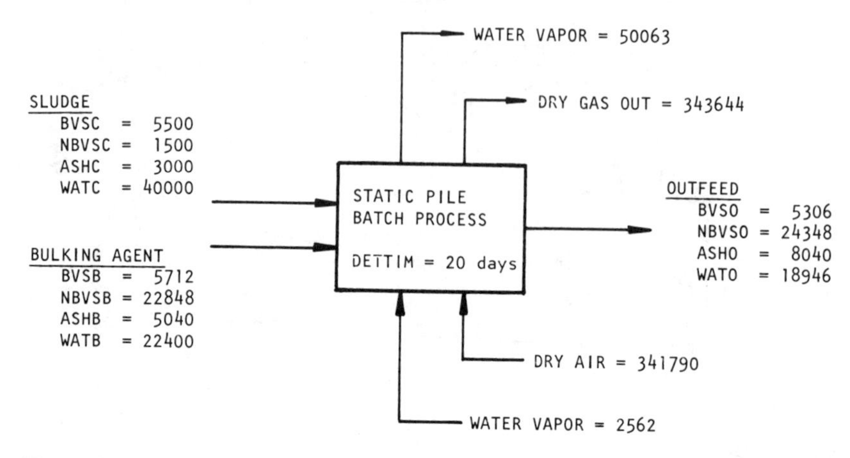

Figure 13-6. BATCH model. Mass balance for solids, water and gas components during static pile composting. Raw sludge, SC = 0.20, TC = TB = TAIR = 20°C, specific air supply = 1.0 m^3/min-dry ton, specific oxygen supply = 7.0 g O_2/g mixture BVS. Units on mass terms are kg and refer to the total input or output over the 20-day compost cycle.

To explain the protracted lag periods it might be assumed that the thermodynamic balance is adversely affected by the low infeed temperatures. Referring to Figure 13-7, however, energy supplied by solids, water and infeed air is relatively insignificant regardless of infeed temperature. Thus, the thermodynamic balance is not significantly affected by the reduction in infeed temperatures considered here. What is affected is the kinetics of the process, and not so much the thermodynamics. Bacterial rates of reaction are markedly reduced as solids temperature is decreased from 20 to 0°C. This is why the lag period is not significantly protracted when only air temperature is reduced to 0°C (Figure 13-3). Microbes are still reacting in response to sludge and bulking agent temperature and only at high air flowrates is an effect observed. When solids themselves are reduced to 0°C, the effect on microbial kinetics is acutely observed, and long lag periods result at all air supply rates. In effect, the energy to perform the composting task is thermodynamically present, but the spark is slow to ignite the process. Note that once pile temperatures reach 20°C, the remainder of the temperature elevation profile is similar to those observed at higher infeed temperatures. It is simply a problem of getting things started.

The above phenomenon has occasionally been observed in actual operations in cold climates. When all feed temperatures are low (i.e., near 0°C), newly formed piles have been observed to sit with little temperature development. Published data on this effect are lacking, however, because process

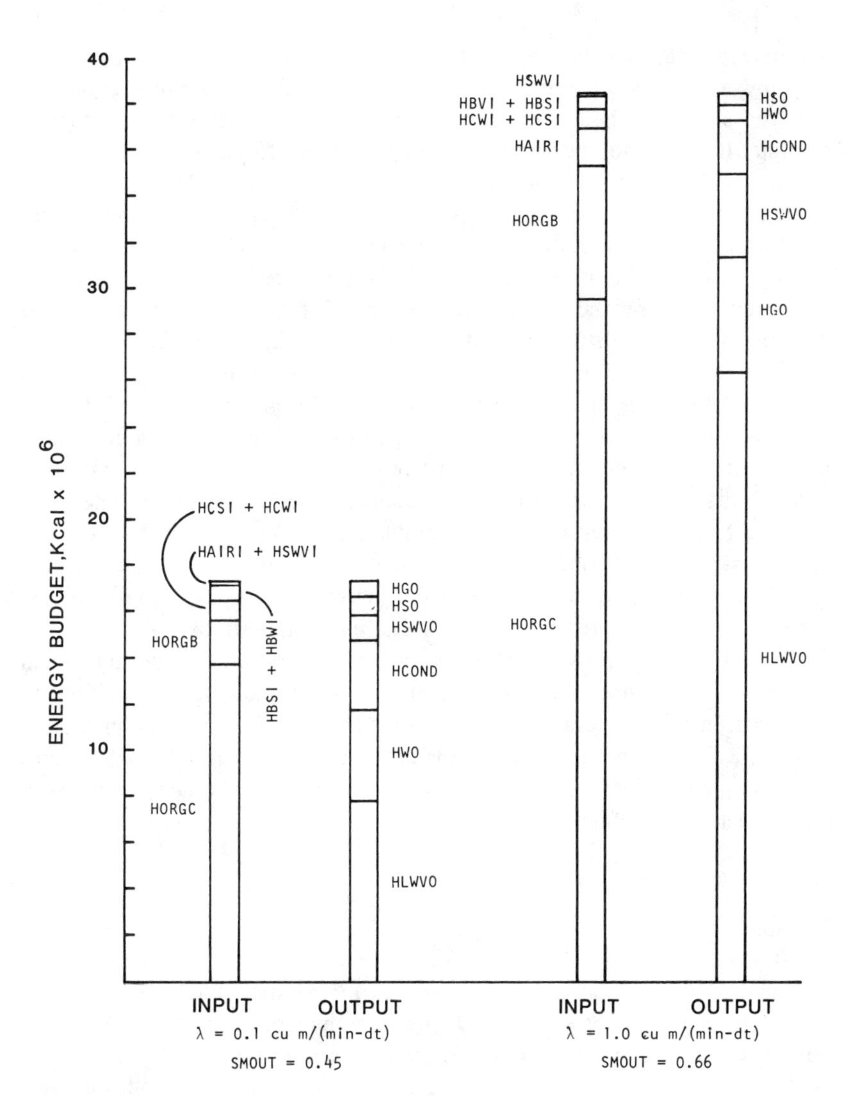

Figure 13-7. BATCH model. Predicted energy balances for static pile composting with low and high air supply rates. Raw sludge, SC = 0.20, TC = TB = TAIR = 20°C.

failures are not commonly reported in scientific journals. Recourse in such situations has often been to apply additional insulating layers to the pile and reduce the air flowrate (the latter will be explored further in the following section). These would be effective measures if the problem were one of process thermodynamics. Unfortunately, it is not and the above procedures will have only a marginal impact. Because it is a kinetic limitation caused by very low starting temperatures, it would be far better to assure that heat in the feed substrates is conserved to assure as high a starting temperature as possible. If this is not sufficient it may be necessary to preheat the air supply during the initial lag period until the process "ignites." Once started any further heating should be unnecessary. Air preheat could be supplied from conventional fuel sources or by heat exchange between exhaust air from an already started pile and feed air to the pile being started.

As stated before, the compost mixture in a batch process must be capable of supporting composting from the inception of the cycle. To this end we have stated that free airspace (FAS) and moisture content must be within the acceptable range of values. Now another condition can be applied. In cold climates the temperature of input solids and water must be high enough to avoid extended lag periods. It is interesting to note that FAS, moisture content and temperature all relate to potential kinetic limitations. Indeed, all these factors have been used in the simulation model to adjust the reaction rate constant (see Chapter 11). The rule for batch systems can then be stated more concisely as follows: for a batch process the initial compost mixture must be adjusted to remove excessive rate limitations.

The distinction between thermodynamic and kinetic limitations again points to differences between batch and continuous feed, well-mixed processes. Infeed temperature in a well-mixed system has very little impact on biochemical reaction rates because it is the reactor temperature which governs process kinetics. This would appear to be a significant advantage for compost operations in cold climates.

Variable Air Supply

The long lag period observed with low-temperature feeds prompted an investigation of the effect of variable air supply rates in reducing the lag period. Could a very low air supply rate be used in the initial phase of the cycle to conserve what little heat is being released? What if it is followed by a higher air flowrate to supply the higher oxygen demand when reaction rates increase? Results for such an operational mode are depicted in Figure 13-8. The initial air supply is only 0.02 m^3/min-dry ton, about 20% of the lowest rate shown previously in Figure 13-4. At day 4 the rate is increased to 0.5 m^3/min-dry ton, and remains constant thereafter. Surprisingly, the temperature profile differs little from those shown in Figure 13-4. Apparently, low air flowrates do little to reduce the lag period associated with batch operation and low infeed temperatures.

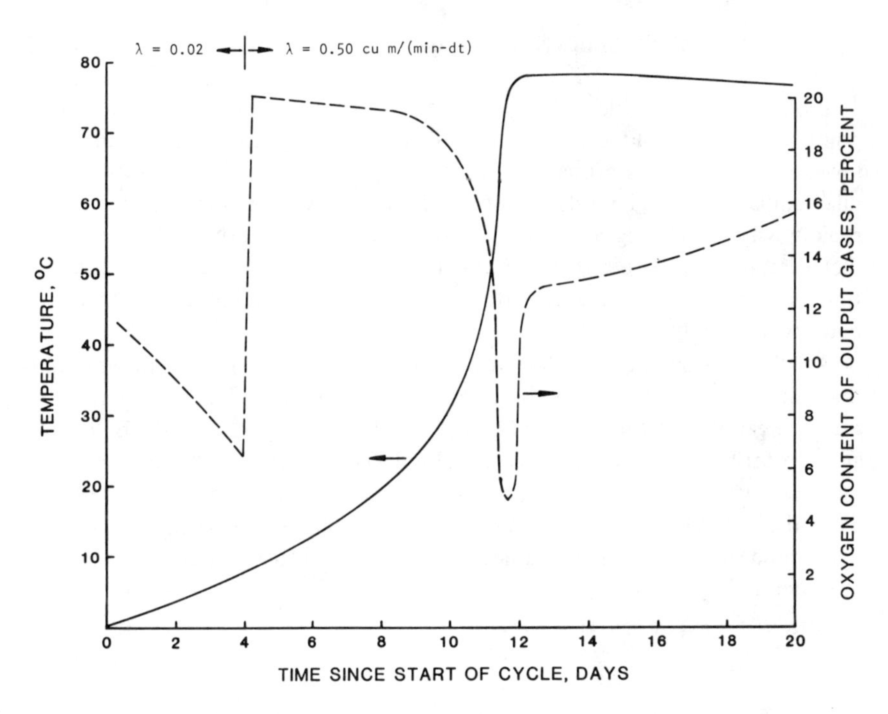

Figure 13-8. BATCH model. Effect of variable air supply rate on the predicted temperature profile and oxygen concentration during static pile composting. Raw sludge, SC = 0.20, TC = TB = TAIR = 0°C.

Average oxygen concentration within the pile is also shown in Figure 13-8 and it is worthy of comment. As would be expected, oxygen concentration decreases in the initial stage as composting temperature increases. Very little air is needed to maintain adequate oxygen levels because of the low reaction rate. A jump in oxygen content results when the step change in air flowrate is made. Oxygen content then remains at an elevated level until the temperature rises steeply. The corresponding decrease in oxygen content is, of course, caused by the flush of biological activity at the higher temperatures. Interestingly, the oxygen content then reaches a minimum point, rises as steeply as it once fell and gives way to a more gradual increase. The reason for this rather unusual display is that once thermophilic conditions are achieved the system is driven to a very high temperature where reaction rates are limited. Just as oxygen concentration is reduced in response to more favored reaction temperatures, it is then increased as rate-limiting temperatures are achieved.

Biodegradable Volatile Solids Stabilization

Biodegradable volatile solids (BVS) reductions for cases examined to this point are shown in Figure 13-9. Low BVS reductions are predicted at the lower air supply rates primarily because of development of excessively high pile temperatures and resultant rate limitations. BVS stabilization increases rapidly with increasing air supply. Peak reductions of about 55% are predicted at air supplies between 0.5 and 1.5 m^3/min-dry ton. Further stabilization is prevented by development of low moisture rate limitations in this range of air supply.

Another reason for the rather low organic stabilization is the assumption that the BVS oxidation rate for bulking agent organics is less than that for sludge organics (see Equations 11-5 and 11-6). From Figure 13-6 the bulking agent contributes a comparable level of biodegradable organics to the starting mixture. With the reduced rate constant a rather large percentage of BVS from the bulking agent can be expected in the final product. On the other hand, biodegradable sludge organics will be stabilized to a greater extent than indicated in Figure 13-9.

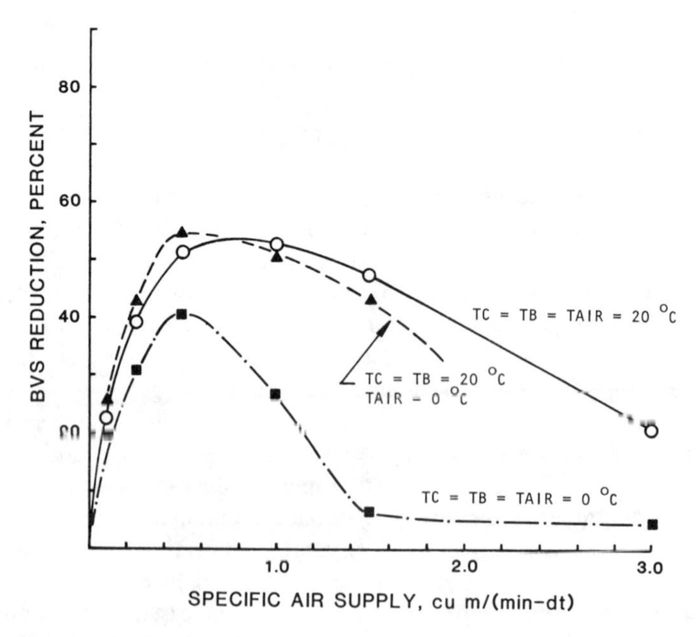

Figure 13-9. BATCH model. Effect of specific air supply and infeed temperature on BVS reduction. Raw sludge, SC = 0.30.

Effect of Cake Solids

Before leaving this section on raw sludge, temperature profiles for the case of 30% sludge solids and 20°C infeed temperatures are shown in Figure 13-10. Compared to Figure 13-2 for the case of a 20% cake, very little difference is evident in the temperature profiles. However, temperature profiles do not tell the entire story and this subject will be discussed later.

TEMPERATURE PROFILE RESPONSES–DIGESTED SLUDGE

Predicted temperature profiles for the case of digested sludge at 20% cake solids are shown in Figures 13-11 and 13-12 for various infeed temperatures. In Figure 13-11, all input temperatures were assumed to be 20°C. Comparing

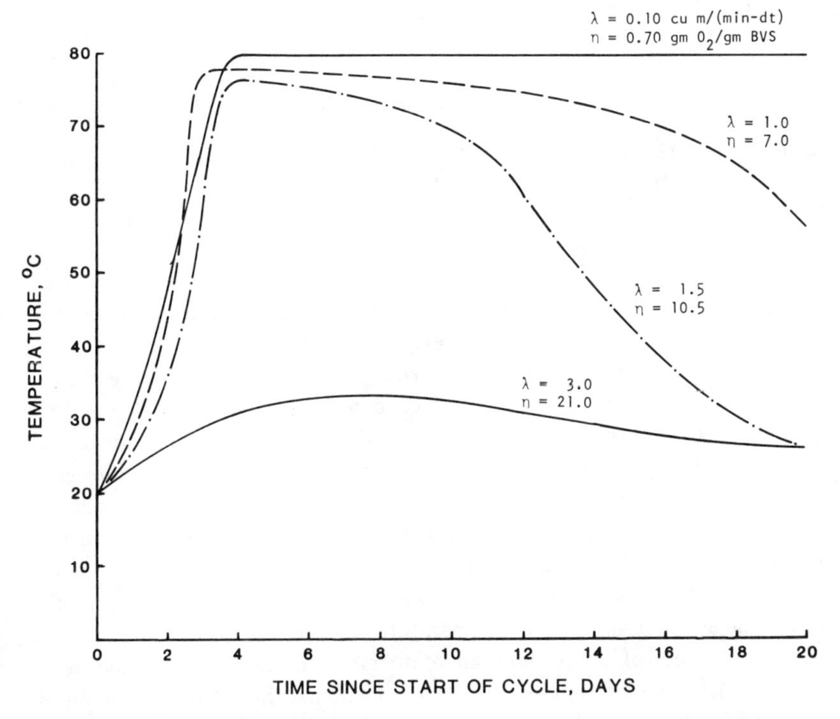

Figure 13-10. BATCH model. Predicted temperature profiles for static pile composting. Raw sludge, SC = 0.30, TC = TB = TAIR = 20°C.

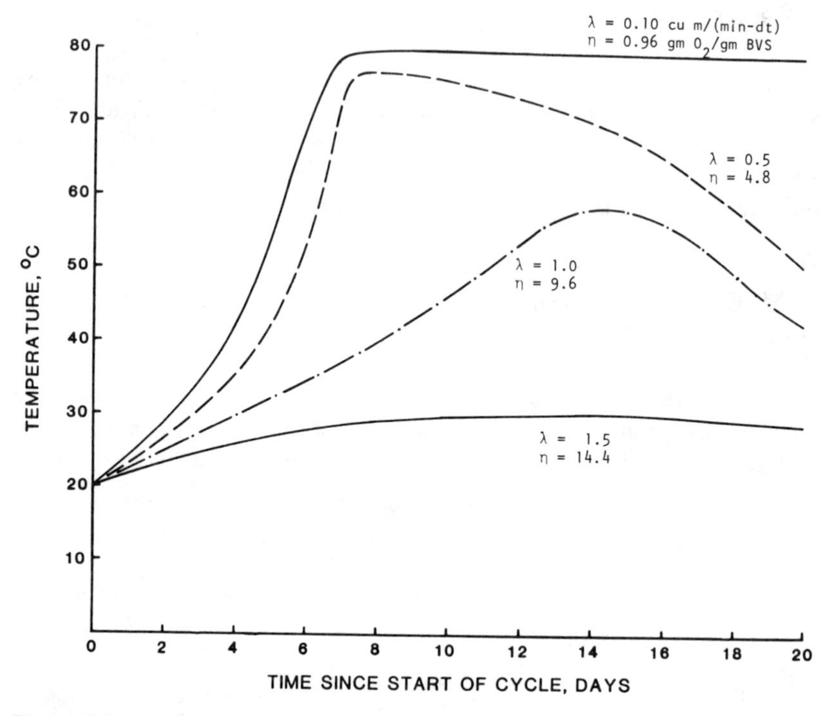

Figure 13-11. BATCH model. Predicted temperature profiles for static pile composting. Digested sludge, SC = 0.20, TC = TB = TAIR = 20°C.

temperature profiles with those for raw sludge at the same infeed temperature (see Figure 13-2) at least two differences are immediately evident. First, digested sludges exhibit a longer lag period before temperatures exceed 60°C. This is caused largely by the lower BVS content in digested sludge and the assumption of first-order kinetics. Second, the range of allowable air supplies is reduced. On a practical basis the maximum air supply is probably limited to about 0.5-1.0 m^3/min-dry ton. At 1.5 m^3/min-dry ton, no significant temperature elevation is predicted. All of these air flowrates resulted in good temperature elevations for the case of raw sludge.

The effect of decreasing inlet air temperature to 0°C while maintaining initial sludge and bulking agent temperatures at 20°C is shown in Figure 13-12. The major effect of the lowered air temperature is another reduction in the range of operating air supplies. On a practical basis the maximum air supply is probably limited to about 0.25-0.5 m^3/min-dry ton. This is about half the maximum rate with an inlet air temperature of 20°C.

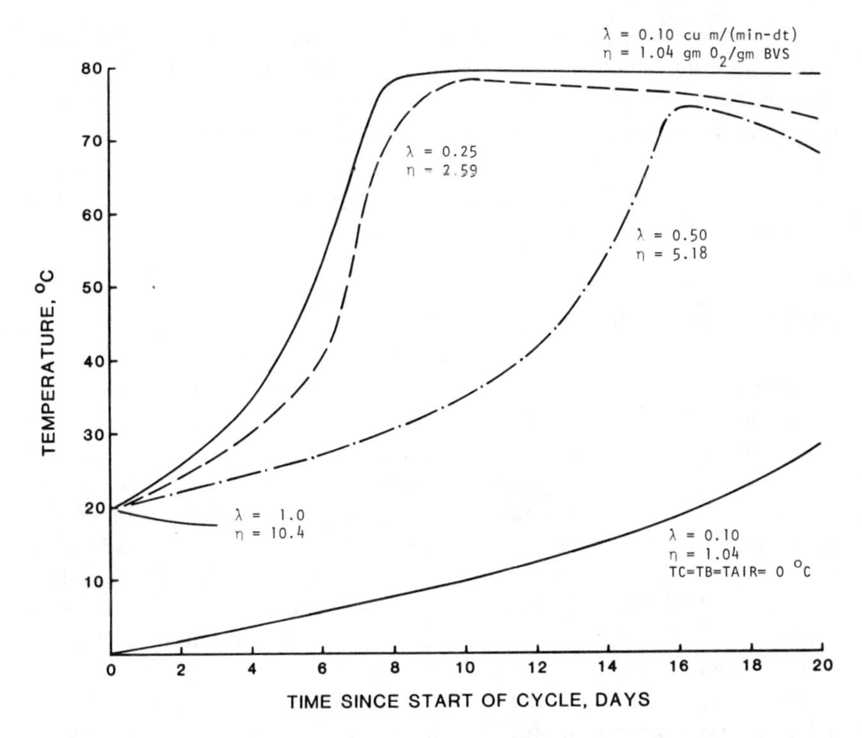

Figure 13-12. BATCH model. Predicted temperature profiles for static pile composting. Digested sludge, SC = 0.20, TC = TB = 20°C with TAIR = 0°C, and TC = TB = TAIR = 0°C.

If all infeed materials enter the process at 0°C (also shown in Figure 13-12), no useful temperature elevation is predicted over the 20-day period. Long lag periods were also predicted with raw sludge under similar conditions. However, adequate temperatures eventually did develop 10-12 days into the cycle.

The question has often been asked whether anaerobically digested sludge can be composted by the static pile system in cold climates. This question often arises because most static pile work has been conducted with raw sludge and there are limited field data to guide those who produce digested sludge. Based on the simulation model, the answer is that digested sludge can be composted with 0°C inlet air temperature, provided the initial temperatures of sludge and bulking agent average about 20°C. Because anaerobic digesters are conventionally maintained at about 35°C, every effort should be made to conserve this heat. This should help overcome initial kinetic limitations. If the air supply is kept within acceptable limits, constrained as they may be, thermodynamic limits should not be exceeded and composting should proceed. If problems develop in maintaining proper temperatures during severely cold weather, the operator should consider bypassing a fraction of raw sludge around the digester and dewatering it with remaining digested sludge. This will increase the energy level of input sludge. The amount of bypassing can be adjusted until a proper thermodynamic balance is reestablished.

OUTPUT SOLIDS CONTENT

Predicted output solids content as a function of specific air supply is shown in Figures 13-13 and 13-14 for conditions of infeed temperature and cake solids previously discussed. Solids contents shown are for the output mixture at the end of the 20-day compost cycle. Referring to Figure 13-13 for the case of raw sludge, output solids content increases in a nearly linear manner with increasing air supply. A maximum solids value is eventually reached and further increases in air supply result in a decrease in output solids. The air supply at which maximum output solids occur is about 1.0 m^3/min-dry ton for the case of 20°C infeed temperature, decreasing to about 0.5 m^3/min-dry ton with 0°C infeed.

Maximum output solids content is a function of infeed cake solids and infeed temperature conditions. Here we can see the advantage of producing drier cake solids from dewatering. Output solids content is consistently greater at a given value of air supply for the drier infeed cake. Recall that temperature profiles were similar for 20 and 30% infeed cakes.

It is interesting to note that output solids contents are very similar for the cases where all infeed temperatures are 20°C and where infeed air temperature is 0°C and cake and bulking agent are maintained at 20°C. Significant drying is possible even with very cold infeed air. The key is that long lag periods must be avoided to the extent possible. When the temperature of all infeed materials is reduced to 0°C, which results in long lag periods, drying is significantly reduced. Maximum output solids content also occurs at a lower air flowrate compared to other curves in Figure 13-13.

For the case of digested sludge, as shown in Figure 13-14, the reduced output solids content compared to raw sludge becomes immediately apparent. With 20% digested cake and 20°C input conditions, maximum output solids are predicted to be only about 52%. Under similar conditions with raw sludge, output solids content was >65%. Thus, with a 20% digested cake, adequate compost temperature can be achieved but with reduced drying potential and over a smaller range of operating variables. This is entirely consistent with the analysis of Chapter 9, which pointed to

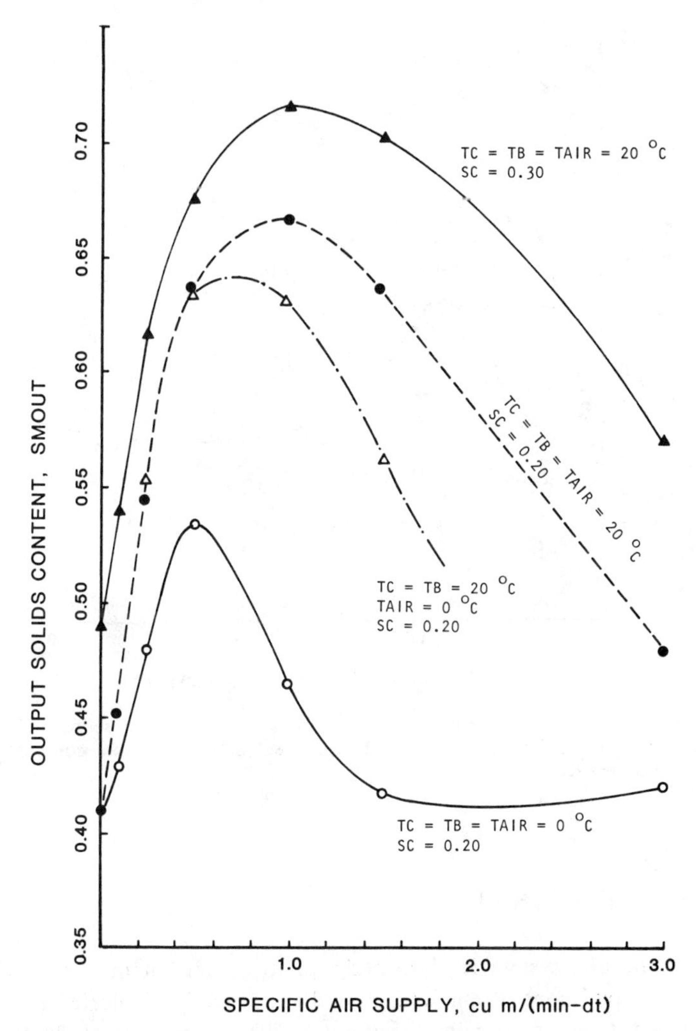

Figure 13-13. BATCH model. Effect of specific air supply, cake solids and infeed temperature conditions on output solids content after static pile composting of raw sludge.

two distinct thermodynamic regions; one in which sufficient energy is available for both composting and drying and another in which energy is sufficient only for composting with limited drying. With 20% digested cake, energy supplies were predicted to be sufficient for only limited drying, a result verified by the present analysis.

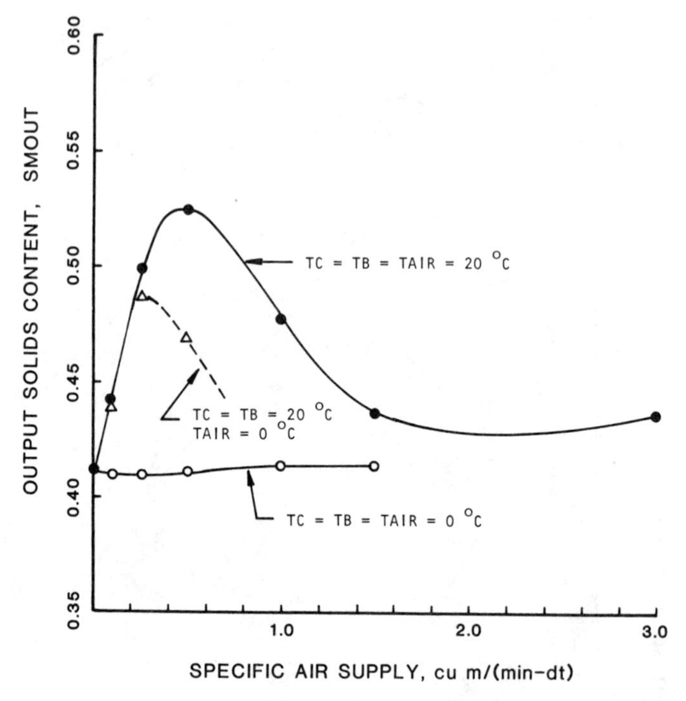

Figure 13-14. BATCH model. Effect of specific air supply, cake solids and infeed temperature conditions on output solids content after static pile composting of digested sludge. SC = 0.20.

EFFECT OF RAINFALL

The static pile system is often designed to operate in the open, subject to ambient conditions of rain and snow. Experience has indicated that such systems are tolerant to rainfall, although difficulties have at times been encountered with screening of the final product. A similar tolerance to rainfall has been observed during windrow composting of dry starting materials such as refuse. In the latter case water is often added to avoid low-moisture rate-limitations, so moderate water additions from rainfall would not be cause for concern. With wet materials such as dewatered sludge, however, water is always a concern and the possible impact of rainfall must be addressed seriously.

The aerated static pile system was simulated under a variety of rainfall conditions. Raw sludge at 20% solids was assumed in all cases. Air temperature was fixed at 10°C, rainfall temperature at 5°C, and relative humidity at 100%. These conditions were selected to simulate a moderately cold and wet climate. Temperature profiles corresponding to one of the more severe rainfall patterns examined is shown in Figure 13-15. Air supply was constant at 0.5 m^3/min-dry ton and two 3-day periods of rainfall were assumed, one near the beginning and one in the latter half of the compost cycle. In each 3-day period the rainfall rate was 2.54 cm/day for a total rainfall of 15.24 cm. A temperature profile under similar conditions but with no rainfall is also shown.

It is apparent from Figure 13-15 that 15 cm of rainfall is predicted to have almost imperceptible impact on the temperature profile. Any effect of added cold water is balanced by increased reaction rates from the added moisture. This is illustrated by the moisture adjustment factor F1, shown for the cases with and without rain. In each case, F1 decreases during the compost cycle as moisture is removed from the pile. However, the decrease is significantly less when rainfall is added. A plateau effect is actually predicted during the second rainfall period.

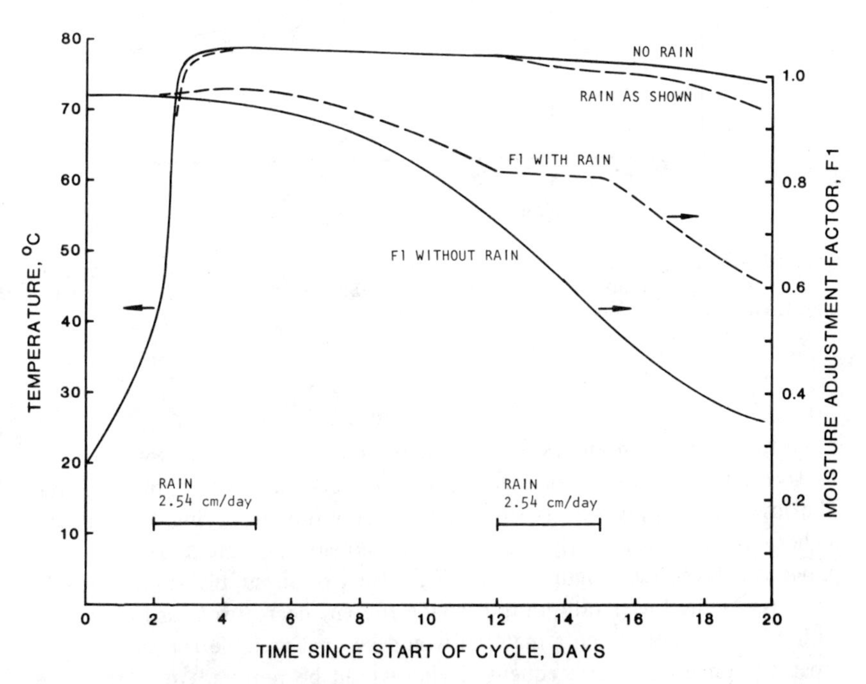

Figure 13-15. BATCH model. Effect of rainfall on predicted temperature profiles and moisture adjustment factor during static pile composting. Raw sludge, SC = 0.20, TC = TB = 20°C, TAIR = 10°C, TRAIN = 5°C, specific air supply = 0.5 m^3/min-dry ton, total rainfall = 15.24 cm.

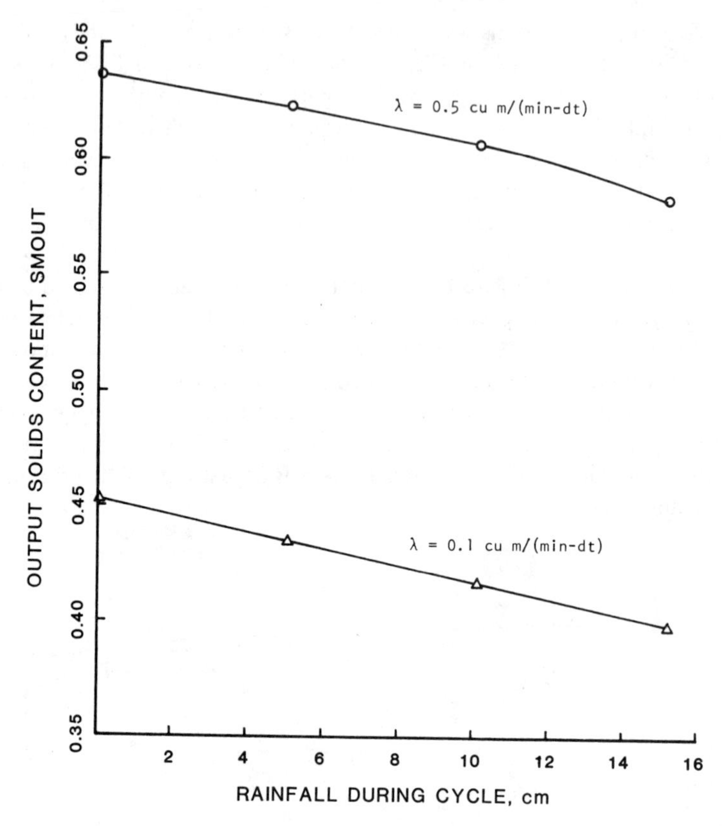

Figure 13-16. BATCH model. Effect of rainfall and air supply on output solids content from static pile composting. Raw sludge, SC = 0.20, TC = TB = 20°C, TAIR = 10°C, TRAIN = 5°C.

Output solids content as a function of total rainfall is shown in Figure 13-16 for two conditions of specific air supply, 0.1 and 0.5 m^3/min-dry ton. The lower air supply rate is sufficient for only limited drying, whereas the higher rate achieves nearly maximum output solids content based on the previous analysis (see Figure 13-13). The effect of air supply is immediately obvious. At the lower rate, output solids content decreases to as low as 40% with 15 cm of rain. Such a material would be unacceptable for screening on most equipment and subsequent drying would be required to produce an acceptable product. Obviously, air-drying would be impractical until wet weather conditions subside.

The situation is considerably improved at the higher air flowrate. In this case output solids content remains above 60% with rainfall up to about 12 cm. Such a situation is far more attractive from an operational standpoint because the final product can be screened and is suitable for direct reuse. Once again, the air flowrate is shown to be a major parameter which can be used to control the compost process.

At this point it would be proper to discuss some of the limitations of the previous analysis. The BATCH simulation model assumes that material which enters the process is evenly distributed throughout the composting material. Thus, rainfall added to the static pile is assumed to be distributed evenly throughout the pile. Possible spatial variations are not considered. All water added can, therefore, act to reduce moisture limitations and increase the rate of reaction. Because a static pile is normally not turned during the compost cycle, spatial variations should be expected. Penetration of water from rainfall may be limited to outer layers of the pile. Thus, impacts on reaction rate constants may be limited to only certain zones of the pile.

What effect would spatial variations have on the predictions of this section? In all likelihood, average temperature conditions in the pile would be similar to predicted profiles. Variations in temperature would be accentuated, however, with outer zones cooled somewhat by the heavy water burden. Average output solids content would be reduced below values shown in Figure 13-16. If spatial variations detract from average performance of the pile, why not remix the pile periodically during the compost cycle? This would redistribute materials such as rainfall which are added after the initial mixing and would also compensate for any errors in the initial mixing. It would also allow materials in colder temperature zones, such as near the toe of the pile, to be redistributed into the pile interior. Performance should then begin to approach that predicted in this section. The practice of remixing during the compost cycle has been incorporated into at least one static pile design and would appear to offer very real benefits.

SUMMARY

The BATCH model was used to simulate the aerated static pile compost process in which all compost materials are assembled at time = 0. Composting then proceeds on a batch basis with no further input of material other than air and possibly water.

The simulation model predicts temperature profiles which in general closely approximate profiles measured under similar field conditions. The predicted shape and duration of the initial temperature elevation period, peak temperatures achieved, subsequent shape of the temperature profile and maximum air flowrates before serious temperature depressions occur correspond well with actual measurements.

An initial lag period of about 2-4 days is predicted for maximum temperature development if all infeed temperatures are 20°C (i.e., TC = TB = TAIR = 20°C). This is increased only slightly if air temperature is reduced to 0°C and mixed cake and bulking agent are maintained at 20°C (TC = TB = 20°C, TAIR = 0°C). If all input materials are at 0°C (TC = TB = TAIR = 0°C), significant lag periods can occur before proper temperature development.

For the case where TC = TB = TAIR = 0°C, decreasing the air supply to low levels in the initial period does not significantly shorten the lag period. The process is kinetically limited by slow biological reaction rates at the low temperature. Every effort should be made to conserve heat contained in the original mixed material. Such heat does not significantly affect the thermodynamic balance but does have a considerable bearing on initial process kinetics. For a batch process the initial compost mixture must be adjusted to reduce kinetic rate limitations, whether caused by low moisture content, low FAS or low initial mixture temperature.

There is an upper level of air supply beyond which process failure occurs. For a 20% raw sludge cake, the maximum air supply is limited to about 1.5 m^3/min-dry ton for the case where TC = TB = TAIR = 20°C. Beyond this value peak temperatures are reduced, and no useful temperature development is predicted at a rate of 3.0 m^3/min-dry ton. The maximum air supply is reduced to about 1.0 m^3/min-dry ton if TC = TB = 20°C and TAIR = 0°C. If TC = TB = TAIR = 0°C, the maximum air supply is effectively limited to about 0.5 m^3/min-dry ton.

Anaerobically digested sludge can be composted effectively in the static pile system, but the range of predicted air supplies is reduced compared to the case with raw sludge. In cold climates care must be taken to conserve heat contained in the feed cake and bulking agent to avoid low reaction rates in the early compost stage. Blending a fraction of raw dewatered sludge will improve the thermodynamic balance and can mitigate operational limitations during cold weather.

Considerable quantities of water can be removed as water vapor during the static pile process. For a given infeed cake solids, output solids content is predicted to increase in a nearly linear manner with initial increases in air supply. A maximum output solids content is eventually reached and further increases in air supply result in a decrease in output solids. Output solids content is also influenced by feed cake solids, and at a given air supply is consistently greater for the drier feed solids. For a raw sludge at 20% solids, maximum output solids content of about 66% is predicted at an air supply of 1.0 m^3/min-dry ton (TC = TB = TAIR = 20°C). Significant drying is possible even with 0°C infeed air, provided long lag periods are avoided.

The static pile system appears to be quite tolerant to moderate rainfall. As much as 15 cm of rainfall during the compost cycle is predicted to have little effect on the average temperature profile. Water added to the pile as rainfall can be partially removed by increasing the air flowrate. With an air flowrate of 0.5 m^3/min-dry ton, output solids content is predicted to decrease from about 63% with no rain to about 58% with 15 cm of rainfall.

The BATCH model assumes that materials input to the process are evenly distributed within the static pile. Possible spatial variations resulting from material input after the initial mixing, such as rainfall, are not considered. Periodic remixing of the pile during the compost cycle would redistribute such materials and compensate for any inadequacies in the initial mixing.

CHAPTER 14

PROCESS DYNAMICS IV: SIMULATION OF THE WINDROW SYSTEM

BASIC ASSUMPTIONS

The BATCH model was used to simulate a windrow compost system. Use of recycled product and air-dried sludge as an amendment were considered in the analysis. Sufficient material was added to adjust the starting mixture to 40% solids. Recycled product was assumed to be well stabilized (i.e., VR = 0.35, KR = 0.05). This is a conservative case because it results in a small energy contribution from recycled material. Addition of bulking particles, such as wood chips, was not considered. The quantity of dry sludge solids was again held constant at 10,000 kg. All sludge was assumed to be anaerobically digested to limit the number of variables included in the analysis. Other conditions and values assumed are presented in Table 14-1.

Windrow cross-sectional dimensions used in the analysis are shown in Figure 14-1. A triangular shape was assumed with dimensions consistent with what is presently attainable with mobile turning equipment. Trapezoidal and even rectangular cross sections can be achieved if fibrous materials, such as straw, are added to the windrow. Addition of such materials was not assumed in the analysis, and a triangular cross section is more consistent with that observed when recycled product is used as the sole amendment.

A detention time of 20 days was assumed. As in the analysis for static pile composting in Chapter 13, the detention time was broken into consecutive time steps, each of 0.25 day duration.

In the case of windrow composting, the heat loss factors F3 and F4 take on nonzero values. Referring to Equation 11-71, the factor F3 is used to account for exposure of compost to ambient conditions during turning. Based on information presented in Chapter 11, a value of 0.10 was assumed.

Table 14-1. Conditions and Variables Assumed in BATCH Model Simulation of Windrow Composting[a]

Feed Rates	
Sludge Feed Dry Weight, SC (XC)	10000 kg
Recycle Product Wet Weight, XR	Quantity adjusted to give SM = 0.40.
Amendment Wet Weight, XA	Quantity adjusted to give SM = 0.40.
Bulking Agent Wet Weight, XB	Bulking particles not used.
Feed Characteristics	
Recycled Product Solids Content, SR	0.70
Amendment Solids Content, SA	0.80
Digested Sludge Volatile Solids Fraction, VC	0.50
Recycled Product Volatile Solids Fraction, VR	0.35
Amendment Volatile Solids Fraction, VA	0.50
Digested Sludge Degradability Coefficient, KC	0.50
Recycled Product Degradability Coefficient, KR	0.05
Amendment Degradability Coefficient, KA	0.50
Windrow Dimensions	
Cross-Sectional Area, AREAXS	3.0 m^2
Windrow Width, WIDTH	4.0 m
Pile Dimension for Conductive Heat Loss, DIMCL	4.0 m
Pile Dimension for Surface Heat Loss, DIMSL	5.0 m
Depth for Surface Heat Loss, DEPTH	0.2 m
Operational Characteristics	
Windrow Turnings per Day, TURNUM	0.25 and 1.0 per day
Heat Loss Turning Factor, F3	0.10
Heat Loss Turning Factor, F4	1.0
Detention Time, DETTIM	20 days
Reference Temperature, TREF	0.0°C
Agitator Power, MIXPW	75 kW
Agitator Operation, HRSDAY	0.50 hr/day
Heat Transfer Coefficient, U	1.5 kcal/m^2-hr-°C

[a]Values in this table were held constant for most of the analysis presented in this section. Exceptions are noted on individual figures and graphs where applicable. Values of specific heat, specific gravity, and heat of combustion are as shown in Table 12-1.

The factor F4 is used in Equation 11-73 and accounts for exposure of the surface to ambient conditions after turning. The factor assumes a value of 1.0 for windrow systems subject to periodic turning. Both of these factors are important because they determine in large measure the heat losses from the system. Of the two factors, F3 is the least defined by experimental data.

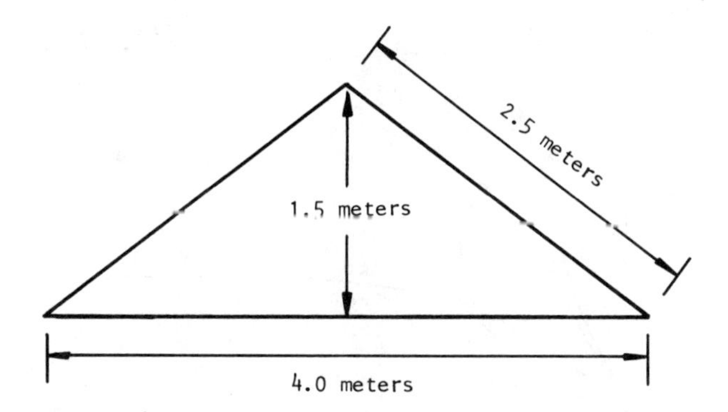

Figure 14-1. BATCH model. Windrow dimensions assumed in the simulation model analysis.

WINDROW DYNAMICS (T = 20°C)

In the analysis of windrow composting a number of additional variables enter the problem which were either not applicable or not considered in the analysis of the CFCM reactor or static pile composting. Important among these is the frequency of windrow turning and the addition of amendments other than recycled product. To make the presentation of results easier, we will begin by discussing process dynamics with all feed temperatures constant at 20°C. The effect of low air temperature is considered later in the chapter. As previously mentioned, the analysis is also limited to the case of digested sludge. Placing these constraints on the analysis will allow us to efficiently explore parameters not considered in Chapters 12 and 13.

Effect Of Air Supply And Cake Solids

Predicted temperature profiles for various air flowrates are presented in Figure 14-2, assuming a cake solids of 30%. Windrow turning was assumed to be on a once-per-day basis. Air flowrates from 0.025 to 1.0 m^3/min-dry ton were considered in the analysis. This spans the range from oxygen supplies less than stoichiometric to those sufficient for both composting and drying. No rainfall or evaporation of surface water was assumed to occur during the compost cycle.

At an air supply of 0.025 m^3/min-dry ton, windrow temperature climbs to about 50°C and remains relatively constant for the remainder of the cycle. Increasing the air supply to 0.05 m^3/min-dry ton results in higher temperature

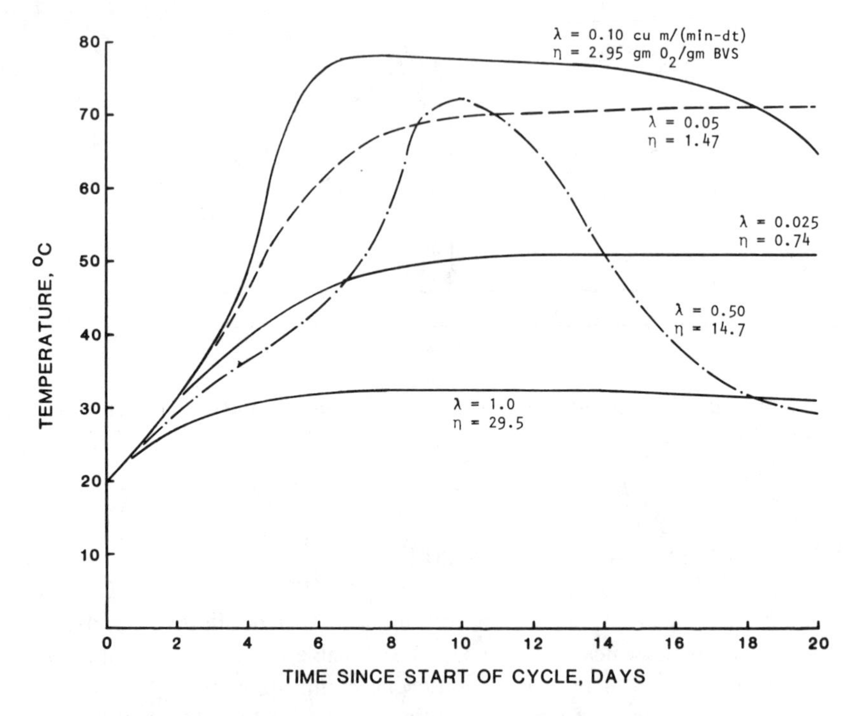

Figure 14-2. BATCH model. Predicted temperature profiles for windrow composting. Digested sludge with recycled product, SC = 0.30, TURNUM = 1.0, TC = TR = TAIR = 20°C, EVAP = RAIN = 0.0.

that also remains relatively constant. In the former case the oxygen content is reduced to <1% during much of the cycle, imposing an oxygen limitation which reduces the rate of reaction. Apparently, the system seeks a dynamic steady-state at about 50°C and can hold this temperature for prolonged periods of time. Only about 34% of the mixture biodegradable volatile solids (BVS) is oxidized in 20 days, so the temperature can probably be prolonged for quite some time. Similar comments apply to the case where λ is 0.05 m^3/min-dry ton. Even though air supply is doubled, the added oxygen increases the rate of reaction such that oxygen content within the windrow is still <1% much of the time. A new steady-state is established at about 70°C, but the system remains largely oxygen-limited. BVS reduction is increased to about 64%, so it is unlikely the temperature will be held much beyond the 20-day period.

Increasing the air supply to 0.10 m^3/min-dry ton results in a higher temperature profile, but one which begins to fall near the end of the cycle.

BVS reduction for this case is about 85%, so the composting conditions are better than at lower air flowrates. The maximum air supply that still maintains an acceptable temperature profile is about 0.5 m^3/min-dry ton. However, the temperature is held for a relatively short period of time and an air supply of 0.25 m^3/min-dry ton may represent a more practical maximum.

Mass and energy balances corresponding to conditions of Figure 14-2 are presented in Figures 14-3 and 14-4. In both cases the balances refer to cumulative input and output over the 20-day compost period. The mass balance corresponds to an air supply of 0.25 m^3/min-dry ton, which is near the peak for output solids content. About 66% of total moisture removal occurs in the compost process. Outfeed material is about 52% solids, which means that subsequent drying is necessary to meet the requirement of 70% solids in product recycle. Even so, considerable moisture removal is possible in the windrow process. And remember, no surface evaporation of water was assumed. Reduction of BVS is nearly complete in the 20-day period. Only a small additional BVS loss remains to be accomplished in the subsequent curing and drying phase.

Energy balances for air supply rates of 0.05 and 0.50 m^3/min-dry ton are shown in Figure 14-4. Input energy terms are similar in relative ranking to those previously observed for the CFCM reactor when using well-stabilized product recycle (i.e., VR = 0.35). Most input energy comes from oxidation of sludge organics with all other terms of relatively minor significance. On the other hand, output energy terms are quite different from what has been predicted to this point. For an air supply of 0.05 m^3/min-dry ton, heat lost from the windrow surface after turning HSURF and heat lost during turning HTURN are the two largest energy losses. Heat lost by conduction HCOND is ranked as fifth largest. Together these three account for about 65% of the total output energy. This is a rather remarkable finding and certainly indicates that heat losses to surroundings are a major term to be considered in windrow composting. It is further interesting that none of the three loss terms dominates the energy picture, each being of comparable magnitude. Note that latent heat of vaporization HLWVO is reduced to third place in the ranking, another interesting finding considering its history of dominance in Chapters 12 and 13. When air flowrate is increased to 0.5 m^3/min-dry ton a realignment of output energy terms along more traditional lines results. HLWVO again assumes a dominating position. However, HSURF, HTURN and HCOND still amount to about 30% of total output energy.

Horvath [9] published the temperature, total solids and volatile solids (VS) profiles shown in Figure 2-11. These were observed under conditions approximating those of the present analysis. Digested sludge dewatered to about 35% solids and recycle product comprised the feed material. Windrows

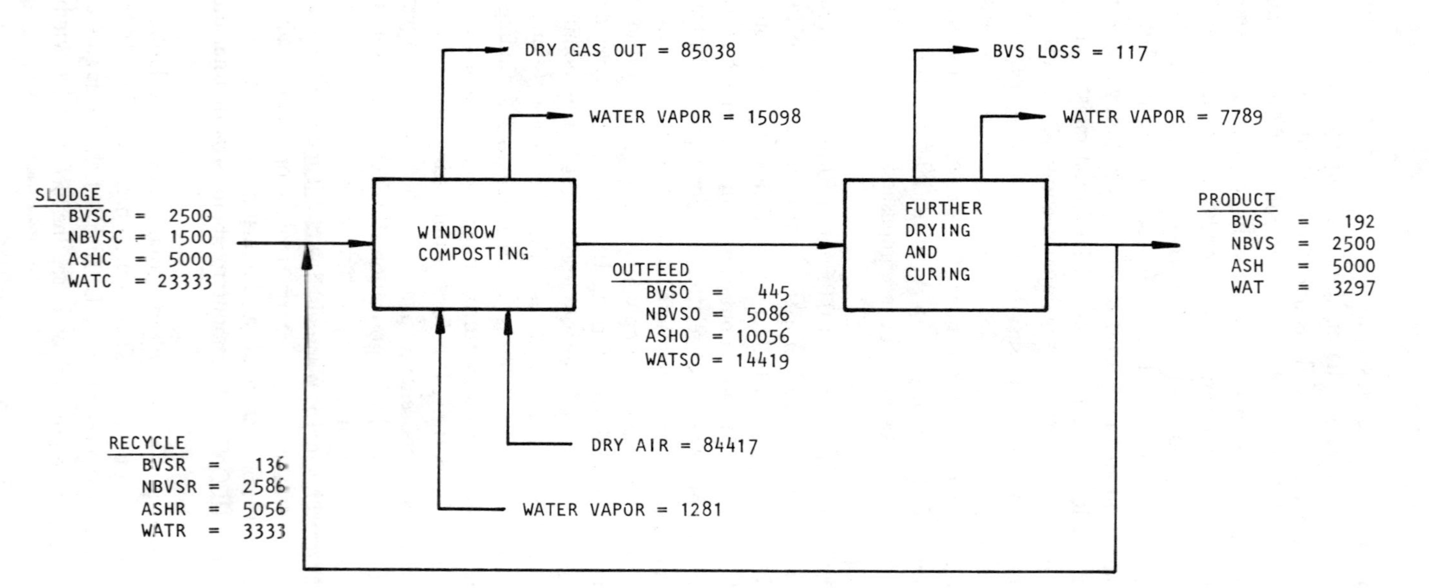

Figure 14-3. BATCH model. Mass balance for solids, water and gas components during windrow composting. Digested sludge with recycled product, SC = 0.30, TURNUM = 1.0, TC = TR = TAIR = 20°C, specific air supply = 0.25 m^3/min-dry ton, specific oxygen supply = 7.36 g O_2/g mixture BVS. Units on mass terms are kg and refer to the total input or output over the 20-day compost cycle.

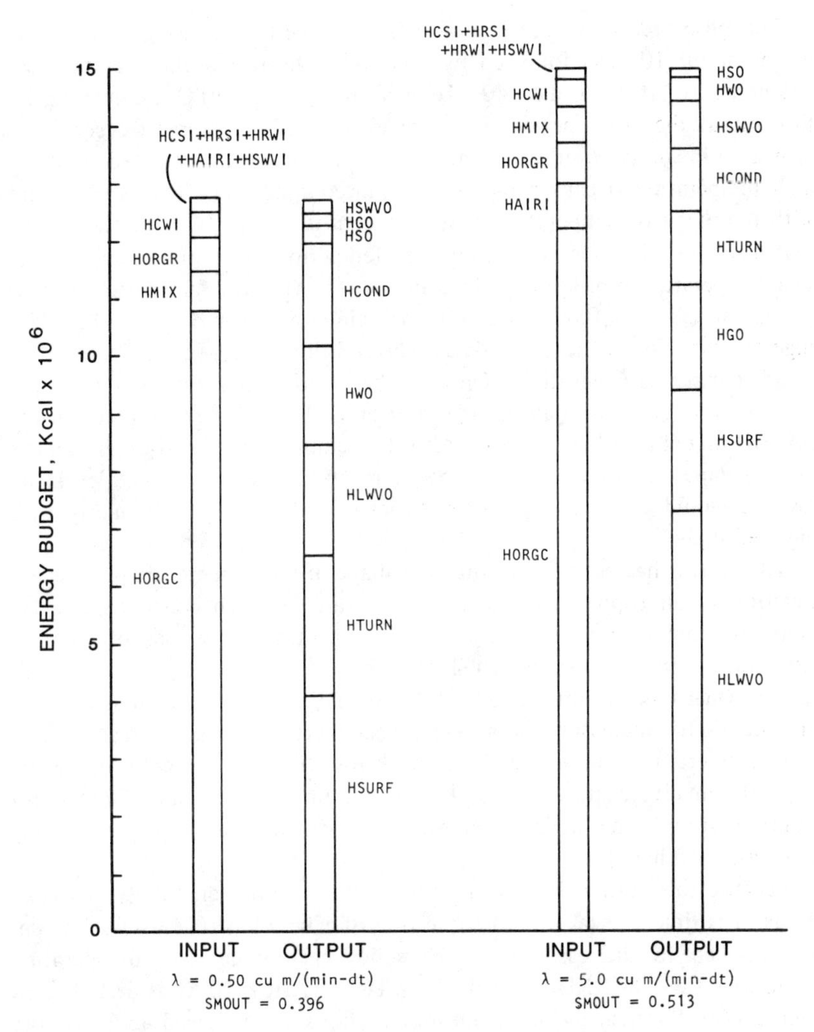

Figure 14-4. BATCH model. Predicted energy balances for windrow composting with low and high specific air supply. Digested sludge with recycled product, SC = 0.30, TURNUM = 1.0, TC = TR = TAIR = 20°C, EVAP = RAIN = 0.0.

were of comparable dimensions to those shown in Figure 14-1 and turning was on a once-per-day basis. Weather conditions were moderate with no rain. Surface evaporation of water would have occurred in the tests reported by Horvath [9], and this has yet to be considered in this analysis. Natural ventilation was relied on to supply oxygen, so the average air supply is unknown, although it was probably low.

The observed temperature profile shows a period of rising temperature lasting about 10 days followed by a period of nearly constant temperature extending to at least day 40. Temperatures above 60°C are maintained throughout the latter period. Iacoboni et al. [112] measured oxygen levels near zero in similar windrows. Thus, it is likely that the protracted period of high temperature is caused by oxygen limited conditions. This is consistent with the shape of temperature profiles shown in Figure 14-2 and the finding that oxygen limitations can cause protracted periods of elevated temperature. The temperature profile maintained in the test windrow falls between those for air supplies of 0.025 and 0.05 m^3/min-dry ton in Figure 14-2. This suggests that forced aeration to supply additional oxygen would result in higher temperature elevations. On the other hand, peak reaction rates for the organisms in question may occur nearer 60°C than the 70°C assumed in the BATCH model (see Figure 11-7). Also, the assumption of first-order kinetics may become an inadequate measure of reaction rates when BVS levels are low, as would occur in the latter stages of composting, particularly with digested sludge.

Thus far, it has been found that the shape of the temperature profiles is a function of air supply. At low air supply rates, oxygen concentration can limit the rate of reaction. This results in protracted periods of elevated temperature, the duration of which is probably controlled by the rate of air supply. High temperature periods of 40-60 days are not inconsistent with the present results. Increasing the air supply can increase the rate of reaction and yield temperature profiles which fall back toward ambient conditions before day 20. Finally, heat losses to the surroundings are a major factor with windrow composting, at least for windrows with dimensions approximating those assumed here.

Because this section is to deal with effects of both air supply and cake solids, it is time to deal with the question of cake solids. The BATCH model was applied to the case of a 20% solids cake under similar operating conditions (i.e., TURNUM = 1.0, TC = TR = TAIR = 20°C) as described in Figure 14-2. Surprisingly, temperature elevations never exceed 45°C over the range of air supplies from 0.025 to 1.0 m^3/min-dry ton. Analysis presented in Chapter 9 indicated that composting a 20% digested cake would be thermodynamically possible if drying were limited to reduce evaporative heat loss.

The reader may question why heat loss is more important with lower cake solids. After all, a dry weight of 10,000 kg was assumed, regardless of feed cake solids. The difference is that the quantity of recycled product (or any amendment for that matter) is greatly increased at the lower cake solids. Thus, windrow volume is increased, which in turn increases windrow length because cross-sectional dimensions were held constant. A comparison of predicted windrow characteristics for both 20 and 30% cake solids is presented in Table 14-2. Volume, length, area subject to conductive loss and the

Table 14-2. Comparison of Initial Windrow Characteristics for Feed Cake Solids of 20 and 30%[a]

	Cake Solids	
	SC = 0.20	SC = 0.30
Bulk Weight of Mixture, g/cm^3	1.00	1.00
Initial Volume of Windrow, VOLUME, m^3	83.3	44.4
Length of Windrow, LENGTH, m	27.8	14.8
Area Subject to Rainfall, AREARN, m^2	111.1	59.3
Area Subject to Conductive Heat Loss, AREACL, m^2	111.1	59.3
Area Subject to Surface Heat Loss, AREASL, m^2	138.9	74.1
Area Subject to Moisture Loss by Surface Evaporation, AREAEV, m^2	138.9	74.1

[a]Windrow cross-sectional dimensions are as shown in Figure 14-1 and the initial mixture of sludge cake and recycled product is 40% solids.

area subject to surface losses are about doubled compared to the case with 30% cake solids. Clearly, the impact of all area-dependent parameters, such as heat loss, rainfall and surface evaporation, will be markedly affected by cake solids.

Effect Of Turning Frequency

One way to reduce heat losses to surroundings is to reduce windrow turning frequency. Heat lost during turning HTURN would be reduced by the lower frequency of turning. Heat lost from the surface after turning HSURF would likewise be reduced. Conductive heat loss HCOND would not be affected as greatly and in the present analysis was assumed to be affected only to the extent that average windrow temperatures are affected (see Equation 11-68).

The effect of reducing the turning frequency to once per four days (TURNUM = 0.25/day) is shown in Figures 14-5 and 14-6 for the cases of 30 and 20% cake solids, respectively. Comparing Figures 14-2 and 14-5 for 30% cake solids, lower turning frequency results in significantly higher temperature elevations at low air flowrates. With an air supply of 0.025 m^3/min-dry ton, temperatures peaked at about 50°C with TURNUM = 1.0, but are predicted to increase above 75°C with TURNUM = 0.25. This is caused entirely by reduced heat losses, which account for 49% of total output energy in the latter case and 73% in the former. The period of initial temperature development is shortened slightly at lower turning frequency and a more acceptable temperature profile results at the high air supply rate of 0.5 m^3/

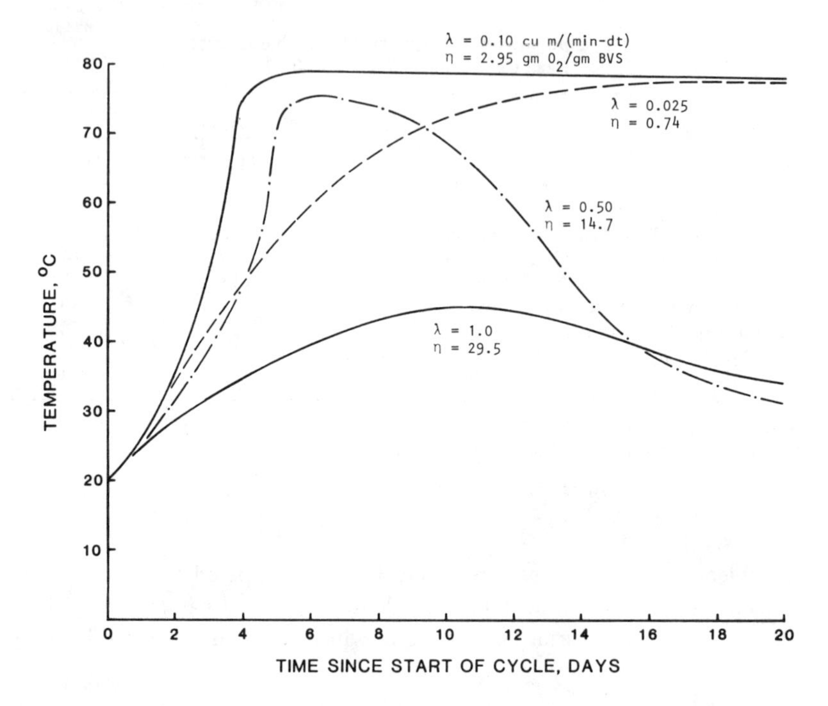

Figure 14-5. BATCH model. Predicted temperature profiles for windrow composting with reduced turning frequency. Digested sludge with recycled product, SC = 0.30, TURNUM = 0.25, TC = TR = TAIR = 20°C, EVAP = RAIN = 0.0.

min-dry ton. However, an air supply of 1.0 m^3/min-dry ton still exceeds available energy resources.

In the case of 20% cake solids, acceptable compost temperatures are predicted with the lower turning frequency. This supports previous comments on the importance of cake solids in determining windrow dimensions and heat losses to surroundings. Reducing these losses through less frequent turning provides the energy conservation needed to support better temperature elevation. It is interesting to note that a high air supply (0.5 m^3/min-dry ton) still provides elevated temperatures, if only for a short duration. The practical maximum air supply is probably about 0.25 m^3/min-dry ton.

It is perhaps a vindication of the thermodynamic analysis in Chapter 9 that proper composting temperatures are predicted to develop when heat losses to surroundings are reduced. Obviously, turning frequency becomes more important as cake solids decrease and the need to conserve energy resources becomes more critical. There is little question that heat losses are reduced by lowering turning frequency. But one eventually has to contend with another

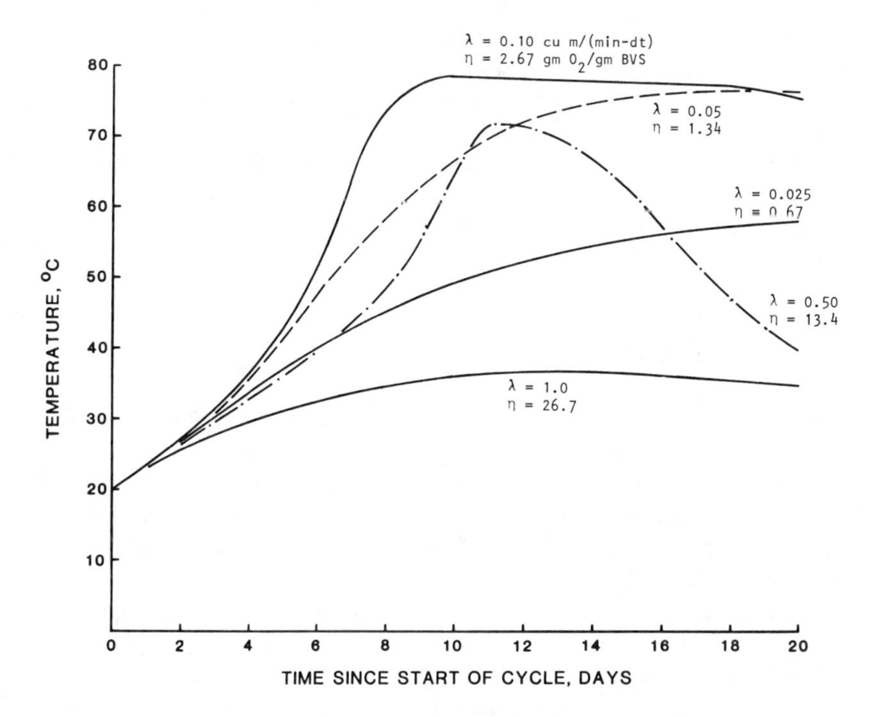

Figure 14-6. BATCH model. Predicted temperature profiles for windrow composting with low cake solids and reduced turning frequency. Digested sludge with recycled product, SC = 0.20, TURNUM = 0.25, TC = TR = TAIR = 20°C, EVAP = RAIN = 0.0.

competing factor, the desire for frequent turning to assure adequate free air-space (FAS), or to provide statistical assurance of pathogen destruction. Even at 40% solids, a cake and recycle product mixture will exhibit some plastic properties and may tend to slump under pressure, thus reducing FAS. Potential effects of such slumping on FAS were not considered in the BATCH model.

Horvath [9] reported that intervals of 2-3 days between turning resulted in temperatures slightly lower than that achieved with once-per-day turning. The initial sludge mixture was 35-40% solids, and air was provided only by turning and natural ventilation. Thus, differing rates of surface drying and air supply could have affected the tests, and results cannot be attributed solely to lack of FAS. LeBrun et al. [22] reported successful aerated static pile tests using digested cake and recycled product, provided the initial starting mixture was adjusted to 50% solids. No turning was practiced in these tests, which lasted more than 20 days. Apparently any effects of slumping can be mitigated by increasing starting mixture solids content to reduce its plastic flow tendencies.

In summary, reducing the turning frequency can minimize heat losses, but care must be exercised to avoid other rate limitations such as those caused by loss of FAS. Solids content can be used to adjust plastic properties of the initial mixture and to reduce the effect of any slumping between turnings. Alternatively, amendments such as straw or bulking particles such as wood chips can be added to the mixture.

Surface Evaporation

Many compost operations are conducted in climates where evaporation exceeds precipitation, at least during certain periods of the year. Moisture will then be lost from the surface of the windrow. As the surface dries, the rate of moisture evaporation should decrease. However, if windrow turning is reasonably frequent, new surface will be exposed to ambient conditions and additional moisture removal can result.

A pan evaporation rate EVAP of 0.50 cm/day was assumed and applied to the case of 20 and 30% cake solids with turning frequencies of 1/day and 0.25/day. All feed temperatures were assumed to be 20°C. The surface evaporation rate is equivalent to about 180 cm/year, which is quite substantial and probably applicable only to more arid regions. The desire was to reveal clearly any effect of surface evaporation and so a significant rate was assumed.

Results of the analysis for a feed cake solids of 20% are presented in Figures 14-7 and 14-8 for turning frequencies of 1/day and 0.25/day, respectively. An analysis was also conducted for 30% cake solids, but predicted temperature profiles were not markedly different from those shown in Figures 14-2 and 14-5. Referring to Figure 14-7, compost temperatures above 60°C are predicted at air flowrates from about 0.05 to 0.25 m^3/min-dry ton even though 1/day turning is assumed. Recall that under similar conditions, but without surface evaporation, compost temperatures never exceeded 45°C regardless of air flowrate. Therefore, surface evaporation can have a very definite beneficial effect on the compost process.

An initial period of slow temperature development extending to about day 10 is predicted for all profiles. A more rapid temperature development then ensues at air flowrates of 0.10 and 0.25 m^3/min-dry ton. By this time surface evaporation has increased the average pile solids content to about 43% from the initial value of 40%. The rate adjustment factor for FAS F2 has correspondingly increased from about 0.74 to 0.87. These values are reasonably constant over the range of air flowrates examined because windrow temperatures are still low and the quantity of vaporized water in

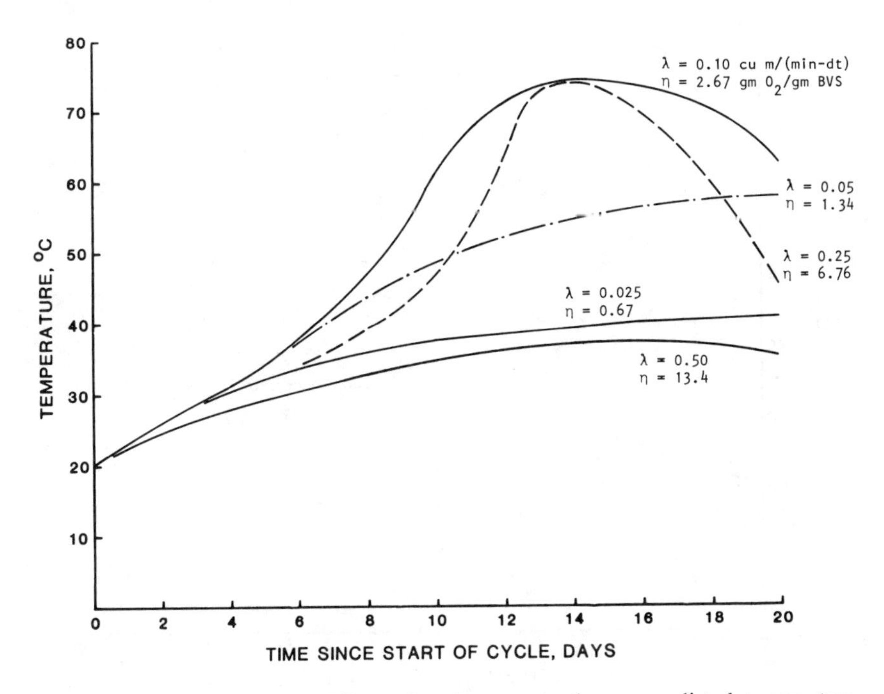

Figure 14-7. BATCH model. Effect of surface evaporation on predicted temperature profiles for windrow composting. Digested sludge with recycled product, SC = 0.20, TURNUM = 1.0, TC = TR = TAIR = 20°C, EVAP = 0.50, RAIN = 0.0.

effluent gas is many times less than surface evaporated water to this point. By comparison and with no surface evaporation, solids content at day 10 actually decreased to about 0.39 and the factor F2 to about 0.71. These rather subtle differences are the underlying reasons for the markedly different temperature profiles for cases with and without surface evaporation.

The effect of reducing turning frequency to 0.25/day is shown in Figure 14-8. Comparing results with those of Figure 14-6, slightly more rapid temperature development and higher peak temperatures at low air flowrates are predicted. The effects of surface evaporation appear to be noticeably beneficial even at reduced turning frequency. However, these results must be viewed with some caution. As pointed out in Chapter 11, the rate of surface evaporation is assumed constant between windrow turnings. This assumption may be stressed somewhat with a low turning frequency of 0.25/day.

The other major effect of surface evaporation is to remove additional water from the windrow. The effect is most keenly felt at low specific air supply and low feed cake solids as the data of Table 14-3 indicate. Under conditions of this analysis, surface evaporation is clearly the dominant

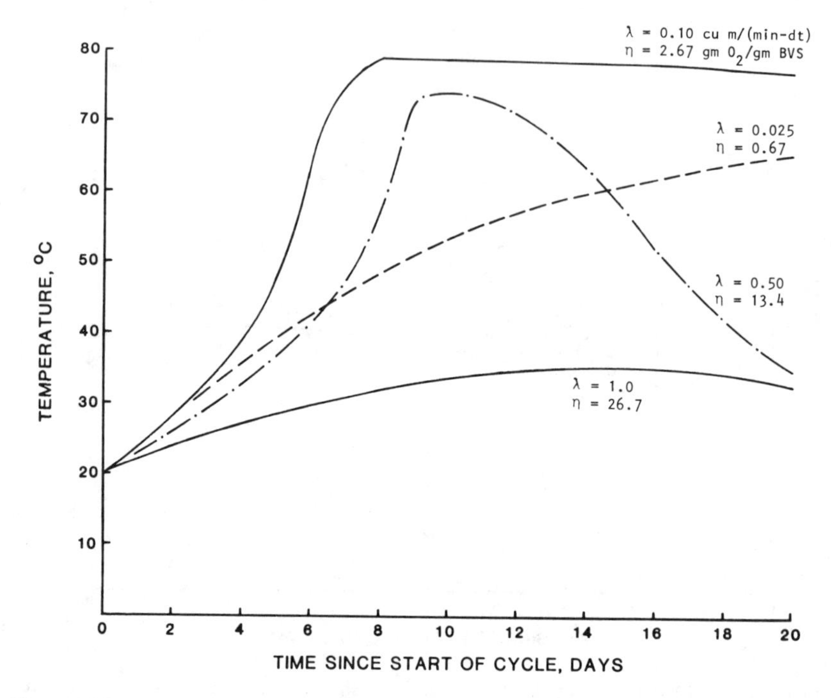

Figure 14-8. BATCH model. Effect of surface evaporation on predicted temperature profiles for windrow composting. Digested sludge with recycled product, SC = 0.20, TURNUM = 0.25, TC = TR = TAIR = 20°C, EVAP = 0.50, RAIN = 0.0.

Table 14-3. Comparison of the Relative Contribution of Surface Evaporation and Vaporization Toward Moisture Removal (EVAP = 0.50, TURNUM = 0.25, TC = TR = TAIR = 20°C)

Specific Air Supply[a] (m^3/min-dry ton)	QAIR[b] (m^3/min)	SC = 0.30		SC = 0.20	
		WATEVP[c]	WATVO	WATEVP	WATVO
0.025	0.25	73	27	94	6
0.05	0.50	54	46	75	25
0.10	1.0	39	61	57	43
0.25	2.5	29	71	44	56
0.50	5.0	26	74	42	58
1.00	10.0	39	61	59	41

[a] Compost temperatures >60°C are predicted with air supplies from 0.025 to 0.50 m^3/min-dry ton for both SC = 0.30 and 0.20.
[b] Values correspond to 10 tons of dry sludge solids at the start of the compost cycle.
[c] Values are expressed as a percentage of total moisture removal.

moisture removal mechanism at air supply rates below 0.05 m^3/min-dry ton. This is caused by limited moisture transport in the exhaust gases at low air supply. Even at an air supply of 0.50 m^3/min-dry ton, surface evaporation can account for about 25% of total moisture removal at SC = 0.30, increasing to 42% at SC = 0.20. Obviously, these results depend on ambient conditions and a rather high pan evaporation rate was assumed here.

In summary, surface evaporation appears to benefit both temperature development and moisture removal, particularly with feed sludges of low cake solids content. The effect is probably lessened as turning frequency is lowered because of the reduction in surface evaporation rate in the interval between turnings.

Output Solids Content

Predicted output solids contents at the end of the 20-day cycle are shown in Figures 14-9 and 14-10. Results with a 30% cake feed are shown in Figure 14-9 for various conditions of turning frequency and surface evaporation rate. Without surface evaporation, output solids content is favored by reduced turning frequency. This is caused by better conservation of heat resources, resulting in higher windrow temperatures for longer duration and, thus, increased moisture loss in output gases. Output solids content is increased further if surface evaporation is available to aid in moisture removal.

At very low air flowrates and with no surface evaporation, output solids content can actually be reduced from that of the initial starting mixture. This is caused by loss of biodegradable solids coupled with low windrow temperature and low air flowrate such that moisture lost in the gas phase is not sufficient to compensate for produced water and lost solids. It is interesting that the general shape of curves in Figure 14-9 is quite similar. With surface evaporation, output solids content is higher at zero airflow because of evaporative loss over the 20-day cycle, even though no biological activity would be predicted. In all cases peak output solids are reached at specific air flowrates between 0.25 and 0.50 m^3/min-dry ton. The shape of the curves between 0.50 and 1.0 m^3/min-dry ton is uncertain because additional data points were not available within this interval.

An interesting comparison can be made between output solids content predicted in this analysis and those predicted by the more simplified thermodynamic analysis of Chapter 9. Referring back to Figure 9-7, a digested sludge with cake solids of 30% was estimated to have sufficient energy to yield a final product of about 65% solids. Note that KR was assumed to be zero in the latter analysis so that recycled product contributed no energy. Nor were heat losses to surroundings considered. The closest conditions to this in the present analysis is for the case of no surface evaporation and a turning frequency of 0.25 to minimize heat losses. In this case, maximum output solids is predicted to be about 60%. The simplified approach based solely on thermodynamics is a good predictive tool under

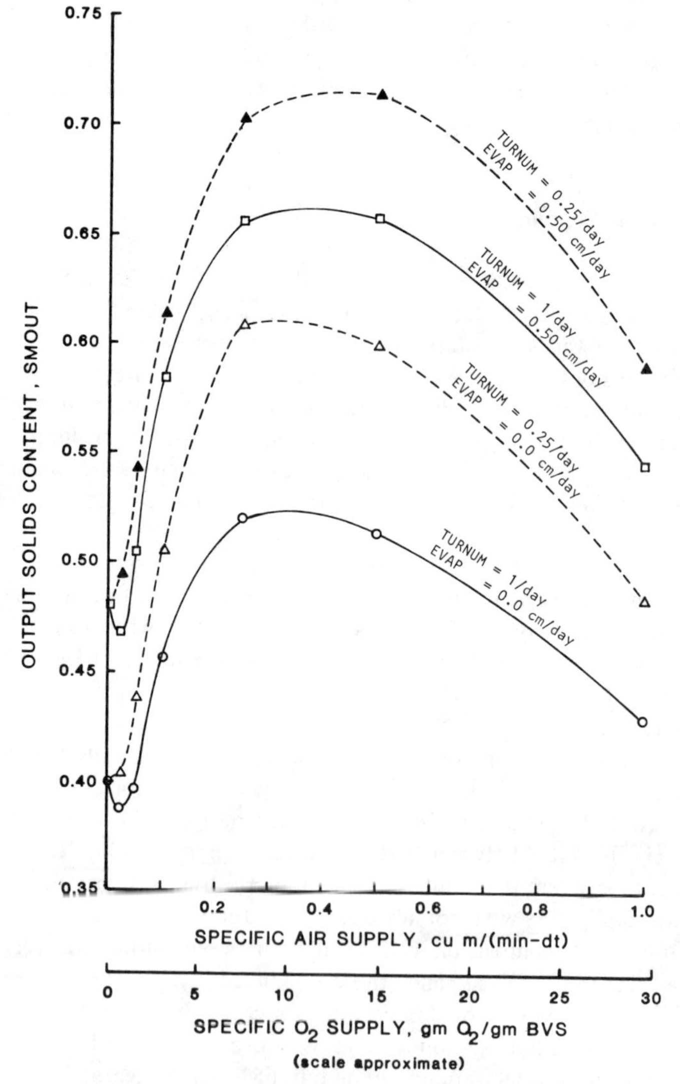

Figure 14-9. BATCH model. Effect of air supply, turning frequency and surface evaporation on output solids content from windrow composting. Digested sludge with recycled product, SC = 0.30, TC = TR − TAIR = 20°C, RAIN = 0.0.

appropriate conditions. However, the effect of heat loss to surroundings becomes evident as turning frequency increases to 1/day and peak output solids decrease to only about 52%.

The predicted results with 20% feed cake are shown in Figure 14-10. The general shape of the curves is not similar for all cases. With no surface evaporation, proper composting temperatures were not predicted with TURNUM = 1.0. This is reflected in the fact that essentially no drying is achieved regardless of air flowrate. Decreasing the turning frequency to 0.25/day resulted in proper composting temperatures, and the effect on output solids content is quite evident. Even so, maximum output solids are only about 47%, significantly less than that achieved under similar conditions

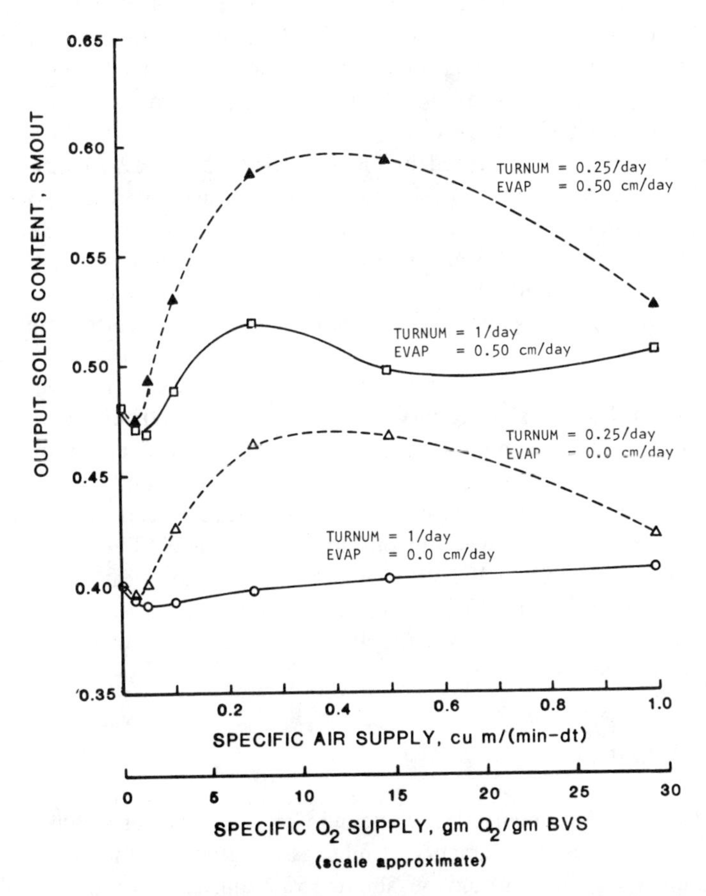

Figure 14-10. BATCH model. Effect of specific air supply, turning frequency and surface evaporation on output solids content from windrow composting. Digested sludge with recycled product, SC = 0.20, TC = TR = TAIR = 20°C, RAIN = 0.0.

with 30% feed cake. With surface evaporation and TURNUM = 1.0, maximum output solids is increased to about 52%, but the range of operating air supply is reduced. Only with surface evaporation and a turning frequency of 0.25/day are output solids approaching 60% predicted. As previously discussed, however, model assumptions become stressed under these conditions and actual practice will likely be less than predicted.

A comparison with the thermodynamic analysis of Chapter 9 can again be made, this time for the case of 20% feed cake. From Figure 9-7 a maximum output solids content of about 47% is estimated on thermodynamic grounds alone. Comparing to the present case with no surface evaporation and TURNUM reduced to 0.25/day, a maximum output solids of about 47% is also predicted, a rather good comparison considering the differences in assumed conditions. Clearly, the conclusion of Chapter 9, that digested cake of 20% solids has insufficient energy for both composting and drying, is supported by the present analysis. Surface evaporation can aid in producing a drier end product, but even so drying is significantly less than that achieveable with a 30% sludge cake.

Mass balances are presented in Figures 14-11 and 14-12 for cases of 30 and 20% cake solids, respectively. Conditions in each case correspond to a surface evaporation rate of 0.50 cm/day, turning frequency of 1/day and a specific air supply of 0.25 m^3/min-dry ton. This air supply is near the peak of output solids content in each case. The balances are presented to illustrate the following point. One might argue that if surface evaporation rates are high, low cake solids may not be a problem because windrow volume will be greater because of the increased recycle rate needed to achieve a 40% solids starting mixture (see Table 14-2). Thus, more surface area will be available, and more water will be removed by surface evaporation, making up for the greater feed water. However, it turns out that this is not quite the case. Water removed by surface evaporation WATEVP is greater for the case of 20% cake solids, but output water vapor WATVO is much less. Total water removed (WATEVP + WATVO) is slightly greater in the 20% solids case, but it is not sufficient to compensate for the increased feed water. Thus, drier product results with the drier feed sludge even with high surface evaporation rates.

BVS Stabilization

Volatile solids content is one measure of final product stability which has often been used in practice. Under assumptions of this analysis, the minimum VS content would be about 35%, equivalent to the nonbiodegradable volatile solids (NBVS) fraction. From a theoretical standpoint it is probably better to concentrate on the BVS fraction of the VS. Predicted BVS

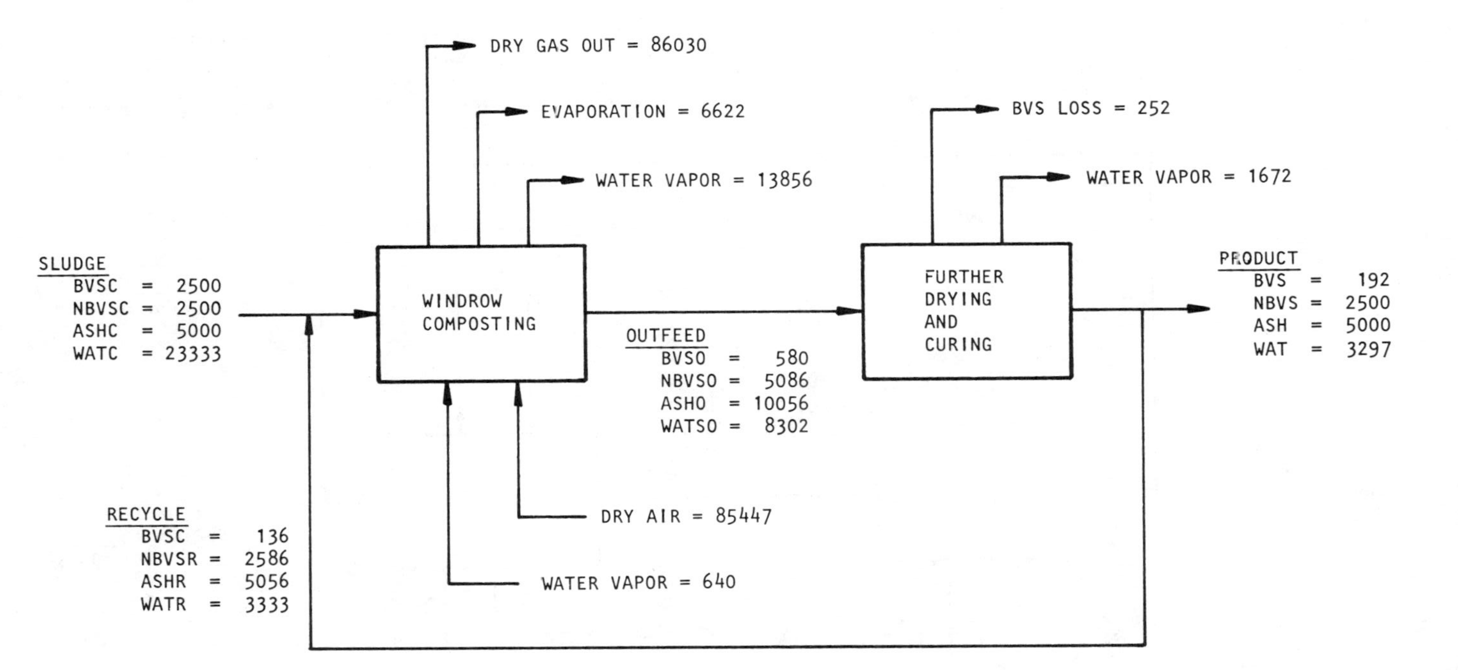

Figure 14-11. BATCH model. Mass balance for solids, water and gas components during windrow composting with high surface evaporation rates. Digested sludge with recycled product, SC = 0.30, TURNUM = 1.0, TC = TR = TAIR = 20°C, specific air supply = 0.25 m^3/min-dry ton, specific oxygen supply = 7.46 g O_2/g mixture BVS, EVAP = 0.50, RAIN = 0.0. Units on mass terms are kg and refer to the total input or output over the 20-day compost cycle.

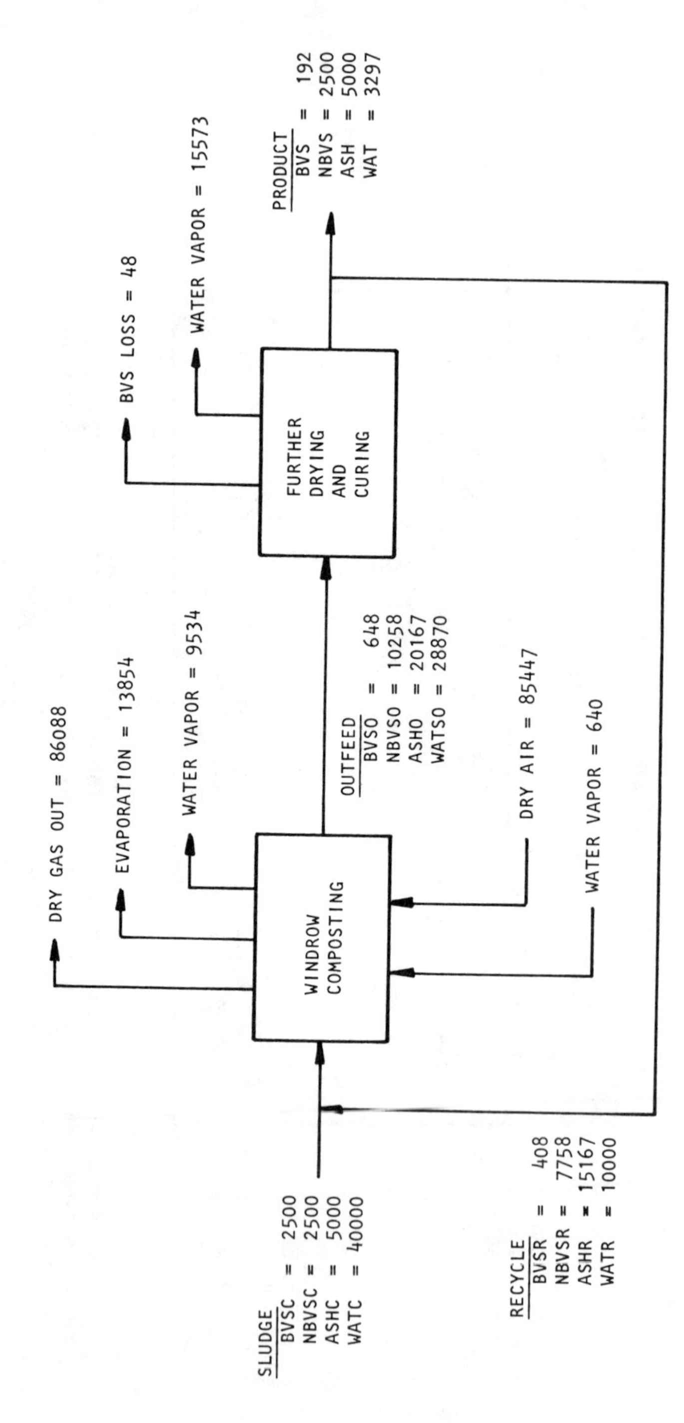

Figure 14-12. BATCH model. Mass balance for solids, water and gas components during windrow composting with high surface evaporation rates and low cake solids. Digested sludge with recycled product, SC = 0.20, TURNUM = 1.0, TC = TR = TAIR = 20°C, specific air supply = 0.25 m^3/min-dry ton, specific oxygen supply = 6.76 g O_2/g mixture BVS, EVAP = 0.50, RAIN = 0.0. Units on mass terms are kg and refer to the total input or output over the 20-day compost cycle.

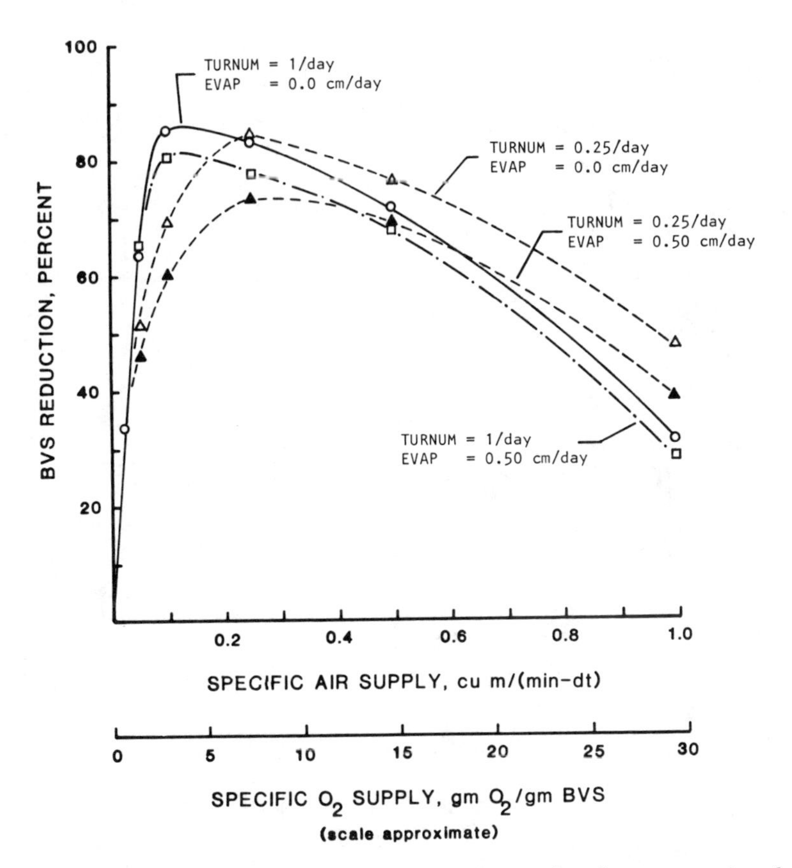

Figure 14-13. BATCH model. Effect of air supply, turning frequency and surface evaporation on BVS reduction. Digested sludge with recycled product, SC = 0.30, TC = TR = TAIR = 20°C, RAIN = 0.0.

reductions for cases examined to this point are shown in Figures 14-13 and 14-14 for SC = 0.30 and 0.20, respectively. With 30% feed sludge, similarly shaped curves result for different turning frequencies and surface evaporation rates. Initially, BVS loss increases rapidly with increasing air supply. With specific O_2 supply less than about 1.5 g O_2/g BVS, reaction rates are oxygen-limited with correspondingly low BVS reductions. In all cases peak BVS reduction exceeds 70% and is usually >80%. With 1/day turning, the peak occurs at about 3 g O_2/g BVS (0.1 m^3/min-dry ton), and with 0.25/day turning at about 7.5 g O_2/g BVS (0.25 m^3/min-dry ton). It is interesting to note that under similar surface evaporation conditions, peak BVS reduction is slightly greater with 1/day turning compared to 0.25/day. Recall that heat

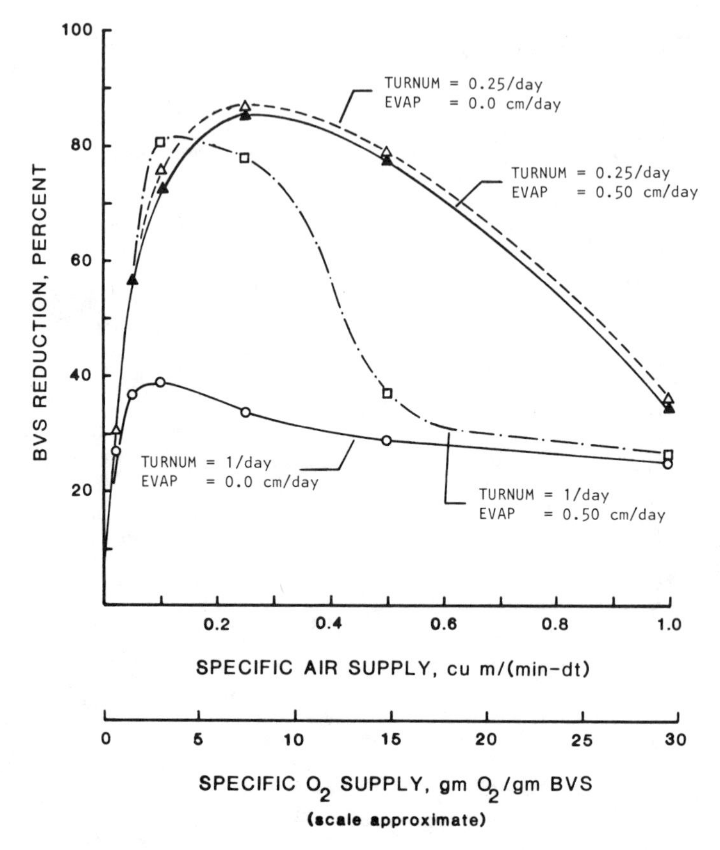

Figure 14-14. BATCH model. Effect of specific air supply, turning frequency and surface evaporation on BVS reduction. Digested sludge with recycled product, SC = 0.20, TC = TR = TAIR = 20°C, RAIN = 0.0.

losses are much reduced with 0.25/day turning. This means that the optimum thermodynamic balance does not assure optimum kinetics rates. We have seen this occur in previous analyses and, although differences in BVS are slight in the present case, it is useful to reiterate this very fundamental distinction.

Corresponding curves for a 20% feed sludge are shown in Figure 14-14. With TURNUM = 0.25/day, similar curves result regardless of surface evaporation rate. Peak BVS reduction of about 85% occurs at about 7.5 g O_2/g BVS (0.25 m^3/min-dry ton). If the turning frequency is increased to 1/day, a significantly different pattern of results is evident. Without surface evaporation, elevated compost temperatures are not predicted and BVS reductions remain low regardless of oxygen supply. With surface evaporation, a peak reduction over 80% is predicted but the range of operating air supply is reduced.

WINDROW DYNAMICS WITH LOW AIR TEMPERATURE

The effect of low ambient temperature was simulated assuming an infeed air temperature of 0°C. Heat losses to surroundings were also calculated assuming a 0°C ambient condition. Analysis was limited to cake solids of 30%. Temperature of infeed solids was assumed to be 20°C for most of the analysis. From the previous study of static pile composting (Chapter 13), it was determined that significant lag periods result if all infeed temperatures are 0°C (i.e., TC = TR = TA = TAIR = 0°C). In fact, for digested sludge no useful temperature elevation was predicted over the 20-day period. Because low starting temperatures should be avoided, further analysis was deemed unnecessary.

Effect Of Air Supply And Turning Frequency

The BATCH model was run assuming 0.25 windrow turnings per day with predicted temperature profiles shown in Figure 14-15. No surface evaporation was assumed. Compared to the corresponding case with all infeed temperatures at 20°C (see Figure 14-5), the major impact of the lower infeed air temperature is a reduction in maximum allowable air supply. With TAIR = 0°C, maximum specific air supply is about 0.25 m^3/min-dry ton, and even here a rather significant lag period is predicted. By comparison, a peak air supply as high as 0.50 m^3/min-dry ton with very little lag period is predicted with TAIR = 20°C. As would be expected the effect is much reduced at lower air flowrates. Temperature profiles for air supplies of 0.025 and 0.10 m^3/min-dry ton are quite similar between the two cases.

When turning frequency is increased to 1/day, no useful temperature elevations are predicted regardless of air flowrate. In fact, temperature profiles immediately decrease below the initial 20°C value and even moderate temperature development is not predicted.

Use Of Initial Preheating

During analysis of the CFCM reactor in Chapter 12, it was observed that the reactor would seek an acceptable, steady-state solution, but only if reactor temperature was above a minimum value. The minimum temperature depended on the severity of operating conditions. It was higher for reduced

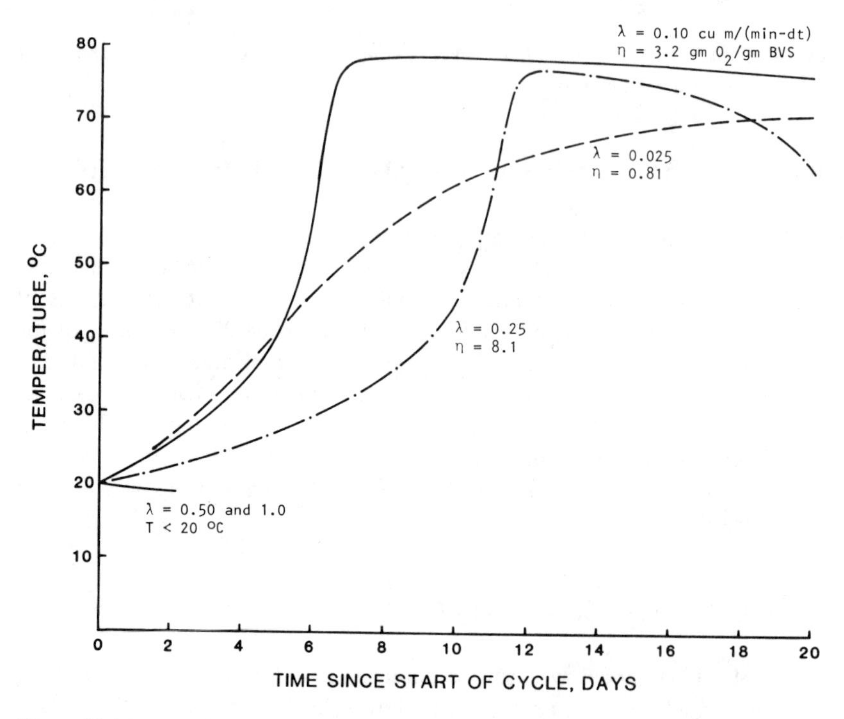

Figure 14-15. BATCH model. Predicted temperature profiles for windrow composting with low ambient air temperatures. Digested sludge with recycled product, SC = 0.30, TURNUM = 0.25, TC = TR = 20°C, TAIR = 0°C, EVAP = RAIN = 0.0.

input mixture solids as shown in Figures 12-37 and 12-41. Some interesting parallels to this can be drawn for the case of windrow composting. Proper temperatures were not predicted with TURNUM = 1.0 and TC = TR = 20°C while TAIR = 0°C. Under similar conditions with TAIR = 20°C, good temperature development was predicted. The difference must be that biological heat production at 20°C cannot keep pace with energy outputs when TAIR = 0°C. But what if initial solids temperatures are higher than 20°C? Rates of biological decomposition, and hence heat production, increase exponentially with temperature in the area of 20°C, whereas heat losses to surroundings increase only linearly. Thus, it may be possible for energy inputs to overtake energy outputs to the point that elevated temperature can be maintained.

This possibility was examined assuming initial cake and recycle temperatures of 50°C. Air temperature remained constant at 0°C and a 1/day turning rate was used. Elevated temperatures are indeed predicted as shown in Figure 14-16. At 50°C the rate of heat production exceeds the rate of

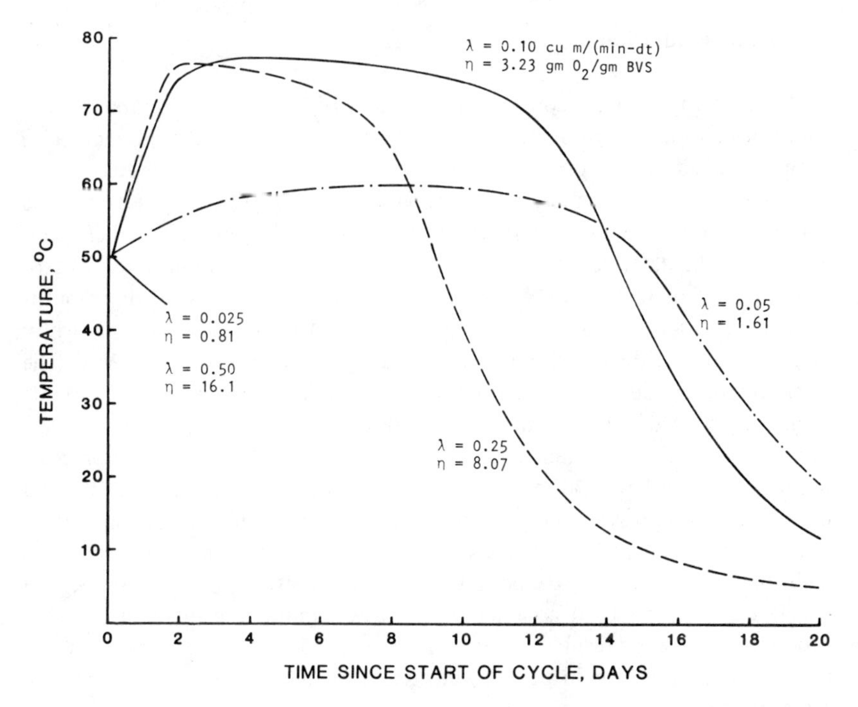

Figure 14-16. BATCH model. Effect of initial preheat on predicted temperature profiles for windrow composting with low ambient air temperatures. Digested sludge with recycled product, SC = 0.30, TURNUM = 1.0, TC = TR = 50°C, TAIR = 0°C, EVAP = RAIN = 0.0.

energy output and temperatures climb to values above 75°C for air flowrates of 0.10 and 0.25 m^3/min-dry ton. Once again it is a case where proper temperature development is not limited by available energy resources as much as by slow reaction rates at lower temperatures. Incidentally, starting temperatures <50°C were not considered so it might be possible to use a lower initial temperature, but one obviously above 20°C.

In practice, this could be taken advantage of by either preheating feed materials used in the windrow or preheating the air supply for the first several days. Referring to Figures 14-2 and 14-15, a period of roughly 4-6 days is required to achieve temperatures >50°C. Using this as a guide, air preheating to 20°C would probably be necessary for roughly 4-6 days. Preheating to higher temperatures would tend to reduce the time requirement. With 50°C air preheating, the time should be no more than that required to bring windrow materials to 50°C.

Amendment Addition

Another approach to counteracting effects of low ambient temperature is addition of other amendments to the composting mixture. Amendments have been discussed at many places in this text, but here we will discuss a very special case of amendment addition. Many locales are characterized by a season of warm, dry conditions followed by wet, cold winters. Even climates which are characterized as arid often experience several months of wet or cold climate. In such cases, digested cake can be air dried during dry months in much the same manner as on a sand drying bed. Spreading the sludge in layers about 1 ft deep with periodic turning facilitates moisture removal. The small piles preclude development of elevated temperatures and solids contents as high as 80-90% can be achieved. The time required for drying is largely a function of the local climate but drying times as low as 6-7 days have been reported for southern California summer conditions [150]. Because of rapid drying and suppressed temperatures, very little organic oxidation occurs during the drying process. Once dried, sludge can be stockpiled with little additional biological activity because of the low moisture content. Stockpiled sludge can then be used as amendment in wet months in place of recycled product. The end result is that organics produced in dry months are conserved for use in cold or wet months when thermodynamic balances are most stressed.

Assuming minimal biological activity during the drying process, air-dried sludge would have the same VS characteristics as the original digested sludge (i.e., VA = 0.50, KA = 0.50). If sufficient dried sludge is added to achieve a 40% solids mixture, then the mixture must have a VS content of 50% and degradability of 50%. These are the same characteristics as would result by directly producing a 40% cake during dewatering. Obviously, the thermodynamic balance will be improved over the use of well stabilized compost product for feed conditioning, because the latter usually contributes only a small fraction of degradable organics. Note also that mixture characteristics become independent of sludge cake solids. In other words, the same mixture characteristics result no matter the value of SC. The quantity of air-dried sludge required to produce the desired mixture would, of course, increase with wetter cake solids, as would total mixture volume.

Predicted temperature profiles, assuming a 30% feed cake with addition of air-dried sludge (termed amendment from here on) to achieve a 40% mixture, are shown in Figure 14-17. The air temperature was assumed to be 0°C and the turning frequency 0.25/day. Comparing to Figure 14-15, the corresponding case using recycled product, some striking differences are noted. First of all, temperature elevations are considerably more rapid when using amendment. Depending on the air supply, peak temperatures develop in as

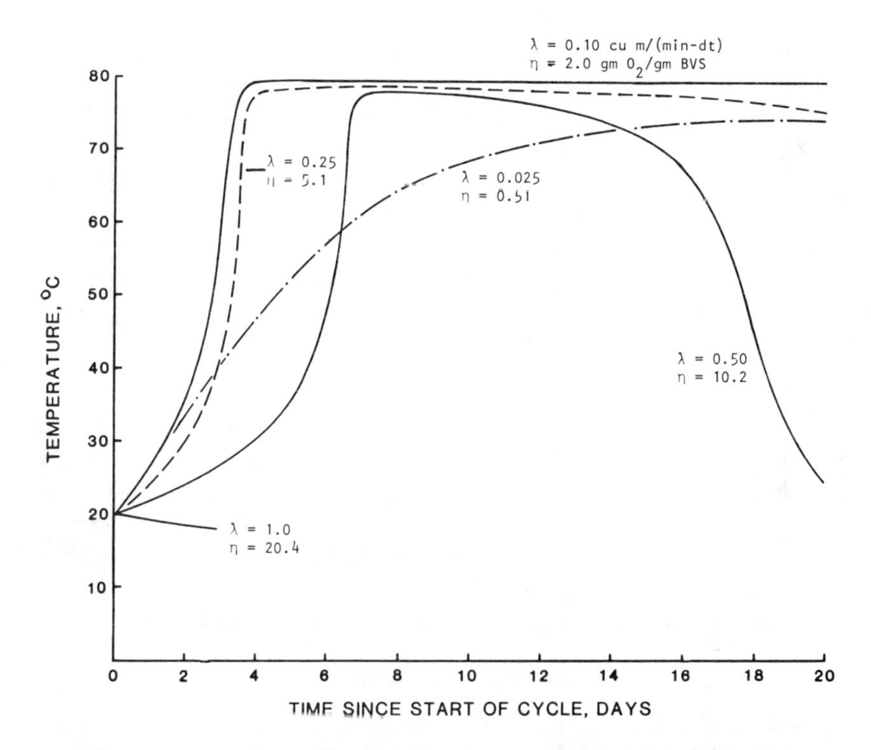

Figure 14-17. BATCH model. Effect of amendment addition on predicted temperature profiles for windrow composting with low ambient air temperatures. Digested sludge with air-dried sludge amendment, SC = 0.30, TURNUM = 0.25, TC = TA = 20°C, TAIR = 0°C, EVAP = RAIN = 0.0.

little as 3-4 days. Second, the range of operating air supply is increased to a maximum somewhat above 0.50 m^3/min-dry ton. It is interesting to note that at 0.50 m^3/min-dry ton the duration of elevated temperature is greater than that predicted using recycled product with all infeed temperatures at 20°C (see Figure 14-5).

Clearly, system thermodynamics have been improved and the impacts of cold weather reduced by use of air-dried sludge as an amendment. But what about process kinetics? Oxygen consumption rate and rate adjustment factors F1, F2 and FO2 are shown in Figures 14-18 and 14-19 for selected air flowrates from Figure 14-17. Figure 14-18 is for an air supply of 0.025 m^3/min-dry ton, 0.51 g O_2/g BVS, which is less than that required for stoichiometric oxidation of all BVS. As a result, oxygen concentrations decrease to a few tenths of a percent and FO2 decreases rapidly. The adjustment factor for moisture F1 and FAS F2 remain relatively constant.

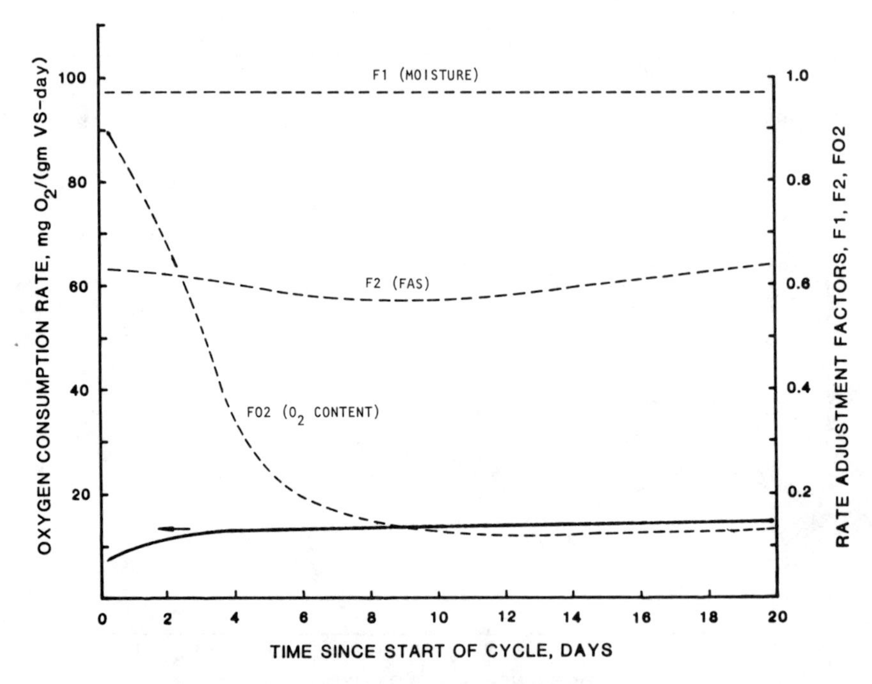

Figure 14-18. BATCH model. Profiles of oxygen consumption rate and rate adjustment factors during windrow composting of sludge and amendment with a low specific air supply. Digested sludge and air-dried sludge amendment, SC = 0.30, TURNUM = 0.25, TC = TA = 20°C, TAIR = 0°C, specific air supply = 0.025 m^3/min-dry ton, specific oxygen supply = 0.51 g O_2/g mixture BVS, EVAP = RAIN = 0.0.

Thus, reaction rates are limited by oxygen supply, and the oxygen consumption rate never exceeds 15 mg O_2/g VS-day during the 20-day cycle. By comparison, the maximum rate (i.e., F1, F2, FO2 = 1.0) at 70°C could be over 200 mg O_2/g VS-day (calculated from Equation 10-1). Thus, the rate limitation is quite severe and leads to a rather slow temperature rise followed by a protracted period of high temperature. In the entire 20-day period only about 25% of the BVS are oxidized. Once again, elevated temperatures for protracted periods do not necessarily imply rapid reaction rates. Quite the converse is true, and severe rate restrictions are often indicated.

The effect of increasing air supply to 0.25 m^3/min-dry ton is shown in Figure 14-19. Oxygen consumption rate increases rapidly to a peak near 110 mg O_2/g VS-day, then rapidly decreases to a plateau of about 60 mg O_2/g VS-day, followed by a period of slow but steady decline. In this case, FO2 remains near 1.0 throughout the cycle, momentarily decreasing to about 0.80 at the time of peak oxygen consumption. Why then the sudden peak and fall in oxygen consumption rate? The explanation lies in the temperature

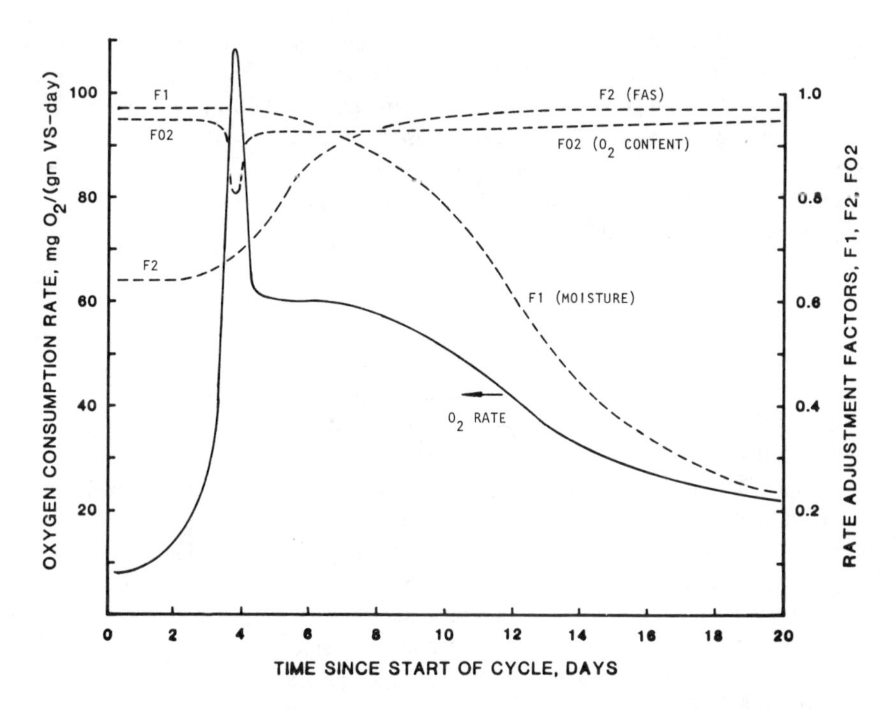

Figure 14-19. BATCH model. Profiles of oxygen consumption rate and rate adjustment factors during windrow composting of sludge and amendment with a high specific air supply. Digested sludge and air-dried sludge amendment, SC = 0.30, TURNUM = 0.25, TC = TA = 20°C, TAIR = 0°C, specific air supply = 0.25 m^3/min-dry ton, specific oxygen supply = 5.1 g O_2/g mixture BVS, EVAP = RAIN = 0.0.

profile of Figure 14-17. The system is driven to very high temperatures such that reaction rates become temperature-limited. A precipitous drop in reaction rate results after the initial peak. Here we see a case where the rate adjustment factors, F1, F2 and FO2, are all near one, but the system finds yet another way to limit itself, this time by temperature. After about 10-12 days, sufficient moisture has been removed to begin imposing an additional rate limitation that becomes more significant as time proceeds. Even with temperature and moisture limitations, however, BVS stabilization is much improved with about 65% oxidation during the 20-day cycle. If it was not obvious before it should be by now. It is difficult to balance all competing factors in a system as complicated as composting.

Even with amendment addition, proper composting temperatures are not predicted with a 1/day turning frequency. Even though BVS in the mixture is increased, kinetic limits with TAIR = 0°C restrict development of the energy resource. Preheating of material to 50°C again solves this problem and it is likely that preheating requirements would be reduced over the previous case using recycled compost. With preheating, adequate temperature profiles are predicted over a range of air flowrates from 0.05 to 0.50 m^3/min-dry ton, corresponding to a specific oxygen supply of about 1-10 g O_2/g BVS. Comparing this to the similar case using recycled product (see Figure 14-16), maximum air flowrate is about doubled whereas the maximum specific oxygen supply remains about the same.

Output Solids Content

Output solids content for the cases examined with TAIR = 0°C are shown in Figure 14-20. The most striking feature is the higher output solids content achieved when air-dried sludge is used in place of recycled compost. Maximum output solids >65% are predicted at specific oxygen loadings about 10 g O_2/g BVS, 0.5 m^3/min-dry ton. Recall that no surface evaporation was assumed as part of this analysis. Using recycled compost, peak output solids between about 50 and 55% are predicted.

At very low oxygen supplies, a slight decrease in output solids is again noted. Further increase in air supply results in a rapid, nearly linear increase in output solids until peak values are approached. Output solids return to essentially infeed values (i.e., SM = 0.40) with oxygen supplies twice that at peak output solids (i.e., about 16 and 20 g O_2/g BVS with recycled product and amendment, respectively). The shape of the curves between these two points is somewhat undetermined because of a lack of intermediate data.

EFFECT OF RAINFALL

Windrow composting of sludge using recycled compost is reported to be adversely affected by wet weather conditions [9]. This susceptibility to wet (and cold) conditions arises from one or more of the following:

1. digested sludge is usually used instead of the more energy-rich raw sludge;
2. cake solids sufficient to assure energy for both composting and drying are not achieved;
3. recycled compost is usually well stabilized and contributes little to the energy budget;
4. windrow dimensions are too small to adequately conserve heat and at the same time project a large surface area for collection of rainfall;
5. turning frequency is too high, resulting in excessive heat loss to surroundings;
6. all weather surfacing of the compost area is not provided resulting in poor traction and inaccessibility of mechanical turning equipment for extended periods; and
7. forced aeration is not provided and moisture removal in output gases cannot be controlled.

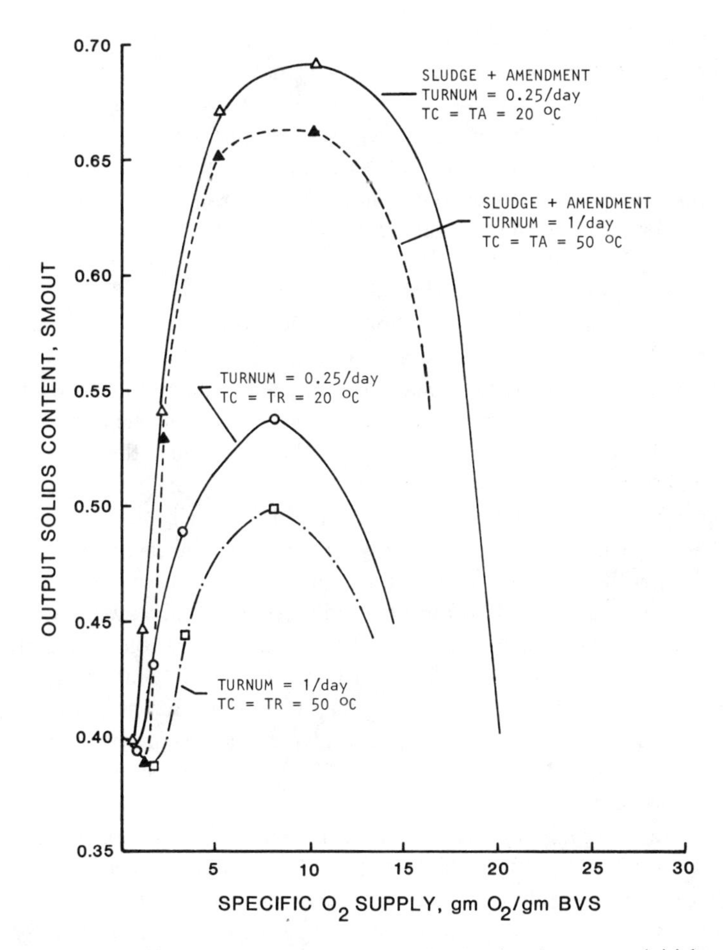

Figure 14-20. BATCH model. Effect of air supply, turning frequency, initial preheat and amendment addition on output solids content from windrow composting. Digested sludge, SC = 0.30, TAIR = 0°C, EVAP = RAIN = 0.0.

The effects of rainfall can be mitigated by fixed roof coverage of a portion of the compost area, providing asphalt or concrete surfacing with a forced aeration system, maximizing windrow dimensions and cake solids achieved during dewatering and, finally, by use of amendments to improve the energy balance.

The effect of roofed coverage has, in a sense, already been evaluated in the previous section on low infeed air temperatures. Rainfall and surface evaporation were both assumed to be zero. This, together with low air temperature, simulates conditions in a covered facility during winter months. Therefore, the present analysis will concentrate on effects of rainfall on uncovered windrows and use of amendments, specifically air-dried sludge.

Persons knowledgeable in the field of composting may be somewhat confused by the statement that windrow composting is susceptible to rainfall. The consensus appears to be that windrow composting is quite resistant to wet weather. The difference arises from the nature of materials being composted and typical windrow dimensions. Windrow composting in large haystack-shaped piles, 5-15 ft high, using periodic turning, has been used with substrates such as refuse, straw and other agricultural wastes. These materials are generally dry and often require moisture addition during composting. Structural strength of material such as straw allows high moisture contents without loss of FAS. Furthermore, the haystack shape and the fibrous nature of the material tend to channel water off the pile. Under these conditions, water additions from intermittent rainfall are usually of little consequence, but heavy rainfall for extended periods may still eventually soak the pile.

These same statements can generally be applied to the case where sludge is composted with large quantities of amendment. For example, digested sludge is composted in Seattle (40-60 in./yr of rainfall) by blending with sawdust in an approximate ratio of 4:1 sawdust: dewatered cake by volume [151]. The mixture is placed in large piles 12-15 ft high. No forced aeration is provided and piles are turned approximately twice in six months. The piles are large and shed water, so that rain is not considered a severe problem. With the proportions used, however, sludge is really the amendment since its solids contribution is so small.

The effects of rainfall were modeled assuming constant rainfall rates of 0.0, 0.25, 0.50 and 0.75 cm/day over the 20-day compost cycle and digested sludge cake solids of 30%. No surface evaporation was assumed and relative humidity was fixed at 100%. Air supply was constant at 0.25 m^3/min-dry ton, which is near the peak for output solids content as shown in Figures 14-9, 14-10 and 14-20. Cake, recycle and amendment temperatures were 20°C with air and rain temperatures of 10°C. In practice, rainfall patterns would likely be intermittent with clear and perhaps moderately dry conditions between storms. However, constant rainfall is a reasonable worst-case situation. Also, it is likely that total water addition is more important than the pattern in which it is applied.

Predicted temperature profiles for cases using recycled product and air-dried sludge amendment are shown in Figures 14-21 and 14-22. Turning frequency was 0.25/day in both cases. A rainfall rate of 0.25 cm/day (5 cm total) had little effect in either case. At 0.50 cm/day (10 cm total) the system using recycled product failed to produce adequate temperatures. At 0.75 cm/day (15 cm total) the system using air-dried sludge for amendment

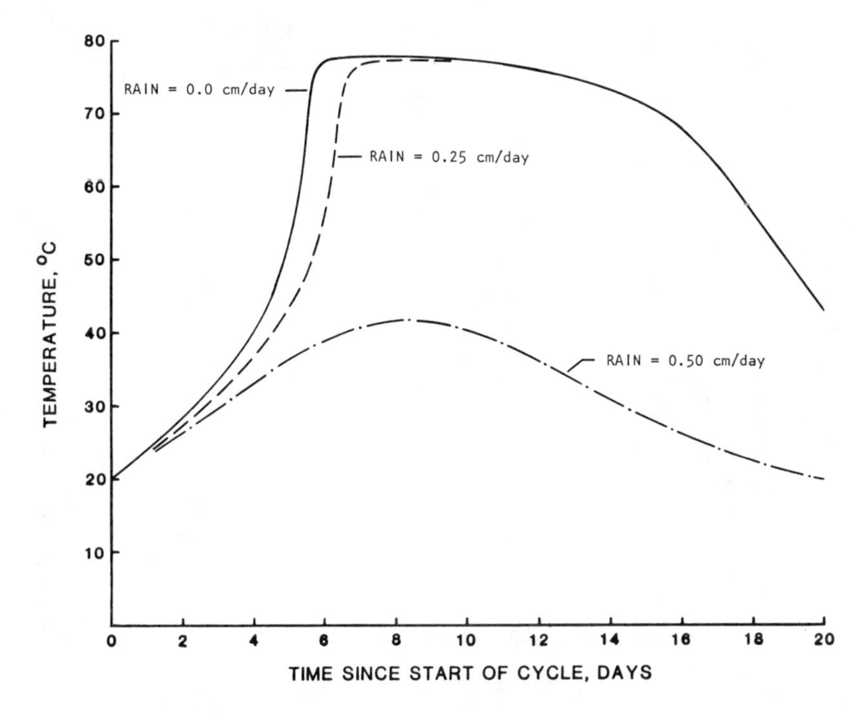

Figure 14-21. BATCH model. Effect of rainfall on predicted temperature profiles for windrow composting. Digested sludge with recycled product, SC = 0.30, TURNUM = 0.25, TC = TR = 20°C, TAIR = TRAIN = 10°C, specific air supply = 0.25 m^3/min-dry ton, specific oxygen supply = 7.71 g O_2/g mixture BVS, EVAP = 0.0.

also failed. Obviously, greater energy resources available in the air-dried material are able to compensate for greater water additions before failure is predicted.

A mass balance for the case using air-dried sludge and a rainfall rate of 0.50 cm/day is shown in Figure 14-23. Rainfall adds about 22% to the water input from all other sources. Nevertheless, almost 80% of total water input is removed in output gases, giving an output solids content of about 62%. Because recycled product is not used, further drying may not be necessary beyond that achieved in composting. BVS oxidation is about 72% complete in the 20-day period. Rate limitations are imposed by high temperatures during much of the cycle and by moisture limitations in later stages.

The ability of a system to withstand rainfall can be improved by reducing the air supply to limit drying and thus extend the available energy. For the case of sludge and amendment with TURNUM = 0.25/day and the air supply

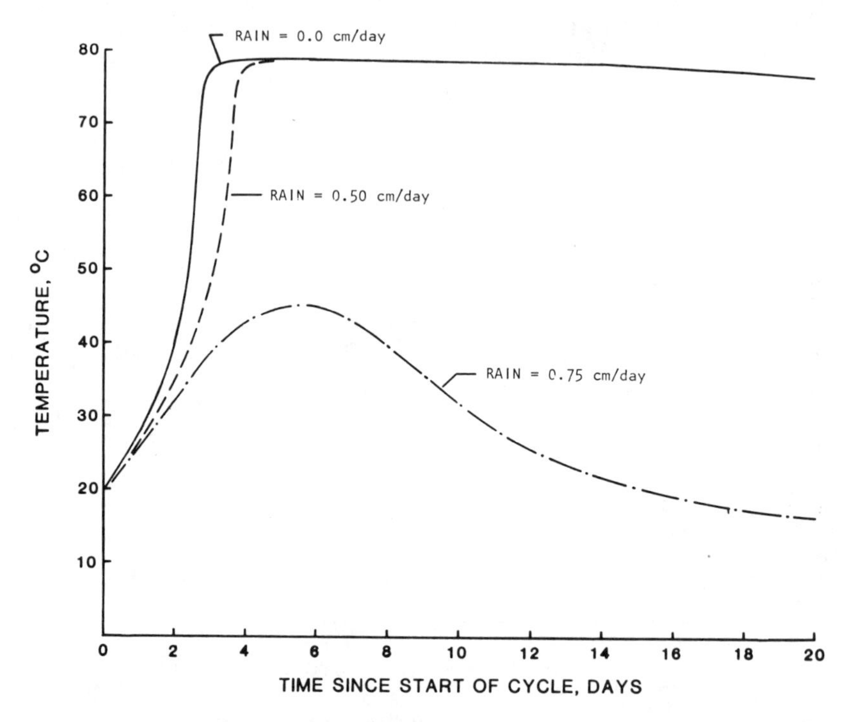

Figure 14-22. BATCH model. Effect of rainfall on predicted temperature profiles for windrow composting. Digested sludge with air-dried sludge amendment, SC = 0.30, TURNUM = 0.25, TC = TA = 20°C, TAIR = TRAIN = 10°C, specific air supply = 0.25 m^3/min-dry ton, specific oxygen supply = 4.88 g O_2/g mixture BVS, EVAP = 0.0.

reduced to 0.10 m^3/min-dry ton, acceptable temperature profiles are predicted at a rainfall rate of 0.75 cm/day (15 cm total). Of course, moisture removal is reduced and the final product may be quite wet. In fact, process failure is predicted at 1.0 cm/day rainfall (20 cm total) primarily from a complete loss of FAS caused by excessive moisture.

Predicted output solids content for the cases examined is shown in Figure 14-24. Output solids content is favored by use of the more biodegradable amendment in place of recycled compost and by reduced turning frequency to lessen heat losses. Maximum rainfall tolerated by the system is increased by use of air-dried amendment, reduced turning frequency and lower air flowrate at least over the range of values tested.

If rainfall rates exceed the system limits, system design must anticipate this eventuality. From the previous analysis, some design and operational approaches to improve the windrow systems tolerance for wet weather can be

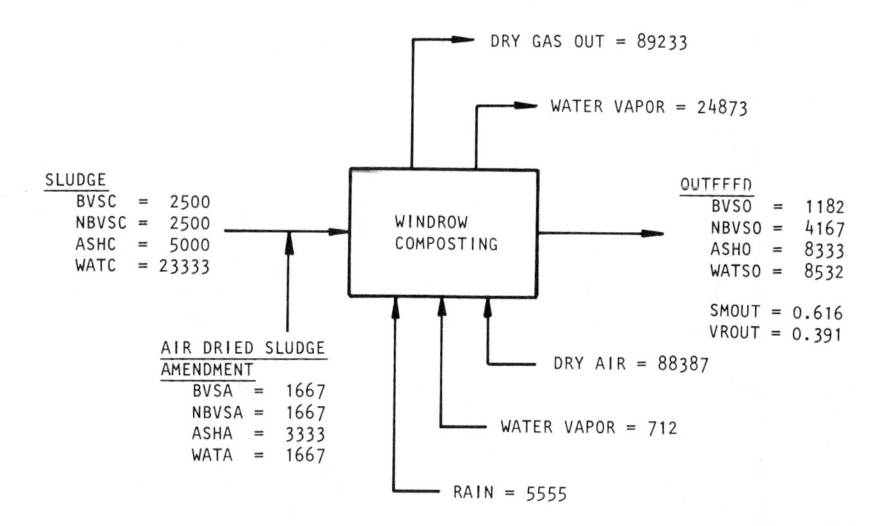

Figure 14-23. BATCH model. Effect of rainfall on the mass balance for solids, water and gas components during windrow composting. Digested sludge and air-dried sludge amendment, SC = 0.30, TURNUM = 0.25, TC = TA = 20°C, TAIR = TRAIN = 10°C, specific air supply = 0.25 m^3/min-dry ton, specific oxygen supply = 4.88 g O_2/g mixture BVS, RAIN = 0.50 cm/day (10 cm total), EVAP = 0.0. Units on mass terms are kg and refer to the total input or output over the 20-day compost cycle.

deduced. Roofed coverage of at least a portion of the compost beds can be provided to eliminate water addition from rainfall. A forced aeration system can be used to provide operational control of the air supply. Windrow dimensions should be maximized consistent with available turning equipment and the structural properties of the mixture. Permanent surfacing, such as asphalt or concrete, should be considered to assure equipment access in all weather conditions. The possibility of amendment addition in winter months should be considered to improve the overall energy balance and reduce the need to produce a very dry final product. Finally, heat drying of the composted material could be considered if a dry product must be produced at all times.

SUMMARY

The BATCH model was applied to the case of windrow composting of digested sludge using recycled product or amendment for moisture control. Windrow dimensions and operating characteristics approximated those of

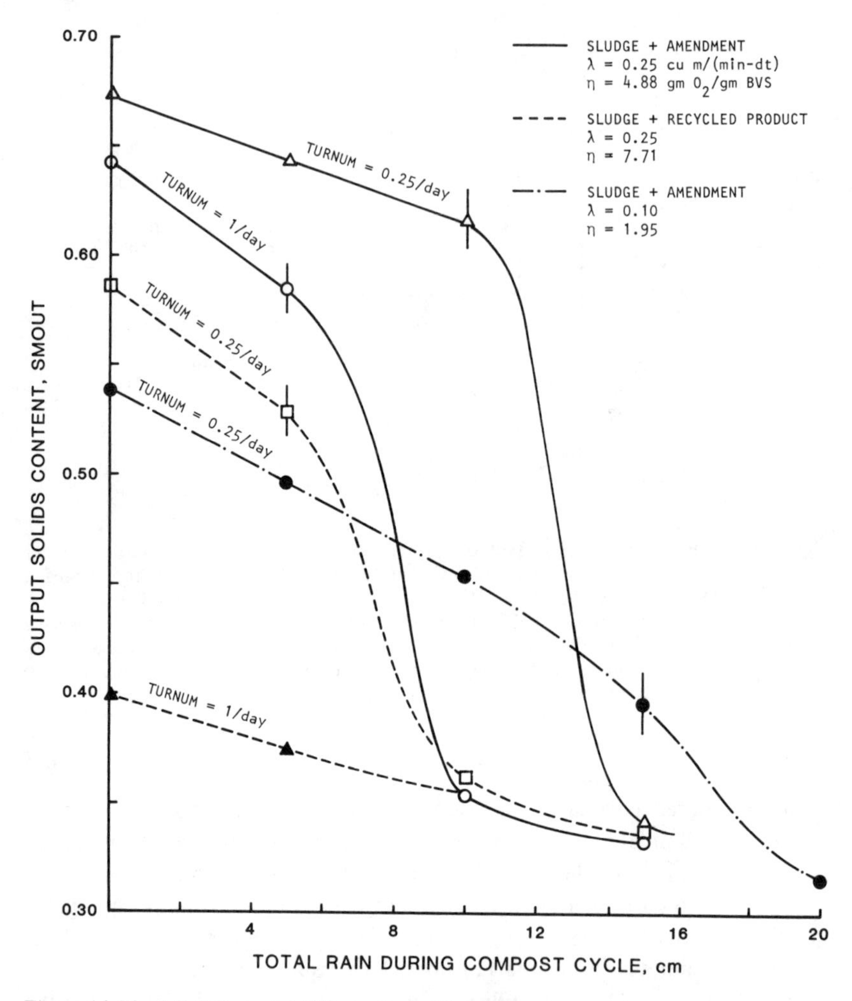

Figure 14-24. BATCH model. Effect of rainfall, turning frequency, amendment addition and specific air supply on output solids content from windrow composting. Vertical line indicates highest rainfall for which acceptable (>60°C) temperature profiles are predicted. Digested sludge, SC = 0.30, TC = TR = TA = 20°C, TAIR = TRAIN = 10°C, EVAP = 0.0.

present sludge-only compost systems. The following conclusions are applicable only to such conditions and should not be extended to cases using very large piles or large quantities of amendment.

Heat losses to surroundings were found to be of much more significance than in other compost systems. Heat loss during turning HTURN, from the

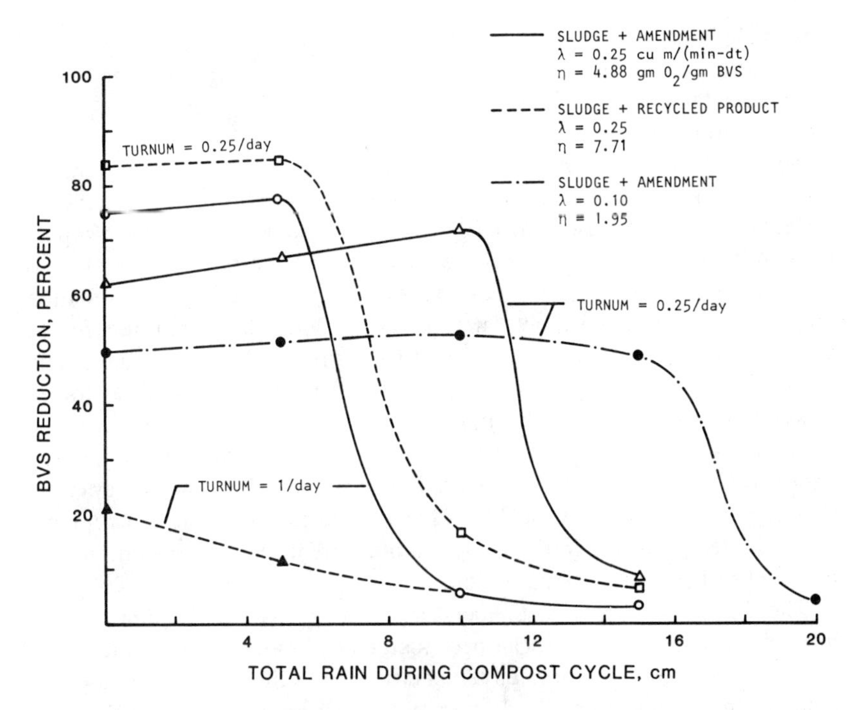

Figure 14-25. BATCH model. Effect of rainfall, turning frequency, amendment addition and air flowrate on BVS reduction. Digested sludge, SC = 0.30, TC = TR = TA = 20°C, TAIR = TRAIN = 10°C, EVAP = 0.0.

surface after turning HSURF and conductive loss to the ground HCOND were all determined to be significant and in many cases of comparable magnitude. In one case with a very low air supply rate, the three terms were predicted to account for 65% of the output energy. Not only did the three terms dominate the energy picture but the latent heat of vaporization HLWVO was displaced from its usual position of dominance. This is perhaps a rather extreme case but it does point to the potential importance of these terms in the total energy balance. At higher air flowrates more heat is removed by vaporization and HLWVO again becomes the dominant term.

Turning frequency is an important parameter in the windrow system because lower turning frequency reduces the heat loss terms HTURN and HSURF. For example, with 30% cake solids using recycled compost product, air and material temperatures of 20°C (i.e. TC = TR = TAIR = 20°C), acceptable temperature profiles are predicted with turning frequencies of both 0.25/day and 1.0/day. Maximum air flowrates are limited to about

0.50 m^3/min-dry ton, 14.7 g O_2/g BVS, in both cases. However, when cake solids are reduced to 20%, only the 0.25/day turning frequency gives acceptable temperature elevations through better heat conservation at the reduced turning frequency.

Surface evaporation of moisture is another important parameter that is generally of minor significance in static pile and enclosed reactor systems. If surface evaporation rates are high (0.50 cm/day used here) and oxygen supply rates near stoichiometric, moisture removed by surface evaporation is generally much greater than that removed by outlet gases. The situation reverses at higher air supply rates. In systems which do not use forced aeration, air supply is probably limited to nearly stoichiometric values, and surface drying is the major moisture removal mechanism. This makes forced aeration necessary in wet climates if maximum drying potential is to be achieved. Not only does surface evaporation affect moisture removal, it can also affect temperature development. For example, with a surface evaporation rate of 0.50 cm/day, a 20% cake is predicted to achieve acceptable compost temperatures with 1/day turning frequency, a condition not obtained without surface evaporation.

The duration of elevated temperatures is a function of air supply. With low air supply rates, oxygen can become the rate-limiting substrate, resulting in protracted periods of elevated temperature. Even though reaction rates are reduced, high-temperature steady-state conditions are predicted, the duration of which is controlled, at least in part, by the rate of air supply. High-temperature periods of 40-60 days are predicted under such conditions. Increasing the air supply removes oxygen limitations, improves BVS stabilization and can yield temperature profiles which fall back toward ambient before day 20.

With cake solids of 30%, significant moisture removal can be achieved even without surface evaporation, although the latter is certainly helpful. With 20% cake solids, drying potential is limited without assistance from surface evaporation. These results are consistent with the analysis of Chapter 9.

Low ambient temperatures can affect the windrow system in particular because of the greater importance of heat loss to surroundings. With an air temperature of 0°C and material temperature of 20°C (i.e., TC = TR = 20°C, TAIR = 0°C), acceptable temperature profiles are predicted with a 30% cake and a turning frequency of 0.25/day, but not at 1/day. Maximum air supply rate is reduced to about 0.25 m^3/min-dry ton. Two approaches were useful in mitigating effects of low air temperature: use of initial preheating and use of degradable amendments in place of recycled product. Preheating of materials allows biological reaction rates to increase to the point where heat production can keep pace with energy outputs. With preheat of solids to 50°C, acceptable temperature elevations are predicted even with 1/day turning.

Addition of degradable amendments can improve energy resources, increase the range of operating air supply, and overcome energy limitations imposed by low cake solids. Air-dried sludge was shown to be effective in this manner. It is an attractive amendment in certain regions because it can be produced in dry months and then stockpiled for later use in wet weather. In effect, organics produced in dry, warm weather are conserved for use when climatic conditions are more severe. Importing other materials to the compost site is also avoided.

The windrow system can withstand limited moisture inputs from rainfall, but process failure is predicted with large inputs. With 30% cake using recycled compost, turning frequency of 0.25/day, and air supply of 0.25 m^3/min-dry ton, maximum rain input of about 5 cm over a 20-day compost period is predicted. Beyond this, adequate temperature profiles are not achieved. Operational approaches to reduce the impact of rainfall include: (1) reduced air supply to limit drying and extend available energy; (2) reduced turning frequency; (3) use of amendments such as air-dried sludge; and (4) maximizing windrow cross-sectional dimensions. Using air-dried sludge, a reduced air flowrate of 0.10 m^3/min-dry ton and 0.25/day turning, the maximum tolerable rain input is extended to about 15 cm. Design approaches to improve the tolerance for wet weather include: (1) roofed coverage to eliminate water input from rain; (2) forced aeration systems to allow operational control of the air supply; and (3) permanent surfacing to assure equipment access in all weather conditions.

CHAPTER 15

CONTROL OF NUISANCE CONDITIONS

INTRODUCTION

It is important to engineer appropriate systems for control of nuisance conditions during design of a compost facility. Potential nuisances most often associated with composting are odors and dusts. Other nuisances such as insects, birds and rodents are of limited concern [15]. Insect eggs are destroyed at the thermophilic temperatures achieved during composting. Also, insects and birds do not appear to be attracted to open compost piles, at least not to those using wood chips or recycled compost. If other amendments are used, such as garbage or refuse, there may be more cause for concern. Some might also suggest visual esthetics and noise as potential nuisances. These are most easily controlled by proper site selection and by matching the type of compost process with the needs of the surrounding community.

This chapter will concentrate on problems of odor and dust, and the alternatives available to control them. The purpose is not to present a complete treatise on the subject, but rather to adapt what is known about various control measures to the needs of a compost system. As will be seen, requirements for nuisance control can have a major influence on the type of compost system selected.

CONTROL OF DUST FORMATION

Dust control in a compost facility can be a major problem, particularly in arid or semiarid climates. Municipal sludge organics are characterized by a small-sized particle distribution. When sludge based compost exceeds about 65–75% solids, the small-sized particles can be easily airborne if agitated. At higher moisture contents dust is generally not a significant problem. Therefore, dust conditions can potentially occur any time dry compost is handled. This would include turning or agitation of dry windrows, screening dry compost from bulking particles, loading of trucks and so on. Another potential source of dust is fugitive emissions that may occur from operation of mobile equipment on dry, unpaved surfaces.

Most of the control measures to avoid excessive dust formation are reasonably straightforward and include the following:

- For open systems, turning or agitation should be discontinued when the material exhibits noticeable dust formation, which should begin at about 65-70% solids.
- Controlled aeration systems can be used to reduce the turning frequency.
- Paved surfaces should be used whenever possible. If unpaved surfaces cannot be avoided, chemical stabilization of the soil and/or generous use of water trucks can relieve the problem.
- Mixing of wet cake and compost product in the open, using front end loaders or similar equipment, should be avoided whenever possible. Mechanical premixing systems can be fully enclosed and also help avoid dry spots when windrows or piles are formed.
- Operations with a high dust potential, such as mixing and screening, should be conducted in an enclosed reactor or building. Enclosures should be equipped with air collection and particulate control devices such as wet scrubbers or fabric filters.
- Compost exposure to ambient conditions can be reduced by use of enclosed reactor composters, dryers and compost storage.
- Good housekeeping should be practiced to avoid accumulation of very dry material. Good housekeeping is aided by use of paved surfaces.
- Compost can be pelleted before drying or storage to increase particle size.

This list does not exhaust all possibilities and the reader can probably think of other dust control measures. The combination of measures required for a given facility depends on the type of compost system used, climatic factors, proximity to neighbors and other site-specific factors. It is possible to provide a very high level of dust control through use of enclosed compost systems, enclosed drying and even enclosed storage. This will be discussed further in Chapter 16. The point here is that dust control should be included in facility design and that a range of options are available to tailor the system to local needs.

One control measure which should receive more consideration is the use of pelleting to increase particle size before the onset of dusting conditions (i.e., low moisture content). Fuller [152] noted the following advantages of pelleted compost: (1) pellets are convenient and easy to handle; (2) pellets are cleaner to handle and cleaner around the private home after applying; (3) pellets are not as subject to dusting if agitated or exposed to high winds; and (4) different sizes and shapes of pellets can be selected for specific purposes. The fine, granular size (1–2 mm) will move down through turf grass until it contacts the soil because of its density and small size. Larger pellets

(2–6 mm) may be more suited for flower beds and mulching. On the other hand, pelleting can be used solely for onsite dust control without regard to end product use. If desired, pellets can be reground before reuse or disposal.

The LA/OMA project experimented with pelleting compost produced from both raw and digested municipal sludges using recycled product for moisture control [23]. The purpose of these studies was to determine the efficiency of the pelleting process and whether pellets would dust or crumble upon drying. Pellets were produced from compost product at about 50% solids content, and were cylindrical in shape, about 10–15 cm long, 5 cm in diameter, with a 1- to 1.5-cm hole through the center to increase the surface area for subsequent drying. The pelleter consisted of a screw that fed compost material at a continuous rate through an extruder as shown in Figure 15-1. Pellets were cut by a rotating blade at the end of the extruder. After formation, pellets were dried in an enclosed dryer using ambient air (see Chapter 16). The large pellet size served to reduce air head loss through the dryer.

Immediately after extrusion the pellets were still somewhat pliable, much like modeling clay. Plasticity limited the initial stacking height to about 2–3 m to avoid compressing the bottom pellets. Pellet strength increased dramatically during the air-drying process. At 80–90% solids, the pellets were characterized as being "rock hard" and exhibited little or no tendency to dust. Even if broken, as by striking with a hammer, the dried pellet would tend to form a number of smaller particles but did not crumble into dust.

In the LA/OMA study, compost product was both dried and stockpiled in the pelleted form. Dry material used for feed moisture control was conveyed to a hammermill, which ground the pellets back to a fine material suitable for blending with feed sludge. The milling process produced considerable dust which was separated from the hammermill exhaust by a dry cyclone and fabric filter.

ENGINEERING FOR ODOR CONTROL

Control of odors is one of the most difficult problems in present sanitary engineering practice. The problem is compounded by the fact that sludge, manure, refuse and other waste organics are inherently odorous and that the human nose is a remarkably sensitive detection instrument. A number of the more common odorants observed in wastewater treatment facilities are presented in Table 15-1. Both inorganic and organic compounds can be malodorous. Hydrogen sulfide (H_2S) and ammonia (NH_3) are the most common inorganic odorants. Organic sources generally come from low-molecular-weight, more volatile compounds such as methyl mercaptans, methyl sulfides and amines. The sensitivity of the human nose from some of these

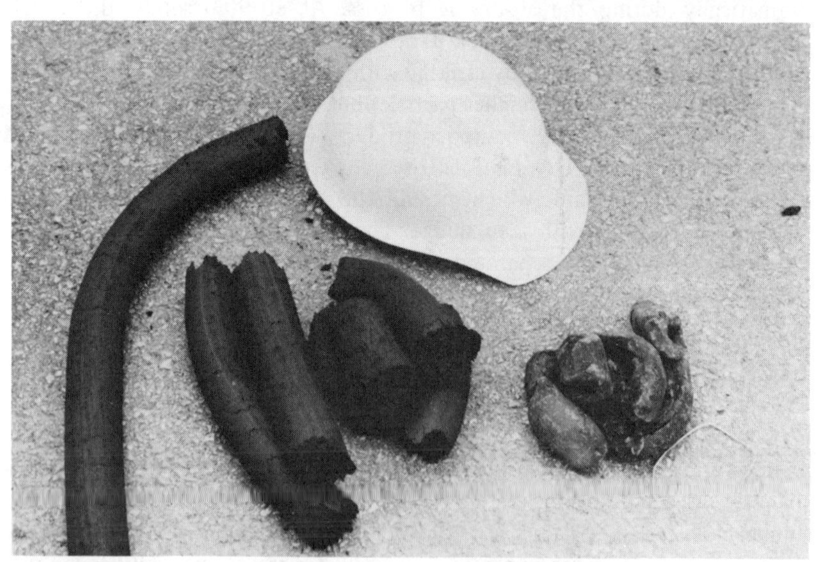

Figure 15-1. Pelleting a compost produced from raw and digested municipal sludges amended with recycled product. The pelleter was part of the Fermentechnik SR-10 system (Euramca) and consisted of a screw which fed compost material at a continuous rate through an extruder. Pellets could be cut at any desired length. Compost was about 50% solids when pelleted. Pellets were then air dried to 80-90% solids. The lower photograph shows both new pellets (left) and dried pellets (right).

compounds is evident in the low-odor threshold concentrations shown in Table 15-1.

Odor is becoming an increasingly sensitive issue as neighbors move closer to existing treatment plants. When new facilities are sited, the potential for odor is invariably one of the first concerns raised by local residents. In fact, odors have been rated as the first concern of the public relative to implementation of wastewater treatment facilities [153]. The designer must be conscious of this fact and familiar with processes used for odor control and the analytical tools used to predict odor transport and dispersion.

Odor Measurement Techniques

To regulate and control odors, it is necessary to quantify and measure them. If the composition of a gas stream is known, it is possible to determine analytically the concentration of odor-causing components using chemical/instrumental techniques. If the odor threshold concentration is known for the various components, the degree of required treatment and/or dilution can be determined. This approach finds its greatest application in certain industrial processes where the number of gas components is limited. Most odor sources, however, are characterized by a wide variety of components whose composition is usually unknown. This is particularly true of odors arising from sewage and sludge processes. In such cases the human nose is still the accepted standard for detecting and determining odor intensity. After all, it is a human receptor that senses the odor, and no machine has yet been able to simulate the human response.

Organoleptic methods for odor detection and measurement are those which use the human olfactory system. The latter system is generally three orders of magnitude more sensitive than currently available chemical/instrumental systems [155]. It is also sensitive to a wide variety of chemical structures. Although the human nose gives only a subjective response to the presence or absence of odor, a number of techniques have been developed to quantify the human response.

A number of organoleptic techniques are in common use for odor quantification, and most involve dilution to a threshold odor concentration. Probably the most commonly used approach is referred to as the "odor panel technique," and is usually conducted in accordance with ASTM procedure D1391-57. A panel of subjects (5–10 members are desirable) are exposed to odor samples that have been diluted with odor-free air. The number of dilutions required to achieve a 50% positive response by panel members is termed the threshold odor concentration (TOC). This is taken to be the

Table 15-1. Odorous Substances Found at

	Formula	Characteristic Odor	Odor Threshold (ppm)
Acetaldehyde	$CH_3 \cdot CHO$	Pungent Fruity	0.004
Allyl Mercaptan	$CH_2 \cdot CH \cdot CH_2 \cdot SH$	Strong Garlic, Coffee	0.00005
Ammonia	NH_3	Sharp Pungent	0.037
Amyl Mercaptan	$CH_3 \cdot (CH_2)_3 \cdot CH_2 \cdot SH$	Unpleasant, Putrid	0.0003
Benzyl Mercaptan	$C_6H_5 \cdot CH_2 \cdot SH$	Unpleasant, Strong	0.00019
Butylamine	$C_2H_5 \cdot CH_2 \cdot CH_2 \cdot NH_2$	Sour, Ammonia-like	
Cadaverine	$H_2N \cdot (CH_2)_5 \cdot NH_2$	Putrid, Decaying Flesh	
Chlorine	Cl_2	Pungent, Suffocating	0.01
Chlorophenol	ClC_6H_5O	Medicinal, Phenolic	0.00018
Crotyl Mercaptan	$CH_3 \cdot CH{:}CH \cdot CH_2 \cdot SH$	Skunk-like	0.000029
Dibutylamine	$(C_4H_9)_2NH$	Fishy	0.016
Diisopropylamine	$(C_3H_7)_2NH$	Fishy	0.0035
Dimethylamine	$(CH_3)_2NH$	Putrid, Fishy	0.047
Dimethyl Sulfide	$(CH_3)_2S$	Decayed Vegetables	0.001
Diphenyl Sulfide	$(C_6H_5)_2S$	Unpleasant	0.000048
Ethylamine	$C_2H_5 \cdot NH_2$	Ammoniacal	0.83
Ethyl Mercaptan	$C_2H_5 \cdot SH$	Decayed Cabbage	0.00019
Hydrogen Sulfide	H_2S	Rotten Eggs	0.00047
Indole	C_2H_6NH	Fecal, Nauseating	
Methylamine	CH_3NH_2	Putrid, Fishy	0.021
Methyl Mercaptan	CH_3SH	Decayed Cabbage	0.0011
Ozone	O_3	Irritating Above 2 ppm	0.001
Propyl Mercaptan	$CH_3 \cdot CH_2 \cdot CH_2 \cdot SH$	Unpleasant	0.000075
Putrescine	$NH_2(CH_2)_4NH_2$	Putrid, Nauseating	
Pyridine	C_6H_5N	Disagreeable, Irritating	0.0037
Skatole	C_9H_9N	Fecal, Nauseating	0.0012
Sulfur Dioxide	SO_2	Pungent, Irritating	0.009
Tert-Butyl Mercaptan	$(CH_3)_3C \cdot SH$	Skunk, Unpleasant	0.00008
Thiocresol	$CH_3 \cdot C_6H_4 \cdot SH$	Skunk, Rancid	0.0001
Thiophenol	C_6H_5SH	Putrid, Garlic-like	0.000062
Triethylamine	$(C_2H_5)_3N$	Ammoniacal, Fishy	0.08

minimum concentration detectable by the average person. Thus, if nine volumes of diluting air added to one volume of odor sample generate a positive response by half the panel, the odor concentration is reported as ten dilutions to TOC. It is common to view the required dilution as a pseudo-concentration and the above sample would be said to have an odor concentration of 10 ou (odor units). An odorous compound at its TOC has, by

Wastewater Facilities [137, 138, 154, 204]

Detection Threshold (ppm)	Recognition Threshold (ppm)	Adsorption of Vapors by Activated Carbon (%)	Boiling Point (°C @ 760 mm Hg)	Molecular Weight
	0.21	7	20.2	44.05
0.016			67	74.15
	46.8	1-2	−33	17.03
			126	104.22
			194	124.21
	0.24		78	73.14
			178	102.18
0.01	0.314		−34	70.91
			214	128.55
0.0077			98	90.19
			159	129.25
	0.085		84	101.19
	0.047		7	45.08
	0.001		37	62.13
	0.0021		296	186.28
	0.83		17	45.08
0.0026	0.001	23	23	62.1
	0.0047	3	−62	34.1
		25	254	117.15
	0.021		−7	31.05
	0.0021	20	8	48.1
0.5			111	48
0.024		25	67	76.16
		25	158	88.15
		25	116	79.10
0.223	0.47	25	266	131.2
		10	−10	64.07
			65	90.19
0.019			195	124.21
0.014	0.28		170	110.18
			89	101.19

definition, a concentration of one odor unit. For example, a 1-m^3 sample with a TOC of 100 ou requires dilution by 99 m^3 of odor free air to reach the threshold concentration.

There are a number of inherent difficulties with the odor panel technique. Samples must be collected and transported to the odor panel site. In some cases, storage of odorant samples has been observed to alter the odor strength. Also, a number of people are required both for sample collection and odor panel analysis, making the procedure somewhat expensive. Human subjects differ in their sensitivity to odors and results can be influenced by the personnel selected for the panel. Despite these shortcomings, the odor panel is a valuable tool for the very difficult task of quantifying odor concentration.

The second major approach to odor quantification uses specialized equipment to supply a range of diluted odor samples to one or more odor judges. Many of these devices, termed direct reading olfactometers (DRO), are portable and avoid the need for sample storage, which is a disadvantage of the traditional odor panel technique. On the other hand, most portable devices use a single individual as a judge, which is statistically less reliable than the odor panel technique.

Descriptions and methods of use of direct reading olfactometers are available in Wilson [156]. Usually, the test subject wears a mask or hood to isolate the olfactory system from ambient air. Carbon-filtered air (assumed to be odorless) is supplied to the subject to prevent fatigue of the olfactory sense or adaptation to the odor. This is followed by a continuous supply of diluted sample air. A number of dilution ratios are supplied to determine the threshold odor concentration. A major advantage of portable DRO units is that they can be used to measure odors at the source without using sam-

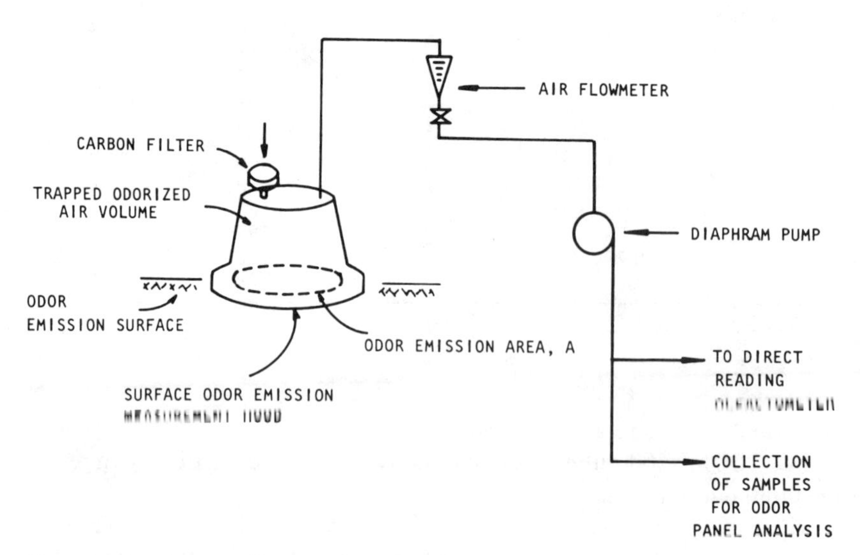

Figure 15-2. Schematic diagram of test equipment used for measurement of unit area SOER. Adapted from Wilson [172].

pling containers. Manpower requirements are also reduced. When using different people with the same odor source, DRO results have been reported to be reproducible to within 15% [157].

The fact that both odor panel and DRO methods are in common use can lead to difficulties in correlating data from different sources. Redner [158] analyzed samples by both techniques and was unable to develop a consistent correlation between the two. In some cases the DRO technique yielded higher concentrations and in other cases it did not. Despite these limitations, the general range of odor emission rates from composting is definable and we will be able to evaluate the effectiveness of various odor control strategies.

Some compost operations are characterized as having a large surface area exposed to ambient air. Certainly the windrow and static pile systems are examples of large area sources of odor. In such cases it is of interest to determine the surface odor emission rate (SOER). Units are usually given as $ft^3/min\text{-}ft^2$ or $m^3/min\text{-}m^2$. The technique for measurement of SOER involves placement of a sampling hood over the surface being analyzed (Figure 15-2). Air is drawn through the hood at a measured flowrate and the odor concentration of exhaust air determined. The product of odor concentration and air flowrate is assumed to equal the odor emission rate from the surface.

Suppose air is drawn at a flowrate of 1 m^3/min across a 0.5 m^2 sampling area. If a sample of the exhaust gas is then determined to contain 10 ou, the SOER would be 20 $m^3/min\text{-}m^2$ (i.e., $1 \times 10/0.5$). The equivalent SOER in English units would be 65 $ft^3/min\text{-}ft^2$. Measurements at various air flowrates indicate that the procedure provides reasonably representative values of the SOER [159].

The units on SOER may be somewhat confusing. One way to interpret their meaning is to envision odorous molecules being released from the material surface at a uniform rate. Consider the above example with a SOER of 20 $m^3/min\text{-}m^2$. Then for each m^2 of surface area a 20-m^3/min flowrate of air would be required to reduce the odor level to the threshold odor concentration (i.e., 1 ou).

Before leaving this section it is appropriate to review the concept of an odor unit again because it is the fundamental measure of odor concentration for organoleptic techniques. An odor unit can be defined as 1 m^3 (or 1 ft^3) of air at the odor threshold. Odor concentration is the number of cubic meters that 1 m^3 of sample (or cubic feet per 1 ft^3 of sample) will occupy when diluted to the odor threshold. Odor concentration is sometimes expressed as ou/m^3 or ou/ft^3, but it is more appropriate to simply state the concentration in ou. The implied units are volume of sample diluted to the odor threshold per volume of original sample. Thus the odor unit can be viewed as a dimensionless number. Although these distinctions are somewhat subtle, they will serve well in later discussions.

Compost Odor Concentrations and Emission Rates

Information on odor emission rates and odor concentrations from compost operations is quite limited. In many cases the available data have been determined by different techniques, making direct comparison somewhat tenuous. Nevertheless, review of the available information will provide the reader with an idea of the general range of odor emissions expected and what can occur if thermodynamic or kinetic constraints are not recognized.

Odor Complaint Criteria

Before examining odor emissions characteristic of compost operations, it is useful to review some of the odor criteria that have been established for ambient air. Huang et al. [160] and Wilson et al. [161] reported that odor concentrations in excess of 5 ou are easily recognized by most people and, regardless of the particular odorant characteristics, represent the threshold level of complaint. This concentration has also been termed the distraction threshold. A concentration of 10 ou usually assures a complaint-level condition and is apparently sufficient to demand conscious attention and to cause psychological stresses commonly associated with odor complaints.

In reviewing the odor concentrations responsible for odor complaints, the Sacramento Regional Wastewater Management District adopted a design ambient air odor criterion of 2 ou maximum at the treatment plant boundary [157]. It is clear from the general range of values that the nearest neighbors should be subjected to no more that 5 ou and prefereably much less. However, such an objective may not be realized 100% of the time because of occasional high odor conditions. In the Sacramento study [157] with full implementation of odor control measures, a plant boundary odor level of 2 ou is expected to be exceeded 6 day/yr, and 5 ou exceeded 0.7 day/yr. Thus, there is a statistical element to evaluation of odor risk and, although the goal may be complete elimination of odors, the best one can usually hope for is management and control to acceptably low levels.

Case History at a Major Compost Facility

Redner et al. [159] published what is probably the first engineered study of odor emissions at a sludge composting facility. The study was prompted by a sudden increase in odor complaints in 1977 at a major treatment plant operated by the Los Angeles County Sanitation Districts. The story behind this study is worth developing because it reveals much about the relationships between odor emissions and constraints placed by thermodynamics and other operating problems.

Before 1977 about 90 dry ton/day of digested sludge at 30–35% solids was windrow-composted at the plant site. Windrow composting was initiated in 1972 and replaced an open-air drying system that had been used for several decades. A benefit of the windrow operation was a reduction in odor emissions as shown in Figure 15-3. Completion of new dewatering facilities in 1977 increased the sludge tonnage to about 270 dry ton/day, but cake solids were decreased to only about 23%. The reader should immediately recognize that the thermodynamic balance was severely stressed, and a resultant lowering of compost temperatures was noted. When combined with wet weather and other operational difficulties, odor emissions and complaints from the local community increased dramatically.

A summary of SOER values determined in 1972 and during the period of odor complaints in 1977 is shown in Table 15-2. For the 1972 data, odor emissions were noted to decrease with compost time. No consistent reduction in the first seven days was noted in the 1977 tests, and SOER values were

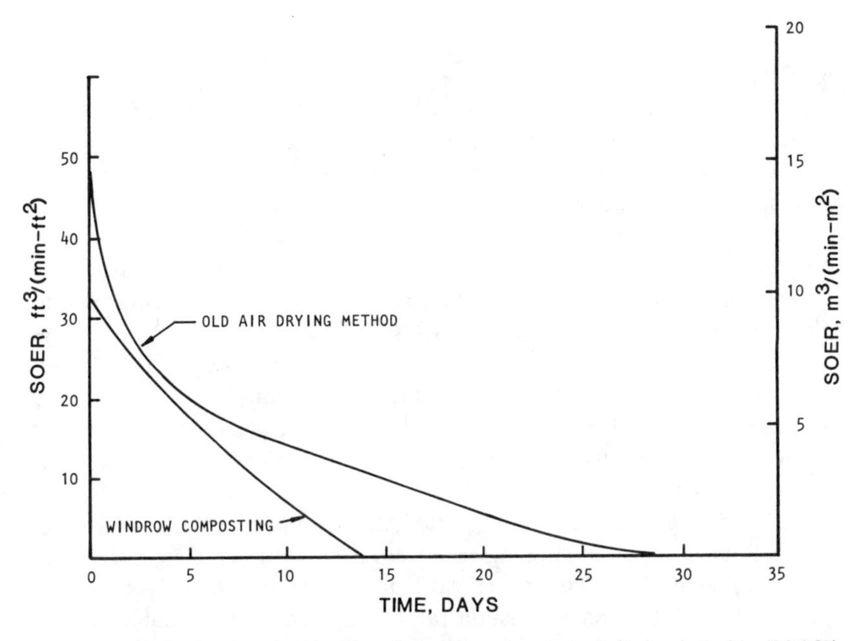

Figure 15-3. SOER observed during windrow composting of digested sludge (30-35% cake solids) blended with recycled compost and compared to open-air drying of the same sludge in piles 40-60 cm deep. Septic conditions often developed in the interior of open-air piles during the six months required for drying. Odor emissions during pile removal were often significantly greater than those shown above. Since this disadvantage was not shared by composting, the actual improvement in the odor situation was greater than that shown above. Odor concentrations were determined by the odor panel technique [159].

Table 15-2. Comparison of Operations and Resultant Odor Emission Rates in 1972 and 1977 at a Major Compost Facility Operated by the Los Angeles County Sanitation Districts [159]

Compost Pile Age (days)	Mean Bed Temperature (°F)		SOER (ft^3/min-ft^2)	
	1972	1977[a]	1972[b]	1977[a,c]
0	90	83	32	43
1	104	85	28	42
2	114	94	25	116
3	122	94	23	148
4	126	100	20	152
5	128	100	18	214
6	129	108	15	455
7	130	89	13	93

[a]Geometric mean.
[b]SOER values for 1972 were determined by the odor panel technique.
[c]SOER values for 1977 were determined by the DRO technique.

significantly increased. Also, windrow turning was observed to have a major effect on odor emissions as shown in Figure 15-4. The equilibrium SOER before turning was about 50–100 ft^3/min-ft^2. The odor emission rate increased to about 1800 ft^3/min-ft^2 immediately after turning and then decayed at an approximately first-order rate. Through integration of the decay curve it was determined that as much as 50% of the odor emission occurred within four hours of turning a windrow. Average odor emission rates from the composting area were estimated to be about 100×10^6 ft^3/min compared to values estimated in 1972 of between 10×10^6 and 15×10^6 ft^3/min. It was also shown that silo storage of dewatered cake for periods of 1–3 days approximately doubled the surface emission rate. It must be clearly pointed out that the data of Figure 15-4 are not typical of a well-operating compost system, but are indicative of what can occur if thermodynamic constraints are not recognized.

During this period a number of alternative compost systems were field-tested. Redner et al. [13] summarized available data for these systems as shown in Figure 15-5. The aerated static pile, windrow and enclosed reactor systems showed low odor emission rates compared to the trouble-plagued windrow system. Exhaust gas from aerated static pile and enclosed reactor processes is also a source of odors, as will be discussed shortly.

To complete this ministory, a number of process modifications were made to reduce the odor impact. A barrier (block wall) was constructed around the compost facility. Input sludge tonnage to the windrow process was reduced and dewatering performance improved to reduce the water tonnage.

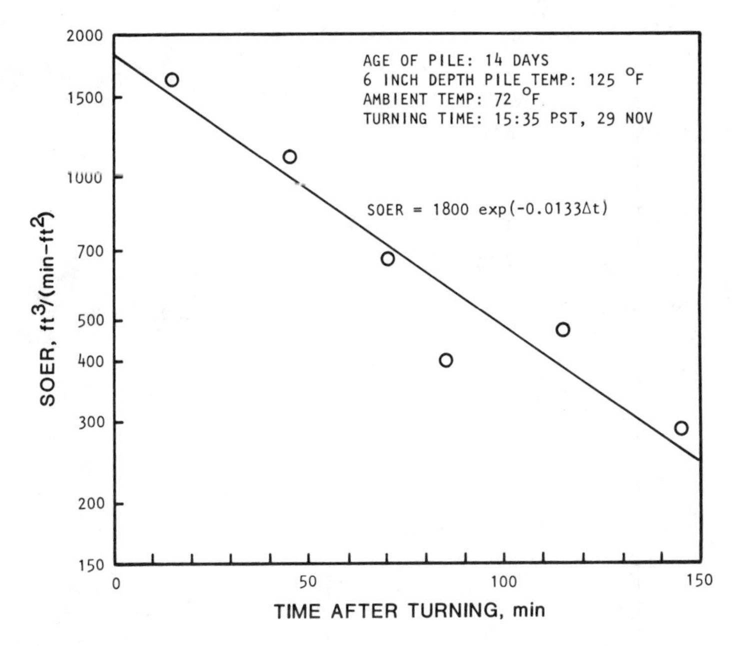

Figure 15-4. Decay of surface odor emission rate after turning of a compost windrow composed of digested sludge and recycled compost. The data were collected during a period of adverse operation and would not be typical of a properly operating windrow. The data are useful in illustrating the potential for odor production that exists if thermodynamic constraints and operating limits are exceeded [162].

Operating hours were restricted to periods of higher atmospheric turbulence to provide greater dilution of odors. The compost area was also reduced in size to provide some buffer between the surrounding community. These measures were effective in reducing the number of odor complaints at least for the short-term. A number of additional odor control measures are available and will be discussed shortly.

Odor Data for Alternative Compost Systems

Iacobini et al. [17] developed the data shown in Figure 15-6 on SOER during windrow composting of digested sludge blended with recycled compost. No forced aeration was used and windrow turning was about once per day. Proper composting temperatures were achieved and SOER values were significantly lower than those observed by Redner et al. [159] during more adverse conditions. The odor emission rate drops significantly during the first few days of composting and then fluctuates somewhat during the

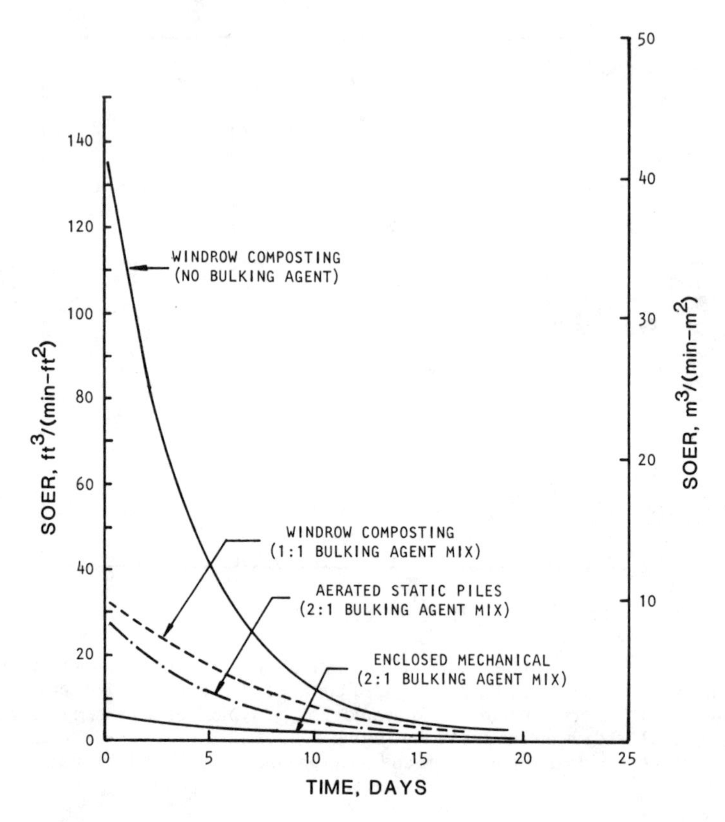

Figure 15-5. Surface odor emission rates for various compost systems. All cases involved composting of digested sludge with recycled compost as a bulking agent. Windrow composting with no bulking agent consisted of periodic turning of dewatered cake to enhance surface evaporation until proper material conditions were achieved. The enclosed mechanical method was a batch-operated, horizontal, agitated solids bed type. Odor concentrations were determined by the odor panel technique [159].

remainder of the cycle. Emission rates during active composting appear to average about one order of magnitude less than the initial rate. The reduction probably corresponds to development of more complete aerobic conditions within the windrow. A frequency distribution of SOER values collected during a number of windrow cycles is shown in Figure 15-7. The median value is about 2 m^3/min-m^2. Most of the high SOER values were observed in the early stages of composting.

Recent studies of the windrow system have confirmed that odor emission rates increase immediately after turning, even with properly operating windrows. This is illustrated by the data of Figure 15-8. Odor emission rates

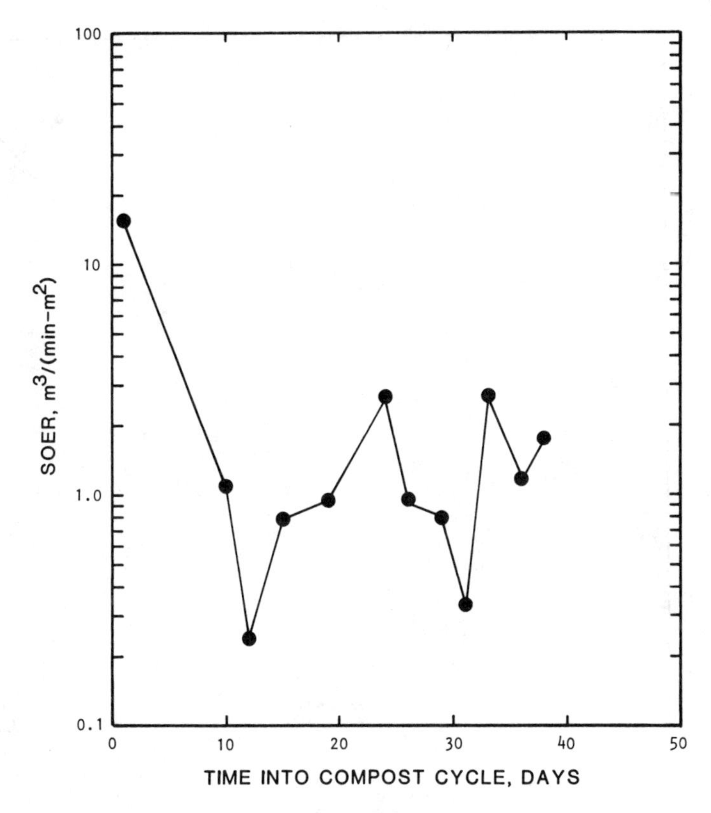

Figure 15-6. SOER during windrow composting of digested sludge blended with recycled compost. Each data point is the average of values obtained on three separate windrows during summer weather conditions. Samples were collected before windrow turning. Odor concentrations were determined by the odor panel technique [17].

increased after turning, but within about one hour began to approach levels evident before turning. Most data in the frequency distribution of Figure 15-7 were collected before windrow turning. Therefore, periodic increases in odor emissions as a result of turning should be considered in the total odor budget in addition to SOER values shown in Figure 15-7.

SOER values observed during static pile composting of digested sludge blended with various amendments are shown in Figure 15-9. Again, the surface emission rate decreases dramatically after the first days of composting. The pile surface was observed to be essentially odor free after the tenth day into the cycle. SOER values did not differ significantly for the various amendments tested. Odor emission rates increased dramatically during pile

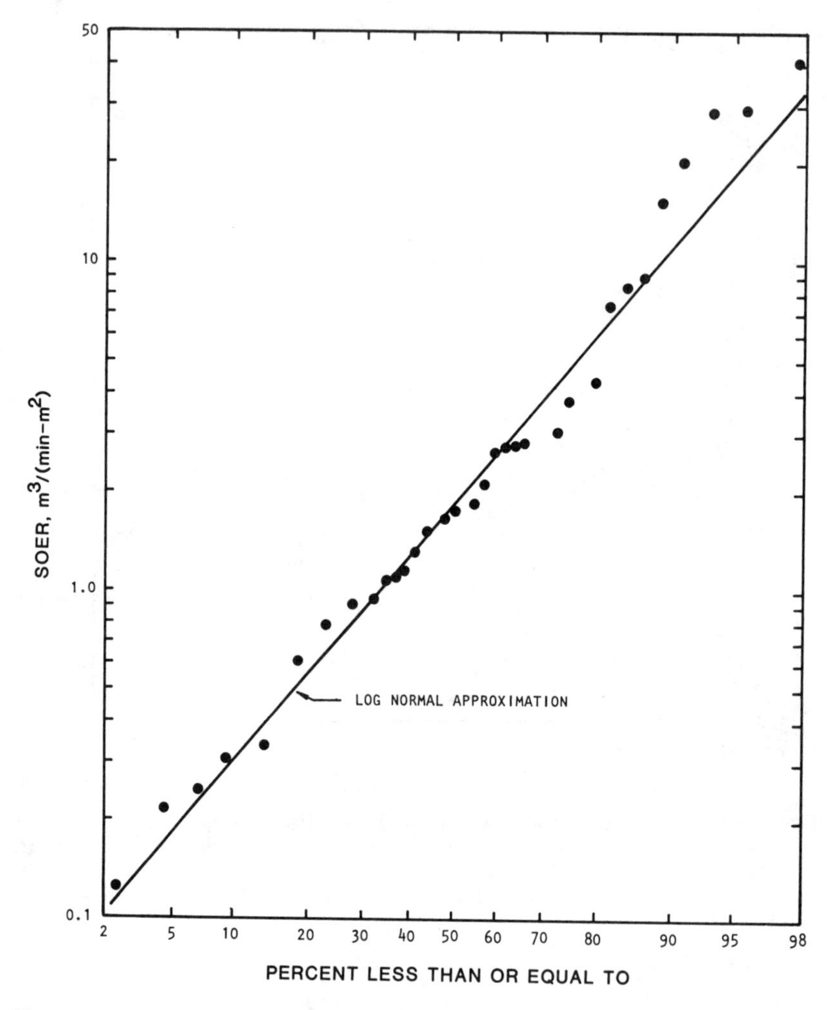

Figure 15-7. Frequency distribution of SOER observed during windrow composting of digested sludge blended with recycled compost. Higher values of SOER occurred within the first few days of the compost cycle. Samples were generally collected before windrow turning. Odor concentrations were determined by the odor panel technique [17,112]

takedown, with higher rates at the end of the pile farthest from the blower. Balls of material ranging from softball to basketball sizes were observed during pile dismantling. The balls exhibited anaerobic interiors and released considerable odor upon breaking, accounting for much of the increased emission rates. Similar balling problems have been observed even when

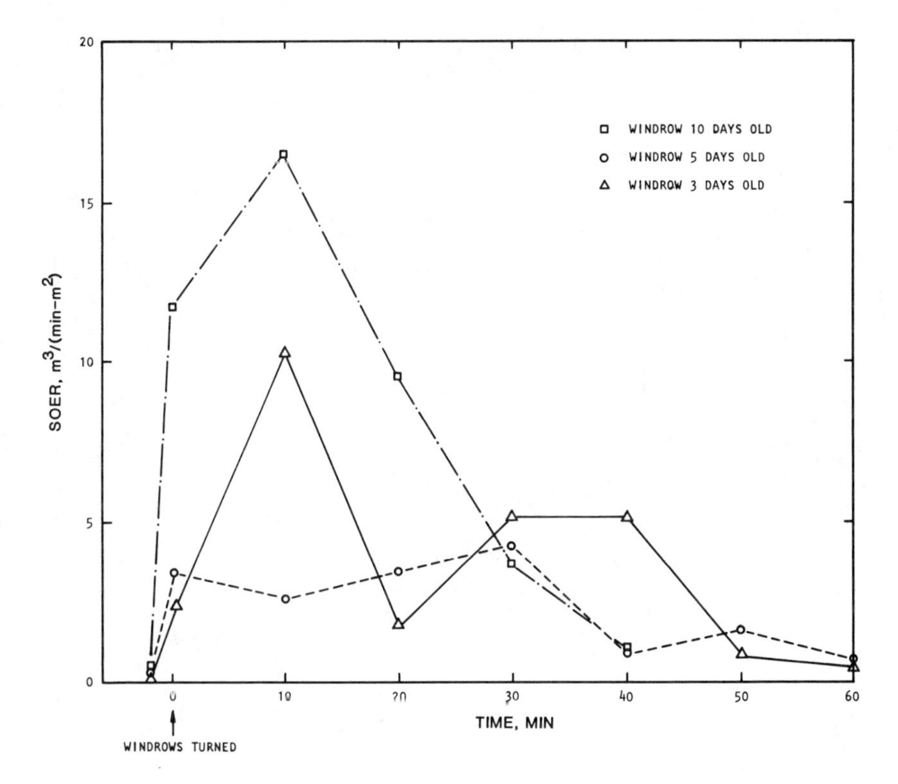

Figure 15-8. Effect of windrow turning on SOER during composting of digested sludge blended with recycled compost. Odor concentrations were determined by the odor panel technique [112].

bulking particles such as wood chips are used. This points to the need for exceptionally good mixing of materials before pile assembly to reduce air channeling.

Exhaust gases from forced aeration systems are another source of odor which must be considered in the total odor inventory. A statistical evaluation of odor concentration from three different forced aeration systems is shown in Figure 15-10. Digested sludge and recycled compost were the primary mixture ingredients, although some tests included rice hulls or redwood shavings as amendments. Systems that provide periodic bed agitation exhibited lower exhaust odor concentrations which may be related to breakup of air channels and anaerobic zones. From the range of values shown it is evident that odor concentrations in exhaust gases can be very high and represent a major contribution to the odor inventory. Concentrations of

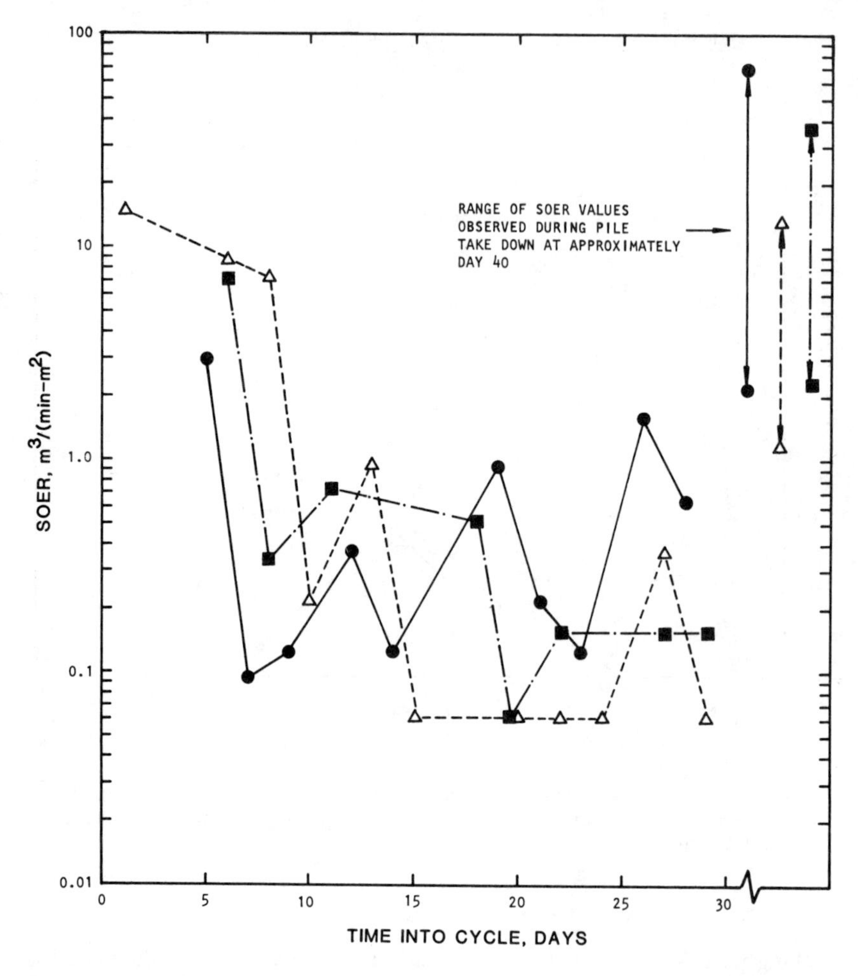

Figure 15-9. SOER during static pile composting of digested sludge blended with recycled compost and other amendments. Odor concentrations were determined by the odor panel technique [17].

several thousand ou may be typical. Fortunately, exhaust gases are collected and, therefore, can be treated and dispersed in a controlled manner.

The LA/OMA Sludge Project conducted field tests of a pilot-scale Fermentechnik SR-10 composter with a process flow diagram as shown in Figure 2-30. The composter can be classified as a batch-feed, moving packed bed, vertical-flow reactor system. Outfeed from the composter was pelleted, placed in an enclosed bin and dried using ambient air. Both digested and raw

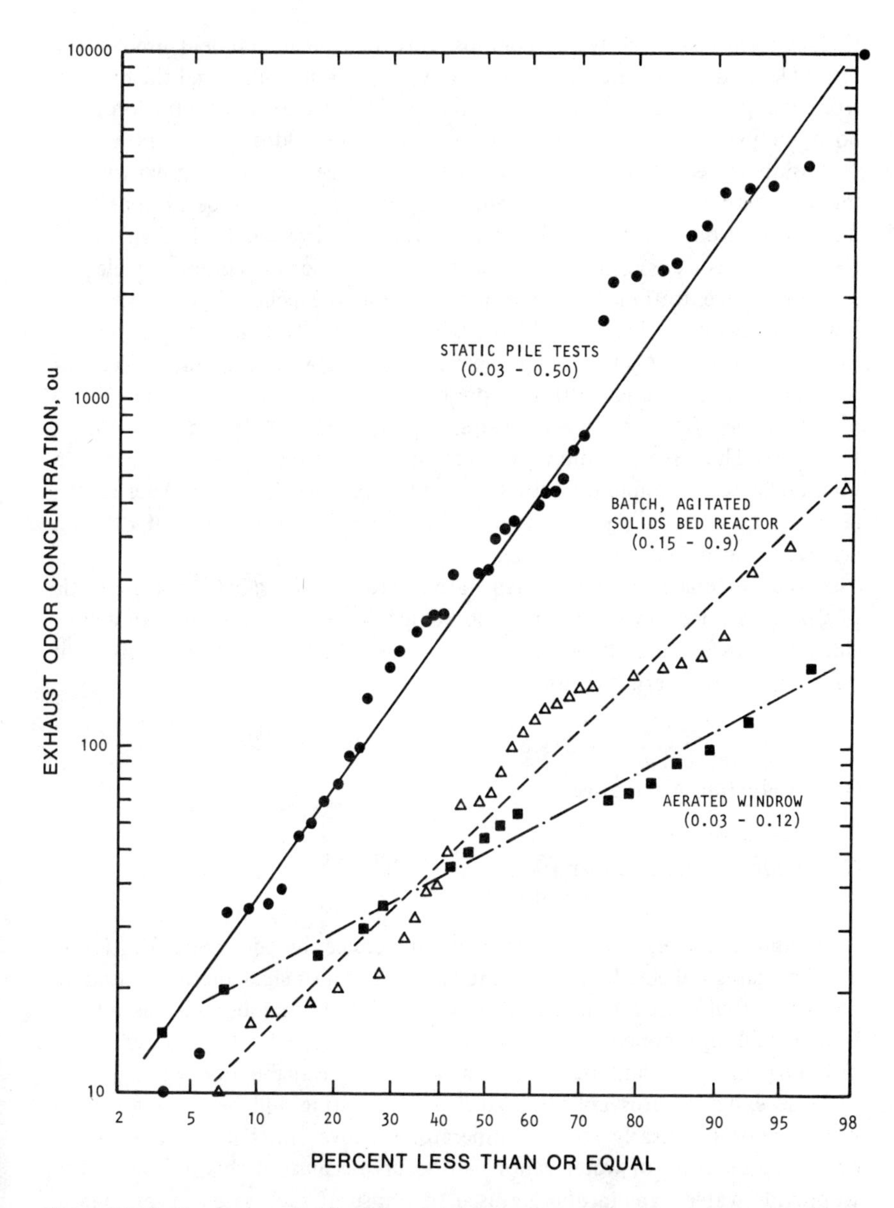

Figure 15-10. Frequency distribution of odor concentrations in exhaust gases from various compost operations. Digested sludge and recycled compost were the primary mixture ingredients in most tests. The range of specific air supply in m^3/min-dry ton is shown in parentheses in each case. Lines shown are log normal approximations of the data. Odor concentrations were determined by the odor panel technique [17,22,112,163].

sludges were tested. Recycled compost was the sole amendment used in the tests. Odor samples were collected from the compost reactor and the exhaust from the pellet dryer. A frequency distribution of odor concentration from both sample locations is shown in Figure 15-11. Odor concentrations in the compost reactor are consistent with the range of values shown in Figure 15-10 for other compost systems. Interestingly, there appears to be little difference in odor levels for the cases of raw and digested feed. Higher odor concentrations were again observed at the beginning of the compost cycle.

Odor concentrations in exhaust gases from the pellet dryer were quite low. The median value from Figure 10-11 is actually below the complaint threshold of 5 ou. Again, no significant difference was noted in the cases of raw and digested sludge. Although odor concentrations were low, the volume of exhaust gas from the dryer was quite high, about 500,000 m^3/dry ton of feed cake. The large air volume is consistent with requirements estimated in Chapter 8 (see Example 8-5) for the case of ambient air drying. Whether the large gas quantity is significant or not will depend on subsequent dilution achieved and the nature of the site.

It is unfortunate that more quantitative data are not available for the static pile process using raw sludge and wood chips. Hopefully, this data gap will be filled in future years allowing for a more reasoned approach to analysis of odor impacts of alternative compost processes.

Odor Reduction Alternatives

Gas Treatment and Deodorization

A number of techniques are available to reduce the odor concentration in exhaust gases collected during composting. The principal methods used to deodorize foul gases are described in Table 15-3. Techniques such as absorption (scrubbing), condensation, adsorption, oxidation by thermal, chemical or biological means, and use of masking agents are available to the designer.

Most exhaust gases collected during composting will be saturated with water vapor (or nearly so) at temperatures above ambient. As these gases cool, condensation of water vapor will occur. Because of this it is necessary to provide water traps in ducting used to transport such gases. As condensation occurs it is likely that water-soluble species in the gas will be absorbed into the condensate. Iacoboni et al. [17] reported the chemical characteristics of condensate collected during aerated static pile composting of digested sludge. Total chemical oxygen demand (COD) ranged from a few hundred to >9800 mg/l (2500 mg/l average) and ammonia from about 100 to 2000 mg/l

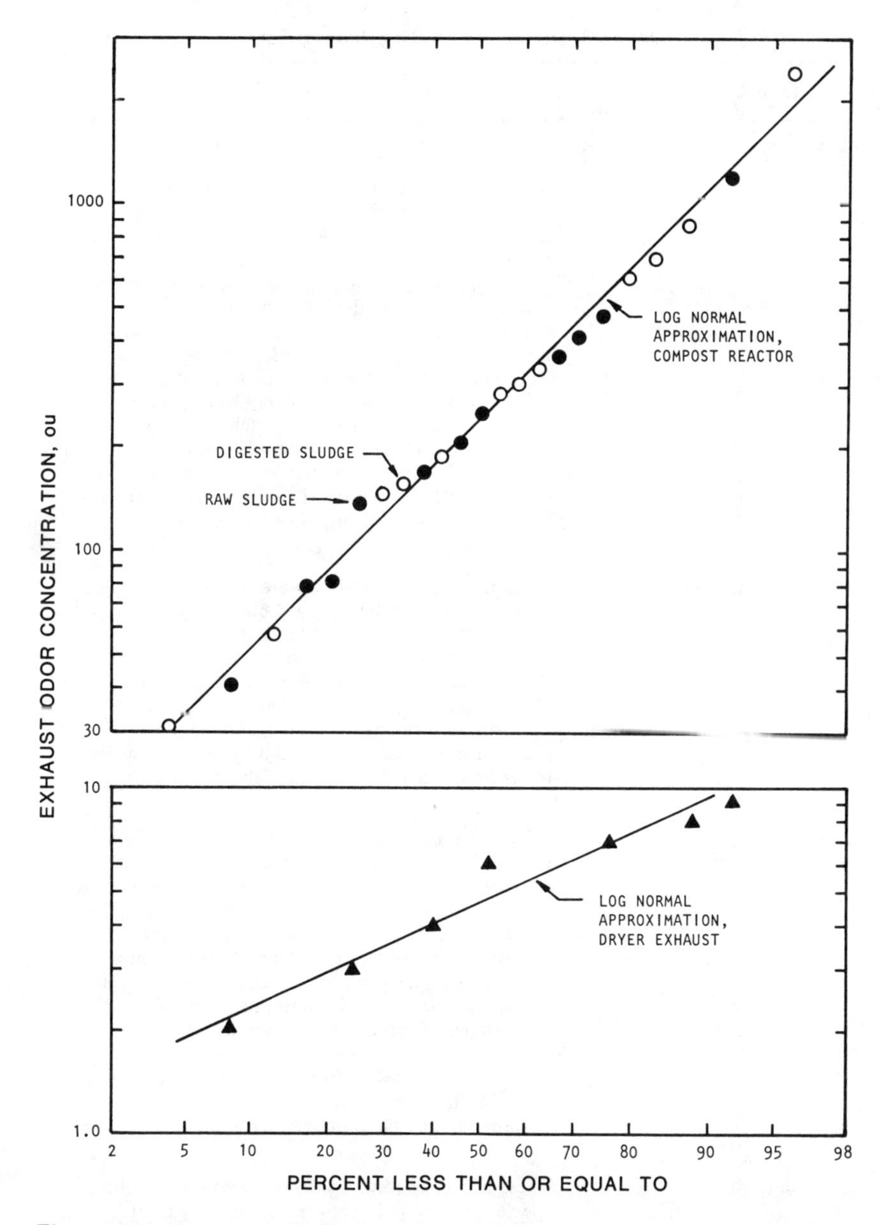

Figure 15-11. Frequency distribution of odor concentrations in exhaust gases from Fermentechnik SR-10 composter (top) and reactor dryer (bottom). Both raw and digested sludges were tested in the study and in each case were blended with recycled compost product. Odor concentrations were determined by the odor panel technique. Tests were sponsored by the LA/OMA project with results reported by Iacoboni [23].

Table 15-3. Methods of Treating Odorous Gases[a]

Method	Description
A. Absorption	
1. Liquid Scrubbing	Gas absorption is a unit operation in which one or more soluble components of a gas mixture are dissolved in a liquid. Absorption may be by simple solubility or may involve solution of the material followed by chemical reaction with constituents in the liquid phase. Specially designed equipment for gas absorption includes spray chambers, packed towers, tray towers and jet or venturi scrubbers. Water is the most common solvent for odor removal and may be combined with chemical oxidizing agents such as potassium permanganate ($KMnO_4$), sodium hypochlorite (NaOCl), alkalies such as lime ($Ca(OH)_2$) or sodium hydroxide (NaOH), or various acids. H_2S, ammonia and certain organics such as mercaptans can be removed efficiently by gas absorption.
2. Condensation	When the temperature of a water-saturated gas is lowered by contact with a cooled surface or a cooling liquid, water condensation results. Water-soluble odor constituents may dissolve in the condensing water and be removed from the gas phase. This is a special case of gas absorption. Condensation can be an important deodorizing mechanism for compost offgases, which are usually moisture-laden at temperatures above ambient.
B. Adsorption	Adsorption involves contacting a fluid phase (gas or liquid) with a particulate phase which has the property of selectively taking up and storing one or more solute species from the fluid. Activated carbon is the most widely used adsorbent and consists largely of neutral atoms which present a nonpolar surface. This makes it effective in humid gas streams because polar water molecules do not compete for the nonpolar surface. Activated carbon is effective in removal of both inorganic and organic species as shown in Table 15-4. However, hydrocarbons may be preferentially adsorbed before compounds such as H_2S are removed.
C. Oxidation	These processes result in the oxidation of organic species to carbon dioxide and water, or partial oxidation to other intermediate compounds. Inorganic species are oxidized to less offensive states. For example, H_2S can be oxidized to sulfur (S^0) or sulfate (SO_4^{2-}). Thermal, chemical and biological processes are available.

Table 15-3., continued

Method	Description
C. Oxidation	
1. Thermal Oxidation	Oxidation of both organic and inorganic species in an odorous gas can be accomplished readily provided sufficient time, temperature, gas turbulence and oxygen are provided. Essentially complete odor destruction can be achieved with operating temperatures of 650-850°C and gas retention times of 0.3-0.5 sec. It is important that complete oxidation take place because intermediate compounds may be more odorous than the original species. A major disadvantage of thermal oxidation is the large energy requirement. Systems with recuperative heat recovery and catalytic oxidation are available to reduce the energy requirement. Another approach is to let an existing boiler do double duty as an incinerator. Nevertheless, applications are generally limited to highly odorous gases of low volume.
2. Chemical Oxidation	Use of chemical oxidants in scrubbing solutions has been described under absorption above. It is also possible to inject gaseous oxidants such as ozone (O_3) directly into an odorous gas. Although ozonation can be effective, few large scale facilities have been installed, and application to odorous gases from composting is limited.
3. Biological Oxidation	Special packed towers have been used to biologically oxidize the odorous components in a foul air stream. Sufficient water (secondary effluent may be used) must be applied to support biological growth on the packed media. The mechanism of removal is probably absorption into the liquid phase followed by biological oxidation. In a similar application, foul sewage air and compost exhaust gases have been used for aeration in activated sludge systems. Odor reduction is probably accomplished by a variety of mechanisms, but biological oxidation is certainly important. Biological oxidation is probably an important contributing mechanism for odor removal from gases passed through soils or compost deodorizing piles.
D. Masking Agents	Perfume scents can be added to the gas stream to mask or combine with an objectionable odor. Results have been erratic and oftentimes the combined odor is worse than the original odor. Masking agents have limited application in compost facilities since more effective techniques are available.

[a]Developed in part from References 148, 164-166.

Table 15-4. Retentivity of Vapors by Activated Carbon [165]

Substance	Formula	Molecular Weight	Normal Boiling Point (°C, 760 mm)	Approximate Retentivity[a] (% at 20°C, 760 mm)
Acetaldehyde	C_2H_4O	44.1	21	7
Amyl Acetate	$C_7H_{14}O_2$	130.2	148	34
Butyric Acid	$C_4H_8O_2$	88.1	164	35
Carbon Tetrachloride	CCl_4	153.8	76	45
Ethyl Acetate	$C_4H_8O_2$	88.1	77	19
Ethyl Mercaptan	C_2H_6S	62.1	35	23
Eucalyptole	$C_{10}H_{18}O$	154.2	176	20
Formaldehyde	CH_2O	30.0	−21	3
Methyl Chloride	CH_3Cl	50.5	−24	5
Ozone	O_3	48.0	−112	Decomposes to oxygen
Putrescine	$C_4H_{12}N_2$	88.2	158	25
Skatole	C_9H_9N	131.2	266	25
Sulfur Dioxide	SO_2	64.1	−10	10
Toluene	C_7H_8	92.1	111	29

[a]Percent retained in a dry airstream at 20°C, 760 mm, based on weight of carbon.

as N (560 mg/l average). Obviously, a large number of volatile, low-molecular-weight, potentially odorous organics along with inorganic ammonia condensed with the moisture. Gaseous odor samples collected downstream of the condensate traps yielded the odor concentrations shown in Figure 15-10. Odor samples were not taken ahead of the condensate traps so the efficiency of removal cannot be determined. Judging from the high concentrations shown in Figure 15-10, however, condensation may partially remove certain odor constituents but cannot be relied on for complete odor control. In fact, careful handling and treatment of the condensate will be necessary to prevent desorption of the odor constituents.

LeBrun et al. [22] and Iacoboni et al. [112] conducted deodorization experiments on exhaust gases from an experimental reactor composting digested sludge blended with recycled compost. A number of deodorization systems were evaluated and data for water scrubbing and water scrubbing followed by activated carbon adsorption are shown in Figure 15-12. The scrubbing tower consisted of a plastic-medium bed. Scrubbing solution was applied to the top of the medium by a spray nozzle. Air flowrates ranged from 10 to 28 $m^3/min\text{-}m^2$ and the liquid-to-air ratio from 3.3 to 5.1 liter/1000 liter. Simple water scrubbing provided an average 44% removal and, although there is some scatter in the data, a consistent trend is evident. More limited data with other

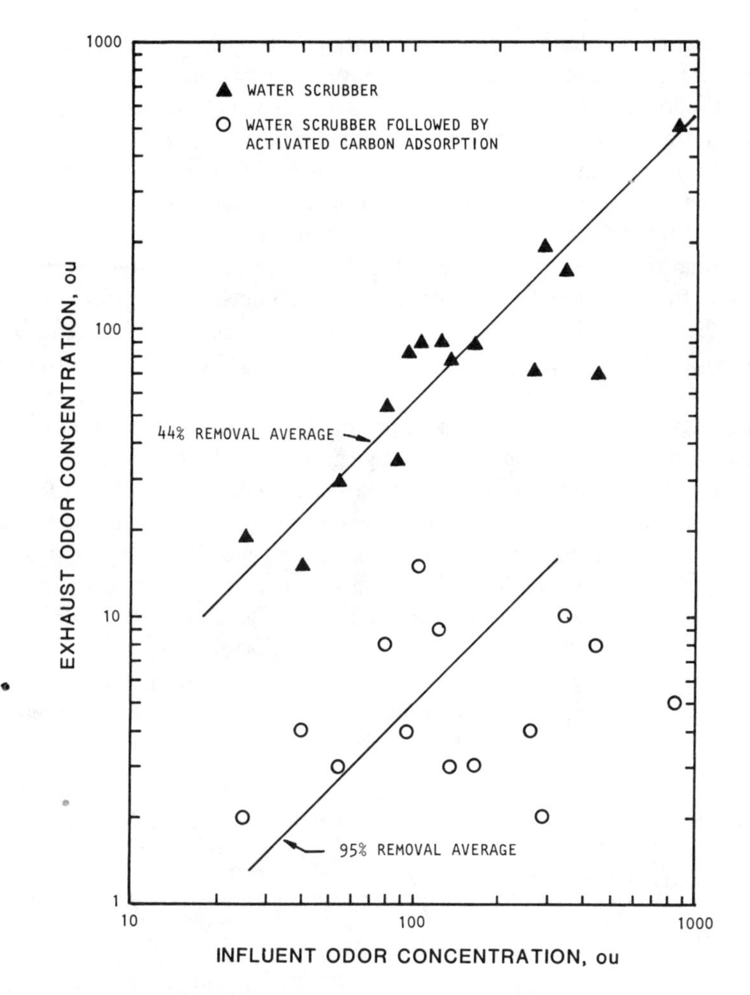

Figure 15-12. The effectiveness of absorption/scrubbing and adsorption for odor removal from compost exhaust gases. Compost exhaust gases were collected from an experimental enclosed composter treating digested sludge blended with recycled product. Odor concentrations were determined by the odor panel technique [22,112].

scrubbing solutions suggested the following removals: potassium permanganate (0.4-0.7 lb/gal of scrubbing solution), 61%; sulfuric acid (0.1 *N*), 57%; and sodium hypochlorite (1% solution), 45%.

The carbon filter consisted of a packed bed filled to about a 2-ft depth with granular carbon. The gas supply rate was about 14 m^3/min-m^2. Carbon adsorption, either alone or preceded by water scrubbing, was quite effective in odor removal. An average 95% removal is shown in Figure 15-12, but effluent odor concentration appears to be largely independent of influent concentration. In all tests but one, effluent concentration was less than 10 ou regardless of the incoming load. The advantage of water scrubbing before adsorption is to possibly extend the life of the carbon bed. The researchers did not determine the adsorption capacity of the carbon.

Information shown in Figure 15-12 is reasonably consistent with data available for deodorization of sewage-related, noncompost gases. For example, Eutek [167] reported average removal efficiencies of 47 and 96% for NaOCl scrubbing and carbon adsorption, respectively, when treating an exhaust gas from vacuum vaporization of digested sludge. With activated carbon adsorption, effluent odor concentrations were generally less than 50 ou with infeed concentrations ranging from 1000 to 7000 ou. Eutek also developed the odor breakthrough curves shown in Figure 15-13 for the case of carbon adsorption of sewer offgas. If the gas was consistent in concentration and composition, breakthrough curves would be expected to be independent of bed depth. Influent odor concentrations actually ranged from 10 to 200 ou. These results indicate a range of carbon life of 1.5×10^5 to 4.0×10^5 vol gas/vol carbon for the particular gas tested.

Carbon adsorption is actually a method for concentration and storage of odor constituents. Therefore, consideration must be given to the method

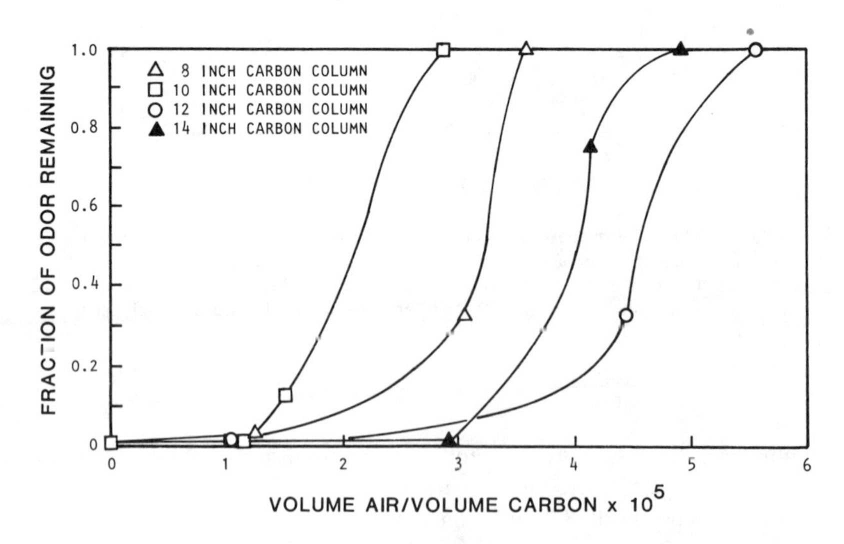

Figure 15-13. Odor breakthrough curves developed for activated carbon adsorption of sewer offgas. Influent odor concentrations ranged from about 10 to 200 ou [167].

of regeneration to assure that odorous vapors are not again released to the atmosphere. Thermal regeneration of carbon will usually assure destruction of organics and oxidation of inorganics. In some cases, adsorbed material can be desorbed in a more concentrated gas stream and then thermally oxidized. The lower gas volume greatly reduces the cost of thermal oxidation.

Natural materials such as soil and compost have been used for odor control in a number of applications. For example, the odor removal capacity of soil is put to practical use in modern sanitary landfills in the form of daily coverage. Compost beds have been used for many years to treat malodorous air streams. Bohn [168] reported their use in Germany, Israel, Switzerland and, of course, the recent use of compost deodorization piles in the aerated static pile process (see Figure 2-13). Bohn [168] also described the useful properties of compost, including its high surface area, air and water permeability, water-holding capacity, microbial populations, and low cost. The mechanism of odor removal is likely a combination of absorption, condensation, adsorption, biological oxidation and atmospheric dispersion.

A number of compost bed designs have been used in practice. A cone-shaped pile of compost has generally been used with the aerated static pile system. About 0.8–1.0 m^3 of deodorization pile is provided per dry ton of sludge [169]. Bohn [168] described a compost filter used at a Duisberg, West Germany, composting facility. The system was designed with an area of 70 m^2/m^3/sec of airflow. A minimum residence time of 30 sec for removal of biodegradable gases in a compost bed was also recommended.

Limitations in the use of compost beds include temperature and moisture constraints to maintain biological activity, limited buffering capacity and, most importantly, a general lack of engineering data by which system performance can be predicted. There appears to be a considerable difference of opinion as to the effectiveness of compost deodorization piles, particularly those used in the aerated static pile process. Numerous references are made to the effectiveness of the piles, but statements are usually subjective and not supported by actual data.

Iacoboni [22] and LeBrun [112] examined the effectiveness of a packed-bed reactor loaded with compost product for deodorization of compost exhaust gases. The bed was operated at an air loading of about 6.1 m^3/min-m^2 with a bed depth of 0.61 m, resulting in an empty bed detention time of about 6 sec. Acknowledging that the detention time may have been less than desirable for a biological oxidation, their data are shown in Figure 15-14. Removals were rather erratic and in one case odor was actually added to the exhaust gas. Several tests were conducted with liquid scrubbing preceding the compost filter and this did not result in any significant improvement in odor removal. An average 50% removal was observed excluding all data above the zero removal line. These same researchers noted surface odor emission

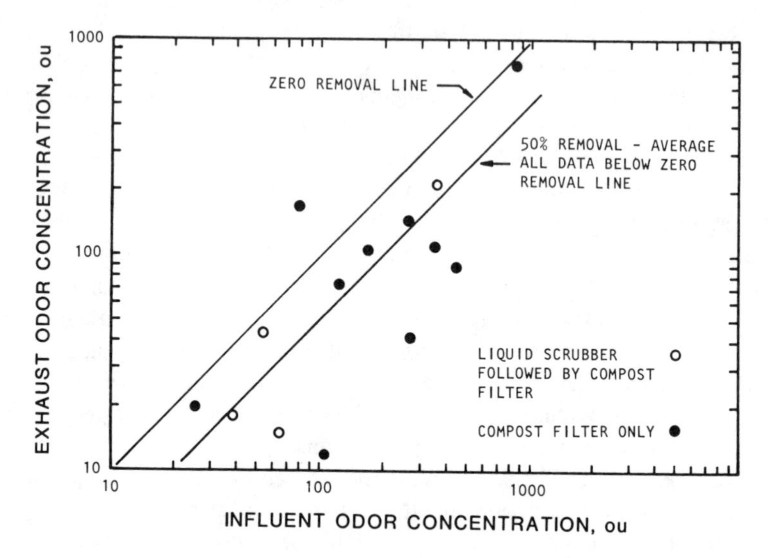

Figure 15-14. Odor reduction across an enclosed bed reactor packed with compost product. Air loadings averaged about 6.1 $m^3/min\text{-}m^2$ with a bed depth of 0.61 m. Odorous gases were collected from an experimental enclosed composter treating digested sludge blended with recycled product. Odor concentrations were determined by the odor panel technique [22,112].

rates as high as 200 $m^3/min\text{-}m^2$ from deodorization piles in the static pile system.

Garber [170] reported on the effectiveness of using odorous air as a feed for the activated sludge process. Contact with biological floc removed most of the odor, and remaining odors were released over the large area of the aeration tank instead of as a point source. Odorous air containing 30–40 ppm by volume of hydrogen sulfide and other wastewater odorants was successfully treated in this manner. Vogt [171] reported on several European compost facilities that are apparently successful in using this approach to control odors from compost exhaust gases. Unfortunately, very few quantitative data are available to predict the odor removal expected as gases are passed through the activated sludge process.

In summary, limited available data suggest that scrubbing with water or chemical solutions can effect a reasonably consistent 45–60% reduction in odor concentration. Activated carbon adsorption can consistently produce an exhaust gas of less than about 10–50 ou, a value which appears largely to be independent of infeed concentration. Compost deodorization piles have accomplished measured odor reductions averaging 50%, but with a wide variation in removal efficiency. It may be that the odor reduction attributed to compost piles is caused by partial odor removal in the pile combined with better atmospheric dispersion from the large surface area. Without question, more data are needed to allow for proper engineering of compost deodorization piles to accomplish the established odor objectives. Similar comments apply to use of odorous air as feed for the activated sludge process. The approach is reportedly effective in a number of full-scale facilities, but quantitative data and predictive tools are generally lacking.

A complete odor control system consists not only of collection and treatment of odorous gases but also their proper dispersal into the atmosphere. Design of systems for treatment and dispersal must be coordinated to assure that ground-level ambient odor standards are not violated. In this regard, deodorization systems such as scrubbing towers and packed bed adsorption reactors have an advantage because treated gases are still contained and can be ducted and dispersed in a controlled manner. Traditional deodorizational piles do not share this advantage because the gases are no longer contained once they enter the pile.

Example 15-1

An odorous gas contains 100 ppm by volume of hydrocarbons and 10 ppm of H_2S. It is estimated that a particular carbon can adsorb 20% by weight of hydrocarbons and 25% by weight of H_2S. Determine the bed volumes of gas that can be deodorized per volume of carbon. Assume a carbon bulk density of 0.55 g/cm^3 and an average hydrocarbon molecular weight of 100.

Solution

1. Determine total carbon life assuming hydrocarbon adsorption is limiting.

$$\text{total hydrocarbon adsorption} = 0.55\ (0.20) = 0.11 \text{ kg/l of carbon}$$

concentration of hydrocarbon in gas stream

$$= \frac{100}{10^6} \times \frac{\text{mol}}{22.4 \text{ liters}} \times \frac{100 \text{ g}}{\text{mol}} \times \frac{1000 \text{ mg}}{\text{g}} = 0.45 \text{ mg/l}$$

$$\text{total carbon life} = 0.11 \text{ kg/l} \times 1 \text{ liter}/0.45 \text{ mg} \times 10^6 \text{ mg/kg}$$

$$= 2.4 \times 10^5 \text{ vol gas/vol carbon}$$

2. Determine total carbon life assuming H_2S adsorption is limiting.

$$\text{total } H_2S \text{ adsorption} = 0.55\ (0.25)$$

$$= 0.138 \text{ kg/l carbon}$$

concentration of H_2S in gas stream

$$= 10/10^6 \times 1 \text{ mol/22.4 liter} \times 34 \text{ g/mol} \times 1000 \text{ mg/g} = 0.015 \text{ mg } H_2S\text{/l gas}$$

$$\text{total carbon life} = 0.138 \text{ kg/l} \times 1 \text{ liter/0.015 mg} \times 10^6 \text{ mg/kg}$$

$$= 9.2 \times 10^6 \text{ vol gas/vol carbon}$$

3. Hydrocarbon adsorption would control the total carbon life in this case.
4. Pilot results shown in Figure 15-13 indicate a total carbon life of 1.5-4.0 $\times$ 10^5 gas vol/vol of carbon when treating an influent gas averaging 92 ppm of hydrocarbon. Therefore, the calculated results are in reasonably close agreement with experimental data.

Feed Conditioning

Another approach to odor control is to reduce odor emission rates from feed substrates themselves. At least two techniques are available including chemical addition and odor stripping.

Chemical Addition. Lime is perhaps the most common inorganic added to sludges for purposes of conditioning before dewatering. Lime addition is sometimes termed lime stabilization because the resulting high pH decreases biological activity. However, organics are not actually decomposed and biological decomposition will eventually produce enough CO_2 to reduce the pH levels, allowing biological rates to return to normal levels. Nevertheless, during the period of high pH, odor emission rates should be reduced by lowered biological activity and chemical ionization of weak acids such as H_2S and low-molcular-weight organic acids such as acetic acid. On the other hand, if the feed substrate had high concentrations of the weak base NH_3, its release would be enhanced by the high pH.

Lime addition is probably most effective with raw sludges that can be very odorous if allowed to decompose in an uncontrolled manner. Control is only temporary, but should be sufficient to mitigate odors in early operations such as sludge storage and mixing before composting. As composting proceeds, pH levels will return to normal and the lime will have little or no additional effect.

Oxidizing agents, such as chlorine and chlorine compounds, have been added to sludge for organic stabilization and odor control. However, concentrations required for effective control are very high because of the large organic concentration. Also, concern has arisen over chlorine addition because of possible formation of halogenated organics. Therefore, addition of oxidizing agents to the sludge itself is not likely to be a common method for odor control.

Odor Stripping. Odor-causing compounds are present in sludge as dissolved gases or vapors and as volatile liquids [157]. There are two basic unit processes for gas/liquid separation, gas stripping and vacuum vaporization. Each can be effective in partially removing odorous compounds from sludge. The objective is to shift the equilibrium and produce a driving force to transfer odor-causing substances into the gas phase. Gas stripping involves injection of air or other carrier gas to remove gaseous and volatile components from the liquid phase. Vacuum vaporization is a process in which a vacuum is produced above the liquid to vaporize odor-causing compounds.

Equilibrium or saturation concentration of a slightly soluble gas dissolved in a liquid is a function of the partial pressure of the gas above the liquid. The relationship is expressed as Henry's law:

$$P_g = H \cdot X_g$$

where P_g = partial pressure of gas, atm
H = Henry's law constant, which is a function of the gas type, liquid type and temperature
X_g = equilibrium mole fraction of dissolved gas, which in an aqueous solution is given as:

$$\frac{\text{moles gas}}{\text{moles gas + moles water}} = \frac{n_g}{n_g + n_w}$$

During gas stripping or vacuum vaporization the gas partial pressure of the vaporizing component is reduced below the equilibrium value. As a result, molecules leave the liquid and enter the gas phase. This process will continue as long as partial pressure of the particular component is less than the equilibrium value. With vacuum vaporization it is also possible to decrease the pressure below the equilibrium boiling pressure of certain components.

Eutek [167] evaluated both air stripping and vacuum vaporization as methods for removing odorous substances from digested sludge. Both systems were found to remove a significant quantity of odorant. Vacuum vaporization was selected as the preferred method primarily because of lower offgas flowrates (about 10% of that for air stripping) and greater energy efficiency.

Lower gas volume is important because exhaust gases must be deodorized before discharge. Activated carbon adsorption was determined to be the most effective approach for gas deodorization in the Eutek study.

Subsequent work with the vacuum vaporization system yielded reductions in SOER values between odor-stripped and control sludges ranging from 47 to 64% (50% average). Referring back to Figures 15-3 and 15-5, the highest SOER appear to occur in the early stages of composting. Odor stripping should be an effective technique to reduce these surface emissions. At present there are no compost facilities using this technique, but facilities are planned for sludge storage basins in Sacramento.

ATMOSPHERIC DISPERSION OF ODORS

It is rather obvious that surface emission of odors from open compost facilities will directly enter the atmosphere and will be conducted and dispersed according to prevailing atmospheric conditions. Because surface odors are released near ground level, downwind odor concentrations will be determined primarily by micrometeorological conditions in the atmospheric sublayer which typically extends about 20 m above the ground surface. Another characteristic of surface emissions from open compost facilities is that they arise from a large area, as opposed to a single point source.

Exhaust gases collected during processing may be deodorized or not, and then either dispersed at ground level or at higher elevations by means of a stack. These constitute point-source discharges, and large dilutions can be achieved by elevated stack dispersal.

Knowledge of the subsequent transport and dispersion of odors is necessary for effective design of odor control facilities. Fortunately, procedures for estimating downwind dispersion from elevated point-sources are well developed. Analysis of large area sources and dispersion in the atmospheric sublayer is less well defined, but recent advances allow estimation of the magnitude of the problem.

Large-Area Sources

Micrometeorology in the Atmospheric Sublayer

Wilson et al. [161] described three categories of atmospheric sublayer dispersion as shown in Figure 15-15. Transport conditions are broadly categorized as unstable, stable and very stable. Time-averaged plume geometry and resultant average downwind concentrations for the three categories are shown.

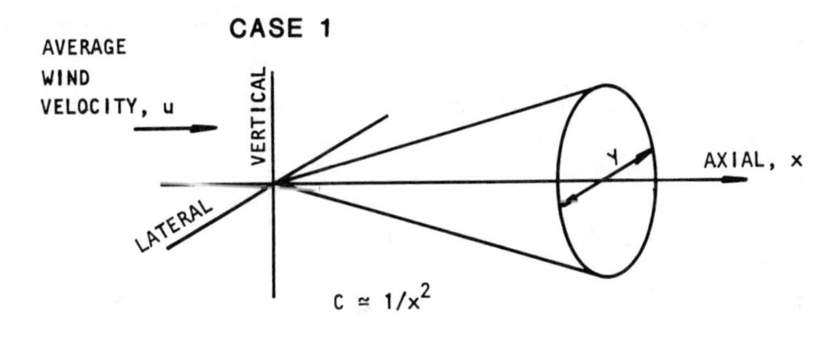

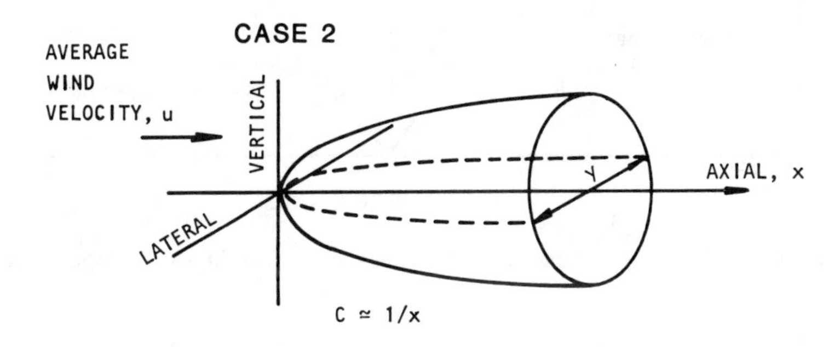

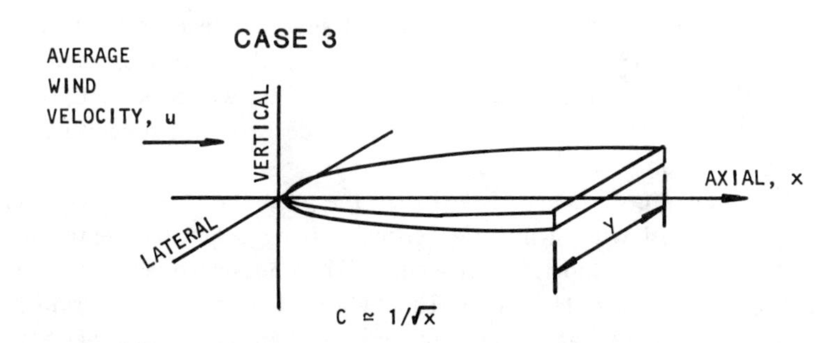

Figure 15-15. Average plume geometry for the three categories of atmospheric sublayer transport: unstable conditions (top), stable (middle) and very stable (bottom) [161].

Atmospheric sublayer stability is determined in large measure by the temperature lapse rate, the rate of temperature change with altitude. Temperature lapse rate in the atmospheric sublayer (see Figure 15-16) can be of greater magnitude than that in the upper atmosphere because of radiant heating and cooling of adjacent ground and water surfaces. Temperature gradients influence the vertical motion of odorant air masses. As a mass of

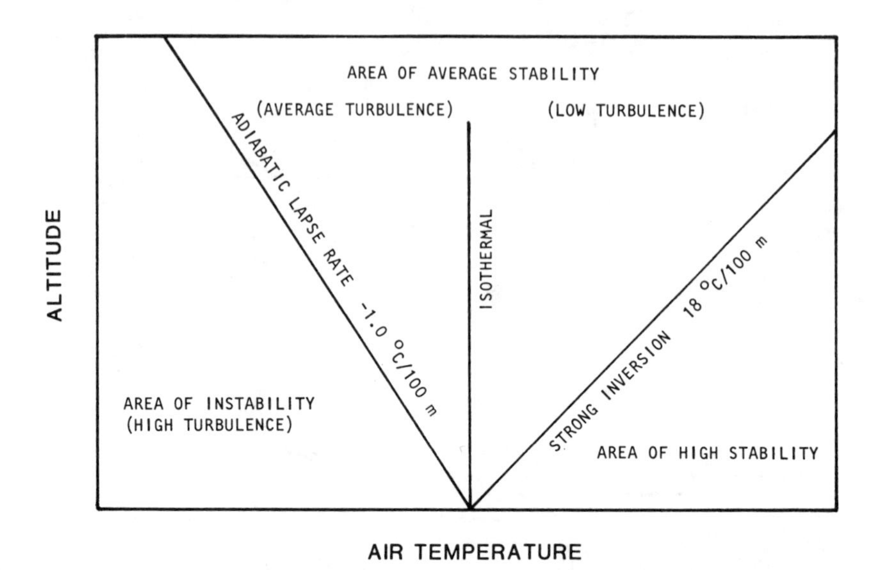

Figure 15-16. Temperature lapse rates and zones of stability in the atmospheric sublayer.

air is moved upward, it will tend to expand and cool as a result of the decrease in pressure. If no heat transfer occurs with the surroundings, dry air will cool at a rate termed the adiabatic lapse rate, which is −1°C/100 m (−5.4°F/1000 ft). The negative sign indicates a decrease in temperature with increasing elevation.

If the temperature gradient is less than the adiabatic lapse rate, unstable air conditions exist with significant vertical mixing. A positive temperature gradient is termed a temperature inversion. When an air parcel rises it tends to cool at the adiabatic lapse rate. Thus, with a temperature inversion, the raising air parcel will be at a lower temperature than its surroundings but at the same pressure. Consequently, it will be heavier than its surroundings and tend to fall to its original position. Vertical mixing will be limited depending on the strength of the inversion.

Temperature gradients less than the adiabatic lapse rate are characteristic of unstable conditions. Dispersion is very rapid and the dispersion coefficient is usually assumed to be a function of the plume scale. Downwind concentrations decrease inversely as the square of distance from the source. Such conditions are commonly encountered when the atmospheric sublayer is buoyantly mixed, which is usually the case during midday of a clear day [159]. The stable category is commonly encountered when the atmospheric sublayer is characterized by a nearly constant temperature (isothermal conditions). Both vertical and horizontal mixing can occur under stable conditions; however, dispersion is not scale-dependent, and downwind concentrations decrease inversely with distance. The third category of micrometeorological transport occurs under conditions of positive temperature gradients that result in very stable or strong inversion transport conditions. Under such conditions, vertical mixing is restricted by the very stable atmospheric conditions. Concentration decays inversely as the square root of distance from the source. Transport under very stable conditions has been termed "puff" transport and, because it results in the lowest rate of dispersion, is usually the critical odor transport condition.

Based on smoke release experiments made in the Sacramento area, Wilson et al. [161] determined that puff transport occurred when the temperature inversion measured between 25 and 5 ft (7.62 and 1.52 m) above ground exceeded 2°F (1.1°C) with wind speeds of 2 mph (0.89 m/sec) or less. During the year these conditions were met for periods of one hour or more on 40% of the evenings in Sacramento.

Positive temperature gradients in the atmospheric sublayer usually occur as a result of radiant cooling of the ground surface. The gradient becomes positive about 2 hr before sunset and usually remains positive until 1.5-2 hr after sunrise [161]. A strong ground-level inversion is likely to occur on a calm, clear night. Thus, ground-level stable conditions are characteristic of evening hours and unstable conditions characterize daylight hours. This is verified by analysis of odor complaints received by the Los Angeles County Sanitation Districts in 1977 as shown in Table 15-5. Most odor complaints occurred in the late afternoon and early evening hours. Even though stable conditions may persist, few complaints were received in the early morning hours presumably because most human receptors were asleep.

Another type of critical odor situation can occur after calm conditions with little or no wind velocity. Absence of wind allows accumulation of odors over the source. If the calm is followed by a period of steady wind, the high odor concentration can be transported under puff conditions. Such a case is termed intermittent puff transport.

Modeling the Puff Transport Condition

A schematic diagram illustrating development of odor over a large-area source and its subsequent dispersion downwind is shown in Figure 15-17. As air moves over the large-area source, odor concentration increases until the downwind edge of the odor source is reached. From this point on, atmospheric turbulence will tend to disperse the air, diluting the odor concentration.

Table 15-5. Number of Odor Complaints vs Time[a] [159]

Time	Number of Odor Complaints	Number of Verified Complaints
1:00 AM	1	0
2:00	0	0
3:00	1	0
4:00	0	0
5:00	0	0
6:00	1	0
7:00	0	0
8:00	1	0
9:00	6	1
10:00	3	0
11:00	4	2
12:00 noon	2	1
1:00 PM	5	1
2:00	4	1
3:00	9	2
4:00	17	5
5:00	10	5
6:00	5	2
7:00	7	5
8:00	2	2
9:00	5	3
10:00	3	2
11:00	3	1
12:00 midnight	2	2
Totals	91	35

[a]Odor complaints analyzed for April to November 1977. These complaint conditions occurred during a period of adverse operating conditions and as such are not typical for this particular facility.

Odor concentration at the downwind edge of a large-area source can be estimated using the idealized model of Figure 15-18. Odor emission rate is assumed to be constant as are the mixing height h, wind speed u and wind direction. Despite the idealized nature of the model, it would provide reasonable estimates under stable, puff-transport conditions. A linear increase in odor concentration over the area source is shown in Figure 15-17, which is consistent with the idealized model of Figure 15-18.

Referring to Figure 15-18, odor emission rate from the area source can be defined as:

$$Q_o = k_S A = k_S LW \qquad (15\text{-}1)$$

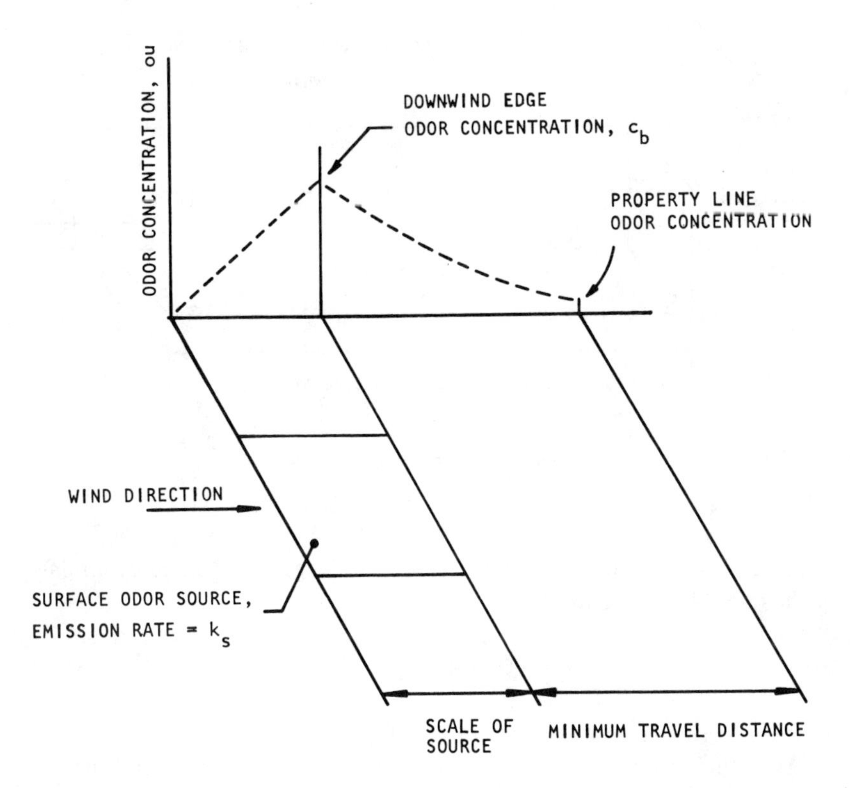

Figure 15-17. Odor concentration downwind from a large emission area [157].

where Q_0 = odor input rate, m^3/min or ft^3/min
k_S = average SOER, m^3/min-m^2 or ft^3/min-ft^2
L = emission source length, m or ft
W = emission source width, m or ft

The odor input rate refers to the flowrate of air required to dilute odor emissions to the threshold odor concentration. Actual air flowrate across the area source is given by:

$$Q_{air} = u \cdot h \cdot W \tag{15-2}$$

where Q_{air} = volumetric flowrate of air moving across the odor source, m^3/min or ft^3/min
u = average wind velocity, m/min or ft/min
h = average puff height over the emission area, m or ft

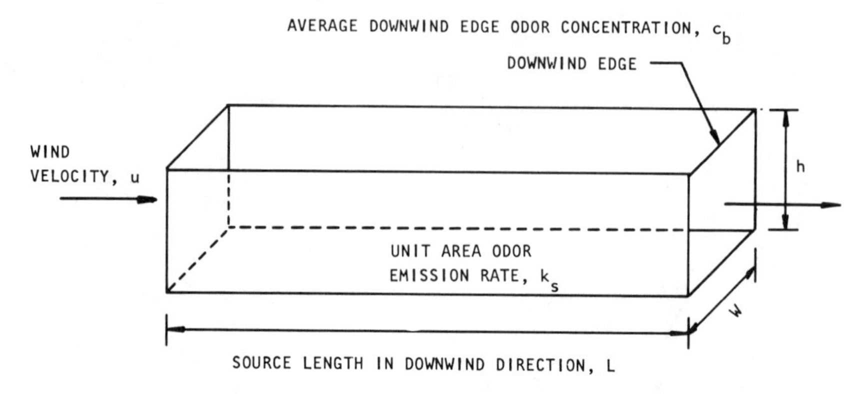

Figure 15-18. Idealized model of air movement over a large area odor source. Model is used to estimate the downwind edge odor concentration.

Average odor concentration at the downwind edge can then be determined as the ratio of Q_0 to Q_{air}:

$$c_b = Q_o/Q_{air} = k_s L/uh \tag{15-3}$$

where $\bar{c}_b$ = average odor concentration at the downwind edge, ou.

After the air mass has moved across the odor source, turbulent dispersion will begin to reduce the odor concentration. In modeling the puff transport condition, let us take a vertical element in the plume as shown in Figure 15-19. Vertical dispersion is assumed to be zero so that the element has a fixed thickness h. Dispersion along the length of the plume is assumed to be zero (i.e., plug flow in the axial direction). Dispersion is limited to the lateral direction and is modeled using Fick's law (Equation 10-2) with a scale-independent dispersion coefficient (i.e., $D \neq f(y)$). The wind velocity u and source odor emission rate Q_0 are assumed to be constant.

With these assumptions the steady-state equation of continuity is given by:

$$u\frac{\partial c}{\partial x} - D\frac{\partial^2 c}{\partial y^2} = 0 \tag{15-4}$$

Integration of Equation 15-4 will yield a Gaussian distribution of concentration in the lateral direction. Instead let us define $\bar{c}$ as the average concentration in the lateral direction given by:

$$\bar{c} = \frac{Q_o}{uyh} \tag{15-5}$$

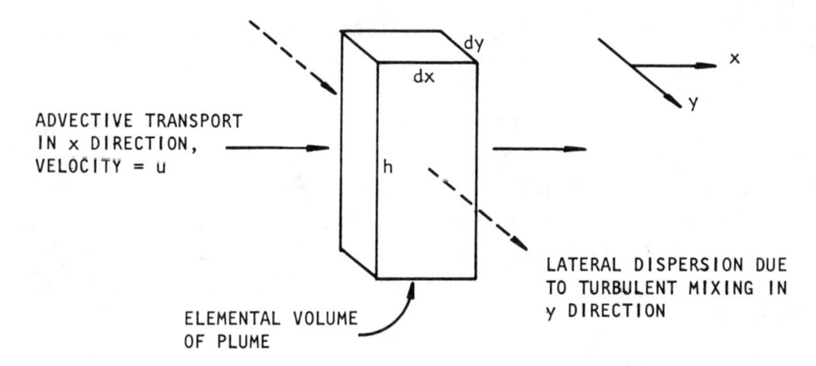

Figure 15-19. Elemental slice from a plume moving by advective transport in the x direction with turbulent dispersion in the y direction. A mass balance over the elemental volume will yield Equation 15-4.

Substituting Equation 15-5 into 15-4 and integrating yields:

$$\bar{c} = \frac{Q_0}{\mu h}\left[\frac{u}{4Dx}\right]^{1/2} \quad x > 0 \tag{15-6}$$

and

$$y = \left[\frac{4Dx}{u}\right]^{1/2} \tag{15-7}$$

where $\bar{c}$ = average downwind odor concentration, ou
D = dispersion coefficient, assumed to be constant; values on the order of 10^5 ft^2/min (10^4 m^2/min) have been reported for the puff transport condition [172]
y = lateral spread of the plume, m or ft

Average downwind odor concentration is shown to be inversely proportional to the square root of distance, which is consistent with Figure 15-15.

Strictly speaking, Equation 15-6 applies to a point source of odor emitted at a rate Q_0. An approximation to an area source can be made by working from the downwind edge of the emission area. A virtual distance X_b can be defined which gives the downwind edge odor concentration c_b, such that:

$$c_b = \frac{Q_0}{uh}\left[\frac{u}{4DX_b}\right]^{1/2} \tag{15-8}$$

The value of c_b can be determined from Equation 15-3, which allows solution of Equation 15-8 for X_b. Equation 15-6 can then be used to solve for the downwind concentration assuming a point source discharge a distance X_b upwind of the downwind edge of the emission area. These relationships are illustrated in Figure 15-20.

One of the problems with the model represented by Equation 15-6 is that wind speed u, puff height h and dispersion coefficient D are treated as independent variables when actually they are not. For example, the dispersion coefficient is a function of observed wind variables including wind speed and the constancy of wind direction. A modified form of Equation 15-6 was developed in the Sacramento odor studies to partially overcome this problem [172]:

$$\bar{c} = \frac{Q_0}{u(\Phi x)^{1/2}} \tag{15-9}$$

where Φ = a modified dispersion coefficient = $4\ h^2D/u$ and x = distance measured from the virtual point source. A frequency distribution for the dispersion coefficient Φ observed under puff transport conditions in Sacramento is shown in Figure 15-21. The distribution would be expected to vary depending on local climatic conditions. However, the distribution in Figure 15-21 can be used for illustrative purposes.

The problem of interdependent variables is not completely resolved by use of Equation 15-9 because the modified dispersion coefficient Φ is still a function of wind speed. Other correlations were developed in the Sacramento studies to overcome these difficulties. However, development of these

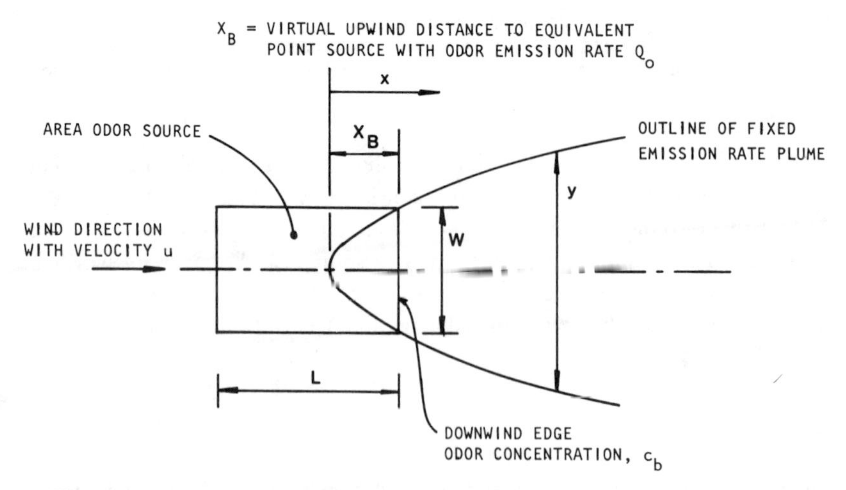

Figure 15-20. Relationship between large area source and equivalent point source.

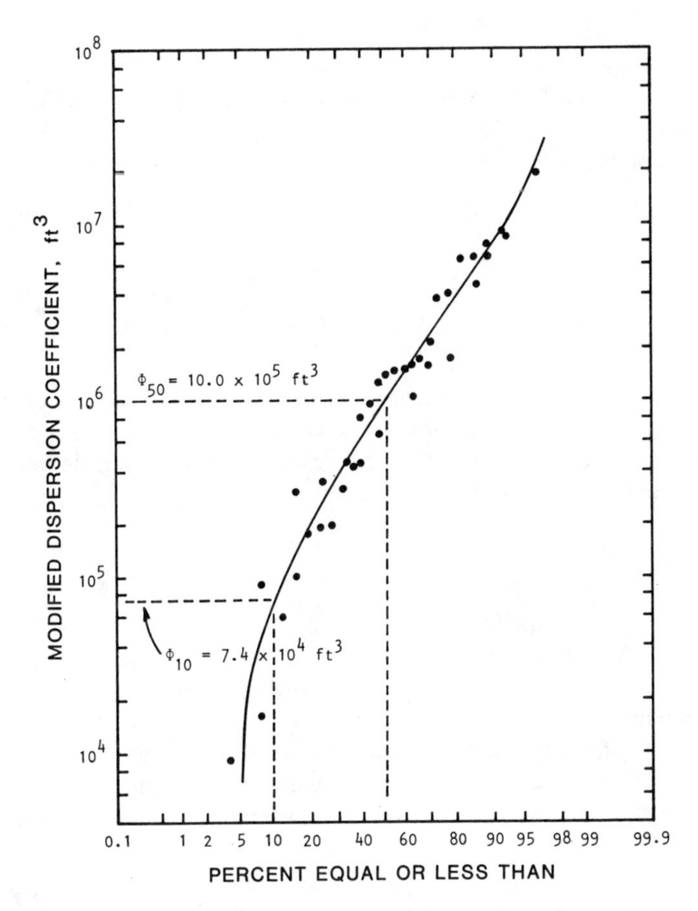

Figure 15-21. Frequency distribution for puff transport dispersion coefficient Φ observed at Sacramento, CA [172].

correlations is beyond the scope of this work and the interested reader is directed to the cited references. Sufficient information has been developed to assess the magnitude of impacts posed by large-area sources.

Example 15-2

An odor criterion limit of 5 ou is set at a plant boundary 2000 ft from the edge of an area odor source. Critical wind velocity is determined to be 220 ft/min. Determine the allowable source emission rate Q_0 to assure that the odor criterion is not exceeded more than 10% of the time during puff transport conditions.

Solution

1. From Figure 15-21, 10% or less of the time the dispersion coefficient has a magnitude of 7.4×10^4 ft^3.
2. The allowable odor emission rate can be estimated from Equation 15-9 as:

$$5 \text{ ou} = \frac{Q_0}{220\,[7.4 \times 10^4 (2000 \text{ ft})]^{1/2}}$$

$$Q_0 = 1.3 \times 10^7 \text{ ft}^3/\text{min}$$

The virtual source distance was ignored in this example because of uncertainties in estimating Φ. This example serves only to illustrate the application of Equation 15-9 and caution should be exercised in applying the frequency distribution of Figure 15-21.

Example 15-3

A windrow compost facility is operated on a square parcel of land, 300 m on a side. A 300-m-wide buffer strip surrounds the compost facility. The surface odor emission rate from the composting site is estimated to average 2 m^3/min-m^2. During puff transport conditions, a mean puff height of 2 m, wind speed of 45 m/min and dispersion coefficient D of 10^4 m^2/min are estimated. Wind direction during puff conditions is assumed to be perpendicular to the downwind edge of the site. Determine the odor concentration at the plant boundary.

Solution

1. Using Equation 15-3, determine the odor concentration c_b at the downwind edge of the source area:

$$c_b = \frac{k_S L}{uh} = \frac{2\,(300)}{45\,(2)} = 6.67 \text{ ou}$$

2. The total source emission rate is given as:

$$Q_0 = k_S A = 2\,(300)(300) = 1.8 \times 10^5 \text{ m}^3/\text{min}$$

3. The virtual distance X_b for an equivalent point source is determined from Equation 15-8:

$$6.67 = \frac{1.8 \times 10^5}{45\,(2)} \left[\frac{45}{4\,(10^4)\,X_b}\right]^{1/2}$$

$$X_b = 101 \text{ m}$$

Therefore, total distance between the virtual point source and the plant boundary is 300 + 101 = 401 m.

4. The lateral spread of a point source plume from the virtual source to the downwind edge of the area source can be checked using Equation 15-7:

$$y = \left(\frac{4Dx}{u}\right)^{1/2}$$

$$y = \left[\frac{4\,(10^4)\,101}{45}\right]^{1/2}$$

$$y = 300 \text{ m}$$

This is equal to the width of the downwind edge of the area source.

5. The odor concentration at the plant boundary under puff conditions can now be calculated from Equation 15-6:

$$\bar{c} = \frac{1.8 \times 10^5}{45\,(2)} \left[\frac{45}{4\,(10^4)\,401}\right]^{1/2}$$

$$\bar{c} = 67.1 \left(\frac{1}{401}\right)^{1/2}$$

$$\bar{c} = 3.35 \text{ ou}$$

6. If the buffer width were increased to 600 m, the resulting odor concentration would be:

$$\bar{c} = 67.1 \left(\frac{1}{600 + 101}\right)^{1/2} = 2.53 \text{ ou}$$

7. It is apparent from this example that odor concentrations decrease very slowly with distance under puff transport conditions. Site conditions used in this problem are similar to those in the case history reported by Redner [159]. During the adverse conditions of 1977, total emission rates as high as 3×10^6 m³/min were determined. Based on this example problem it is not surprising that the odor complaints shown in Table 15-5 were recorded.

Methods to Increase Atmospheric Dispersion

Methods are available to mitigate critical odor concentrations which develop with stable transport conditions in the atmospheric sublayer. Artificial barriers and wind machines have both been used in agricultural practice to modify local micrometeorological conditions. Barriers have been used for wind control and wind machines for frost control [157]. Tests were recently conducted on the effectiveness of these devices in controlling odor transport from sludge storage basins adjacent to the city of Sacramento. Results should be reasonably applicable to odor sources from large-area compost facilities.

Barriers were shown to increase turbulence and vertical mixing, which increased the height of the downwind puff under strong inversion conditions. Increased puff height was found to be a function of barrier height and independent of wind velocity. Puff height downwind of a barrier can be estimated as [157]:

$$h_b = B \cdot H \tag{15-10}$$

where h_b = puff height downwind of the barrier
H = barrier height
B = constant = 2.4

It was assumed that puff transport would persist downwind of the barrier but with the increased puff height. A 12-ft barrier surrounding sludge storage basins in Sacramento has been recommended [157]. The barrier will be "sharp-edged" and impermeable. Wind tunnel tests showed that a sharp-edged barrier increases vertical mixing to a greater extent than a curved or streamlined barrier. Because the barrier will surround the site, a minimum of two points of barrier mixing are provided along each wind line.

Wind machines will be used to maintain air movement across the area odor source during calm wind periods. This is designed to reduce high odor levels which result during intermittent puff transport conditions. The objective is to convert intermittent puff transport to the lower-risk continuous puff transport. Wind machines will be turned on only during calm periods and are designed to maintain a minimum velocity of 2 mph over the area source [157].

One other method of reducing odor risk that should be discussed is the proper timing of more odorous operations. In this case the idea is not to increase atmospheric dispersion per se but to conduct more odorous operations during periods of higher atmospheric turbulence. Fortunately, high turbulence in the atmospheric sublayer is characteristic of daylight hours and it is relatively easy to conduct operations during this period. Redner [159] reported that windrow turning at a major compost facility is restricted to hours of greater vertical mixing, between 8:00 AM and 4:00 PM. This is not means to imply that any amount of odor can be released during the daylight period. Obviously, odors will still be transported, although dispersion may be increased over the puff transport case.

Elevated Point Sources

It is not the purpose of this work to provide an elaborate treatise on the subject of dispersion of elevated discharges, but to provide an understanding of the mechanisms of transport and the more commonly used engineering tools to evaluate dispersion. This will allow estimation of ground-level odor concentrations resulting from stack discharge of odorous gases. The designer will then be in a position to judge the tradeoffs between ground level emissions vs collection, treatment and stack dispersal. Fortunately, a number of excellent summaries of practical procedures for estimating plume dispersion have been written in recent years. If the reader desires further information beyond that provided here, the works by Turner [173] and Budney [174] are recommended as a starting point.

The Gaussian Plume Formula

The Gaussian plume formula is most often used for estimation of downwind concentrations from a point-source discharge. The following assumptions are made: (1) the plume spread has a Gaussian distribution in both the horizontal and vertical planes; (2) dispersion along the length of the plume is negligible; (3) the emission is continuous and of uniform concentration; (4) steady-state conditions prevail; (5) wind speed and direction are constant; (6) the effective emission height H is the sum of stack height h and plume rise ΔH; (7) no deposition of material occurs from the plume and there is no reaction at the ground surface; (8) the emissions are conservative (i.e., no reactions occur to remove contaminants from the plume). Under these conditions the concentration at coordinates x,y,z from a continuous point source of effective height H is given by:

$$c(x,y,z;H) = \frac{S}{2\pi\sigma_y\sigma_z u} \exp\left[-\frac{1}{2}\left(\frac{y}{\sigma_y}\right)^2\right] \cdot \left\{\exp\left[-\frac{1}{2}\left(\frac{z-H}{\sigma_z}\right)^2\right] + \exp\left[-\frac{1}{2}\left(\frac{z+H}{\sigma_z}\right)^2\right]\right\} \qquad (15\text{-}11)$$

where c = pollutant concentration at coordinates x,y,z, g/m^3
S = mass emission rate of pollutant, g/sec
u = mean wind velocity effecting the plume transport, m/sec
H = effective emission height, m
σ_y = standard deviation of plume concentration in the horizontal direction, m
σ_z = standard deviation of plume concentration in the vertical direction, m
x,y,z = coordinate distances, m

Development of Equation 15-11 is beyond the scope of this text but is similar to the mass balance approach used in development of Equation 15-4. The form of plume predicted by Equation 15-11 and the assumed coordinate system are shown in Figure 15-22. Any consistent set of units may be used but it is common practice to use metric units as indicated above.

For the case of odor emissions we are most interested in ground-level concentrations (i.e., $z = 0$) for which Equation 15-11 simplifies to:

$$c(x,y,0;H) = \frac{S}{\pi \sigma_y \sigma_z u} \exp\left[-\frac{1}{2}\left(\frac{y}{\sigma_y}\right)^2\right] \exp\left[-\frac{1}{2}\left(\frac{H}{\sigma_z}\right)^2\right] \tag{15-12}$$

Further simplification can be achieved by considering the centerline concentration of the plume (i.e., $y = 0$). This is of particular significance because maximum concentrations lie along the plume centerline.

$$c(x,0,0;H) = \frac{S}{\pi \sigma_y \sigma_z u} \exp\left[-\frac{1}{2}\left(\frac{H}{\sigma_z}\right)^2\right] \tag{15-13}$$

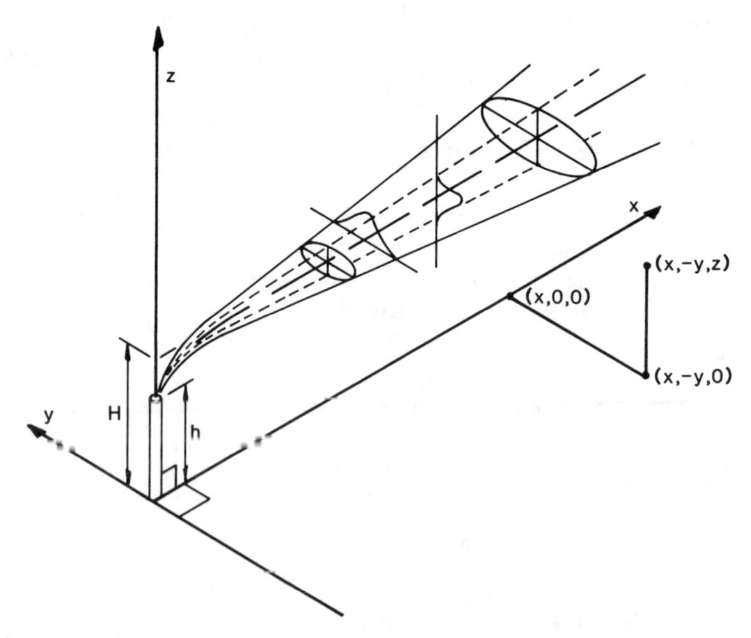

Figure 15-22. Coordinate system showing Gaussian distributions in the horizontal and vertical [173].

If the point source is at ground level with no effective plume rise (H = 0):

$$c(x,0,0;0) = \frac{S}{\pi \sigma_y \sigma_z u} \tag{15-14}$$

Depending on the assumed values of σ_x and σ_y, Equation 15-14 can be applied to Case I and Case II meteorological conditions of Figure 15-15. It has not generally been applied to the case of puff transport.

Values of σ_y and σ_z vary with the turbulent structure of the atmosphere, height above the surface, surface roughness, sampling time over which the concentration is to be estimated, wind speed and distance from the source [173]. Most practical studies employ a set of correlations of Pasquill [175] and Gifford [176] to estimate atmospheric dispersion. Six categories of atmospheric stability are defined and used to determine values of σ_y and σ_z as a function of downwind distance. The relationships are shown in Figures 15-23 and 15-24. Stability class is generally estimated by a method given by Turner [173] which requires information on solar elevation angle, cloud cover, ceiling height and wind speed (Table 15-6).

These methods will give representative indications of dispersion over open country or rural terrain. They are less reliable for urban areas because of the greater surface roughness and different heat transfer characteristics. Figures 15-23 and 15-24 probably underestimate plume dispersion from low-level sources in developed areas [173]. Budney [174] gives further discussion of urban dispersion patterns.

In using the Gaussian plume formulas, the height is assumed to be the lowest several hundred meters of the atmosphere. Concentrations estimated using values of σ_y and σ_z from Figures 15-23 and 15-24 correspond to a sampling time of about 10 min. Concentrations downwind from a source will decrease with increasing sample time mainly because of a larger σ_y caused by increased meander of wind direction. Turner [173] presented the following relationship for estimating concentrations for time intervals greater than a few minutes:

$$c_s = c_k \left(\frac{t_k}{t_s}\right)^p \tag{15-15}$$

where c_s = concentration estimate for the desired sampling time t_s
c_k = concentration estimate for the shorter sampling time t_k (assumed to be 10 min if relationships presented in Figures 15-23 and 15-24 are used)
p = constant, values between 0.17 and 0.20.

In using Equation 15-15 it is assumed that wind speed and stability remain relatively constant for the time duration t_s.

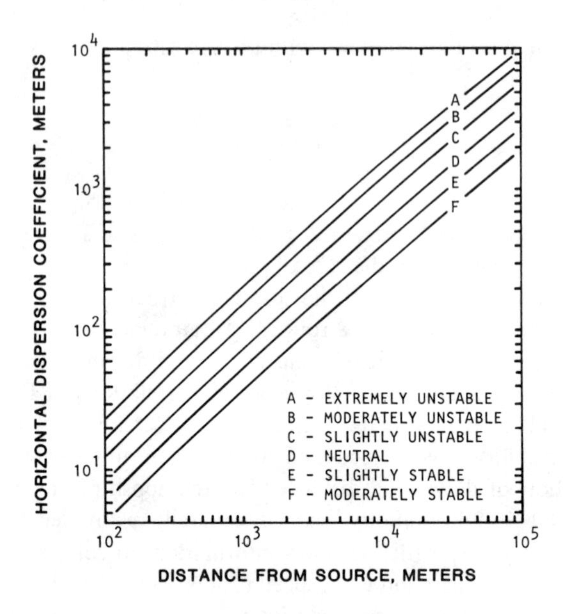

Figure 15-23. Horizontal dispersion coefficient σ_y as a function of downwind distance and stability class; rural terrain [173].

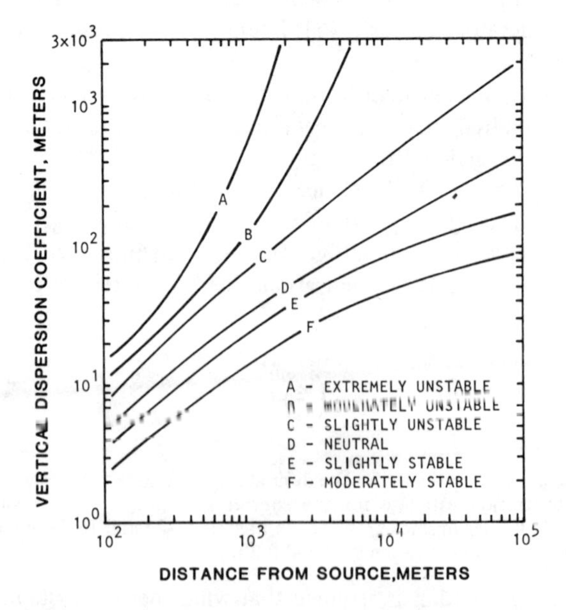

Figure 15-24. Vertical dispersion coefficient σ_z as a function of downwind distance and stability class; rural terrain [173].

Table 15-6. A Key to Atmospheric Stability Classes[a] [173,174]

Surface Wind Speed at a Height of 10 m (m/sec)	Day: Incoming Solar Radiation[b] (Insolation)			Night	
	Strong	Moderate	Slight	Thinly Overcast or ⩾4/8 Low Cloud Cover	⩽3/8 Cloud Cover
<2	A	A-B	B		
2-3	A-B	B	C	E	F
3-5	B	B-C	C	D	E
5-6	C	C-D	D	D	D
>6	C	D	D	D	D

[a]The neutral class (D) should be assumed for all overcast conditions during day or night.
[b]Appropriate insolation categories may be determined through the use of sky cover and solar elevation information as follows:

Sky Cover	Solar Elevation Angle >60°	Solar Elevation Angle ⩽60° But >35°	Solar Elevation Angle ⩽35° But >15°
4/8 or Less or Any Amount of High Thin Clouds	Strong	Moderate	Slight
5/8 to 7/8 Middle Clouds (7000-16,000-ft base)	Moderate	Slight	Slight
5/8 to 7/8 Low Clouds (less than 7000-ft base)	Slight	Slight	Slight

Gaussian plume formulas presented so far are arranged to estimate downwind concentration given the mass emission rate from the point source. In the case of odor discharges it is more appropriate to consider dilutions achieved between the source and downwind distance. This is particularly true because odor concentration is a measure of dilutions required to reach the threshold odor concentration. Consider Equation 15-13 and divide both sides by the stack gas volume flowrate V (m^3/sec):

$$\frac{1}{V}\,c(x,0,0;H) = \frac{S}{\pi \sigma_y \sigma_z u V} \exp\left[-\frac{1}{2}\left(\frac{H}{\sigma_z}\right)^2\right]$$

The term S/V is equal to the concentration c_s (g/m^3) in the source. Substituting and rearranging:

$$\frac{c_s}{c(x,0,0;H)} = D_l = \pi\sigma_y\sigma_z u \Bigg/ \left\{ V \exp\left[-\frac{1}{2}\left(\frac{H}{\sigma_z}\right)^2\right]\right\} \qquad (15\text{-}16)$$

where D_l = dilutions achieved between the source and the ground level, centerline of the plume.

Plume Rise

The effective height of plume H is taken as the sum of stack height h and plume rise ΔH, and is defined as the height at which the plume becomes passive and follows the ambient air motion. Plume behavior is a function of a number of parameters, including exit gas velocity (momentum), temperature difference between the plume and ambient air (buoyancy), density stratification of the atmosphere, wind speed and wind velocity gradient du/dz. Buoyancy forces produced by a temperature difference and momentum forces from the gas velocity are both important to plume rise. Given the complexity of the problem, numerical integration techniques have been used to increase the number of parameters which can be considered. Unfortunately, measurements of all parameters are seldom available. A number of semiempirical equations have been developed which are widely used and provide reasonable estimates of the plume rise. The equation of Holland [177] will be used here because buoyant and momentum forces are considered and both are likely to be important in the discharge of odorous gases from composting:

$$\Delta H = \frac{V_s d}{u}\left[1.5 + (2.68 \times 10^{-3})p\,\frac{T_s - T_a}{T_s}\,d\right] \qquad (15\text{-}17)$$

where ΔH = rise of the plume above the stack, m
V_s = exit velocity of stack gas, m/sec
d = inside stack diamter, m
u = wind speed, m/sec
p = atmospheric pressure, mbar
T_s = stack gas temperature, °K
T_a = air temperature, °K

The value 2.68×10^{-3} is a constant with units of mbar^{-1}m^{-1}. The reader is referred to standard texts on air pollution for more complete coverage of plume rise calculations.

If the effective height of emission were the same under all atmospheric conditions, highest ground-level concentrations would occur with the lightest winds (see Equation 15-11). However, plume rise is generally an inverse function of wind speed (see Equation 15-17). Thus, maximum ground-level concentrations generally occur at an intermediate wind speed where a balance is reached between dilution from wind speed and the effective height of emission [173].

Higher exit velocities will give more plume rise, but will also increase the pressure drop in the ducting and stack system and, thereby, increase power requirements. Tradeoffs must be made between the needs for odor reduction, the cost of increasing stack height and operating costs. As a minimum, stack-to-wind velocity should be greater than 1.5 to allow exhaust gases to break cleanly from the stack and avoid plume downwash behind the stack. The limit on maximum velocity is about 35 m/sec for masonry stacks and 45 m/sec for steel stacks [155]. Typical exit velocities are between 10 and 25 m/sec.

The following series of example problems will illustrate the calculation procedures for elevated point sources. They will also demonstrate the effectiveness of stack dispersal in a properly engineered odor control system.

Example 15-4

A compost facility is processing 10 dry ton/day of raw sludge by the static pile process. A maximum air supply of 1.5 m^3/min-dry ton is anticipated over a 21-day compost cycle. Collected exhaust gases are expected to have an average odor concentration of 300 ou (see Figure 15-10) and an average temperature of 50°C. Gases will be discharged through a 50-m stack at a discharge velocity of 20 m/sec. Wind speed is 2 m/sec and the ambient air temperature 20°C. Calculate the ground-level odor concentration at the plume centerline at a downwind distance of 1 km under class D (neutral) stability conditions.

Solution

1. Determine the average gas volume to be discharged:

$$1.5 \text{ m}^3\text{/min-dry ton} \times 10 \text{ ton/day} \times 21 \text{ days} \times \text{min/60 sec} = 5.25 \text{ m}^3\text{/sec}$$

2. Determine the required stack diameter:

$$\text{area of stack} = \pi d^2/4 = V/V_S = 5.25/20 = 0.263 \text{ m}^2$$

$$d = 0.58 \text{ m}$$

3. Estimate the plume rise using Holland's formula, Equation 15-17, assuming atmospheric pressure at 1000 mb:

$$\Delta H = \frac{20\,(0.58)}{2}\left[1.5 + (2.68 \times 10^{-3})(1000)\left(\frac{323 - 293}{323}\right)0.58\right]$$

$$\Delta H = 9.5\ \text{m} \quad (\text{say } 10\ \text{m})$$

The effective plume height, H, is then 50 + 10 = 60 m.

4. Dilution at the plume centerline and ground level is estimated using Equation 15-16. From Figures 15-23 and 15-24, values of σ_y and σ_z for a downwind distance of 1 km and class D stability are estimated as 68 and 32 m, respectively.

$$D_l = \pi(68)(32)(2)\Big/\left\{5.25\ \exp\left[-\frac{1}{2}\left(\frac{60}{32}\right)^2\right]\right\}$$

$$D_l = 1.5 \times 10^4$$

5. Downwind odor concentration can then be estimated as:

$$\frac{300}{1.5 \times 10^4} = 0.02\ \text{ou}$$

Because this is far below the threshold level, no odor should be detectable at the downwind location.

Example 15-5

Using the conditions of Problem 15-4, estimate the downwind odor concentration if the compost facility is designed to process 100 dry ton/day.

Solution

1. Repeat the calculation procedures of Problem 15-4.

$$\text{gas volume, } V = 1.5\,(100)\,21\,\frac{1}{60} = 52.5\ \text{m}^3/\text{sec}$$

$$\text{required stack diameter, } d = \left(\frac{4}{\pi}\cdot\frac{52.5}{20}\right)^{1/2} = 1.83\ \text{m}$$

$$\text{plume rise, } \Delta H = \frac{20\,(1.83)}{2}\left[1.5 + (2.68 \times 10^{-3})(1000)\left(\frac{323 - 293}{323}\right)(1.83)\right]$$

$\Delta H = 36$ m

effective plume height, $H = 50 + 36 = 86$ m

dilution at centerline, ground level =

$$D_l = \pi(68)(32)(2) \Big/ \left\{52.5 \exp\left[-\frac{1}{2}\left(\frac{86}{32}\right)^2\right]\right\}$$

$$D_l = 9.6 \times 10^3$$

downwind odor concentration $= 300/(9.6 \times 10^3)$

$= 0.03$ ou

2. Note that downwind odor concentration remains below the threshold odor level even if the gas discharge rate is increased by a factor of 10.

Example 15-6

For the conditions of Problem 15-4, determine the maximum ground-level centerline concentration.

Solution

1. Using the calculation procedure shown in Problem 15-4, the following table can be constructed. Note that average gas volume, stack diameter and effective plume height remain the same as in Problem 15-4.

x (km)	σ_y (m)	σ_z (m)	H/σ_z	$\exp\left[-\frac{1}{2}\left(\frac{H}{\sigma_z}\right)^2\right]$	D_l
0.25	19	10	6.00	1.52×10^{-8}	1.49×10^{10}
0.50	36	18	3.33	3.91×10^{-3}	1.98×10^{5}
0.75	52	25	2.40	5.61×10^{-2}	2.80×10^{4}
1.00	68	32	1.88	0.172	1.50×10^{4}
1.25	85	37	1.62	0.269	1.39×10^{4}
1.50	100	43	1.40	0.375	1.37×10^{4}
1.75	115	46	1.30	0.430	1.47×10^{4}
2.00	130	50	1.20	0.487	1.60×10^{4}
3.00	190	64	0.938	0.644	2.26×10^{4}
4.00	240	77	0.779	0.738	3.00×10^{4}
5.00	300	89	0.674	0.797	4.00×10^{4}

2. The maximum dilution occurs about 1.5 km downwind of the point source. Odor concentration at this point would be about:

$$300/(1.37 \times 10^4) = 0.022 \text{ ou}$$

which is significantly below the odor threshold.

3. Similar tables can be constructed for different wind speeds to determine the critical wind speed resulting in the maximum downwind concentration. In the same manner, other stability classes can be assumed. Thus, system characteristics can be examined over a range of atmospheric parameters to assure that design is adequate in all cases.
4. Given the conditions of this problem and the range of stability classes from A to F, maximum ground level concentrations will range from about 0.011 ou (class F) to 0.035 ou (class A). Thus, changes in stability class generally effect less than a factor of 10 change in ground level concentration. A factor of 2-5 is typical for effective heights less than 70 m. Distance to the point of maximum concentration is affected more significantly, however.

It is apparent from these example problems that a properly engineered stack dispersal system is an effective method for handling odorous gases. Significant dilution can be achieved before the plume reaches ground level. In combination with gas collection and treatment, an effective system of odor control can be provided. Sufficient information has been developed here for preliminary planning of odor control facilities. However, the help of trained meteorologists is recommended during final design because there are many subjects of atmospheric transport and dispersion beyond the scope of this text.

A number of limitations of stack dispersal should be noted. When odorous air is diluted with odor-free air, the perceived odor may decrease less rapidly than the concentration of odorant molecules. In other words, dilution of an odorant can yield less than a proportional reduction in odor intensity. For example, a tenfold reduction in the concentration of amyl butyrate is needed to reduce its perceived odor intensity by half. On the other hand, such factors are accounted for, at least in part, if the odor unit is the basis of measurement since it is based on required dilutions to the threshold concentration.

Perhaps more important is the fact that public reaction to odors is not so much based on average concentration as it is to occasional peak or excess concentrations. Recall also that odor panel measurements determine the TOC at which 50% of the panel members respond positively. A certain percentage of panel members will be sensitive to lower odorant concentrations, i.e., higher TOC values. Thus, worst-case conditions should be included in the dispersion analysis. A factor of safety of 30 applied to calculated ground-level concentrations has been suggested to assure that threshold concentrations are not exceeded [154].

SUMMARY

Appropriate systems for control of nuisance conditions must be included in the design of compost facilities. Nuisances of most concern include odors and dusts, both of which can be controlled through proper design and operation.

Dust control can be a major problem particularly in arid or semiarid climates, and is aggravated by the small size-distribution characteristic of most sludges. Proper housekeeping and enclosure of operations with a high dust potential are necessary for effective management. Pelleting has been demonstrated to be an effective dust control procedure but has seen only limited application to date.

Control of odors is one of the most difficult problems in present sanitary engineering practice. Odors will be emitted from surfaces of open piles and will also be present in exhaust gases collected from controlled aeration systems. Organoleptic techniques which use the human olfactory system remain the standard method for quantification of odors. The odor unit (ou) is defined as the number of dilutions with odor-free air required to achieve the minimum detectable odor concentration. Odor concentration in ou is usually determined by supplying a number of diluted samples to one or more individuals until the odor is either not detected or detected by only 50% of panel members.

Another parameter of importance with open facilities is the SOER, usually expressed in $m^3/min\text{-}m^2$. The latter is determined by placing a sample hood over the surface being analyzed. Air is drawn through the hood at a controlled rate and the odor concentration of exhaust gases determined.

Based on a number of studies, an odor concentration of about 5 ou represents the threshold level of complaint. A concentration of 10 ou usually assures a complaint-level condition. A design odor criterion of 2 ou maximum at the plant boundary has been used in practice.

Available data on odor emission rates from compost facilities are rather limited. A case history at a major compost facility has demonstrated that odor emissions can increase significantly if thermodynamic and other operational constraints are exceeded. During normal operation the SOER for windrow composting of digested sludge should range from about 0.3 $m^3/min\text{-}m^2$ (10% less than value, $SOER_{10}$) to about 10 $m^3/min\text{-}m^2$ (90% less than value, $SOER_{90}$). SOER values from aerated static pile and enclosed mechanical systems have been observed to be in the same general range or slightly lower. In most cases the SOER is greatest at the beginning of the compost cycle, decreasing with time into the cycle.

Odor concentrations in exhaust gases from controlled aeration systems have been observed to vary over a wide range. With aerated windrow and enclosed reactor systems operating on digested sludge, ou_{10} and ou_{90} values of about 20 and 250 ou, respectively, have been observed in collected gases. With other systems, odor concentrations of several thousand ou have been recorded. Limited data obtained from an enclosed reactor system suggest that odor concentrations are comparable when using raw or digested sludge.

Controlled aeration systems allow collection of exhaust gases which makes treatment and deodorization possible before atmospheric discharge. A number of deodorization techniques are available but those which have received most attention include water or chemical scrubbing, activated carbon adsorption, deodorization piles consisting of previously composted material and use as feed air in activated sludge systems. Based on limited data, wet scrubbing is moderately effective with an average removal efficiency of about 45%. Activated carbon adsorption is more effective and can generally produce an effluent concentration less than 10-50 ou even with influent concentrations of several thousand ou. Few engineering data are available to judge the effectiveness of compost deodorization piles and considerable difference of opinion exists in engineering circles. One controlled study indicated an average 50% reduction with considerable variability in the data. Use of exhaust gases as feed air in the activated sludge process is reportedly effective in a number of full-scale facilities, but quantitative data are generally lacking. Deodorization systems such as scrubbing towers and packed-bed adsorption reactors have an advantage because treated gases are still contained and can be dispersed in a controlled manner.

A complete odor control system consists not only of treatment of odorous gases but also their proper dispersal into the atmosphere. Odors emitted from the surface of open compost piles will be dispersed according to prevailing conditions in the atmospheric sublayer. The latter typically extends about 20 m above the ground surface. A critical transport condition defined as "puff" transport, has been observed to occur during periods of high atmospheric stability. Dispersion is limited to the lateral direction and downwind odor concentration is inversely proportional to the square root of distance from the source. Puff transport conditions are characteristically observed during evening hours, and can result in transport of odor for considerable distance with little attenuation in concentration. Barriers and wind machines have been used to increase dispersion during such conditions. More odorous operations should be conducted during periods of higher atmospheric turbulence to avoid puff transport conditions.

Discharge of collected exhaust gases from an elevated stack is an effective means of dispersal into the atmosphere. Elevated discharge avoids many of the problems encountered in the atmospheric sublayer. With typical engineering designs, dilutions on the order of 10^3-10^4 can be achieved before the plume reaches ground level. In combination with gas collection and deodorization, stack dispersal is one approach to providing an engineered, controlled system of odor management.

CHAPTER 16

CLOSING THOUGHTS

A FEW COMMENTS

Before drawing this book to a close I would like to take a measure of editorial license and present a personal view on the future course of composting, particularly sludge composting. The future of any unit process depends largely on the advantages and disadvantages offered compared to competing processes and other management alternatives. The advantages of composting are many. It can convert putrescible organics to a stabilized form, destroy pathogenic microbes and provide significant drying of wet substrates such as sludge. Drying reduces the cost of subsequent handling and increases the attractiveness of composted product for reuse. All of these advantages are obtained with minimal outside energy input; the major energy resource is the substrate organics themselves. Even if subsequent heat-drying is required to polish the final product, requirements can be greatly reduced by maximizing drying achieved during composting. Furthermore, composting is compatible with a variety of feedstocks, ranging from raw to digested sludge, conditioned by heat, organic polymers or inorganic materials such as lime, and can use a variety of amendments as may be locally available.

With all these advantages, it begins to sound as if nothing can compete with composting, at least for sludge management. However, this is not the true picture. Most experienced engineers know that panacea solutions are very rare. The invention of the wheel may have been the last recorded by man. All unit processes in sanitary and chemical engineering seem to have both advantages and disadvantages, and selection of a particular process depends on many site-specific factors. It has been my experience that very few decisions in sludge management are so clear and straightforward that one process is the obvious choice. After the technical analysis is complete and all quantitative information has been exhausted, qualitative factors often tip the scale toward one process or another. Will nuisance conditions, such as odor and dust be increased? Is there a reasonable likelihood of obtaining a processing site outside existing plant boundaries? What are the political, social and environmental consequences of the proposed action? Are there regulatory unknowns or new regulations pending? In today's world the answer is usually yes. What are the technological uncertainties?

It is in these areas that composting reveals most of its disadvantages. Although composting itself is a very old process, its application to sludge is quite recent. Perhaps more important has been the lack of a rational, engineered approach to design and operation of compost facilities. The major purpose of this book is the development of a basic unified approach to analysis.

Composting can also be susceptible to nuisance conditions such as odors and dusts. These factors are becoming increasingly important in process selection. Most treatment facilities are eventually surrounded by neighbors who demand essentially complete elimination of nuisance conditions or guarantees that there will be "no odor at any time" in the case of new facilities. Composting can also require large land areas compared to other processes such as heat-drying, combustion or direct landfilling. Many facilities will not have sufficient land and will face the decision of obtaining an offsite location (a difficult task at best) or adopting other processes which allow one to stay onsite.

So far, I have referred several times to composting as a process. Many view composting as a complete management alternative by and of itself. I feel that it is more appropriately thought of as a process, a part of a total management system. A repeated theme of this book is that composting depends on feedstock characteristics that are largely determined by the nature of upstream processes. Reuse or disposal of the final product also depends on the success of the compost process. Therefore, it is important to understand the underlying principles of composting and to know what can and cannot be achieved, so that composting can be coordinated within the treatment train. Composting must act in concert with upstream and downstream processes to develop a successful management system.

It appears to be the nature of government to seek universal solutions to problems. From my previous comments it is obvious that I do not subscribe to this approach. Recent experiences on the East Coast seem to bear this out. Development of the static pile system at Beltsville, MD, coupled with federal requirements to cease ocean disposal of sludge, resulted in nearly feverish interest in sludge composting. This had led many in government to view the process as a panacea, applicable to all no matter the situation. Pressure was brought to bear on many communities to implement this system without a proper assessment of other composting processes or other management alternatives. In my opinion, this has led to implementation of several ill-conceived compost facilities either through improper design, unwise process selection or location of facilities at sites not well suited to composting.

This presents a very real and imminent danger to the future of composting. A few bad facilities can so taint a process that it becomes difficult if not impossible to implement again. It may be that all the problems encountered are solvable, but just try to convince a local official of that if he has visited a facility plagued by odors, dusts and financial problems, or one that is visually unpleasing. All the engineering in the world may not overcome that initial impression or the "everybody knows that won't work" syndrome.

The author has many colleagues instrumental in development of the static pile system. The above remarks are not intended to cast any aspersions on their work or on the static pile system. Those who have read this book know the many advantages this process can offer and the significant contributions these researchers have made to the understanding and advancement of composting. But like its cousins, the windrow and enclosed reactor systems, it is not the appropriate process for all situations. Composting at the wrong place or implementation of the wrong type of composting can be just as damaging as no composting at all.

One only has to review the history of refuse composting to realize that the above scenario is indeed possible. Refuse composting was characterized by periods of extreme interest and other times when the literature was essentially silent on the subject. The 1950s and 1960s saw intense interest in refuse composting only to see eventual abandonment of nearly all U.S. facilities by the late 1960s. The reasons are many, but overly optimistic economic analysis, lack of markets, competition with less expensive disposal alternatives and overstated claims by developers coupled with a lack of scientific understanding of the process by many engineers are certainly paramount. A number of technologically sound facilities, designed and operated by knowledgeable people, were made to suffer by the general disfavor eventually applied to refuse composting in general.

This then is a plea for reasoned analysis of composting and alternative compost processes, along with full evaluation of other management alternatives on a case-by-case basis. Only by recognizing its advantages and disadvantages, applying the tools of engineering analysis to allow for educated decisions and resisting mandated "universal" solutions, will the future of composting be assured. There are enough advantages to composting that it will still be the system of choice in many cases. But where it is not, it is best that other management alternatives be implemented.

A PROSPECTUS ON THE FUTURE

It is my personal conviction that reasoned and fundamentally sound analysis will prevail, and that composting will attain its proper status among the unit processes of engineering. Assuming this conviction is realized, what direction is composting likely to take? What types of systems will be in use ten years from now? If technical and economic considerations are the main guiding forces, I believe a reasonable prospectus can be developed. Six subject areas will be considered in this analysis: (1) bulking agent or amendment selection; (2) agitation and forced aeration, (3) moisture removal and wet weather operation; (4) nuisance conditions; (5) raw vs digested sludge; and (6) system costs.

Bulking Agents and Amendments

Bulking agent or amendment selection is a critical factor in the engineering of any compost facility. Commenting on the use of wood chips in the static pile method, Shea [178] noted that they are a commodity as well as a by-product, subject to price escalation far greater than average rates of inflation. Actual bids received during design of facilities were $12-13/m^3$ (1979 dollars), 2-3 times greater than previously estimated in project economic planning. At that rate wood chip purchase alone accounted for 60% of total system cost. Further, that estimate assumed use of screening to recover and reuse a large fraction of the chips. Shea [178] concluded that on economic grounds alone alternatives to wood chips must be found if static pile composting is to be cost-effective in the long run.

Situations have developed in several eastern U.S. cities where there has been actual competition for the available wood chip supply. The imbalance of supply and demand has led to greatly increased costs. In some cases, compost operations have been jeopardized and periodically curtailed by a lack of supply.

Another interesting fact comes to light when the energy content of wood chips is examined. Heat of combustion available in wood chips is presented in Table 16-1, assuming 10 dry ton/day of sludge at 20% solids, a volumetric mixing ratio of 2.5, and wood chip recoveries ranging from 60 to 80%. A recovery of 80% is somewhat optimistic, whereas values as low as 60% have been observed in practice. Heats of combustion of new chips range from about 20×10^6 to 40×10^6 kcal/day, depending on the efficiency of recovery and reuse. By comparison, if wet sludge at 20% solids is simply heat dried to 90% solids in a conventional indirect steam dryer, the energy requirement is about 33×10^6 kcal/day. Thus, we are

Table 16-1. Comparison of Energy Available in Wood Chips to Energy Required for Heat Drying

Sludge Quantity (dry ton/day)	Sludge Quantity[a] (wet ton/day)	Sludge Volume[b] (m^3/day)	Total Chip Volume[c] (m^3/day)	Chip Recovery (% by vol)	New Chips Required (m^3/day)	Heat of Combustion New Chips[d] (10^6 kcal/day)	Energy to Heat Dry Sludge[e] (10^6 kcal/day)
10	50	48.2	120.5	60	48.2	40.7	33
10	50	48.2	120.5	70	36.2	30.6	33
10	50	48.2	120.5	80	24.1	20.3	33

[a] SC = 0.20.
[b] Based on Equation 7-3 and assuming VC = 0.70, GV = 1.0, GF = 2.5.
[c] Volumetric mixing ratio M_{bc} = 2.5.
[d] HB = 3739 kcal/kg, SB = 0.80, VB = 0.95.
[e] Energy required to heat dry sludge from 20 to 90% solids in an indirect steam dryer. Includes heat losses and boiler efficiency (80%), 825 kcal/kg feed water.

left with the rather incredible conclusion that under some conditions new wood chips could be burned in a boiler and supply sufficient heat to completely dry the sludge! It is also interesting to note that at $13/m^3 dry wood chips are being valued at about $12/Mkcal ($3/MBtu), essentially equivalent to the cost of OPEC crude oil (1979 dollars). So not only are we adding nearly sufficient energy to heat dry the sludge but we are paying a price comparable to the use of petroleum.

I would conclude from the above that use of wood chips for bulking will be limited to smaller treatment facilities where the impact on available supply is not great, and also limited to those regions where wood products are in plentiful supply. If the use of wood chips is phased out, two other options are available. First, other bulking particles could be used. Nondegradable particles such as shredded rubber tires and plastic chips have been suggested, but there is little practical experience with either. Shredded tires have the added disadvantage of being very dense, which greatly increases material handling costs. Garden debris, certain agricultural wastes and shredded refuse can be used. But here we begin to encounter problems of structural stability. These materials have more tendency to compact under pressure, and periodic turning may be required to maintain free airspace. The designer should determine the availability of any such materials as well as their ability to structurally maintain the static pile.

If the designer is willing to provide occasional turning, the option of using recycled compost and/or other amendments becomes available. The latter materials do not have to possess the structural integrity of bulking particles. Furthermore, when used in conjunction with compost, amendment requirements can be significantly reduced or entirely eliminated. If sufficient energy is available in the sludge to satisfy thermodynamic demands and a high level of agitation can be provided, complete reliance on compost product is possible. Another advantage to use of compost product is that the quantity of final product is not increased as occurs when bulking particles or amendments are used.

Agitation and Controlled Aeration

Agitation or turning of material during the compost process can accomplish a number of purposes including (1) averaging out initial errors in mixing; (2) prevention of consolidation and air channeling; (3) increased statistical assurance that all material has achieved a minimum time-temperature condition; and (4) allowed use of compost product and/or amendments for moisture control which might not be suitable in the absence of agitation. If agitation is coupled with uniform feeding, conditions approximating those of a continuous-feed, complete-mix (CFCM) reactor can be achieved. The latter has several advantages over batch processes, including reduced detention time and less severe moisture limitations on the infeed material. But whether batch or CFCM processes are used, agitation is likely to be a rather common feature in future designs.

Controlled aeration is another feature which should be considered. Forced aeration provides improved operational control over the rate of oxygen supply, which is in turn related to the thermodynamic balance and is a major controlling parameter in the compost process. Almost all enclosed reactor systems provide controlled aeration as does the static pile system. There are a few windrow systems which do not. These generally use small pile dimensions, large quantities of porous materials such as straw or sawdust, or very long detention times. It is advantageous to maximize pile dimensions to conserve heat, reduce land requirements and reduce susceptibility to wet weather. As pile dimensions increase, however, the ability of natural ventilation to supply adequate oxygen becomes stressed, particularly if compost product is used for moisture control. Therefore, controlled aeration should be considered seriously in design of any compost system. Control over both rates of agitation and aeration provides the maximum in operational flexibility to achieve the desired end product.

Moisture Removal and Wet Weather Operation

High rates of aeration can be used to effect significant moisture removal from wet feed material such as sludge. The extent of moisture removal will be limited either by thermodynamics or by kinetics if moisture content is reduced below about 40%. Even with an "energy-rich" sludge and high aeration rates, drying from biologically produced heat is limited to about 65% solids content. If a drier product is required, if the sludge is energy-deficient or if the system is exposed to rainfall, additional drying will usually be required.

Two basic approaches can be used to achieve the desired drying: air drying, where reliance is on surface evaporation into ambient air of <100% relative humidity, or heat drying. Options are then available within these two basic approaches as shown in Table 16-2. Systems which rely on air drying are energy-efficient, but are most effective in arid climates or dry periods in wet climates. This usually means that large storage piles are necessary to stockpile compost in wet weather for later

Table 16-2. Drying Methods Applicable to Compost Material

Air Drying

Drying is accomplished by natural evaporation promoted by contact with ambient air of relative humidity <100%. Generally effective in arid climates or dry months in wetter climates.

Open-Air Drying

Compost is spread into windrows or thin piles and periodically turned to improve contact with ambient air. Simple and energy efficient method in arid climates.

Reactor Drying

Compost is contained in a reactor and ambient air supplied by forced aeration. Exhausted air is easily collected and can be treated and/or dispersed via stack.

Heat Drying

Heated ambient air, hot combustion gases or fluids such as steam are used to remove moisture. Systems can range from simply heating ambient air 10-20°C to decrease relative humidity, to use of conventional heat drying systems. Heat is usually supplied from fossil fuels, alternative fuel sources or waste heat sources. At least two basic types.

Reactor Drying Systems

Direct Contact of wet material with heated ambient air or other gases.
Indirect Contact of wet material with heating fluid across a heat exchange surface. Steam is usually used for heating.

Heated Curing Piles

Some recent designs have included provision for forced aeration of curing piles using heated ambient air. Heat can be provided from heat exchange with hot exhaust gases from the compost process.

drying when weather conditions permit. Open-air drying systems are more prone to odor and dust nuisances and require relatively large land investments. If site conditions are appropriate, however, it can be a very effective approach. Reactor drying decreases the land requirement and allows for collection and control of potentially odorous exhaust gases. Energy requirements are increased over the open-air drying case, but can be kept within reasonable limits by pelleting the compost. This increases the size of void passages and decreases head loss across the reactor.

Heat drying can remove all dependence on ambient air conditions and provide a facility capable of producing a dry product in all weather conditions. The major disadvantage of heat drying has traditionally been the large energy requirement. By maximizing drying potential of the compost process itself, subsequent energy requirements can be reduced substantially below those required for conventional heat drying of dewatered cake. Composting followed by heat drying provides an orchestrated system which can produce the desired end product in all climatic conditions and with minimal dependence on outside energy sources. Another measure of operational flexibility is provided since the composter can run under conditions more conducive to biodegradable volatile solids (BVS) reduction, delegating more or less moisture removal to the heat-drying step, while always under control of the operator.

With heat requirements reduced by the compost process, a number of heretofore wasted heat sources can be used. Although availability of waste heat sources is very site-specific, a number are commonly found in treatment plants. For example, digester gas is often used in internal combustion engines. Both combustion gases and jacket cooling water are sources of waste heat potentially useful for compost drying. Recent compost facility designs [11,180] have included provisions for forced aeration of curing piles using ambient air preheated by heat exchange with hot gases exhausted from the compost process. Heating ambient air by only 10-20°C greatly increases its moisture-carrying capacity (see Figure 8-2). Direct use of compost exhaust gases is not practical because they are near saturation and will likely cool during passage through the curing pile, resulting in moisture condensation rather than evaporation. The point here is that the design engineer should investigate possible sources of waste heat before deciding on a drying approach.

The subject of moisture removal leads directly to problems of wet-weather operation. Process selection, design and operation of the compost facility will usually be determined by wet weather requirements. If the designer is not careful, a simple compost system, adequate in arid climates, may become very complex and costly when requirements for wet weather operation are considered. Previous chapters discussed a number of design options to mitigate effects of wet weather. These can be summarized as follows: (1) maximize dewatered cake solids to provide sufficient energy for both composting and drying; (2) compost raw sludge or a raw/digested blend to increase available energy resources; (3) add biodegradable amendments, preferably dry ones, to control moisture and to improve energy resources compared to use of well-stabilized compost; (4) maximize windrow or pile dimensions to reduce the area subject to rainfall; (5) provide forced aeration to allow operational control of the drying rate; (6) provide all weather surfacing of the compost area; (7) enclose the compost area by either roofed coverage or use of an enclosed reactor composter; (8) provide large stockpiles of dry material (e.g., compost product, amendments and bulking particles) for use during wet weather; and (9) provide heat drying to remove all dependence on ambient air conditions. Exactly which combinations of these measures are appropriate will depend on the severity and duration of wet weather conditions and other site-specific factors.

There still remains much for the design engineer to consider because the range of possibilities for wet weather design and operation is large. On the one hand, open-air systems such as the windrow and static pile may require little if any wet weather provisions in arid climates with infrequent rainfall. On the other, use of enclosed reactor systems or open-air systems with roofed coverage followed by heat drying of compost product can provide a system virtually immune to effects of wet weather. Obviously, decisions in these matters are very site-specific.

Nuisance Conditions

Major nuisance problems associated with composting are odors and dusts. Concern over odors has in recent years become of paramount importance in siting waste treatment facilities and even in process selection. Municipal agencies and industrial facilities are often finding themselves in a position of having to guarantee no odor impact on the surrounding community. Of course, this is virtually impossible to guarantee and as a result many agencies are shunning any process with high odor potential. Unfortunately, composting is a process that can have a high odor and nuisance potential if the system is improperly designed or the wrong system is selected for a particular location.

It may seem somewhat contradictory to speak about the problems of obtaining dry product and wet weather operation, and then turn around and talk about problems of dust formation from having too dry a product. Nevertheless, dust control can be a major problem, particularly in arid climates. Municipal sludge organics are characterized by a small-size particle distribution. When sludge based compost exceeds about 65-75% solids, the small-size particles are easily airborne if agitated. Dust problems can be further aggravated if composting is conducted on an unpaved surface.

Fortunately, the design engineer has tools available to control odors and dusts and virtually eliminate any nuisance impact on the surrounding community. It's a question of tailoring the system to the needs of the situation. A generalized diagram relating the nuisance problem with the degree of control (and hence cost?) is shown in Figure 16-1. The diagram is provided for purposes of discussion and simply states that nuisance problems should be reduced as more controls are included in the system design. This appears to be a reasonable and rather obvious conclusion. There are no scales on the axes, and this is one of the present problems with compost facility design. It is difficult for the designer to tailor nuisance

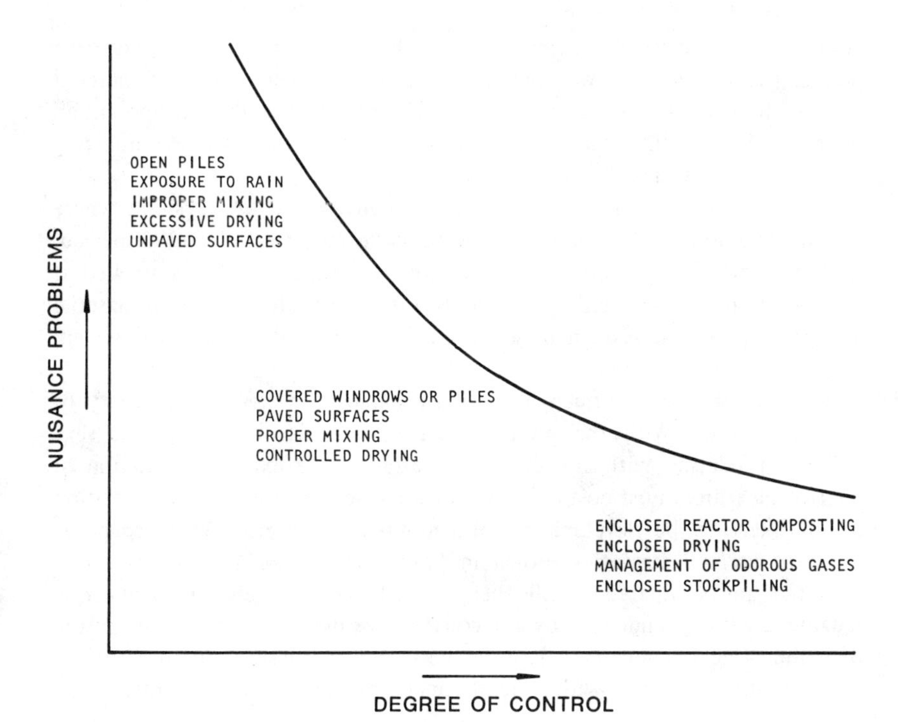

Figure 16-1. Generalized diagram relating nuisance problems with the degree of control exercised over them.

controls to the particular situation. For example, quantifying odor emissions and prediction of odor levels at plant boundaries is a new science whose tools have yet to stand the test of time. As a result we see engineers still relying on qualitative assessments and hear statements such as "this process smells and I can't have any odors at my plant." In fact, all sludge processes result in odor emissions to some extent. It is a question of the rate of odor emissions, their perceived characteristics to the human receptor and the application of engineered systems for collection, treatment and/or dispersal.

I have taken the liberty of labeling sections of the curve in Figure 16-1. In light of above comments the labeling represents my subjective opinion of the degree of control offered by various systems. A minimum of control is provided by open-air systems, such as the windrow and to a lesser extent the static pile. Odor emissions from the surface cannot be contained or controlled. If forced aeration is provided, that portion of the air is collected and can be controlled. However, surface emissions will still remain.

Potential for nuisances increases if windrows are exposed to excessive rainfall (i.e., inadequate wet weather design). Improper mixing of material before windrow or static pile formation will also increase the odor potential. Excessive drying will increase potential for dust formation, as will composting on an unpaved surface.

The maximum level of control is provided by enclosed systems where all air in contact with compost can be collected, treated and dispersed. Nuisances which may arise from subsequent drying can be controlled if reactor drying is used, either with ambient air or heated gas. Even material and product stockpiles can be enclosed. Thus, a totally controlled system can be achieved if desired.

Between the totally open and totally enclosed systems is a variety of process options. With the present state of the art, however, it is very difficult to define with any degree of certainty the exact combination of control measures most cost-effective for a particular case. Such information is not likely to be developed until a number of systems, which span the range of control strategies shown in Figure 16-1, have been constructed, operated and evaluated. Until then we will have to rely on engineering judgment with a tendency toward conservative design. As far as the future of composting is concerned, it is far better to overdesign nuisance control measures than to end up with a facility unacceptable to local residents.

Raw vs Digested Sludge

Anaerobic digestion has often been compared to composting with the idea that only one or the other should be implemented at a given facility. This is most unfortunate since the processes do not accomplish the same purposes. It is true that both processes will biologically stabilize organics, but from that point on they differ considerably. Anaerobic digestion produces a useful and valuable fuel and reduces odor potential of the sludge for subsequent processing, but leaves the sludge in a liquid form and is only moderately effective in destroying pathogens. Composting is more effective in destroying pathogens and can produce a reasonably dry final product suitable for reuse, but produces no energy and can have a higher odor potential, particularly if raw sludges are processed. As pointed out in this text, digested sludge contains less available energy than raw sludge, and any compost system using digested sludge will be more constrained thermodynamically. Nevertheless, composting of digested sludge can be achieved as long as thermodynamic constraints are recognized and not exceeded.

I am certainly not proposing that all sludges be digested before composting. Each situation must be evaluated on its own site-specific factors, and there are many cases where raw sludge composting is the system of choice. The point is that anaerobic digestion and composting are not mutually exclusive. An attempt to make them mutually exclusive ignores the fact that decisions regarding use of anaerobic digestion involve many factors including economics, energy production and previous agency experience. Given the national energy situation and the number of existing treatment plants using anaerobic digestion, restricting composting to only raw sludge could limit future applications of composting in sludge management. Fortunately, composting can be accomplished regardless of decisions concerning anaerobic digestion, provided thermodynamic constraints are not exceeded.

System Economics

The subject of costs is very difficult to address properly in any text, and indeed it is not really the subject of this book. Costs are very site-specific and a complete chapter would have to be written to do justice to the subject. All that can be said is composting is cost-competitive with other sludge management alternatives. If this were not so, we would not see such a large number of systems being implemented. Notice that I used the term cost-competitive. Composting is likely to be cost-competitive in most cases, but whether it is the most cost-effective depends on a site-specific analysis.

My purpose for bringing up the subject concerns Figure 16-1. Is the cost of enclosed composting or other nuisance mitigation measures so prohibitively high as to eliminate them from serious consideration? This would have the effect of placing an upper bound on the degree of nuisance control practicable in a compost system. Obviously, this would have serious implications on the future direction of composting.

To answer the question I need to present only a few general cost figures. Windrow composting using recycled compost product can usually be conducted for less than $20-40/dry ton, including land purchase, site preparation, mobile equipment, a concrete pad, fixed roof cover, controlled aeration system, and operation and maintenance (O&M) expenses including labor, maintenance and fuels [10,101,181,198]. Static pile composting with wood chips is considerably more expensive, primarily because of the cost of wood chips. Referring to the mass balance shown in Table 2-3 and assuming chip costs of $13/m^3, the cost of new chip replacement alone is about $42/dry ton of sludge. Total system costs are usually in the range of $60-120/dry ton; the major cost items are bulking agents and labor [15]. I realize the inadequacies of stating cost figures without a detailed listing of all assumptions and design criteria used in developing the costs. However, I feel these generalized values will be adequate for the present purposes.

A number of preliminary proposals for enclosed reactor composting were reviewed [180,182,183] as well as costs developed by the LA/OMA Sludge Project (for which I was senior engineer for five years). In all cases the systems included enclosed reactors with forced aeration and provision for periodic and frequent agitation. Recycled compost product was used for moisture control in most cases. Wood chips or other expensive amendments were not used. Subsequent drying was provided either by open-air or enclosed heat drying systems. Amortized capital costs were generally in the range of $10-25/dry ton of sludge feed and included reactors, buildings, conveyance and drying systems. O&M costs were generally in the same range ($10-25/dry ton), and included labor, fuels, maintenance and electricity. Total system costs were then in the area of $20-50/dry ton of sludge.

I present the above cost data with reservations because economics are so much a function of site-specific factors that generalized values have little meaning beyond the conditions for which they were developed. However, it does appear that fully enclosed systems can be constructed and operated competitively with other compost processes. In particular, there seems to be a significant competitive edge over the traditional static pile method if local wood chip costs are high. As previously discussed, total system costs are greatly influenced by the cost of bulking particles and amendments. Therefore, if local cost for such materials is high, a competitive advantage is given to systems which can use compost recycle either alone or in conjunction with other low-cost amendments.

This analysis also implies that any practical upper bound to nuisance control measures in Figure 16-1 is far to the right on the diagram. Enclosed composting and enclosed drying, with collection, treatment and dispersal of all odorous gases and with a high degree of dust control can be provided at cost-competitive prices. Because nuisance factors are an ever-increasing concern, it is reasonable to assume that they will continue to have a major role in process selection and design. Therefore, pressures to control nuisance conditions will tend to move composting in the direction of more enclosed systems, particularly because costs do not appear to present an overwhelming hurdle to such movement.

A GENERALIZED SYSTEM DESIGN

In summarizing the above analysis, I offer this personal view of future facility designs. It is likely that future systems will be designed with more attention to the costs and availability of bulking particles and amendments. The trend will be away from the use of high-cost materials and more toward use of lower-cost amendments, such as recycled compost product and/or locally available waste materials. As this occurs there will be an increasing need to provide periodic agitation of material during the compost process. Controlled aeration will be provided in most systems to provide greater operational flexibility. Requirements for wet weather operation and control of nuisance conditions will largely determine process selection and design. Enclosed systems which provide agitation, forced aeration, less susceptibility to wet weather and better control of odors and dusts will be increasingly favored. This conclusion is supported by the fact that costs for more enclosed systems are not prohibitive. In fact, they may be even more economical than systems which rely on expensive bulking particles or amendments.

I do not wish to imply that all future systems will be enclosed. This would contradict earlier statements concerning the need for rational analysis of all options before process selection. However, I do see a larger share of the compost market going to enclosed systems because of the difficulty in obtaining remote offsite locations and the proximity of many treatment plants to large population centers.

A block diagram showing major components of an enclosed composting system is shown in Figure 16-2. It is assumed that the system is to have a high degree of operational flexibility and stability, control over nuisance conditions such as odors and dusts, and resistance to wet weather impacts. Storage, metering and premixing of all feed components is provided ahead of the composter. Without a metered and controlled flow of material the operator's ability to produce a consistent feed substrate is severely hampered. It is also considered good engineering practice to provide premixing of all components even if the composter itself can provide a high level of agitation. Mixing energies in most compost systems are not sufficient to rapidly blend different feed components. Material inhomogeneity in the compost system can cause obvious operational problems, and it is far better to simply avoid the entire problem with good premixing.

The enclosed compost system should provide both agitation and controlled aeration. A number of available designs satisfy this criterion. Batch processes and reactor systems which can approximate continuous-feed, complete mix conditions are both available, as well as hybrid intermediates. The reader should refer to Chapter 2 for further discussion of systems meeting these criteria.

The capability of adding water to the composting material should be provided, particularly with high-energy feed substrates. High temperature rate limitations can often result with such feed materials. Increasing the

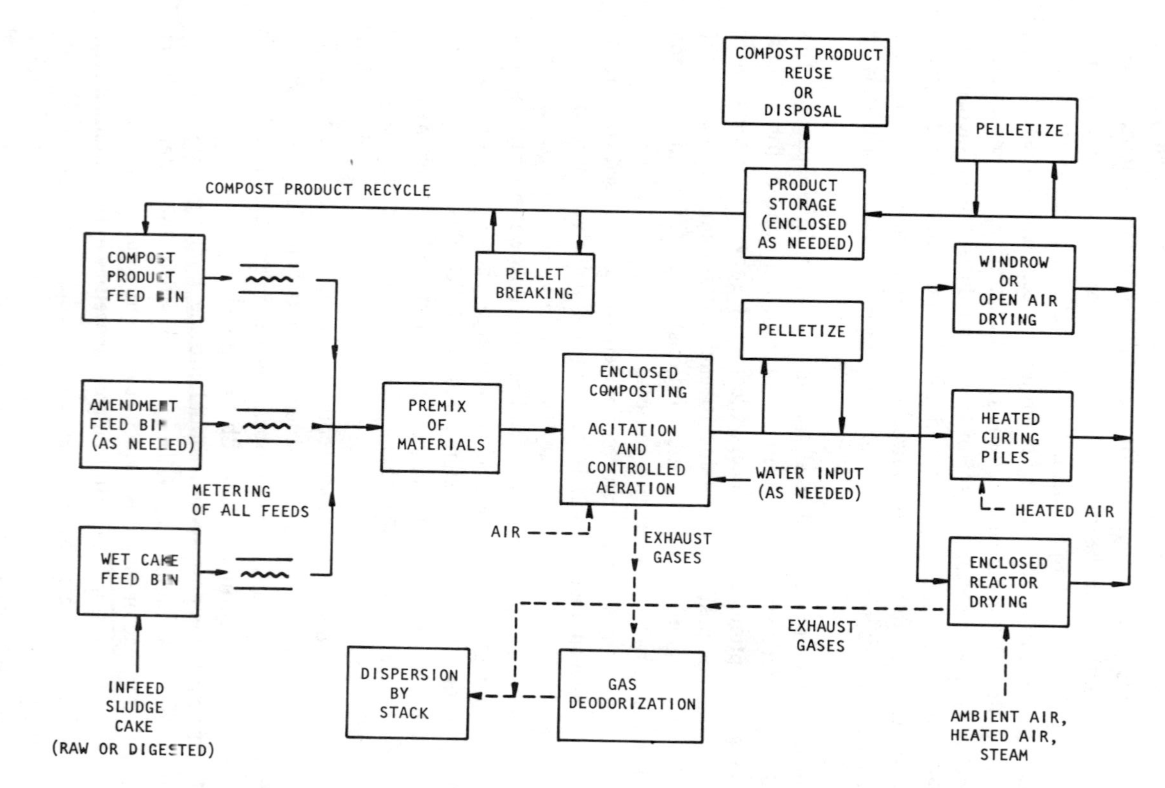

Figure 16-2. Process diagram of a generalized sludge compost system designed for a high level of operational flexibility and stability, control of odors and dusts, and resistance to wet weather.

air flowrate increases the rate of heat removal and can lower the reactor temperature to a more favorable range for biological oxidation. With high air flowrates, however, moisture lost in the exhaust gases can increase to the point where excessive drying results with attendant low moisture rate limitations. Designing the system with the capability for supplemental water addition provides flexibility to control low moisture rate limitations if desired.

A number of options are shown for product drying after composting. The options vary in their control of exhaust gases and susceptibility to wet weather conditions. A site-specific analysis is required to determine the approach or combination of approaches most suited to a particular situation. It is important to note, however, that a high level of odor and dust control and a high resistance to wet weather can be provided if necessary. An all-weather facility is possible even under the worst of circumstances.

Options to pelletize the composter outfeed or the product after drying are included in the block diagram. Pelleting is effective in reducing head losses in reactor dryers using ambient air or slightly heated air. This can be important because gas volumes can be quite high when using ambient air. If conventional heat drying is used, pelleting is generally not necessary because gas volumes are greatly reduced. Pelleting is also effective as a dust control measure and should be considered, particularly if open-air drying or stockpiling is intended.

Exhaust gases from the composter and enclosed dryer (if used) should be collected, treated and dispersed. A number of gas deodorization techniques are available which range in their effectiveness for odor removal. The deodorization system must be designed in concert with the dispersal system. If a high level of gas dispersion is provided, little or no gas treatment may be necessary. Stack dispersion of collected gases should be considered in any compost facility design. Techniques for predicting subsequent dilution and ground-level concentrations of released gases are well developed. Therefore, a deodorization and stack dispersal system can be engineered to achieve stringent standards for ground-level odor concentrations. This is far better than allowing uncontrolled odor emissions at ground level.

It is within the present state-of-the-art to design and operate sludge compost facilities using a reasoned, engineering approach. Systems can be tailored to different feed characteristics, adjusted to the severity of weather conditions and designed to meet requirements for control of odors and dusts. However, actual operating experience with such systems is limited at the present time. Also, municipal agencies are generally reluctant to implement systems which lack a long established history of operation. Therefore, advanced sludge composting systems are presently in a difficult stage of development. If rational analysis of the process continues to expand and if a number of systems can be implemented incorporating various features of the process diagram in Figure 16-2, sludge composting should advance to the point where it is an accepted, engineered alternative for sludge management.

APPENDIX I: NOTATION

The following symbol notation has been adopted in this text. Where applicable, the units most commonly used are shown. A symbol may sometimes be used to define more than one variable. In such cases, the reader should refer to the appropriate section of the text to determine the context in which the symbol is used.

A = area perpendicular to direction of heat or mass transfer, cm^2

A_v = available surface area per unit volume

a = thermal diffusivity, cm^2/hr

a = number of free enzyme adsorption sites per unit volume

a_0 = total number of enzymes adsorption sites per unit volume

B = constant

C_i = concentration of gas in the liquid phase at the gas/liquid interface

C_l = concentration of gas in the liquid phase

c = pollutant concentration at coordinates x, y, z, g/m^3

$\bar{c}$ = average downwind odor concentration, ou

c_b = avarage odor concentration at the downwind edge of a large area odor source, ou

c_k = concentration estimate for the shorter sampling time, t_k

c_p = heat capacity at constant pressure per unit mass, cal/g-°C

c_s = concentration estimate for the desired sampling time, t_s

c_v = heat capacity at constant volume per unit mass, cal/g-°C

D = atmospheric dispersion coefficient, ft^2/min or m^2/min

D_f = diffusion coefficient in the biofilm, cm^2/sec

D_g = diffusion coefficient in the gas phase, cm^2/sec

D_l = diffusion coefficient in the liquid phase, cm^2/sec

D_l = dilutions achieved between the point source and the ground level, centerline of the plume

D_r = decimal reduction factor, time required to achieve a tenfold reduction in cell population, T_{90}

d = inside stack diameter, m

d_c = depth of sludge cake, cm

d_w = depth of water to be evaporated, cm

E = evaporation rate, cm/day

ΔE = change in internal energy within a system

E_a = activation energy, cal/mol or kcal/mol

E_d = inactivation energy, cal/mol or kcal/mol

e = number of free enzymes per unit volume of the reaction mixture

F = association factor for solvent, equal to 2.6 for water

F_a = configurational factor to account for the relative position and geometry of two heat radiating bodies

F_e = emissivity factor to account for nonblack-body radiation

f = free airspace, ratio of gas volume to total volume

f_b = free airspace within the interstices of a bulking agent before sludge addition

f_c = free airspace within the interstices of a sludge cake; usually assumed to be zero

f_h = fraction of compost material in the high temperature zone of a compost pile

f_l = fraction of compost material in the low temperature zone of a compost pile

f_m = free airspace within the interstices of a mixture of composting materials

ΔG = free energy change between two states of a system, kcal/mol or cal/mol

ΔG^0 = free energy change measured under standard state conditions, kcal/mol or cal/mol

G_a = mass flowrate of air, g air/g cake solids

G_f = specific gravity of the fixed or ash fraction of the total solids

G_m = specific gravity of mixture solids

ΔG_R = reaction free energy change based on activities of chemical constituents, cal/mol or kcal/mol

ΔG_R^0 = reaction free energy change under standard state conditions, cal/mol or kcal/mol

G_s = specific gravity of the total solids

G_v = specific gravity of the volatile fraction of the total solids

g = acceleration of gravity, 980 cm/sec^2

H = Henry's law constant, atm/mol fraction

H = effective point source emission height, m or ft

H = barrier height, m or ft

ΔH = height of rise of a plume above the discharge stack, m or ft

ΔH = enthalpy change between two states of a system, cal/mol or kcal/mol

ΔH^0 = enthalpy change measured under standard state conditions, cal/mol or kcal/mol

ΔH_{fg} = enthalpy change from liquid to vapor at temperature TNEW, kcal/kg

h = average puff height over an odor emission area, m or ft

h = point source stack height, m or ft

h_b = puff height downwind of a barrier, m or ft

K_{eq} = equilibrium constant

K_s = half-velocity coefficient, also referred to as the Michaelis-Menten coefficient, mass/volume

K_x = half-velocity coefficient equal to the microbial concentration where $ds/dt = k/2$

k = reaction rate constant, $time^{-1}$

k = thermal conductivity, cal/hr-cm^2-°C/cm

k = maximum rate of solid substrate hydrolysis which occurs at high microbial concentration

k_a = fraction of amendment volatile solids degradable under composting conditions

k_b = fraction of bulking agent volatile solids degradable under composting conditions

k_c = fraction of sludge cake volatile solids degradable under composting conditions

k_d = thermal inactivation or death coefficient, time^{-1}

k_d = rate constant, time^{-1}

k_{da} = actual rate constant for sludge, amendment and recycled organics corrected for effects of moisture, free airspace and oxygen content, time^{-1}

k_{da}^b = actual rate constant for bulking agent organics corrected for effects of moisture, free airspace and oxygen content, time^{-1}

k_{dm} = maximum rate constant uncorrected for effects of moisture, free airspace and oxygen content, time^{-1}

k_e = endogenous respiration coefficient, mass of microbes respired/mass of microbes-time

k_m = maximum utilization coefficient, maximum rate of substrate utilization at high substrate concentration, mass substrate/mass microbes-day

k_r = fraction of the compost recycle volatile solids degradable under composting conditions

k_s = average surface odor emission rate, SOER, m^3/min-m^2 or ft^3/min-ft^2

L = odor emission source length, m or ft

L = length of pore tubes which equals the height of the composting mixture; used in pore tube model

M = moisture content, percent of total solids

M = solvent molecular weight

M_{bc} = volumetric mixing ratio, ratio of volume of bulking agent to volume of sludge cake

M_{mb} = volume ratio of mixed materials to bulking agent

m = mass, kg

m_a = mass of dry gas

m_v = mass of water vapor

N = number of pile turnings

n = porosity, ratio of void volume to total volume

n = viable cell population

n_f = final viable cell population

n_0 = initial viable cell population

n_t = viable cell population after time, t

P = absolute pressure

P = precipitation rate, cm/day

P_c = percent of inorganic conditioning chemicals in sludge

P_g = partial pressure of gas in the gas phase

P_i = partial pressure of gas in the gas phase at the gas/liquid interface

$P_0(t)$ = extinction probability, probability that all organisms are inactivated

P_v = percent volatile solids in sludge

p = pressure

p = weight of water which must be evaporated to reach the desired cake solids per weight of original sludge cake, g water/g sludge cake

Q = heat of combustion or fuel value, Btu/lb or cal/g

Q = ventilation rate, $cm^3/sec\text{-}m^2$

Q_{air} = volumetric flowrate of air moving across the odor source, m^3/min or ft^3/min

Q_0 = odor input rate, m^3/min or ft^3/min

q = heat flow into (+) or out of (−) a system

q_a = heat to air, cal or kcal

q_p = heat flow in a constant pressure system, cal or kcal

q_s = heat flow to solids, cal or kcal

q_v = evaporative heat burden, cal or kcal

q_v = heat flow in a constant volume system, cal or kcal

q_w = heat flow to water, cal or kcal

R = universal gas constant, 1.99 cal/deg-mol

R = universal gas constant, 287 J/kg-°K for air, 461.6 J/kg-°K for water vapor

R = radius of a pore tube used in the pore tube model

R = radius of a spherical particle

R_d = dry weight recycle ratio, ratio of dry weight of compost product recycled to dry weight of sludge cake

R_w = wet weight recycle ratio, ratio of wet weight of compost recycled to total wet weight of sludge cake

r = radius of a composting particle, used in pore tube model

S = mass emission rate of pollutant, g/sec

S = surface area of pore tubes which equals the surface area of particles in the composting mixture; used in pore tube model

S = concentration of the rate limiting substrate, mass/volume

ΔS = entropy change between two states of a system

S_a = fractional solids content of amendment

S_b = fractional solids content of bulking agent

S_{bm} = fractional solids content of bulking agent in sludge/bulking agent mixture after moisture absorption

S_{bm}^m = minimum fractional solids content of bulking agent achievable by absorption of moisture from sludge to bulking agent

S_c = fractional solids content of dewatered sludge cake

S_{cm} = fractional solids content of sludge in a sludge/bulking agent mixture after moisture absorption

S_{cm}^m = maximum fractional solids content of sludge achievable by absorption of moisture from sludge to bulking agent

T = temperature

ΔT = temperature change

T_a = absolute temperature, °K

T_a = air temperature, °K or °C

T_K = temperature, °K

T_s = stack gas temperature, °K

t = time

t_a = time required for air drying, days

t_{90} = time required to achieve a tenfold reduction in cell population

U = overall heat transfer coefficient which includes effects of both conductive and convective heat transfer, cal/hr-cm^2-°C

u = average wind velocity over a large area odor source, m/min or ft/min

u = mean wind velocity effecting plume transport, m/sec

V = volume of a thermodynamic system

V = volume of reactor, m^3

V = total volume of pore tubes which equals the void volume in the composting mixture; used in pore tube model

V = volume, m^3

V = stack gas volume flow rate, m^3/sec

V_a = volatile solids content of amendment, fraction of dry solids

V_b = volatile solids content of bulking agent, fraction of dry solids

V_c = volatile solids content of sludge cake, fraction of dry solids

V_m = volatile solids content of mixture, fraction of dry solids

V_0 = solute molal volume at normal boiling point, equal to 25.6 cm^3/g mol for oxygen

V_r = volatile solids content of compost product and recycle, fraction of dry solids

V_s = exit velocity of stack gas, m/sec

V_s = volatile fraction of the total solids

v = average fluid velocity in a pore tube; used in pore tube model

v = rate of product formation in an enzyme catalyzed reaction

v_g = volume of gas phase

v_s = volume of solids

v_t = total volume of solids, water and gas in a composting matrix

v_w = volume of water phase

W = water to degradable organic ratio, weight of water to weight of degradable organic in a composting mixture

W = odor emission source width, m or ft

W_s = weight of dry solids

W_w = weight of water

w = work done on (−) or by (+) a system

w = specific humidity, mass of water vapor per mass of dry gas

w_{O_2} = rate of oxygen consumption, mg O_2/g VS-hr

X = concentration of microbes, mass/volume

X_a = total wet weight of organic amendment, other than sludge cake or compost recycle, added to mixture per day

X_B = virtual upwind distance to equivalent point source with an odor emission rate Q_o

X_b = total wet weight of bulking agent added to mixture per day

X_c = total wet weight of dewatered sludge cake produced per day

X_g = equilibrium mole fraction of dissolved gas in an aqueous solution

X_m = total wet weight of mixed material entering the compost process per day

X_p = total wet weight of compost produced per day

X_r = total wet weight of compost recycled per day

x = downwind distance measured from the virtual point source

Y_m = growth yield coefficient, mass of microbes/mass of substrate

y = lateral spread of the plume, m or ft

z = particle thickness

γ = specific weight of fluid

γ_b = unit bulk weight of bulking agent, wet weight per volume, g/cm^3

γ_c = bulk weight of sludge cake, g/cm^3

γ_c = unit bulk weight, total wet weight per unit volume, g/cm^3

$\gamma_c(dry)$ = unit dry weight, dry weight per unit volume, g/cm^3

γ_g = thickness of laminar gas film as used in the two-film model of gas transfer

γ_1 = thickness of laminar liquid film as used in the two-film model of gas transfer

γ_m = unit bulk weight of the mixed material to be composted, wet weight per volume, g/cm^3

γ_w = bulk weight of water, g/cm^3

η = solvent viscosity, cP

η = specific oxygen supply, g O_2/g BVS in the feed mixture

θ = detention time, days

θ = temperature coefficient for a chemical or biochemical reaction

λ = specific air supply, m^3/min-dry ton/day of sludge feed in the CFCM reactor, m^3/min-dry ton sludge feed in the BATCH model

μ = net specific growth rate, g cells grown/g cells-day

μ = fluid viscosity, g/cm-sec or N-sec/m^2

μ_m = maximum net specific growth rate, g cells grown/g cells-day

ρ = mass density, g/cm^3

σ = Stefan-Boltzmann constant, 4.87×10^{-8} kcal/hr-m^2-°K^4

σ_y = standard deviation of plume concentration in the horizontal direction, m

σ_z = standard deviation of plume concentration in the vertical direction, m

Φ = modified dispersion coefficient

APPENDIX II: NOMENCLATURE USED IN SIMULATION MODELS

MASS TERMS

Except where noted mass terms have units of kg/day in the CFCM model and kg in the BATCH model.

ASHA = ash or inert solids contributed by the amendment

ASHB = ash contributed by the bulking agent

ASHC = ash contributed from the sludge cake

ASHR = ash contributed from recycle material

BVSA = biodegradable volatile solids (BVS) contributed by the amendment

BVSB = BVS contributed by the bulking agent

BVSC = BVS contributed from the sludge cake

BVSR = BVS contributed from recycle material

BVSO = sum of BVSAO, BVSBO, BVSCO and BVSRO

BVSAO = amendment BVS remaining in the output from the CFCM model or in the output from a BATCH model time step

BVSBO = bulking agent BVS remaining in the output

BVSCO = sludge cake BVS remaining in the output

BVSRO = recycle BVS remaining in the output

DAIRI = input dry air

DBVSA = amendment BVS oxidized in the reactor or time step

DBVSB = bulking agent BVS oxidized in the reactor or time step

DBVSC = sludge cake BVS oxidized in the reactor or time step

DBVSR = recycle BVS oxidized in the reactor or time step

DELBVS = sum of DBVSA, DBVSB, DBVSC and DBVSR

DGASO = output dry gases

DWTIN = input dry weight to time step I; equal to output dry weight from time step I − 1

DWTOUT = output dry weight from time step I; equal to input dry weight to time step I + 1

EVAP = evaporation rate, cm/day

FAS = free airspace, fraction of total volume occupied by air voids

F1 = correction factor applied to the rate constant to account for moisture rate limitations; also used to adjust water vapor pressure in outfeed gas from saturation conditions

F2 = correction factor applied to rate constant to account for rate limitations from insufficient free airspace

F3 = factor to account for relative exposure of compost to ambient conditions during turning

F4 = factor to account for relative exposure of compost to ambient conditions after pile turning

F5 = factor to correct measured pan evaporation rates to evaporation rates from a compost surface

FO2 = correction factor applied to the rate constant to account for the rate limitations from insufficient oxygen in the free airspace

GAMMAB = bulk weight of bulking agent, g/cm^3

GAMMAM = bulk weight of mixed materials, g/cm^3

KA = amendment degradability coefficient, fraction of total volatile solids that are degradable under composting conditions

KB = bulking agent degradability coefficient, fraction of total volatile solids that are degradable under composting conditions

KC = sludge cake degradability coefficient, fraction of total volatile solids that are degradable under composting conditions

NBVSA = nonbiodegradable volatile solids (NBVS) contributed by the amendment

NBVSB = NBVS contributed by the bulking agent

NBVSC = NBVS contributed by the sludge cake

NBVSO = sum of NBVSA, NBVSB, NBVSC and NBVSR

NBVSR = NBVS contributed by the product recycle

PV = actual water vapor pressure in infeed gas, mm Hg

PVS = saturation water vapor pressure in infeed gas, mm Hg

PVO = actual water vapor pressure in outfeed gas, mm Hg

PVSO = saturation water vapor pressure in outfeed gas, mm Hg

QAIR = air supply flowrate, m^3/min

RAIN = rainfall intensity in time step DELTAT, cm/day (BATCH model only)

RATEK = rate constant for sludge cake, amendments and product recycle components adjusted by correction factors F1, F2 and FO2, day^{-1}

RATEKB = rate constant for bulking agent adjusted by correction factor F1, F2 and FO2, day^{-1}

RATEKM = maximum rate constant for compost decomposition as a function of temperature, day^{-1}

RCYCF1 = factor to account for change in dry gas weight from organic decomposition of recycle material, based on a weighted average of sludge cake and amendment components in the outfeed material

RCYCF2 = factor to account for biologically produced water as a result of organic decomposition of recycle material, based on a weighted average of sludge cake and amendment components in the outfeed material

RCYCF3 = factor to account for oxygen consumption as a result of organic decomposition of recycle material, based on a weighted average of sludge cake and amendment components in outfeed material

RCYCF4 = factor to account for carbon dioxide production from organic decomposition of recycle material, based on a weighted average of sludge cake and amendment components in outfeed material

RHAIR = relative humidity of inlet air, fraction of saturation vapor pressure

SA = solids content of amendment; fraction of total wet weight

SB = solids content of bulking agent, fraction of total wet weight

SC = solids content of sludge cake, fraction of total wet weight

SM = solids content of mixed material, fraction of total wet weight

SMOUT = solids content of outfeed material, fraction of total wet weight

SMNEW = new fractional solids content calculated during iterative loop; final value put into SMOUT or SM once iterative procedure is completed

SR = solids content of product recycle, fraction of total wet weight

SURFWT = wet weight of material in enclosed reactor composter subject to surface heat loss, kg (CFCM model only)

VA = amendment volatile solids, fraction of total dry solids

VB = bulking agent volatile solids, fraction of total dry solids

VC = sludge cake volatile solids, fraction of total dry solids

VLPCO2 = volume percentage of carbon dioxide in the outfeed gas

VOLPN2 = volume percentage of nitrogen in the outfeed gas

VOLPO2 = volume percentage of oxygen in the outfeed gas

VR = recycle volatile solids, fraction of total dry solids

VRNEW = volatile solids of recycle component, fraction of total dry solids

VROUT = volatile solids of recycle component, fraction of total dry solids

VSO = sum of BVSO and NBVSO

WATA = water input from the amendment

WATB = water input from bulking agent

WATC = water input from sludge cake

WATP = water produced from organic decomposition

WATR = water input from product recycle

WATRN = water supplied by rain in time step DELTAT, kg (BATCH model only)

WATSI = water associated with solids and input to time step I; equal to output water from time step I − 1

WATSO = output water associated with solids from CFCM reactor or from time step in BATCH model

WATVI = input water vapor

WATVO = output water vapor

WATEVP = water removed by surface evaporation

WEIGHT = wet weight of composting material, kg

XA = input wet weight of amendment

XB = input wet weight of bulking agent

XC = input wet weight of sludge cake

XR = input wet weight of recycle material

ENERGY AND RELATED TERMS

Except where noted, energy terms have units of kcal/day in the CFCM model and kcal in the BATCH model.

CPGAS = specific heat of dry gases, usually 0.24 cal/g-°C

CPSOL = specific heat of solids, usually 0.25 cal/g-°C

CPWAT = specific heat of water, usually 1.00 cal/g-°C

CPWATV = specific heat of water vapor, usually 0.444 cal/g-°C

DELHFG = enthalpy change from liquid to vapor at temperature TNEW, kcal/kg

HA = heat of combustion of amendment, cal/g

HAIRI = heat content of input dry air

HASI = heat content of infeed amendment solids

HAWI = heat content of infeed amendment water

HB = heat of combustion of bulking agent, cal/g

HBSI = heat content of infeed bulking agent solids

HBWI = heat content of infeed bulking agent water

HC = heat of combustion of sludge cake, cal/g

HCOND = heat loss by conduction from a surface not exposed to ambient air

HCSI = heat content of infeed sludge cake solids

HCWI = heat content of infeed sludge cake water

HGO = heat content of output dry gases

HLWVO = latent heat of output water vapor less the latent heat of the input water vapor

HMIX = mixing energy input

HORGA = heat release from amendment decomposition

HORGB = heat release from bulking agent decomposition

HORGC = heat release from sludge cake decomposition

HORGR = heat release from product recycle decomposition

HRAIN = heat input from rainfall or snowfall

HRSDAY = hours per day in which the agitator and/or aeration equipment is operated

HRSI = heat content of infeed product recycle solids

HRWI = heat content of infeed product recycle water

HSO = heat content of the output solids

HSURF = heat loss from surface after windrow or pile turning

HSWEV = sensible heat in surface evaporated water

HSWVI = sensible heat content of input water vapor

HSWVO = sensible heat in output water vapor

HTOTI = total input energy, summation of input energy terms

HTOTO = total output energy, summation of output energy terms

HTURN = heat loss in turning a windrow or pile

HWO = heat content of output water associated with output solids, WATSO

MIXPW = agitator and/or aerator power, kW

RAINTP = average rainfall temperature in time step DELTAT, °C

TA = temperature of amendment infeed, °C

TAIR = temperature of input air, °C

TB = temperature of bulking agent infeed, °C

TC = temperature of sludge cake infeed, °C

TIN = temperature of infeed components to time step I, °C; equal to temperature of outfeed from time step I − 1

TNEW = temperature of outfeed solids, water and gases, °C

TR = temperature of product recycle infeed, °C

TREF = reference temperature used in energy balance, usually taken as 0.0°C

TURNUM = number of windrow or pile turnings per day

U = overall heat transfer coefficients, kcal/m^2-hr-°C

TIME AND DIMENSION TERMS

AREACL = surface area over which conductive heat transfer occurs, m^2

AREAEV = area of windrow or pile subject to surface evaporation, m^2

AREARN = area of windrow or pile subject to rainfall, m^2

AREASL = area of windrow or pile subject to surface losses with ambient air, m^2

AREAXS = cross-sectional area of windrow or pile, m^2

DELTAT = time step used in BATCH model, days

DEPTH = depth of pile surface influenced by ambient convective currents, assumed to be about 0.20 m

DETTIM = detection time in CFCM reactor or total compost cycle time in BATCH model

DIMCL = characteristic dimension for conductive heat transfer, m

DIMSL = characteristic dimension for surface losses with ambient air, m

HEIGHT = height to which compost is placed in CFCM reactor, m

LENGTH = length of windrow or pile, m

MAXT = number of last iteration step in BATCH model, (DETTIM/DELTAT) + 1

VLREAC = volume of enclosed reactor composter, m^3

VOLUME = volume of pile or windrow, m^3

WIDTH = maximum width of pile, m

APPENDIX III: ANALYSIS OF NATURAL VENTILATION USING A PORE TUBE MODEL

The pore tube model was used in Chapter 8 to estimate rates of oxygen supply from natural ventilation as a result of density differences between hot, moist gases within an actively composting mass and outside ambient air. Uniformly sized pore tubes replaced the random pore spaces in an actual composting mixture but retained the same total void volume and surface area. Composting solids are assumed to be discrete particles and the void volume (porosity) is assumed to equal the free airspace (FAS).

For a cylindrical pore tube the following equations hold:

$$V = \pi R^2 L \tag{1}$$

$$S = 2\pi R L \tag{2}$$

$$R = 2V/S \tag{3}$$

where V = total volume of pore tubes which equals the FAS space in the composting mixture
R = radius of pore tube
S = surface area of pore tubes which equals the surface area of particles in the composting mixture
L = height of pore tubes which equals the height of the composting mixture

Considering spherical particles of radius r and randomly packed with a FAS f, the following equations can be developed:

$$\text{no. of particles} = \frac{3(1-f)}{4\pi r^3} \tag{4}$$

$$\text{area of particles} = \frac{3(1-f)}{4\pi r^3}(4\pi r^2) = \frac{3(1-f)}{r} \tag{5}$$

$$\text{radius of pore tube} = R = \frac{2V}{S} = \frac{2}{3}\left(\frac{f}{1-f}\right)r \tag{6}$$

The number of pore tubes per m^2 of mixture N is given by:

$$N = \frac{\text{pore volume}}{\text{volume/tube}} = \frac{f\,(10^6)}{\pi R^2\,(100)}$$

$$N = 10^4 \frac{f}{\pi R^2} = \text{number/m}^2 \tag{7}$$

where R = pore tube radius (cm).

The velocity of flow in the pore tube is determined from the Hagen-Poiseuille relationship:

$$v = \frac{\Delta\rho g R^2}{8\mu} \tag{8}$$

$$Q = Av = \pi R^2 N v = \pi R^2 \frac{f\,(10^4)}{\pi R^2} \frac{\Delta\rho g R^2}{8\mu} \tag{9}$$

where Q = ventilation rate
v = average flow velocity in pore tube
$\Delta\rho$ = density difference between fluid inside and outside the pore tube
g = acceleration of gravity
μ = fluid viscosity

Substituting Equation 6 for R^2:

$$Q = 1250\, f \left(\frac{f}{1-f}\right)^2 \frac{\Delta\rho g}{\mu} \left(\frac{2}{3}\right)^2 r^2 \tag{10}$$

In the temperature range under consideration μ is about 2×10^{-4} g/cm-sec (2×10^{-5} N-sec/m^2) and g = 980 cm/sec^2. Substituting and dividing by 1000 to convert $\Delta\rho$ to units of g/1:

$$Q = 2.72 \times 10^6\, f \left(\frac{f}{1-f}\right)^2 \Delta\rho r^2 \tag{11}$$

where Q = ventilation rate, cm^3/sec-m^2
$\Delta\rho$ = density difference, g/1
r = particle radius, cm
f = FAS, fraction of total mixture volume

Equation 11 was used to construct the curves shown in Figure 8-9.

APPENDIX IV: SAMPLE OUTPUT FROM THE CFCM SIMULATION MODEL

The following sample output is for the case of 10 dry ton/day of raw sludge at 20% cake solids composted in a CFCM reactor. Product recycle is used for feed conditioning, and it is assumed that no further organic stabilization occurs after material leaves the reactor (VR = VROUT). Specific air supply is 20 m^3/min-dry ton/day and all infeed temperatures are 20°C. A steady-state reactor temperature of 75.4°C and output solids content of 63% are predicted by the model. Initial problem conditions and steady state mass are energy balances are included in the program output.

SIMULATION MODEL

ENCLOSED REACTOR COMPOSTING OF SLUDGE, AMENDMENT, RECYCLE MIXTURES

COMPLETE MIX REACTOR

INITIAL PROBLEM CONDITIONS

	TOTAL WEIGHT KILOGRAMS/DAY	SOLIDS CONTENT	VOLATILE SOLIDS FRACTION	DEGRADABILITY COEFFICIENT	INPUT TEMPERATURE
SLUDGE	50000.00	0.2000	0.7000	0.7857	20.00
AMENDMENT	0.0	0.0	0.0	0.0	0.0
RECYCLE	50000.00	0.7000			20.00
MIXTURE	100000.00	0.4500			20.00

```
DETENTION TIME =          7.00  DAYS
HEIGHT OF COMPOST =          2.00  METERS
HEAT TRANSFER COEFFICIENT =        1.5000  KCAL/M2-HR-C

AIR FLOWRATE =      200.00  CU. METERS PER MIN
AIR TEMPERATURE =          20.00  DEGREES C
RELATIVE HUMIDITY =          0.50
REFERENCE TEMPERATURE =         0.0   DEGREES C

AGITATOR POWER =        75.00  KILOWATTS
AGITATOR OPERATION =          6.00  HRS/DAY

SPECIFIC HEAT OF SOLIDS =       0.2500  CAL/GM-C
SPECIFIC HEAT OF WATER =     1.0000  CAL/GM-C
SPECIFIC HEAT OF DRY GASES =       0.2400  CAL/GM-C
SPECIFIC HEAT OF WATER VAPOR =       0.4440  CAL/GM-C

SPECIFIC GRAVITY OF VOLATILES =       1.0000
SPECIFIC GRAVITY OF ASH =       2.5000

HEAT OF COMBUSTION OF SLUDGE ORGANICS =     6766.00  CAL/GM OF ORGANIC OXIDIZED
HEAT OF COMBUSTION OF AMENDMENT ORGANICS =     3739.00  CAL/GM OF ORGANIC OXIDIZED

SOLIDS CONTENT OF INFEED MIXTURE =       0.4500

HEAT LOSS TURNING FACTOR F3        0.0
HEAT LOSS SURFACE FACTOR F4        0.0
PILE TURNINGS PER DAY      10.00
```

PRINTOUT OF INTERMEDIATE VALUES

ITERATION NUMBER	TEMPERATURE DEGREES C	HEAT IN KCAL/DAY	HEAT OUT KCAL/DAY	SMOUT	KNT3
5	25.4226	15456956.	6440182.	0.45366	5
10	33.3622	20894144.	9999789.	0.46517	12
15	42.4100	27048640.	15160333.	0.49106	18
20	51.4834	31344320.	21084624.	0.53442	17
25	59.2986	33250016.	25512496.	0.57411	10
30	65.6316	33843152.	28174800.	0.59875	6
35	70.5143	33609504.	29921424.	0.61571	5
40	73.6390	32670080.	30897792.	0.62621	4
45	74.9936	31775344.	31266752.	0.63084	1
50	75.3544	31449424.	31375280.	0.63243	3
55	75.3952	31379200.	31303136.	0.63118	3
60	75.4029	31401104.	31387840.	0.63262	3
65	75.4106	31366272.	31314352.	0.63137	3
70	75.4183	31382976.	31384624.	0.63255	3

MASS BALANCE

	DEGRADABLE VOLATILE SOLIDS KGM/DAY	NONDEGRAD VOLATILE SOLIDS KGM/DAY	ASH KGM/DAY	DRY WEIGHT KGM/DAY	WATER KGM/DAY	TOTAL WEIGHT KGM/DAY
(MASS INPUTS)						
SLUDGE	5499.89	1500.10	3000.00	10000.00	40000.00	50000.00
AMENDMENT	0.0	0.0	0.0	0.0	0.0	0.0
RECYCLE	7919.01	9027.38	18053.61	35000.00	15000.00	50000.00
AIR FEED				341790.69	2561.60	344352.25
SUM INPUTS	13418.91	10527.48	21053.60	386790.63	57561.61	444352.25
(MASS OUTPUTS)						
OUTFEED	9235.04	10527.48	21053.60	40816.12	23709.73	64525.84
GAS OUT				342977.13	36849.28	379826.38
SUM OUTPUT	9235.04	10527.48	21053.60	383793.19	60559.01	444352.19

```
VOLATILE FRACTION OF INFEED MIXTURE          0.5321
VOLATILE FRACTION OF OUTFEED SOLIDS          0.4842
SOLIDS CONTENT OF OUTFEED SOLIDS             0.6326
REQUIRED REACTOR VOLUME                      714.34   CUBIC METERS
REACTOR AREA CONDUCTIVE HEAT LOSS            491.16   SQ METERS
REACTOR AREA SURFACE HEAT LOSS               357.17   SQ METERS
BULK WEIGHT OF OUTFEED                       0.6330   GMS/CC
INITIAL O2 CONSUMPTION RATE, 25 C              27.05   MG O2/GM VS-DAY
FINAL O2 CONSUMPTION RATE, 25 C                16.09   MG O2/GM VS-DAY
SPECIFIC OXYGEN SUPPLY                       5.8583   GM O2/GM MIXED FEED BVS


OUTPUT GAS CHARACTERISTICS - PERCENT BY VOLUME, DRY BASIS
        NITROGEN             79.60
        OXYGEN               18.60
        CARBON DIOXIDE        1.80
        WATER VAPOR PARTIAL PRESSURE =   111.9408   MM HG


                 ENERGY BALANCE
     REFERENCE TEMPERATURE =        0.0    DEGREES C
TEMPERATURE OF OUTPUT SOLIDS AND GASES =    75.4183  DEGREES C
 RATE CONSTANT FOR OUTPUT TEMPERATURE =   0.064714  PER DAY
  ADJUSTMENT FACTOR FOR MOISTURE CONTENT,F1 =       0.36
   ADJUSTMENT FACTOR FOR FREE AIR SPACE, F2 =       0.97
```

```
ENERGY INPUTS        SLUDGE - SENSIBLE HEAT IN SOLIDS                49999.98  KCAL/DAY
                     SLUDGE - SENSIBLE HEAT IN WATER                800000.06
                     AMENDMENT - SENSIBLE HEAT IN SOLIDS                 0.0
                     AMENDMENT - SENSIBLE HEAT IN WATER                  0.0
                     RECYCLE - SENSIBLE HEAT IN SOLIDS              174999.88
                     RECYCLE - SENSIBLE HEAT IN WATER               300000.06
                     DRY AIR INPUT - SENSIBLE HEAT                 1640593.00
                     WATER VAPOR INPUT - SENSIBLE HEAT               22747.00
                     SLUDGE ORGANIC DECOMPOSITION                 11601646.00
                     AMENDMENT ORGANIC DECOMPOSITION                     0.0
                     RECYCLE ORGANIC DECOMPOSITION                16706371.00
                     AGITATION ENERGY                               386632.69
                     TOTAL INPUT ENERGY                           31382976.00

ENERGY OUTPUTS       MIXED OUTFEED SOLIDS - SENSIBLE HEAT           769570.56  KCAL/DAY
                     OUTFEED WATER - SENSIBLE HEAT                 1788147.00
                     OUTPUT DRY GASES - SENSIBLE HEAT              6208016.00
                     WATER VAPOR OUTPUT - SENSIBLE HEAT            2671696.00
                     HEAT OF VAPORIZATION                         18967312.00
                     CONDUCTIVE HEAT LOSS TO SURROUNDINGS           979892.19
                     HEAT LOSS FROM SURFACE                              0.0
                     HEAT LOSS FROM PILE TURNING                         0.0
                     TOTAL OUTPUT ENERGY                          31384624.00

   PROGRAM RUN INFORMATION
   KNT1 =     70
   KNT2 = 11255

   KNT3 =    423
```

REFERENCES

1. Klass, D. L. "Wastes and Biomass as Energy Resources: An Overview," Symposium papers Clean Fuels from Biomass, Sewage, Urban Refuse, Agricultural Wastes, Institute of Gas Technology, Chicago, IL (1976).
2. Suler, D. "Composting Hazardous Industrial Wastes," *Compost Sci.* (July/August 1979).
3. "Multimedium Management of Municipal Sludge," National Academy of Science, Committee on Multimedium Approach to Municipal Sludge Management (1978).
4. Meek, B., L. Chesnin, W. Fuller, R. Miller and D. Turner. "Guidelines for Manure Use and Disposal in the Western Region, USA," Bulletin 814, College of Agriculture Research Center, Washington State University (1975).
5. General Electric Co. *Solid Waste Management Technology Assessment* (New York: Van Nostrand Reinhold Co., 1975).
6. Briedenbach, A. W. "Composting of Municipal Solid Wastes in the United States," U.S. EPA, Washington, DC, SW-47R (1971).
7. Spohn, E. "Recent Developments in Composting of Municipal Wastes in Germany," *Compost Sci.* (March/April 1977).
8. Compton, C. R., and F. R. Bowerman. "Composting Operation in Los Angeles County," *Compost Sci.* (Winter 1961).
9. Horvath, R. W. "Operating and Design Criteria for Windrow Composting of Sludge," *Proceedings of the National Conference on Design of Municipal Sludge Composting Facilities* (Rockville, MD: Information Transfer, 1978).
10. Robbins, M. H. "Solids Disposal at the Upper Occoquan Sewage Authority," *Proceedings of the National Conference on Municipal and Industrial Sludge Composting-Materials Handling* (Rockville, MD: Information Transfer, 1979).
11. Toups and Loiderman, Engineers. "Composting Site Evaluation and Preliminary Design Report for Montgomery County, Maryland, Composting Facility," report to Washington Suburban Sanitary Commission (1977).
12. Colacicco, D., E. Epstein, G. B. Willson, J. F. Parr and L. A. Christensen. "Costs of Sludge Composting," Agricultural Research Service, USDA, ARS-NE-79 (1977).

13. Epstein, E., G. B. Willson, W. D. Burge, D. C. Mullen and N. Enkiri. "A Forced Aeration System for Composting Wastewater Sludge," *J. Water Poll. Control Fed.* 48:4 (1976).
14. Willson, G. B. "Equipment for Composting Sewage Sludge in Windrows and in Piles," *Proceedings of the 1977 National Conference on Composting of Municipal Residues and Sludges* (Rockville, MD: Information Transfer, 1977).
15. Ettlich, W. F., and A. E. Lewis. "A Study of Forced Aeration Composting of Wastewater Sludge," U.S. EPA Municipal Environmental Research Laboratory, Cincinnati, OH, EPA-600/2-78-057 (1978).
16. Willson, G. B., E. Epstein and J. R. Parr. "Recent Advances in Compost Technology," *Proceedings of the Third National Conference on Sludge Management, Disposal and Utilization* (Rockville, MD: Information Transfer, 1977).
17. Iacoboni, M., T. LeBrun and D. L. Smith. "Composting Study," Technical Services Dept., Los Angeles County Sanitation Districts (September-December 1977).
18. "Washington Suburban Sanitary Commission Compost Facility–Site 2 Proposed Design Criteria and Plans," Toups and Loiederman Engineers (1978).
19. Houser, J. E., Fairfield Engineering Co., Marion, OH. "Test Report for the Reduction of Organic Matter in Municipal Waste Water Under Aerobic Conditions in the Thermophilic Phase by the Fairfield Digester System at Altoona, PA," Unpublished.
20. Houser, J. E., Kramer, Comer, Passe, Racher Engineers. Personal communication (1979).
21. Senn, C. L. "Dairy Waste Management Project–Final Report," University of California Agricultural Extension Service, Public Health Foundation of Los Angeles County (1971).
22. LeBrun, T., M. Iacoboni and J. Livingston. "Composting Study," Technical Services Dept., Los Angeles County Sanitation Districts (1978).
23. Iacoboni, M. "Summary Report–Fermentechnik SR-10 Composter Study," LA/OMA Sludge Project (1980).
24. Lehninger, A. L. *Bioenergetics* (New York: W. A. Benjamin, Inc., 1965).
25. Perry, R. H., and C. H. Chilton, Eds. *Chemical Engineer's Handbook*, 5th ed. (New York: McGraw-Hill Book Co., 1973).
26. *Handbook of Chemistry and Physics*, 50th ed. (Cleveland, OH: CRC Press, 1970).
27. Lehninger, A. L. *Biochemistry* (New York: Worth Publishers, Inc., 1970).
28. Sawyer, C. N., and P. L. McCarty. *Chemistry for Environmental Engineering* (New York: McGraw-Hill Book Co., 1978).
29. Burd, R. S. "A Study of Sludge Handling and Disposal," report to FWPCA, Dept. of Interior, by Dow Chemical Co., Pub. WP-20-4 (1968).

30. Olexsey, R. A. "Thermal Degradation of Sludges," paper presented at Symposium on Pretreatment and Ultimate Disposal of Wastewater Solids, Rutgers University, May 1974.
31. Bailey, J. E., and D. F. Ollis. *Biochemical Engineering Fundamentals* (San Francisco, CA: McGraw-Hill, 1977).
32. Asimov, I. *Life and Energy* (New York: Avon Books, 1962).
33. Poindexter, J. S. *Microbiology–An Introduction to Protists* (New York: The Macmillan Co., 1971).
34. Lee, J. F., and F. W. Sears. *Thermodynamics* (Reading, MA: Addison-Wesley Publishing Co., Inc., 1963).
35. Fair, G. M., J. C. Geyer and D. A. Okun. *Water and Wastewater Engineering, Vol. 2, Water Purification and Wastewater Treatment and Disposal* (New York: John Wiley & Sons, Inc., 1968).
36. Spoehr, H. A., and H. W. Milner. "The Chemical Composition of Chlorella; Effect of Environmental Conditions," *Plant Physiol.* 24:120 (1949).
37. Corey, R. C., Ed. *Principles and Practices of Incineration* (New York: Wiley-Interscience, 1969).
38. Loran, B. I. "Burn that Sludge," *Water Wastes Eng.* (October 1975).
39. McCarty, P. L. "Anaerobic Treatment of Soluble Wastes," in *Advances in Water Quality Improvement* (Austin: University of Texas Press, 1968).
40. Buffaloe, N. D., and D. V. Ferguson. *Microbiology* (Boston, MA: Houghton Mifflin Co., 1976).
41. *Bergey's Manual of Determinative Bacteriology*, 7th ed.
42. Peppler, H. J., Ed. *Microbial Technology* (New York: Reinhold Publishing Corporation, 1967), p. 421.
43. Aiba, S., A. E. Humphrey and N. F. Millis, Eds. *Biochemical Engineering*, 2nd ed. (New York: Academic Press, Inc., 1973), p. 29.
44. Golueke, C. G. *Biological Reclamation of Solid Waste* (Emmaus, PA: Rodale Press, 1977).
45. Poincelot, R. P. "The Biochemistry of Composting," in *Composting of Municipal Residues and Sludges, Proceedings of the 1977 National Conference* (Rockville, MD: Information Transfer, 1977).
46. McKinney, R. E. *Microbiology for Sanitary Engineers* (New York: McGraw-Hill Book Co., Inc., 1962).
47. Alexopoulos, C. J. *Introductory Mycology* 2nd ed. (New York: John Wiley & Sons, Inc., 1962).
48. Frobisher, M., R. D. Hinsdill, K. T. Crabtree and C. R. Goodheart. *Fundamentals of Microbiology* (Philadelphia: W. B. Saunders Co., 1974).
49. Kane, B. E., and J. T. Mullins. "Thermophilic Fungi in a Municipal Waste Compost System," *Mycologia* 65:1087-1100 (1973).
50. Love, G. J., E. Tompkins and W. A. Galke. "Potential Health Impacts of Sludge Disposal on the Land," paper presented at 2nd National Conference on Sludge Management and Disposal, Anaheim, CA, 1975.

51. "Intestinal Parasite Surveillance, Annual Summary 1976," (Atlanta, GA: Center for Disease Control, 1977).
52. Emmons, C. W. "Natural Occurrence of Opportunistic Fungi," *Lab. Invest.* 11(2):1026-1032 (1962).
53. Millner, P. D., P. B. Marsh, R. B. Snowden and J. F. Parr. "Occurrence of Aspergillus fumigatus During Composting of Sewage Sludge," *Appl. Environ. Microbiol.* 34(6) (1977).
54. Pahren, H., J. Lucas and N. E. Kowal. "An Assessment of the Health Risks from the Oxon Cove Compost Piles Resulting from Aspergillus fumigatus," letter report to EPA Region III (1978).
55. Solomon, W. R. "Assessing Fungus Prevalence in Domestic Interiors," *J. Allergy Clin. Immunol.* 53(2):71 (1974).
56. Austwick, P. K. "Ecology of Aspergillus fumigatus and the Pathogenic Phycomycetes," in *Recent Progress in Microbiology* (Toronto: University of Toronto Press, 1963).
57. Austwick, D. K. "The Role of Spores in the Allergies and Mycoses of Man and Animals," *Proceedings of the 18th Symposium, Colston Research Society* (London: Butterworths, 1966).
58. Lacey, J., and M. E. Lacey. "Spore Concentrations in the Air of Farm Buildings," *Trans. Brit. Mycol. Soc.* 47:547-552 (1964).
59. LeBrun, T. Los Angeles County Sanitation Districts. Memo to the LA/OMA project on status of *Aspergillus* monitoring (1979).
60. McCarty, P. L. "Thermodynamics of Biological Synthesis and Growth," in *Adv. Water Poll. Res.* 2(1) (1964).
61. Servizi, J. A., and R. H. Bogan. "Free Energy as a Parameter in Biological Treatment," *J. San. Eng. Div., ASCE* 89(SA3):17 (1963).
62. Payne, W. J. "Energy Yields and Growth of Heterotrophs," *Ann. Rev. Microbiol.* 24:17 (1970).
63. Sykes, R. M. "Theoretical Heterotrophic Yields," *J. Water Poll. Control Fed.* 47(3) (1975).
64. McCarty, P. L. "Energetics and Bacterial Growth," paper presented at the Fifth Rudolf Research Conference, Rutgers University, July 1969.
65. O'Rourke, J. T. "Kinetics of Anaerobic Treatment at Reduced Temperatures," PhD Thesis, Stanford University (1968).
66. "Criteria for Classification of Solid Waste Disposal Facilities and Practices," U.S. EPA *Federal Register* 44:179 (1979).
67. Haug, R. T., L. D. Tortorici and S. K. Raksit. "Sludge Processing and Disposal–A State of the Art Review," publication of the LA/OMA project, Whittier, CA (1977).
68. Farrell, J. B., J. E. Smith, S. W. Hathaway and R. B. Dean. "Lime Stabilization of Primary Sludges," *J. Water Poll. Control Fed.* 46(1) (1974).
69. Wiley, J. S. "Pathogen Survival in Composting Municipal Wastes," *J. Water Poll. Control Fed.* 34:1 (1962).
70. Roediger, H. J. "The Technique of Sewage-Sludge Pasteurization: Actual Results Obtained in Existing Plants; Economy," International

Research Group on Refuse Disposal Information, Bulletin 21-31 (1964–1967).
71. Gotaas, H. B. "Composting–Sanitary Disposal and Reclamation of Organic Wastes," World Health Organization, Mono. Ser. No. 31 (1956).
72. Stern, G. "Pasteurization of Liquid Digested Sludge," *Proceedings of National Conference on Municipal Sludge Management* (Rockville, MD: Information Transfer, 1974).
73. Ward, R. L., C. S. Ashley and R. H. Moseley. "Heat Inactivation of Poliovirus in Wastewater Sludge," *Appl. Environ. Microbiol.* 32(3) (1976).
74. Ward, R. L., and C. S. Ashley. "Identification of the Virucidal Agent in Wastewater Sludge," *Appl. Environ. Microbiol.* 33(4) (1977).
75. Ward, R. L., and J. R. Brandon. "Effect of Heat on Pathogenic Organisms Found in Wastewater Sludge," in *Composting of Municipal Residues and Sludges, Proceedings of the 1977 National Conference* (Rockville, MD: Information Transfer, 1977).
76. "Progress Report–Beneficial Uses Program, Period Ending Sept 30, 1977," Sandia Laboratories, NM, SAND78-0242.
77. Brandon, J. R., and K. S. Neuhauser. "Moisture Effects on Inactivation and Growth of Bacteria and Fungi in Sludges," Sandia Laboratories, NM, SAND 78-1304 (1978).
78. Kruse, C. W. "Sludge Disinfection," in *Evaluation of Current Developments in Municipal Waste Treatment*, Technology Information Center, Energy Research and Development Administration (1977).
79. Chick, H. "Investigations of the Laws of Disinfection," *J. Hyg.* 8:92-158 (1908).
80. Burge, W., D. Colacicco, W. Cramer and E. Epstein. "Criteria for Control of Pathogens during Sewage Sludge Composting," *Proceedings of National Conference on Design of Municipal Sludge Compost Facilities,* (Rockville, MD: Information Transfer, 1978).
81. Moats, W. A. "Kinetics of Thermal Death of Bacteria," *J. Bacteriol.* 105(1) (1971).
82. Wei, J. H., and S. L. Chang. "A Multi-Poisson Distribution Model for Treating Disinfection Data," in *Disinfection: Water and Wastewater*, J. D. Johnson, Ed. (Ann Arbor, MI: Ann Arbor Science Publishers, Inc., 1975).
83. Carslaw, H. S., and J. C. Jaeger. *Conduction of Heat in Solids* 2nd ed. (Oxford: Clarendon Press, 1959).
84. Brandon, J. R. "Parasites in Soil/Sludge Systems," Sandia Laboratories No. 77-1970 (1978).
85. Brandon, J. R. "Sandia's Sludge Irradiation Program," in *Sludge Management, Disposal and Utilization* (Rockville, MD: Information Transfer, 1976).
86. Selna, M., and D. Smith. "Pathogen Inactivation During Composting," Technical Services Department, Los Angeles County Sanitation Districts (September 1975–December 1976).

87. Hornick, R. B., T. E. Woodward, F. R. McCrumb, M. J. Snyder, A. T. Dawkins, J. T. Bulkeley, F. DeLaMacorra and F. A. Carozza. "Study of Induced Typhoid Fever in Man. I. Evaluations of Vaccine Effectiveness," *Trans. Assoc. Am. Physicians*, 79:361-367 (1966).
88. Hornick, R. B., S. E. Greisman, T. E. Woodward, H. L. Dupont, A. T. Dawkins and M. J. Snyder. "Typhoid Fever: Pathogenesis and Immunologic Control," *New England J. Med.* 283(13):686-691 (1970).
89. Mechalas, B. J., K. K. Hekimian, L. A. Schinazi and R. H. Dudley. "An Investigation into Recreational Water Quality–Water Quality Criteria Data Book, Vol. 4," Office of Research and Monitoring, U.S. EPA (1972).
90. McCullough, N. B., and C. W. Eisele. "Experimental Human Salmonellosis. I. Pathogenicity of Strains of Salmonella meleagridis and Salmonella anatum obtained from Spray-dried Whole Egg," *J. Infect. Dis.* 88:278-289 (1951).
91. McCullough, N. B., and C. W. Eisele. "Experimental Human Salmonellosis. II. Immunity Studies Following Experimental Illness with Salmonella meleagridis and Salmonella anatum," *J. Immunol.* 66:595-608 (1951).
92. McCullough, N. B., and C. W. Eisele. "Experimental Human Salmonellosis, III. Pathogenicity of Strains of Salmonella newport, Salmonella derby, and Salmonella bareilly Obtained from Spray-Dried Whole Egg," *J. Infect. Dis.* 89:209-213 (1951).
93. McCullough, N. B., and C. W. Eisele. "Experimental Human Salmonellosis, IV. Pathogenicity of Strains of Salmonella pullorum Obtained from Spray-dried Whole Egg," *J. Infect. Dis.* 89:259-265 (1951).
94. Kaprowski, H. *Am. J. Trop. Med. Hyg.* 5:440 (1956).
95. Akin, E. W., H. P. Pahren, W. Jakubowski and J. B. Lucas. "Health Hazards Associated with Wastewater Effluents and Sludges: Microbiological Consideration," Proceedings of the Conf. on Risk Assessment and Health Effects of Land Application of Municipal Wastewater and Sludges, Center for Applied Research and Technology, University of Texas at San Antonio (1977).
96. McGauhey, P. H., and H. G. Gotaas. "Stabilization of Municipal Refuse by Composting," *Trans. Am. Soc. Civil Eng.* Paper No. 2767.
97. Shell, G. L., and J. L. Boyd. "Composting Dewatered Sewage Sludge," Public Health Service Publication No. 1936, U.S. HEW (1969).
98. Haug, R. T., and L. A. Haug. "Sludge Composting–A Discussion of Engineering Principles," *Compost Sci. J. Solid Wastes Soil* (November/December 1977; January/February 1978).
99. Haug, R. T. "Engineering Principles of Sludge Composting," *J. Water Poll. Control Fed.* 51(8) (1979).
100. Cooper, G., Golden West Fertilizer Co., Santa Ana, CA. Personal communication (1977).
101. "Quantitative Analysis of Sludge Management Projects," a report of the LA/OMA project, Whittier, CA (1979).

102. Haug, R. T. "Composting Wet Organic Sludges–A Problem of Moisture Control," in *Proc. of the National Conference on Design of Municipal Sludge Compost Facilities* (Rockville, MD: Information Transfer, 1978).
103. Jeris, J. S., and R. W. Regan. "Controlling Environmental Parameters for Optimum Composting, Part II: Moisture, Free Air Space and Recycle," *Compost Sci.* (March/April 1973).
104. Schultz, K. L. "Continuous Thermophilic Composting," *Compost Sci.* (Spring 1962).
105. *Pathogen Inactivation During Sludge Composting–Status Report* (Los Angeles, CA: Los Angeles County Sanitation Districts, 1976).
106. Gossett, J. M., and P. L. McCarty. "Heat Treatment of Refuse for Increasing Anaerobic Biodegradability," Civil Engineering Tech. Report No. 192, Stanford University, Stanford, CA (1975).
107. Golueke, C. G. "Bioconversion of Energy Studies at the University of California (Berkeley)," Proceedings, Bioconversion Energy Research Conference, University of Massachusetts, Amherst, MA (1973).
108. Pfeffer, J. T., and J. C. Liebman. "Biological Conversion of Organic Refuse to Methane, Annual Progress Report," NSF Grant No. GI-39191 (1974).
109. Rich, L. G. *Unit Processes of Sanitary Engineering* (New York: John Wiley & Sons, Inc., 1963).
110. "Laboratory-Scale Tests of the Multiple-Hearth Gasification Process on Selected Dried Sludges for the LA/OMA Project," report to the LA/OMA project by BSP Division, Envirotech Corp. (1979).
111. Smith, D. Los Angeles County Sanitation Districts. Personal communication (1977).
112. Iacoboni, M., T. LeBrun and J. Livingston. "Composting Study," Technical Services Dept., Los Angeles County Sanitation Districts (1979).
113. Maier, P. P., E. R. Williams and G. F. Mallison. "Composting Municipal Refuse by the Aeration Bin Process," *Proceedings of the 12th Purdue Industrial Waste Conference* (West Lafayette, IN: Purdue University, 1957).
114. Golueke, C. G. *Composting–A Study of the Process and Its Principles* (Emmaus, PA: Rodale Press, 1972).
115. Shell, B. J. "The Mechanism of Oxygen Transfer Through a Composting Material," PhD Thesis, Civil and Sanitary Engineering Department, Michigan State University (1955).
116. Snell, J. R. "Some Engineering Aspects of High-Rate Composting," *J. San. Eng. Div., ASCE*, Paper 1178 (1957).
117. Mears, D. R., M. E. Singley, C. Ali and F. Rupp. "Thermal and Physical Properties of Compost," in *Energy, Agriculture and Waste Management*, W. J. Jewell, Ed. (Ann Arbor, MI: Ann Arbor Science Publishers, Inc., 1975).
118. Grant, F. A. "Liquid Composting of Dairy Manure," Proceedings 3rd International Symposium on Livestock Wastes, Univ. of Illinois (1975).

119. Schultz, K. L. "Rate of Oxygen Consumption and Respiratory Quotients during the Aerobic Decomposition of a Synthetic Garbage," *13th Annual Purdue Industrial Waste Conference* (West Lafayette, IN: Purdue University, 1958).
120. Schultz, K. L. "Rate of Oxygen Consumption and Respiratory Quotients During the Aerobic Decomposition of a Synthetic Garbage," *Compost Sci.* (Spring 1960).
121. Schultz, K. L. "Relationship Between Moisture Content and Activity of Finished Compost," *Compost Sci.* (Summer 1961).
122. Wiley, J. S. "Studies of High-Rate Composting of Garbage and Refuse," *10th Annual Purdue Industrial Waste Conference* (West Lafayette, IN: Purdue University, 1955).
123. Wiley, J. S. "Progress Report on High-Rate Composting Studies," *11th Annual Purdue Industrial Waste Conference* (West Lafayette, IN: Purdue University, 1956).
124. Wiley, J. S. "II. Progress Report on High-Rate Composting Studies," *12th Annual Purdue Industrial Waste Conference* (West Lafayette, IN: Purdue University, 1957).
125. Wiley, J. S., and G. W. Pearce. "A Preliminary Study of High-Rate Composting," *Trans. Am. Soc. Civil Eng.* Paper No. 2895.
126. Jeris, J. S., and R. W. Regan. "Controlling Environmental Parameters for Optimum Composting, Part I: Experimental Procedures and Temperature," *Compost Sci.* (January/February 1973).
127. Jeris, J. S., and R. W. Regan. "Controlling Environmental Parameters for Optimum Composting, Part III," *Compost Sci.* (May/June 1973).
128. Lawrence, A. W., and P. L. McCarty. "A Unified Basis for Biological Treatment, Design and Operation," *J. San. Eng. Div., ASCE* 96(SA3) (1970).
129. Mueller, J. A. "Oxygen Diffusion through a Pure Culture Floc of *Zoogloea ramigera*," PhD thesis, University of Wisconsin (1966).
130. Atkinson, D., E. L. Swilley, A. W. Busch and D. A. Williams. "Kinetics, Mass Transfer, and Organism Growth in a Biological Film Reactor," *Trans. Inst. Chem. Eng.* 45:T257 (1967).
131. Kornegay, B. H., and J. F. Andrews. "Kinetics of Fixed Film Biological Reactors," *Proc. 22nd Annual Purdue Industrial Waste Conference* (West Lafayette, IN: Purdue University, 1967).
132. Maier, W. J., V. C. Behn and C. D. Gates. "Simulation of the Trickling Filter Process," *J. San. Eng. Div., ASCE* 93(SA4) (1967).
133. Haug, R. T., and P. L. McCarty. "Nitrification with the Submerged Filter," Tech. Report No. 149, Dept. of Civil Engr., Stanford University (1971).
134. Young, J. C., and P. L. McCarty. "The Anaerobic Filter for Waste Treatment," *J. Water Poll. Control Fed.* 41:R160 (1969).
135. Williamson, K., and P. L. McCarty. "A Model of Substrate Utilization by Bacterial Films," *J. Water Poll. Control Fed.* 48(1) (1976).

136. Williamson, K., and P. L. McCarty. "Verification Studies of the Biofilm Model for Bacterial Substrate Utilization," *J. Water Poll. Control Fed.* 48(2) (1976).
137. *Wastewater Treatment Plant Design, Manual of Practice No. 8* (Washington, DC: Water Pollution Control Federation).
138. *Manual of Engineering Practice No. 36* (New York: American Society Civil Engineers, 1977).
139. Reid, R. C., and T. K. Sherwood. *The Properties of Gases and Liquids* (New York: McGraw-Hill Book Co., Inc., 1958).
140. Bungay, H. R. et al. "Microprobe Techniques for Determining Diffusivities and Respiration Rates in Microbial Slime Systems," *Biotechnol. Bioeng.* 11(5):765 (1969).
141. "Second Interim Report of the Inter-departmental Committee on Utilization of Organic Wastes," *New Zealand Eng.* (November/December 1951).
142. Lossin, R. D. "Compost Studies, Part III," *Compost Sci.* (March/April 1971).
143. Chrometzka, P. "Determination of the Oxygen Requirements of Maturing Composts," International Research Group on Refuse Disposal Information Bulletin 33 (1968).
144. Lossin, R. D. "Compost Studies, Part I," *Compost Sci.* (November/December 1970).
145. McCarty, P. L. "Anaerobic Processes," paper presented at the Birmingham Short Course on Design Aspects of Biological Treatment, Int. Assoc. of Water Poll. Research, England (1974).
146. Metcalf and Eddy, Inc. *Wastewater Engineering* (New York: McGraw-Hill Book Co., 1972).
147. Andrews, J. F., and K. Kambhu. "Thermophilic Aerobic Digestion of Organic Solid Wastes," Office of Research and Development, U.S. EPA NTIS PB-222-396 (1973).
148. McCabe, W. L., and J. C. Smith. *Unit Operations of Chemical Engineering*, 3rd ed. (New York: McGraw-Hill Book Co., 1976).
149. "Reclamation of Municipal Refuse by Composting," Technical Bulletin No. 9, Sanitary Engineering Research Lab. (Berkeley, CA: University of California, 1953).
150. "A Proposal for Sewage Sludge Management at the Coyote Canyon Landfill, County of Orange," Orange County Sanitation Districts (1978).
151. Dominowske, R., Metro. Seattle. Personal communication. (1979).
152. Fuller, W. H. "Pelleted Compost Now Turns Former Waste into Useful Product," *Progressive Agric. Arizona* XVIII:(3) (1966).
153. Sullivan, R. J. "Preliminary Air Pollution Survey on Odorous Compounds: A Literature Review," NAPCA Pub. APTD 66-24 (1969).
154. *Odor Control for Wastewater Facilities, Manual of Practice No. 22* (Washington, DC: Water Pollution Control Federation, 1979).

155. Liptak, B. G., Ed. *Environmental Engineers Handbook* (Radner, PA: Chilton Book Co., 1974).
156. Wilson, G. "Odors, Their Detection and Measurement," EUTEK, Process Development and Engineering, Sacramento, CA (1975).
157. Sacramento Area Consultants. "Draft Environmental Impact Report–Sewage Sludge Management Program Wastewater Solids Processing and Disposal" (1978).
158. Redner, J., Los Angeles County Sanitation Districts. Personal communication (1979).
159. Redner, J., G. E. Wilson, T. W. Schroepfer and J. Y. Huang. "Control of Composting Odors," *Proceedings of the National Conference on Design of Municipal Sludge Compost Facilities* (Rockville, MD: Information Transfer, 1978).
160. Huang, J. Y., G. E. Wilson and T. W. Schroepfer. "Evaluation of Sludge Odor Control Alternatives," *J. Environ. Eng. Div., ASCE* 6:1135-1148 (1978).
161. Wilson, G. E., J. Y. Huang and T. W. Schroepfer. "Atmospheric Sublayer Transport and Odor Control," ASCE preprint 3493 (1979).
162. Eutek, Inc. "Preliminary Evaluation of Alternatives for Scaling and Odor Control–JWPCP," Los Angeles County Sanitation Districts (1978).
163. Livingston, J. "Mechanically Aerated Windrow Study, Phase I (Nov 1978) and Phase II (Dec 1978)," Los Angeles County Sanitation District.
164. Metcalf and Eddy. *Wastewater Engineering: Treatment, Disposal, Reuse*, revised by G. Tchobanoglous (New York: McGraw-Hill Book Co., 1979).
165. Cross, F. L. *Air Pollution Odor Control Primer* (Westport, CT: Technomic Publishing Co., 1973).
166. Perry, R. H., and C. H. Chilton. *Chemical Engineers' Handbook* 5th ed. (New York: McGraw-Hill Book Co., 1973).
167. Eutek, Inc. "Solids Storage Basins Source Odor Control Study," Sacramento Area Consultants (1977).
168. Bohn, H. L. "Compost Scrubbers of Malodorous Air Streams," *Compost Sci.* (Winter 1976).
169. Toups and Loiederman Engineers. "Washington Suburban Sanitary Commission, Compost Facility–Site 2, Proposed Design Criteria and Plans," Rockville Pike, MD (1978).
170. Garber, W. F. "Odors and Their Control in Sewers and Treatment Facilities," paper presented at the IAWPR Workshop, Vienna, Austria (1979).
171. Vogt, D. R. "Synopsis of Observations, European Tour of Mechanical Enclosed Composting Systems," Washington Suburban Sanitary Commission (1979).
172. Eutek, Inc. "Development of Odor Criteria for Large Emission Areas," Sacramento Reg. Wastewater Management Program, Sacramento Area Consultants (1977).

173. Turner, D. B. "Workbook of Atmospheric Dispersion Estimates," Office of Air Programs, Pub. No. AP-26, U.S. EPA (1970).
174. Budney, L. J. "Guidelines for Air Quality Maintenance Planning and Analysis, Vol. 10: Procedures for Evaluating Air Quality Impacts of New Stationary Sources," U.S. EPA Pub. No. EPA-450/4-77-001 (1977).
175. Pasquill, F. "The Estimation of the Dispersion of Windborne Material," *Meteorol. Mag.* 90(1063):33-49 (1961).
176. Gifford, F. A. "Use of Routine Meteorological Observations for Estimating Atmospheric Dispersion," *Nucl. Safety* 2(4):47-51 (1961).
177. Holland, J. Z. "A Meterorological Survey of the Oak Ridge Area," Atomic Energy Commission, Report ORO-99, Washington, DC (1953).
178. Shea, T. G., J. Braswell and C. S. Coker. "Bulking Agent Selection in Sludge Compost Facility Design," *Compost Sci./Land Util.* (Nov/Dec 1979).
179. Marsh, P. B., P. D. Millner and J. M. Kla. "A Guide to the Recent Literature on Aspergillosis as Caused by Aspergillus fumigatus," Agricultural Reviews and Manuals, USDA, ARM-NE-5 (Sept 1979).
180. Brown, V., and K. L. Donnelly. "Advancement in the Mechanized Sludge Composting System," paper presented at the Annual Meeting of the Water and Waste Operators Association and the Chesapeake Water Pollution Control Association, June 1979.
181. "Proposed Sludge Management Program for the Los Angeles/Orange County Metropolitan Area, Facilities Plan and Management Program," LA/OMA project, Whittier, CA (1980).
182. Houser, J. E. Preliminary proposal to the City of Los Angeles for Composting Facilities at the Terminal Island Treatment Plant (1979).
183. Houser, J. "Engineering Study for Composting Paper Mill Sludge," Kramer, Comer, Passe, Racher Engineers (1979).
184. Klein, S. A. "Anaerobic Digestion of Solid Wastes," *Compost Sci.* (Jan/Feb 1972).
185. Krough, A. "The Rate of Diffusion of Gases Through Animal Tissues, with Some Remarks on the Coefficient of Invasion," *J. Physiol.* 52:391-408 (1919).
186. Greven, K. "Über die Bedeutung der Zellmembranen für die O_2-Diffusion im Gewebe," *Pfluger's Arch. Gesamte Physiol.* 273:353-366 (1961).
187. Longmuir, I. S., and A. Bourke. "The Measurement of the Diffusion of Oxygen Through Respiring Tissue," *Biochem J.* 76:225-229 (1960).
188. Thews, G. "Oxygen Diffusion in the Brain," *Pfluger's Arch. Gesamte Physiol.* 271:197-226 (1960).
189. Greenwood, D. J., and D. Goodman. "Effect of Shape on Oxygen Diffusion and Aerobic Respiration in Soil Aggregates," *J. Sci. Food Agric.* 15:781-790 (1964).
190. Wise, D. L. "The Determination of the Diffusion Coefficients of Ten Slightly Soluble Gases in Water and a Study of the Solution Rate of Small Stationary Bubbles," PhD Thesis, University of Pittsburg (1963).

191. *International Critical Tables* (New York: McGraw-Hill Book Co., 1929).
192. Brannen, J. P., D. M. Garst and S. Langley. "Inactivation of Ascaris Lumbricoides Eggs by Heat, Radiation, and Thermoradiation," Sandia Laboratories, NM (1975).
193. Regan, R. W., and J. S. Jeris. "A Review of the Decomposition of Cellulose and Refuse," *Compost Sci.* 11:17 (1970).
194. Ettlich, W., Culp, Wesner, Culp/Clean Water Consultants, El Dorado Hills, CA. Personal communication (1977).
195. Brown, V., Resource Conversion Systems, Inc., Houston, TX. Personal communication (1977).
196. Meyer, J., Biological Waste Management Systems, New York. Personal communication (1977).
197. CH2M-Hill. "Preliminary Computer Analysis of Sludge Management Alternatives, Unit Process Design, Cost, Energy, and Emission Factors," report to the LA/OMA Project, Whittier, CA (1978).
198. Haug, L. A. "Windrow Composting Analysis for a Ten Dry Ton per Day Plant," *Proceedings of the Municipal and Industrial Sludge Handling and Disposal, Equipment and Services Show* (Rockville, MD: Information Transfer, 1980).
199. Merritt, F. S., Ed. *Standard Handbook for Civil Engineers* (San Francisco, CA: McGraw-Hill Book Co., 1968).
200. Schultz, K. L. "The Fairfield-Hardy Composting Pilot Plant at Altoona, Pennsylvania," *Compost Sci.* (Autumn-Winter 1965).
201. Clark, C. S., C. O. Buckingham, D. H. Bone and R. H. Clark. "Laboratory Scale Composting Techniques," *J. Environ. Eng. Div., ASCE* 5:893-906 (1977).
202. "Static Pile Composting of Wastewater Sludge," EPA Technology Transfer, EPA-625/2-71-014.
203. "Direct Environmental Factors at Municipal Wastewater Treatment Works," Office of Water Program Operations, U.S. Environmental Protection Agency (1976).
204. "Threshold Limit Values for Chemical Substances in Workroom Air," *Nat. Safety News* (September 1976), p. 116.
205. Tsuk, A. G., and G. Oster. "Determination of Enzyme Activity by a Linear Measurement," *Nature* 190:721 (1961).
206. McLaren, A. D., and L. Packer. "Some Aspects of Enzyme Reactions in Heterogeneous Systems," *Adv. Enzymol. Rel. Sub. Biochem.* 33:245 (1970).
207. Niese, G. "Experiments to Determine the Degree of Decomposition of Refuse by Its Self-Heating Capability," International Research Group on Refuse Disposal Information Bulletin 17 (1963).
208. Hudson, H. J. "Thermophilous and Thermotolerant Fungi in the Airspora at Cambridge," *Brit. Mycol. Soc. Trans.* 60:596-598 (1973).
209. Motegi, K., Department of Sewage Works, Tokyo Metropolitan Government. Personal communication and unpublished data (1980).

210. Comar, C. L. "Risk: A Pragmatic De Minimis Approach," *Science* 203: 4378 (1979).
211. Epstein, E. "Bulking Materials," *Proceedings of National Conference on Municipal and Industrial Sludge Composting-Materials Handling* (Rockville, MD: Information Transfer, 1979).

INDEX